Justice and Tourism

Research related to justice and tourism is at an early stage in tourism studies. Challenges abound due to the complex scope and scale of tourism, and thus the need to transcend disciplinary boundaries to inform a phenomenon that is intricately interwoven with place and people from local to global. The contributors to this book have drawn from diverse knowledge domains including but not limited to sociology, geography, business studies, urban planning and architecture, anthropology, philosophy and management studies, to inform their research.

From case-based empirical research to descriptive and theoretical approaches to justice and tourism, they tackle critical issues such as social justice and gender, discrimination and racism, minority and worker rights, indigenous, cultural and heritage justice (including special topics like food sovereignty), while post-humanistic perspectives that call us to attend to non-human others, to climate justice and sustainable futures. A rich array of principles are woven within and between the chapters. The various contributions illustrate the need for continuing collaboration among researchers in the Global North and Global South to enable diverse voices and worldviews to inform the pluralism of justice and tourism, as arises in this book.

The chapters in this book were originally published as a special issue of the *Journal of Sustainable Tourism* on Justice and Tourism, Volume 29, Issues 2–3.

Tazim Jamal is Professor in the Department of Recreation, Park and Tourism Sciences at Texas A&M University, USA. Her research focuses on sustainable tourism and collaborative tourism planning. She is the author of *Justice and Ethics in Tourism* (2019) and co-editor of *The SAGE Handbook of Tourism Studies* (2009).

James Higham is Professor in the Department of Tourism at the University of Otago, New Zealand. His research addresses tourism and environmental change at the global, national and local scales of analysis. He is the co-editor of the *Journal of Sustainable Tourism*.

Justice and Tourism

Principles and Approaches for Local-Global Sustainability and Well-Being

Edited by
Tazim Jamal and James Higham

Routledge
Taylor & Francis Group
LONDON AND NEW YORK

First published 2022
by Routledge
2 Park Square, Milton Park, Abingdon, Oxon OX14 4RN

and by Routledge
605 Third Avenue, New York, NY 10158

Routledge is an imprint of the Taylor & Francis Group, an informa business

British Library Cataloguing in Publication Data
A catalogue record for this book is available from the British Library

ISBN: 978-0-367-69739-6 (hbk)
ISBN: 978-0-367-69740-2 (pbk)
ISBN: 978-1-003-14305-5 (ebk)

DOI: 10.4324/9781003143055

Typeset in Myriad Pro
by Newgen Publishing UK

Publisher's Note
The publisher accepts responsibility for any inconsistencies that may have arisen during the conversion of this book from journal articles to book chapters, namely the inclusion of journal terminology.

Contents

Citation Information vii
Notes on Contributors xi

1 Justice and ethics: towards a new platform for tourism and sustainability 1
Tazim Jamal and James Higham

2 Overtourism, place alienation and the right to the city: insights from the historic centre of Seville, Spain 16
Iban Diaz-Parra and Jaime Jover

3 Who has the right to the rural? Place framing and negotiating the Dungog festival, New South Wales, Australia 34
Judith Mair and Michelle Duffy

4 Locally situated rights and the 'doing' of responsibility for heritage conservation and tourism development at the cultural landscape of Honghe Hani Rice Terraces, China 51
Jun Gao, Hongxia Lin and Chaozhi Zhang

5 Indigenous tourism and cultural justice in a Tz'utujil Maya community, Guatemala 72
Lucy C. Harbor and Carter A. Hunt

6 Becoming common plantain: metaphor, settler responsibility, and decolonizing tourism 92
Michela J. Stinson, Bryan S. R. Grimwood and Kellee Caton

7 Heritage justice, conservation, and tourism in the Greater Caribbean 111
Brent R. Fortenberry

8 Representation of "mill girls" at a UNESCO World Heritage Site in Gunma, Japan 135
Naho Ueda Maruyama and Kyle Maurice Woosnam

9 Beyond accessibility: exploring the representation of people with disabilities in tourism promotional materials 153
Stefanie Benjamin, Ethan Bottone and Miranda Lee

10 Tourism, animals and the scales of justice 172
David A. Fennell and Valerie Sheppard

11 Intergenerational rights to a sustainable future: insights for climate justice and tourism 194
Dawn Jourdan and Jani Wertin

12 Megaliths, material engagement, and the atmospherics of neo-lithic ethics: presage for the end(s) of tourism 204
Mick Smith, Siobhan Speiran and Peter Graham

13 Tourism, inclusive growth and decent work: a political economy critique 220
Raoul V. Bianchi and Frans de Man

14 Humanism, dignity and indigenous justice: the Mayan Train megaproject, Mexico 239
Blanca A. Camargo and Mario Vázquez-Maguirre

15 Indigenous food sovereignty and tourism: the Chakra Route in the Amazon region of Ecuador 259
Verónica Santafe-Troncoso and Philip A. Loring

16 Roots tourism: a second wave of Double Consciousness for African Americans 279
Alana Dillette

17 Heritage tourism, historic roadside markers and "just representation" in Tennessee, USA 295
Candace Forbes Bright, Kelly N. Foster, Andrew Joyner and Oceane Tanny

18 Resisting marginalisation and reconstituting space through LGBTQI+ events 315
Oscar Vorobjovas-Pinta and Anne Hardy

19 Slow food justice and tourism: tracing Karakılçık bread in Seferihisar, Turkey 334
İlkay Taş Gürsoy

20 The dialogic negotiation of justice 355
Becky Shelley, Can-Seng Ooi and Lisa Denny

21 Conceptualizing justice tourism and the promise of posthumanism 370
Jaume Guia

22 World heritage and social justice: insights from the inscription of Yazd, Iran 388
Raymond Rastegar, Zohreh (Zara) Zarezadeh and Ulrike Gretzel

23 Smart Korea: governance for smart justice during a global pandemic 408
Jiyoung Choi, Seunghoon Lee and Tazim Jamal

Index 418

Citation Information

The chapters in this book, except Chapter 11, were originally published in the *Journal of Sustainable Tourism*, volume 29, issue 2–3 (2021). Chapter 11 was originally published in volume 28, issue 8 of the same journal. When citing this material, please use the original page numbering for each article, as follows:

Chapter 1

Justice and ethics: towards a new platform for tourism and sustainability
Tazim Jamal and James Higham
Journal of Sustainable Tourism, volume 29, issue 2–3 (2021), pp. 143–157

Chapter 2

Overtourism, place alienation and the right to the city: insights from the historic centre of Seville, Spain
Iban Diaz-Parra and Jaime Jover
Journal of Sustainable Tourism, volume 29, issue 2–3 (2021), pp. 158–175

Chapter 3

Who has the right to the rural? Place framing and negotiating the Dungog festival, New South Wales, Australia
Judith Mair and Michelle Duffy
Journal of Sustainable Tourism, volume 29, issue 2–3 (2021), pp. 176–192

Chapter 4

Locally situated rights and the 'doing' of responsibility for heritage conservation and tourism development at the cultural landscape of Honghe Hani Rice Terraces, China
Jun Gao, Hongxia Lin and Chaozhi Zhang
Journal of Sustainable Tourism, volume 29, issue 2–3 (2021), pp. 193–213

Chapter 5

Indigenous tourism and cultural justice in a Tz'utujil Maya community, Guatemala
Lucy C. Harbor and Carter A. Hunt
Journal of Sustainable Tourism, volume 29, issue 2–3 (2021), pp. 214–233

Chapter 6

Becoming common plantain: metaphor, settler responsibility, and decolonizing tourism
Michela J. Stinson, Bryan S. R. Grimwood and Kellee Caton
Journal of Sustainable Tourism, volume 29, issue 2–3 (2021), pp. 234–252

Chapter 7

Heritage justice, conservation, and tourism in the Greater Caribbean
Brent R. Fortenberry
Journal of Sustainable Tourism, volume 29, issue 2–3 (2021), pp. 253–276

Chapter 8

Representation of "mill girls" at a UNESCO World Heritage Site in Gunma, Japan
Naho Ueda Maruyama and Kyle Maurice Woosnam
Journal of Sustainable Tourism, volume 29, issue 2–3 (2021), pp. 277–294

Chapter 9

Beyond accessibility: exploring the representation of people with disabilities in tourism promotional materials
Stefanie Benjamin, Ethan Bottone and Miranda Lee
Journal of Sustainable Tourism, volume 29, issue 2–3 (2021), pp. 295–313

Chapter 10

Tourism, animals and the scales of justice
David A. Fennell and Valerie Sheppard
Journal of Sustainable Tourism, volume 29, issue 2–3 (2021), pp. 314–335

Chapter 11

Intergenerational rights to a sustainable future: insights for climate justice and tourism
Dawn Jourdan and Jani Wertin
Journal of Sustainable Tourism, volume 28, issue 8 (2020), pp. 1245–1254

Chapter 12

Megaliths, material engagement, and the atmospherics of neo-lithic ethics: presage for the end(s) of tourism
Mick Smith, Siobhan Speiran and Peter Graham
Journal of Sustainable Tourism, volume 29, issue 2–3 (2021), pp. 337–352

Chapter 13

Tourism, inclusive growth and decent work: a political economy critique
Raoul V. Bianchi and Frans de Man
Journal of Sustainable Tourism, volume 29, issue 2–3 (2021), pp. 353–371

Chapter 14
Humanism, dignity and indigenous justice: the Mayan Train megaproject, Mexico
Blanca A. Camargo and Mario Vázquez-Maguirre
Journal of Sustainable Tourism, volume 29, issue 2–3 (2021), pp. 372–391

Chapter 15
Indigenous food sovereignty and tourism: the Chakra Route in the Amazon region of Ecuador
Verónica Santafe-Troncoso and Philip A. Loring
Journal of Sustainable Tourism, volume 29, issue 2–3 (2021), pp. 392–411

Chapter 16
Roots tourism: a second wave of Double Consciousness for African Americans
Alana Dillette
Journal of Sustainable Tourism, volume 29, issue 2–3 (2021), pp. 412–427

Chapter 17
Heritage tourism, historic roadside markers and "just representation" in Tennessee, USA
Candace Forbes Bright, Kelly N. Foster, Andrew Joyner and Oceane Tanny
Journal of Sustainable Tourism, volume 29, issue 2–3 (2021), pp. 428–447

Chapter 18
Resisting marginalisation and reconstituting space through LGBTQI+ events
Oscar Vorobjovas-Pinta and Anne Hardy
Journal of Sustainable Tourism, volume 29, issue 2–3 (2021), pp. 448–466

Chapter 19
Slow food justice and tourism: tracing Karakılçık bread in Seferihisar, Turkey
İlkay Taş Gürsoy
Journal of Sustainable Tourism, volume 29, issue 2–3 (2021), pp. 467–487

Chapter 20
The dialogic negotiation of justice
Becky Shelley, Can-Seng Ooi and Lisa Denny
Journal of Sustainable Tourism, volume 29, issue 2–3 (2021), pp. 488–502

Chapter 21
Conceptualizing justice tourism and the promise of posthumanism
Jaume Guia
Journal of Sustainable Tourism, volume 29, issue 2–3 (2021), pp. 503–520

Chapter 22
World heritage and social justice: Insights from the inscription of Yazd, Iran
Raymond Rastegar, Zohreh (Zara) Zarezadeh and Ulrike Gretzel
Journal of Sustainable Tourism, volume 29, issue 2–3 (2021), pp. 521–540

Chapter 23

Smart Korea: Governance for smart justice during a global pandemic
Jiyoung Choi, Seunghoon Lee and Tazim Jamal
Journal of Sustainable Tourism, volume 29, issue 2–3 (2021), pp. 541–550

Notes on Contributors

Stefanie Benjamin, Department of Retail, Hospitality & Tourism Management, The University of Tennessee, Knoxville, Knoxville, TN, USA.

Raoul V. Bianchi, School of Business & Law, Docklands Campus, University Way, London, UK.

Ethan Bottone, Department of Humanities and Social Sciences, Northwest Missouri State University, Maryville, MO, USA.

Candace Forbes Bright, Department of Sociology and Anthropology, East Tennessee State University, Johnson City, TN, USA.

Blanca A. Camargo, Business School, Universidad de Monterrey, San Pedro Garza Garcia, Mexico.

Kellee Caton, Department of Tourism Management, Thompson Rivers University, Kamloops, British Columbia, Canada.

Jiyoung Choi, Department of Recreation, Park and Tourism Sciences, Texas A&M University, College Station, TX, USA.

Lisa Denny, College of Arts, Law and Education, University of Tasmania, Hobart, Australia.

Iban Diaz-Parra, Department of Human Geography, Facultad de Geografıa e Historia, University of Seville, Sevilla, Spain.

Alana Dillette, L. Robert Payne School of Hospitality and Tourism Management, San Diego State University, San Diego, CA, USA.

Michelle Duffy, School of Environmental and Life Sciences, The University of Newcastle, Callaghan, Australia.

David A. Fennell, Department of Geography & Tourism Studies, Brock University, St. Catharines, Canada.

Brent R. Fortenberry, Department of Architecture, Texas A&M University, College Station, TX, USA.

Kelly N. Foster, Department of Sociology and Anthropology, East Tennessee State University, Johnson City, TN, USA.

Jun Gao, School of Tourism Management, Sun Yat-sen University, Guangzhou, China.

Peter Graham, Community and Public Affairs, Concordia University, Montreal, Canada.

Ulrike Gretzel, Center for Public Relations, University of Southern California, Los Angeles, CA, USA.

Bryan S. R. Grimwood, Department of Recreation and Leisure Studies, University of Waterloo, Waterloo, Ontario, Canada.

Jaume Guia, Department of Business Administration, Faculty of Tourism, University of Girona, Girona, Catalonia, Spain; School of Tourism and Hospitality, College of Business and Economics, University of Johannesburg, Johannesburg, South Africa.

İlkay Taş Gürsoy, Reha Midilli Foca Tourism Faculty, Dokuz Eylul University, Izmir, Turkey.

Lucy C. Harbor, Department of Recreation, Park and Tourism Management, The Pennsylvania State University, State College, PA, USA.

Anne Hardy, School of Social Sciences, University of Tasmania, Hobart, Tasmania, Australia.

James Higham, Department of Tourism, University of Otago, Dunedin, New Zealand.

Carter A. Hunt, Department of Recreation, Park and Tourism Management, The Pennsylvania State University, State College, PA, USA.

Tazim Jamal, Department of Recreation, Park and Tourism Sciences, Texas A&M University, College Station, Texas, USA.

Dawn Jourdan, Landscape Architecture and Urban Planning, Texas A&M University, College Station, TX, USA.

Jaime Jover, Department of Human Geography, University of Seville, Sevilla, Spain.

Andrew Joyner, Department of Geosciences, East Tennessee State University, Johnson City, TN, USA.

Miranda Lee, The University of Tennessee, Knoxville, Knoxville, TN, USA.

Seunghoon Lee, Department of Recreation, Park and Tourism Sciences, Texas A&M University, College Station, TX, USA.

Hongxia Lin, School of Business Administration, Southwestern University of Finance and Economics, Chengdu, China.

Philip A. Loring, Arrell Food Institute and the Department of Geography, Environment, and Geomatics, University of Guelph, Guelph, Ontario, Canada.

Judith Mair, Tourism Discipline, UQ School of Business, The University of Queensland, St Lucia, Australia.

Frans de Man, Retour Foundation, Rotterdam, Netherlands.

Naho Ueda Maruyama, Department of Tourism Policy, Takasaki City University of Economics, Takasaki, Gunma, Japan.

Can-Seng Ooi, College of Arts, Law and Education, University of Tasmania, Hobart, Australia.

Raymond Rastegar, Business School, University of Queensland, Brisbane, Australia.

Verónica Santafe-Troncoso, School of Environment and Sustainability, University of Saskatchewan, Saskatoon, Saskatchewan, Canada.

Becky Shelley, Peter Underwood Centre, University of Tasmania, Hobart, Australia.

Valerie Sheppard, Independent Research Scholar.

Mick Smith, Environmental Studies/Philosophy, Biosciences, Queen's University, Kingston, Canada.

Siobhan Speiran, Environmental Studies, Biosciences, Queen's University, Kingston, Canada.

Michela J. Stinson, Department of Recreation and Leisure Studies, University of Waterloo, Waterloo, Ontario, Canada.

Oceane Tanny, Department of Sociology and Anthropology, East Tennessee State University, Johnson City, TN, USA.

Mario Vázquez-Maguirre, Business School, Universidad de Monterrey, San Pedro Garza Garcia, Mexico.

Oscar Vorobjovas-Pinta, School of Business and Law, Edith Cowan University, Joondalup, Western Australia, Australia.

Jani Wertin, Landscape Architecture and Urban Planning, Texas A&M University, College Station, TX, USA.

Kyle Maurice Woosnam, Warnell School of Forestry and Natural Resources, Parks, Recreation and Tourism Management Program, University of Georgia; The School of Tourism and Hospitality Management, University of Johannesburg.

Zohreh (Zara) Zarezadeh, Business School, University of Queensland, Brisbane, Australia.

Chaozhi Zhang, School of Tourism Management, Sun Yat-sen University, Guangzhou, China.

Justice and ethics: towards a new platform for tourism and sustainability

Tazim Jamal and James Higham

ABSTRACT

In times of upheaval and uncertainty justice has come to the fore as a key principle to guide tourism development and policy. Justice is being sought by individuals, groups and societies, as vulnerable individuals and populations seek to be able to protect themselves from longstanding and deeply embedded historical, and new emerging forms of, injustice. In response, this special issue presents a range of theoretical perspectives and empirical insights into justice and tourism. Given the early stage of theory building in this field, tourism researchers are well served by exploring the multi-/inter-/trans-disciplinary knowledge domains in which theories of justice are being addressed. In this paper frame some emerging principles and approaches to justice and tourism as addressed in the special issue, including social justice, equity and rights; inclusiveness and recognition; sustainability and conservation; well-being, belonging and capabilities; posthumanistic justice; and governance and participation. In doing so we outline the wide range of issues and insights for 'just' tourism that demand urgent and rigorous scholarly attention. A nascent new platform of research on justice and ethics is emerging to guide tourism and sustainability. This special issue offers valuable insights and guidance towards this timely and important research agenda.

Introduction

The World Health Organized declared a global pandemic on 11 March, 2020. Cruise lines, air travel and mobilities of goods and services came to a standstill as border closures, lockdowns and other containment, mitigation and, in some cases, elimination strategies were attempted at local, regional and national levels. Black Lives Matter (BLM) sparked anti-racism protests over police brutality following the death of George Floyd at the hands of a Minneapolis police officer in the USA in May 2020. BLM protests against the systemic injustices of colonialism erupted worldwide. Justice and care is being sought at all levels, in all places, from the individual to the group and societal level, for front-line workers (including in tourism and hospitality) as well as first responders at high risk without adequate personal protective equipment, and for vulnerable individuals and populations least able to protect themselves (Jamal & Budke, 2020).

The COVID-19 pandemic has revealed deep inequities that stand to be further exacerbated by the existential threat of climate change. Extreme weather, rising sea level rise and loss of food security are displacing populations and contributing to increasing human rights issues. Up to a million species are threatened with extinction, many very imminently (IPBES, 2019), as the biodiversity crisis accelerates. A recent study published in *Nature Climate Change* indicates that tourism contributes 8% of total GHGs from related transport and food consumption (Lenzen et al., 2018).

Tourism is deeply implicated in these challenges, but it also offers hope for more sustainable futures. In light of the unprecedented pandemic crisis, many are calling for new strategies and approaches for *environmentally and socially responsive* tourism. Will destinations struggling for swift recovery pay attention to the climate crisis and develop restorative measures towards ecological and social resilience? Will it be 'business-as-usual', racing towards a return to neoliberal globalization, or will places seek new way to address *overtourism* and look towards *degrowth* and individual as well as communal *well-being* rather that economic growth as a measure of modernist notions of 'progress' (Escobar, 2018)? What ethical principles guide 'just' and 'sustainable' tourism in an Anthropocene where climate change and pandemics are interwoven with other global issues such as large-scale migrations and refugees, water and food crises, geopolitical conflicts, systemic oppression of minority groups, and acts of domestic and global terrorism?

Justice as a key principle to guide tourism development and policy is the focus of this special issue. The two volumes of the special issue present a range of perspectives on justice and tourism through practical cases and theoretical directions provided by scholars researching in a range of disciplinary fields. It speaks to their concerns, their critical knowledge, their hopes and aspirations, and their insights on the subjects of their scholarly attention. They illustrate that issues of justice are interwoven with social-ecological worlds and everyday life in the places and spaces of travel and tourism. Authors approach this interrelated domain from their situated perspectives, their positionality in relation to different disciplinary or inter-/trans-/post-disciplinary viewpoints.

We have grouped their efforts broadly under some general categories that illustrate the complex scope and scale of this research domain. The social-ecological and technological systems in which travel, tourism and hospitality are enacted weave together the local and the global. The issues that arise span different levels and modes of engagement—from individual well-being, belonging and caring for land, building capabilities for a good life, to social justice at the societal level; from governance and (public) participation, to inclusion and recognition of diverse and disadvantaged groups; and from mainstream (and modernist) notions of sustainability and conservation to posthumanistic approaches that eschews anthropocentrism for relational ways of being and belonging with human and non-human others.

Many other principles and values also weave through the twenty-two papers in the special issue. Some are grounded in empirical or theoretical support, others are fleeting, tantalizing invitations to engage in much-needed empirical research, theory building and praxis in justice and tourism. All are crucial reminders of the young stage of research in justice and tourism, as we note below. Inter-/trans-/post-disciplinary approaches to justice and an ethic of care are needed as the various papers in the special issue illustrate. The special issue contributors offer critical insights on a wide range of tourism phenomena in local to global social-ecological systems, together with principles of justice and ethics to help navigate the journey towards building the knowledge base for just tourism.

It's a long and winding road

In 1990, Jafar Jafari succinctly summarized the emerging field of tourism studies in four platforms: the Advocacy platform (advocating tourism as an effective development strategy to provide employment and income), the Cautionary platform (summarizing a growing base of concern and criticism of tourism's adverse effects), the Adaptancy platform (calling for alternative approaches to increase tourism benefits and reduce its environmental and social costs), and the

Knowledge Based Platform (moving descriptive and applied research towards more scientific approaches and theoretical constructs) (Jafari, 1990, 2001). Indeed, academics have been challenging inequalities and injustices in tourism since the 1970s (e.g., Hills and Lundgren, 1977), for instance, economic leakage leaving destinations carrying the burden of environmental degradation, resource depletion, human rights violations (e.g., on service workers in the labor market, using children in sex tourism, etc.) and other social and cultural disbenefits (e.g.,Mowforth & Munt, 2016). Britton's (1982) well-cited work on the inequalities evident in the political economy of tourism has been echoed on criticism of globalization by various researchers, with some calling for alternative globalizations and more 'sustainable' approaches to tourism development (e.g., Higgins-Desbiolles, 2008, Sharpley & Telfer, 2016).

Critiques of mass tourism, neocolonialism and dependency gathered momentum in the 1980s and 1990s, spurring growing literatures on 'alternative tourism' and 'responsible tourism' approaches, and various types or forms of tourism, such as community-based tourism (CBT), volunteer tourism, fair trade tourism and pro-poor tourism (e.g., see Gohori & van der Merwe, 2020; McGehee, 2014). *Environmental justice* and *social justice* address critical issues related to equity, fairness and rights in relation to environmental health and social well-being from the local to the global systems in which tourism occurs. A number of studies explore inequalities and discriminatory practices being experienced by, women and children, LGBTQI+and people of color, Indigenous, minority and diverse populations, the poor, the elderly and people with disabilities, as well as migrant populations and climate refugees (e.g., Cole, 2017, Bianchi & Stephenson, 2014). Political ecology studies, for example, reveal inequalities in the distribution of costs and benefits, loss of access and use of local commons, and disproportion allocation of scarce resources like water to tourists while disadvantaged local populations suffer (Stonich, 1998, Nepal & Saarinen, 2016).

Cultural injustices also abound, as evident in misrepresentation, stereotyping, systemic racism and various forms of cultural imperialism. Speaking to the forced dislocation of the San (also known as Basarwa) from their traditional land, Marobela (2010) notes that the Basarwa, the First People of the Kalahari, feel that the social oppression and economic injustice they face relate to making way for diamond mining and tourism by dispossessing them of access and belonging to their ancestral land. Of course examples of 'good' tourism also exist, as in places that are typically community-driven and inclusive of local and traditional knowledge. Such cases seek equitable distribution and caring relationships, respectful and enriching cultural experiences, and offering ways to disidentify from unjust practices (e.g., Hollinshead, 1992, Dredge, 2019; Kaminski et al., 2014). *Inclusiveness* and *recognition* (political and personal), especially of those who stand to be most affected by tourism development and marketing, have been acknowledged as critical priorities for a 'just' and fair tourism.

Yet, while such issues of justice and rights, well-being and the "good" play out widely in tourism and hospitality studies, they often remain implicit or poorly theorized, with rare exception (see, for example, Fennell, 2006, Smith & Duffy, 2003; Luh Sin et al., 2015). The *capabilities approach* forwarded by development theorists Sen and Nussbaum, for instance, offers useful theoretical and practical concepts for advancing social justice and well-being in responsible tourism, pro-poor tourism, and community-based tourism. Feminist theorists including Gilligan (1982) and Noddings (1999) offer essential insights into an ethic of care. Too little is evident of such insights to help inform social justice and other important companions like restorative justice and care ethics. It is not surprising that Jafar Jafari makes yet another attempt to spur greater advances on the 'knowledge-based platform' by proposing a fifth platform: the 'Public platform', where he calls on tourism researchers to recognize the relationship between tourism and other areas and to therefore bring interdisciplinary perspectives to inform empirical studies and knowledge building (Jafari, 2005).

Another important occurrence in ongoing efforts toward fairer and more ethical outcomes of tourism development arrived in the 1980s and set the stage for what Macbeth (2005) felt should be a separate platform: 'Tourism and sustainability'. Prolific growth in tourism research and critical insights to progress justice relate priorities arose as the United Nations World Tourism Organization (UNWTO) assumed a comprehensive North-South global agenda for tourism and conservation, following the Brundtland Commission's report *Our Common Future* (UNWCED, 1987). Equity and fairness for present and future generations of visitors and host destinations, environmental conservation and sustaining "cultural integrity" are highlighted in the UN's approach to sustainable tourism development:

> Sustainable tourism development meets the needs of present tourists and host regions while protecting and enhancing opportunity for the future. It is envisaged as leading to management of all resources in such a way that economic, social, and aesthetic needs can be fulfilled while maintaining cultural integrity, essential ecological processes, biological diversity, and life support systems. (UNWTO (United Nations World Tourism Organization), 1995, p. 30)

Offsetting macro-global perspectives as outlined above, are micro-local approaches that offer a different discourse to sustainability and a more holistic approach to justice that is situated in notions of well-being, belonging and building capabilities that enable living a good life. Moscardo and Murphy (2014) offer a trenchant critique of sustainable tourism and argue that its narrow, rule-driven framing omits consideration of the wider domain and can disempowerment local residents. Furthermore, it "ignores the value-laden, subjective, cultural nature of sustainability, and thus fails to address the justice issues that are critical to understanding the origins of unsustainability" (Moscardo & Murphy, 2014: 2540) They call for more explicit consideration of ethical approaches to tourism and social justice, and argue that the idea of well-being or quality of life (QoL) is increasingly being used by researchers as a better indicator to assess sustainability in tourism. In their usage, QoL applies to the individual level and well-being to a broader level (e.g., community well-being):

> QoL refers to the idea that a good and satisfying life depends on meeting key basic needs... These needs are physiological, including food, water, good health, and security, opportunities for education, productive activity and involvement in relevant decisions, belongingness, and self-esteem... Well-being is often seen as the broader term ... (Moscardo & Murphy, 2014: 2541).

But what ethical principles guide such notions of a good life that sustainability and tourism should endeavor to provide? Hultsman (1995: 561) was an early proponent of "just tourism" and sought a "paradigm of ethicality" for service providers and tourism marketers, arguing that "[s]hould tourism reach the point of being considered by service providers as first a business and second an experience, it is no longer "just tourism"; it is industry" (*ibid*, p. 561). What constitutes "just tourism" remains unclear. Hultsman states 'just tourism' counsels service providers to act in a manner that reflects ethicality."

A critical examination of the tourism knowledge base at the start of the 21st century shows a predominance of modernist, primarily growth-driven, anthropocentric and Eurocentric approach to tourism, obfuscating "other" discourses (e.g., Indigenous, critical, Marxist discourses). Animal welfare and nature's rights have languished (see Holden, 2019), but the new century has brought greater diversity in tourism scholarship and new challenges. Caton (2016), for instance, describes the human-centric approach to humanism that ascribes positive value and moral worth to human beings, but notes also the rise of posthumanism since the 1980s, which is non dualist, immanent (not transcendental), and relational with human and non-human others (Keeling & Lehman, 2018). Modernist and anthropocentric values are challenged and decentered by posthumanist approaches, as human-centered privilege is displaced. New opportunities arise to address identity politics, colonial histories (including settler colonialism), Indigenous perspectives and ways of being and caring for land, alternative economies like community social enterprise (i.e., other than market capitalism), human-environmental/human-animal relationships and nature's rights (e.g. Holden, 2019; Puriri and McIntosh, 2019) Situated place-based research and decolonizing methodologies aid the task of advancing justice, self-determination and autonomy for Indigenous and other historically oppressed groups (e.g., Tuck & McKenzie, 2015).

Yet, while new spaces of understanding injustices in tourism are dawning as the 21st century advances, what constitutes "good governance" for tourism bound up in complex social ecological systems, dominated by neoliberal globalization, lacking global regulation and oversight while industry discourses favor "voluntary" certification programs? Some hope is afforded by new form of tourism where greater responsibility is assumed by residents and visitors, such as in justice tourism. Justice tourism is "both ethical and equitable," according to Scheyvens (2002: 104), and social change is facilitated by political activism by organizations such as NGOs (e.g., the reality tours offered by Global Exchange), but also by "tourists of conscience" (Scheyvens 2011: 107-8). It advocates a just form of travel that aims to achieve mutual understanding, solidarity and equality among participants, as well as "social and environmental respect, economic equity, and support for people's struggles for self-determination in relation to tourism policy" (Scheyvens, 2002: 104). Planning and policy are political, after all, often involving multi-stakeholder contestation over self-determination, autonomy, voice and control over tourism development (see Wardana, 2018, 2019). Examining contemporary destinations like Singapore, Fainstein (2017) argues that justice is the first principle to guide policy evaluation in spatial and urban planning—justice principles must clearly be situated in the policy domain (see also Jamal & Camargo, 2014).

Justice and ethics in tourism is receiving increasing attention and study, but much work awaits to clearly conceptualize and identify key principles and approaches to justice to navigate through the challenges that await tourism in a (post)covid world. It is a long and winding road with many discoveries lying ahead to weave just tourism futures. The early stage of theory building in the complex multi-/inter-/trans-disciplinary context of tourism studies means that hospitality and tourism researchers must extend themselves into the knowledge domains in which theories of justice are being addressed. The contributors to the special issue on justice and tourism have embarked on this challenging task in various ways. It is important to note that the special issue was well advanced when the pandemic was declared by the World Health Organisation on 11 March, 2020. This special issue was not driven by the paralysis of global tourism due to COVID-19. However, its timing has been amplified, for without concerted action the cumulative threats of climate change alongside current and future pandemics will continue to exacerbate economic inequalities and socio-political as well as ecological and cultural injustices. As the Special Rapporteur on extreme poverty and human rights, Philip Alston noted in his report to the United Nations (UN), lack of action on climate change will not only undermine basic rights to sustenance for hundreds of millions of people, it will also threaten democracy and the rule of law (Alston, 2019).

Approaches to justice and tourism in the special issue

The subject of justice in tourism and hospitality studies is indeed slowly being advanced by the academic community, alongside closely related areas of ethics, but growing pains are apparent. Issues of equity and rights, well-being and the "good" play out widely in tourism and hospitality studies, but they often remain implicit and poorly theorized. There is clearly a large body of historical and contemporary perspectives worldwide that inform justice and ethics, ranging from Eastern religious and Indigenous sacred worldviews to the contemporary perspectives offered by civil rights leaders such as Martin Luther King and Nelson Mandela. The European/Western tradition in which many of the papers in this special issue are situate include philosophers like Plato and Aristotle in Antiquity, liberal Enlightenment philosophers like John Locke, classical utilitarian philosophers like Jeremy Bentham and John Stuart Mill, modern moral philosophers like Immanuel Kant, leading up to contemporary political theorists like John Rawls and critical feminist theorists like Nancy Fraser. The *Stanford Encyclopedia of Philosophy* (https://plato.stanford.edu/) is a helpful online site for scholarly resources on various philosophers and their theoretical perspectives. Indigenous, feminist and post-development approaches have also emerged to challenge Eurocentric modernist views (see, for example, Sachs 2019).

So, what is justice and how have leading authors in the field of justice and tourism taken it up? Justice, according to the preeminent Harvard professor and philosopher, Michael Sandel, is giving people what they deserve (Sandel, 1998). For some, this is based on rights and duties, and notions of fairness and equality (equity being a related aspect). For example, the contemporary political philosopher John Rawls uses a contractarian approach to social justice. He puts forward a highly influential, liberal egalitarian paradigm of 'justice as fairness' (Rawls, 1971), where people have rights and should have equal opportunity to societal goods and opportunities (equal rights). Rawls' rights-based approach to social justice draws upon Immanuel Kant's influential moral philosophy (including notions of autonomy and respect) to address the fair distribution of societal goods, using appropriate institutional structures. Institutional rules should reduce inequalities or at least ensure that inequities favor the most disadvantages. His rights-based approach to social justice has dominated liberal pluralist societies in the west since *A Theory of Justice* was published (Rawls, 1971).

Social justice, equity and rights are taken up in various ways in the special issue by various leading scholars specialising in a range of disciplines and fields. Some authors have focused specifically on issues related to social well-being (including workers' rights, diversity and gender rights, and people with disabilities) culture and heritage (including Indigenous Tourism, fair food and sovereignty); others on sustainability and conservation (including heritage conservation as well as animal justice and intergenerational rights). In addition to research articles that present cases from various countries and regions, there are a few descriptive and theoretical articles related to posthumanist and Indigenous worldviews and methodologies. Each provides valuable insights on a variety of justice issues. Rather than attempt to summarize them systemically, the focus of the articles that appear in this special issue are considered in relation to interrelated emergent themes that weave through the special issue. They illustrate a multitude of theoretical and applied approaches to justice from the diverse knowledge domains of the special issue contributors.

The categories in Figure 1 represent some key principles and topics that weave through the articles in the special issue. Numerous other principles are nested within these overall categories as the papers reveal, such as respect, care, autonomy and sociality. The boundaries are invisible and porous in this emergent picture of the diverse approaches to justice the authors undertake, and the principles and values they identify are interwoven within and between the articles. In Figure 1 we move away from the use of boxes, solid lines and arrows, so typical in the presentation of conceptual models. Instead, the concepts and approaches that emerge in this special issue, as illustrated in Figure 1, should be viewed from a holistic and interrelated perspective rather than from an either/or binary standpoint.

The picture that emerges is partial and processual. Justice is in the process of 'becoming' part of the knowledge base in tourism studies. Important concepts and considerations for justice and tourism emerge through the struggles, challenges and experiences of the authors as they grapple with complex and critical issues for local-global sustainability and well-being. In their articles, too, you see diverse methodologies and reflexive voices, some sharing tentative propositions, others forwarding firm convictions. They are supported by empirical insights, case studies or descriptive/theoretical efforts as summarized below.

Social justice, equity and rights

In a globalized world of porous borders and increasing mobile travelers, residents and visitors have increasingly 'collided' in neighborhoods and public spaces that became quickly popular, and one question that accompanies *overtourism* is: Who has the rights to the social? Ibán-Diaz-

Figure 1. Emerging principles and approaches to justice and tourism in the special issue.

Parra and Jaime Jover draw on Henri Lefebvre's work of the right to city (Lefebvre, 1968) to address overtourism and gentrification in Seville, Spain. Here, both long-time residents and more recent incomers have had to *appropriate* the rights to their city and overcome *place-alienation*. By contrast, Judith Mair and Michelle Duffy explore rights to the rural, specifically, the rural community of Dugong, Australia, and the Dugong Festival held there. While drawing inspiration from Lefebvre, they use Pierce *et al.*'s (2016) framework of rights in places to guide their investigation of diverse interests, inclusion and exclusion (see also LGBTQI+ rights and appropriation of the Australian Outback further below).

Fairness and equity in distribution of benefits and burdens are principles that run through the special issue. Rights and responsibilities of diverse stakeholders groups are taken up by Jun Gao, Hongxia Lin and Chaozhi Zhang at the Cultural Landscape of Honghe Hani Rice Terraces, China (a UNESCO World Heritage Site). The authors examine issues related to the rights of minority groups whose traditional agricultural practices maintain these heritage landscapes, but from which they fail to realize benefit compensation and a fair and equitable share of tourism benefits. Systematic unequal power relations between the minority local resident population and the state exacerbate the inequalities, a theme that is seen in other articles as well. By contrast, Lucy Harbor and Carter Hunt argue that positive benefits are being facilitated by cultural tourism in the Tz'utujil Maya community in Guatemala, despite typical critiques of commodification and 'authenticity' often associated with tourism development. They draw upon and build on principles of Indigenous environmental justice discussed by Whyte (2010) in their discussion of Indigenous cultural justice.

Inclusiveness and recognition

Important principle of inclusiveness and recognition in tourism development and conservation arises in several articles in the two volumes of the special issue, and are interwoven with other justice principles. Michela Stinson, Bryan Grimwood and Kellee Caton use the metaphor of *becoming common plantain* (i.e., *Plantago major*) to illustrate how Indigeneity may be reclaimed through storytelling. Their experimental, more than representational approach endevours to storytell how Settler accountability, decolonization and justice as healing can emerge as historic exclusions and the invisibility of colonization are ameliorated. Brent Fortenberry's article on documentation and heritage conservation in the Greater Caribbean illustrates the tensions of postcolonial exclusion, erasure and disconnect with enslavement history. Heritage justice and tourism are interconnected in the examples from St. George's, Bermuda, and Falmouth, Jamaica. The paper calls for transformation of heritage programming, building local capabilities and greater inclusiveness and equity for under-represented communities in the Greater Caribbean. Exclusion of marginalized groups in a historical context is also taken up in Naho Maruyama and Kyle Woosnam's paper, which presents a case study of young women workers at the Tomioka Silk Mill in Japan between 1870-1940. Economic injustice and recognition justice are interwoven in this study. Touristic representations of the "mill girls" reveal gender injustice where economic exploitation during the mill's operation morph into stereotyping and misrecognition when the mill becomes a cultural attraction, further entrenching recognition injustice (see, too, Fraser, 2010).

Inclusiveness in various articles is closely related to not only recognition justice and distributive justice plus fair distribution of economic costs and benefits, but also fair *representation* in the programs and processes of tourism development, marketing and conservation. Scheyvens and Biddulph (2018) describe inclusiveness in tourism development as a transformative activity beyond economic objectives like employment and income generation. Their focus is on including marginalized groups actively in the development process, constructive and respectful representation, mutual understanding and inclusion in participatory process for joint decision making, transforming power relations that exclude disadvantaged groups. This resonates with Fortenberry's article on conservation in the Greater Caribbean, and also with Stefanie Benjamin, Ethan Bottone and Miranda Lee's exploration of the representation of people with disabilities (PWD) in tourism marketing materials in southeast USA. Their findings call for greater inclusion of PWD and breaking down physical and psychological barriers to PWD participation in travel and tourism. Exclusion is evident also in the study of historic roadside markers in Tennessee, USA, by Candace Bright, Kelly Foster, Andrew Joyner and Oceane Tanny. Their analysis of 313 historic road markers enacted since 1988 reveals lack of "just representation" of non-white male history in these markers that has carried through historically to the present day, as they point out.

Sustainability and conservation

Yet more is missing with respect to the rights of those who cannot speak for themselves in the past and present, as well as the future. Sustainable tourism as advocated by the United Nations World Tourism Organization (UNWTO) draws on an important justice principle (equity) to address inter- and intra-generational equity in tourism with a North-South perspective (see UNWTO (United Nations World Tourism Organization), 1995, p. 30; see also UNWTO, 1985). However, critics of the UNWTO point to a pervasive neoliberal discourse, which has also tended to be anthropocentric and Eurocentric (see Mowforth and Munt, 2003). Raoul Bianchi and Frans de

Man provide a political economy critique of "inclusive growth" and "decent work" as provided in Sustainable Development Goal 8, used in the UNWTO sustainable tourism agenda (https://www.unwto.org/tourism-in-2030-agenda). They hone in on how these concepts reinforce a growth-driven tourism development model, neoliberal market notions of justice, and the primacy of capital.

A number of other important issues tend to be overshadowed by a neoliberal agenda, such as the commercial uses of animals for tourism and the development of approaches to animal justice. Attention to the area of animal welfare is growing slowly, while little is being said on the rights of Nature (Holden, 2019). Noting this, David Fennell and Valerie Sheppard propose a 'scales of justice' framework for animals used in tourism, contrasting two different normative ethics models: a conventional model that includes deontology and consequentialism, versus a model rooted in virtue ethics. Attentiveness to non-human 'others' is taken up further by other contributors to the special issue, in terms of voice, rights and well-being. They also show the interrelatedness of human and non-human others with landscapes, place, past (e.g., in terms of cultural heritage) and future (e.g., in relation to climate change). Dialogic resistance and relational praxis for justice and well-being are also evident in the themes related to well-being and posthumanistic justice further below.

Despite reference to intra- and inter-generational equity in the UNWTO's definition of sustainable tourism, progress is also slow with respect to current generations of youth as well as future generations. Dawn Jourdan and Jani Wertin[1] draw upon the case of Juliana v. United States to draw attention to intergenerational rights, specifically the rights of youth and children to a sustainable future, in the face of climate change. They also discuss climate justice in the Global North and South, noting the challenge faced by Small Island Developing States (SIDS). Mick Smith, Siobhan Speiran and Peter Graham offer another perspective on climate change that illuminates the ontological and ethical presumptions of reducing the world to a 'standing reserve' of human resources, leading to destructive environmental consequences. They draw on Material Engagement Theory (MET) and megaliths to discuss the distribution of agency and how even archetypically inanimate objects (like stones) actively constitute and transform places, thoughts, and lives, in diverse ways, leading to redemptive agentic possibilities.

Well-being, belonging and capabilities

Recognition of industry-driven injustices related to mass tourism internationally and especially in developing countries has paved the way for more responsible, equitable and 'just' forms of tourism to arise. Referring to *responsible tourism*, Goodwin (2011: 16) states that it "recognizes the importance of cultural integrity, ethics, equity, solidarity, and mutual respect, placing quality of life at its core". But what constitutes "quality of life" and what is the role of justice here? Following a perspective of human flourishing or well-being, development theorists Amartya Sen and Martha Nussbaum argue that justice should be about helping people live lives that they consider worthwhile rather than focusing on the appropriate distribution of primary goods as John Rawls does (Sen, 1999, Nussbaum, 2011). Moral values of equal dignity and equal respect for people's dignity guide the concept of well-being in Nussbaum's approach. The basic idea is that "some living conditions deliver to people a life that is worthy of the human dignity that they possess, and others do not … A focus on dignity is quite different, for example, from a focus on satisfaction" (Nussbaum, 2011, p. 30, cited in Jamal, 2019, p. 68). But some basic capabilities (a minimum threshold of capability) are needed that would enable a dignified and minimally flourishing life, as she puts it. She asks:

> Among the many things that human beings might develop the capacity to do, which ones are the really valuable ones, which are the ones that a minimally just society will endeavor to nurture and support? (Nussbaum, 2011, p. 28)

In contrast to mass tourism and modernist notions of "progress" (often measured in terms of economic growth, hence visitor numbers, revenues, etc.), approaches to "alternative tourism" development are oriented towards individual and communal well-being as an end. Community-based tourism (CBT) similarly address capacity building for local well-being. Nussbaum's approach offers valuable guidance to building capabilities, and using guiding principles of dignity and equal respect. Blanca Camargo and Mario Vázquez-Maguirre in this special issue draw upon management literature grounded in humanism to argue for the priority of dignity as a key principle for well-being and justice. They propose guiding principles and stakeholder responsibilities to enable fairer development outcomes and building capabilities for Indigenous Maya in the case of the Maya Train project in southern Mexico. Research by Verónica Santafe-Troncoso and Philip Loring on Indigenous food sovereignty in Ecuador illustrates how the Kichwa and non-Kichwa people on the "Chakra Route" visited by tourists conserve biodiversity through traditional agroforestry methods, enabling cultural flourishing and food security. The article demonstrates how the women's management of the chakra gardens and participation in tourism facilitate self-determination, autonomy, agency, and well-being.

Well-being is addressed explicitly or implicitly in many articles in the special issue. Alana Dillette's study of the lived experience of roots-seeking African-American tourists to Ghana demonstrates the importance of such journeys to self-awareness, identity and belonging, as well as fostering a "double consciousness" that enables civic commitment to social justice. Facilitating roots travel, building capabilities and inclusiveness in heritage conservation and interpretation (see also Fortenberry's article on the Greater Caribbean) and ensuring "just representation" in public history as Bright *et al.* point out in their study of historic markers in Tennessee, are among the opportunities that heritage tourism can undertake to facilitate the well-being of these historically oppressed diasporic populations.

Posthumanistic justice

While papers like those mentioned above do not explicitly take a posthumanistic turn, the close relationships between people and place, objects like cultural sites, heritage buildings, and historic markers, as well as traditional food and cultivation practices, show how human and non-human others are woven together and guided by situated ethical practices and non-moralistic subjectivities that open possibilities for other ways of relating and caring. A posthumanistic approach eschew binaries of ecology "versus" culture and de-centering anthropocentric human dominance and towards non-dualistic agencies and identities, countering stereotypes and enabling generative and transformative ways of being and becoming. Modernist values of instrumental reason and commodification are resisted through situated acts and dialogue, as illustrated by many of the special issue articles.

A number of articles like that of the Chakra women and the roots-seeking tourists noted above illustrate performative acts in which participants exercise agency, appropriate social and cultural rights, and engage actively in acts to advance well-being and sustainability. These are enacted in close relationships to land and belonging in social-ecological systems, i.e., to their cultural heritage. In other articles, justice is "performed" by human as well as non-human actors. In their study of the Broken Heel drag queen festival at Broken Hill, Australia, Oscar Vorobjovas-Pinta and Anne Hardy describe how stereotyping, marginalization and exclusion of LGBTQI+communities are being resisted. Drawing on neo-tribal theory, they discuss how LGBTQI+participants appropriate and reconstitute the Outback landscape and festival space through individual and joint expressions of non-heterosexual identities and sociality. Sociality arises also in the paper by İlkay Taş Gürsoy as she follows the traces of a traditional heritage food, Karakılçık bread, in everyday and touristic spaces in the Cittaslow city of Seferihisar, Turkey. Various performative acts conserve heritage and the ancient grains in this slow food destination.

Actor-Network Theory in her study helps to trace the policies enacted around this ancestral food, the festivalization of the bread as well as co-created experiences and interactions of conviviality between tourists and local residents in places like the farmer's market in Seferihisar. Slow food justice here is being enacted in the performing of the bread in the slow food network of human and non-human actors.

Contesting exclusion and negotiating justice through the performance and voices of those who stand to be most affected is also evident in the case by Becky Shelley, Can-Seng Ooi and Lisa Denny of workers' experiences during a long-standing industrial dispute at Australia's first legal casino. They draw on Bakhtin's *dialogic imagination* to show how, from a Bakhtinian theoretical perspective, justice is a social activity and doing justice is a social process (and a difficult one, as the case illustrates). Like the sociality and conviviality being social constructed around the ancestral bread above, Bakhtin's theoretical perspectives comport well with posthumanist approaches that are attentive to human and non-human others situated in immanent social-ecological and technological networks rather than the transcendental plane (see also the article on *neo-lithic ethics* by Mick Smith, Siobhan Speiran and Peter Graham. Herein lies potential for decommodifying and reclaiming the depoliticizing of various forms of "alternative" tourism that aspire towards justice and fairness. Jaume Guia's paper provides a trenchant critique of the notion of *justice tourism* and deconstructs moralistic approaches from political responsibility as a primary aspect of justice tourism. As per the conceptual argument in this paper, justice tourism offers radical transformative potential when it is free of moralistic precepts and coupled with post-humanist *affirmative ethics* and political responsibility (solidarity and advocacy are complementary principles here).

Governance and participation

Contemporary philosopher Michael Sandel (1998) argues that we cannot make sense of our debates on justice without drawing on ethics, because virtues and notions of the good are woven into our arguments. The idea of who deserves what and why, and what this has to do with the virtue of persons is at play often without us realizing, so we must consider desert, virtue and the common good, in addition to utility and rights (Sandel, 1998). He draws here on Aristotle, for whom justice is supreme among the virtues, and what is just is "whatever produces and maintains happiness and its parts for a political community" (Aristotle, 1999: 68). The aim of the state is to achieve human good, such as by designing institutions and programs that work toward individual as well as collective good. The article by Raymond Rastegar, Zohreb (Zara) Zarezadeh and Ulrike Gretzel on the UNESCO World Heritage Site of Yazd, Iran, notes the challenge of such inclusive participation in governing a local-global heritage for fair and equitable distribution of tourism's costs and benefits, for the communal good of its residents and the good of visitors.

Even greater challenges await as travel and tourism has come under immense pressure with COVID-19, and pandemics are expected to be among future global threats to be anticipated, along with rising economic and social inequalities. Information sharing and preparedness, collaborative planning and coordination of a local to global phenomenon raises news question for governance, from the local to the global. The research note by Jiyoung Choi, Seunghoon Lee and Tazim Jamal describes South Korea's efforts to govern responsibly as it tackles COVID-19 with the help of information and communication technologies, together with the active engagement of its residents and visitors to co-create 'smart' containment and mitigation strategies aided by digital tools. Much awaits future research on 'smart justice' and good governance in this area; surveillance and privacy concerns on contact tracing apps are just one of the issues currently being debated. But it would seem that an ethic of care is an important correlate to an ethic of justice for good destination governance during the pandemic—a "politics of kindness" and empathy (see, for example, BBC, 2020).

Weaving just tourism futures

Research related to justice and tourism is in its infancy and much awaits research and praxis to build a robust knowledge base and weave tighter just tourism futures. The early stage of theory building in tourism studies and in the area of justice and tourism means that contributors have had to reach into multi-/inter-/trans-disciplinary knowledge domains in which theories of justice are being addressed. Authors in this special issue have drawn from sociology, geography, business studies, gender and women's studies, urban planning and architecture, anthropology and philosophy, etc., to theoretically inform their research. The various contributions illustrate not only the range of issues and insights for 'just' tourism that await future exploration, but also the need for collaborative research weaving diverse knowledge domains and researchers in the Global North and Global South to address challenging local to global futures in the Anthropocene.

The articles in this special issue provide diverse examples of the wide range of issues that arise. They tackle critical issues ranging from rights and recognition, inclusion and representation, individual communal capacity and well-being, sustainability and conservation, discrimination and racism, gender, equity and fairness, in the local to the global systems in which tourism occurs. Slowly, they weave a picture of justice including social rights and equity, representation, recognition and inclusion, governance and participation, identity, belonging and well-being. Topics range from social justice and gender, minority and worker rights, building capabilities, indigenous, cultural and heritage justice (including special topics like food sovereignty), posthumanistic perspectives that call us to attend to non-human others, to climate justice and sustainable futures.

Given the complex scope and scale of tourism as a phenomenon that is intricately interwoven with place and people in diverse spaces from the local to the global, they offer a partial and situated picture. Many issues and subject areas that might usefully inform theoretical and empirical inquiry on justice and tourism were regrettably omitted, like political ecology, feminist ethics and ecofeminism, global migrations and the unequal rights to travel across borders, among others. There is also the challenging issue of global justice for a global phenomenon (tourism). In the era of the Anthropocene, this needs to be worked out for present and future generations. As Martha Nussbaum (2019) pointed out, we have well-formed doctrines about duties of justice that ground first-generation rights, but we have not equally worked out doctrines on other duties, such as those of second generations (we can add future generations) and do not seem to know where to begin when we step outside of national boundaries (see also: https://iai.tv/articles/the-cosmopolitan-tradition-a-noble-but-flawed-ideal-auid-1255). Young (2006, p. 103), too, calls for more, saying:

> There are some moral obligations that human beings have to one another as human; these are cosmopolitan obligations or obligations to respect human rights." However: "While the basic moral respect owed to all persons grounds the cosmopolitan obligations ... obligations of justice require more and are based on more than common humanity (Young, 2006, p. 104)

Young (2000) offers an interesting perspective on the challenge of transborder justice and the notion of relational autonomy that is entailed—being with others in relations of solidarity and interdependence, being included in democracy with recognition of difference between and within groups, yet retaining self-determination and cultural difference. For Young, political responsibility is incurred on the part of those who are embedded in the network in which systematic injustices are being facilitated, and change must be advanced through collective action within and across borders. The anti-racism riots that arose worldwide in May following the tragic death of George Floyd demonstrate the potential for such change. History and cultural heritage sites are being revised in many destinations with removal of statues representing colonialism and enslavement. Young was also a philosopher of *difference*—recognition justice entails the *recognition of difference*, she argued, so there must be general rules of justice as well as particular rules suited to context and place—including the excluded "others" who may be erased from history and too often, from public consciousness and civic action (Young, 2011). What, then, constitutes global justice? Future research awaits!

Cultural and heritage justice, recognition (political and personal), equity and inclusiveness, affirmative ethics and posthumanistic 'others', are among the themes that are explored in this special issue. They offer helpful insights to help weave a more hopeful future through solidary and joint actions. Each article in its own unique way reveals the critical importance of exploring diverse approaches to justice, towards individual, communal and societal well-being and planetary flourishing. They demonstrate the need to view justice and tourism from a macro-micro perspective, as occurring within complex local-global systems. As a reflective reading of the papers indicate, it requires understanding the history, context, and interrelatedness of tourism with diverse social-ecological systems. Justice here is a pluralistic concept, the scales of justice extend from the general and universal to a situated, place base "pluriverse" (Escobar, 2018) of diverse meanings and approaches to justice. The contributions of the authors highlight the urgent need for advancing this subject area of justice and tourism through North-South collaborative scholarship in tourism studies and related fields.

Note

1. While, regrettably, their paper was mistakenly published in an earlier issue of the journal, the electronic version can be accessed here.
   See: https://www.tandfonline.com/doi/full/10.1080/09669582.2020.1732992

Acknowledgements

We thank all those who participated in this special issue effort, contributors as well as the patient, constructive and timely reviewers. We would also like to acknowledge those whose voices and work did not make it to publication but whose research on important subjects will hopefully contribute to the knowledge base in future. Our thanks also to Kavitha Kannadasan and Iniya Chitravel for their assistance, and to our families - Tazim to her brother for his constant support and wise assistance.

Disclosure statement

No potential conflict of interest was reported by the author(s).

References

Alston, P. (2019). "Climate change and poverty: report of the special rapporteur on extreme poverty and human rights", Report presented to the UN Human Rights Council Forty-first session, 24 June–12 July, A/HRC/41/39.

Aristotle. (1999). Nicomachean ethics. (T. Irwin, T., Ed.) (2nd ed.). Indianapolis, IN: Hackett Publishing Company.

BBC (2020). Coronavirus: How New Zealand relied on science and empathy. https://www.bbc.com/news/world-asia-52344299).

Bianchi, R. V., & Stephenson, M. L. (2014). *Tourism and Citizenship: Rights, Freedoms and Responsibilities in the Global Order*. Routledge.

Britton, S. (1982). The political economy of tourism in the Third World. *Annals of Tourism Research*, *9*(3), 331–358. https://doi.org/10.1016/0160-7383(82)90018-4

Caton, K. (2016). A Humanist Paradigm for Tourism Studies?. In *Tourism Research Paradigms: Critical and Emergent Knowledges*. A. Munar and T. Jamal (eds.). Published online: 01Jun 2016; 35-56. https://doi.org/10.1108/S1571-504320150000022009

Cole, S. (2017). Water worries: An intersectional feminist political ecology of tourism and water in Labuan Bajo, Indonesia. *Annals of Tourism Research, 67*, 14–24. https://doi.org/10.1016/j.annals.2017.07.018

Dredge, D. (2019). Street Voices. In *Case 4.3 in T. Jamal, Justice and Ethics in Tourism* (p. 129–132). Routledge.

Escobar, A. (2018). *Designs for the Pluriverse*. Duke University Press.

Fainstein, S. (2017). Urban planning and social justice. In M. Gunder, A. Madanipour, & V. Watson (Eds.), *The Routledge handbook of planning theory* (pp. 130–142). Routledge. https://www.routledge.com/guilfordpress-books/details/9781138905016?books/details/9781138905016/

Fennell, D. A. (2006). *Tourism ethics*. Channel View Publications.

Fraser, N. (2010). *Scales of justice: Reimagining political space in a globalizing world*. Columbia University Press.

Gilligan, C. (1982). *In a Different Voice: Psychological Theory and Women's Development*. Harvard University Press.

Gohori, O., & van der Merwe, P. (2020). Towards a Tourism and Community-Development Framework: An African Perspective. Sustainability, *12*(13), 5305. https://doi.org/10.3390/su12135305

Goodwin, H. (2011). *Taking Responsibility for Tourism*. Goodfellow Publishers Limited.

Higgins-Desbiolles, F. (2008). Justice tourism: A pathway to alternative globalisation. *Journal of Sustainable Tourism, 16*(3), 345–364. https://doi.org/10.1080/09669580802154132

Hills, T. L., & Lundgren, J. (1977). The Impact of Tourism in the Caribbean—A Methodological Study. *Annals of Tourism Research, 4*(5), 248–267. https://doi.org/10.1016/0160-7383(77)90098-6.

Holden, A. (2019). Environmental ethics for tourism- The state of the art. *Tourism Review, 74*(3), 694–703. https://doi.org/10.1108/TR-03-2017-0066

Hollinshead, K. (1992). White' gaze, 'red' people — Shadow visions: the disidentification of 'Indians' in cultural tourism. *Leisure Studies, 11*(1), 43–64. https://doi.org/10.1080/02614369100390301

Hultsman, J. (1995). Just tourism: An ethical framework. *Annals of Tourism Research, 22*(3), 553–567. https://doi.org/10.1016/0160-7383(95)00011-T

IPBES (2019). "Global Assessment Report on Biodiversity and Ecosystem Services." Intergovernmental Science-Policy Platform on Biodiversity and Ecosystem Services (IPBES), United Nations report online: https://ipbes.net/ipbes-global-assessment-report-biodiversity-ecosystem-services.

Jafar, J. (2001). The scientification of tourism. In V. L. Smith, and M. Brent (Eds.). *Hosts and Guests Revisited: Tourism Issues of the 21st Century* (pp. 28–41). Cognizant Communication Corporation.

Jafari, J. (1990). Research and scholarship: The basis of tourism education. *Journal of Tourism Studies, 1*, 33–41.

Jafari, J. (2005). Bridging out, nesting afield: Powering a new platform. *Journal of Tourism Studies, 16*, 1–5.

Jamal, T. (2019). *Justice and Ethics in Tourism*. Routledge.

Jamal, T. (2019). Tourism Ethics: A Perspective Article. *Tourism Review, 75*(1), 221–224. https://doi.org/10.1108/TR-05-2019-0184.'

Jamal, T., & Budke, C. (2020). Tourism in a world with pandemics: Local-global responsibility and action. *Journal of Tourism Futures, 6*(2), 181–188. https://doi.org/10.1108/JTF-02-2020-0014

Jamal, T., & Camargo, B. (2014). Sustainable tourism, justice and an ethic of care: towards the Just Destination. *Journal of Sustainable Tourism, 22*(1), 11–30. https://doi.org/10.1080/09669582.2013.786084

Kaminski, J., Benson, A.M. & Arnold, D. (Eds.). (2014). *Contemporary issues in cultural heritage management*. Routledge.

Keeling, D. M., Lehman, M. N. (2018). "Humanism." Entry in Oxfor published online: April 2018. https://doi.org/10.1093/acrefore/9780190228613.013.627. Online: https://oxfordre.com/communication/view/10.1093/acrefore/9780190228613.001.0001/acrefore-9780190228613-e-627.

Lefebvre, H. (1968). *Le droit a la ville*. Anthropos.

Lenzen, M., Sun, Y. Y., Faturay, F., Ting, Y. P., Geschke, A., & Malik, A. (2018). The carbon footprint of global tourism. *Nature Climate Change, 8*(6), 522–528. https://doi.org/10.1038/s41558-018-0141-x

Luh Sin, H., Oakes, T., & Mostafanezhad, M. (2015). Traveling for a cause: Critical examinations of volunteer tourism and social justice. *Tourist Studies, 15*(2), 119–131. https://doi.org/10.1177/1468797614563380

Macbeth, J. (2005). Towards an ethics platform for tourism. *Annals of Tourism Research, 32*(4), 962–984. https://doi.org/10.1016/j.annals.2004.11.005

Marobela, M. (2010). The State, Mining and the Community: The Case of Basarwa of the Central Kalahari Game Reserve in Botswana. *Labour, Capital and Society*/Travail. *Capital Et Société, 43*(1), 137–154. Retrieved September 20, 2020. from http://www.jstor.org/stable/43158366

McGehee, N. G. (2014). Volunteer tourism: Evolution, issues and futures. *Journal of Sustainable Tourism, 22*(6), 847–854. https://doi.org/10.1080/09669582.2014.907299

Moscardo, G., & Murphy, L. (2014). There is no such thing as sustainable tourism: Reconceptualizing tourism for sustainability. Sustainability, *6*(5), 2538–2561. https://doi.org/10.3390/su6052538

Mowforth, M., & Munt, I. (2016). *Tourism and Sustainability: Development, globalisation and new tourism in the Third World*. Routledge.

Nepal, S. and Saarinen, J. (Eds.). (2016). *Political ecology and tourism*. Routledge.

Nepal, S. and Saarinen, J. (Eds.). (2016). *Political ecology and tourism*. Routledge.

Noddings, N. (1999). *Care, justice and equity*. In M. S. Katz, N. Noddings, & K. A.

Nussbaum, M. (2011). *Creating capabilities: The human development approach*. The Belknap Press of Harvard University Press.
Nussbaum, M. (2019). *The cosmopolitan tradition: A noble but flawed ideal*. Harvard University Press.
Pierce, J., Williams, O., & Martin, D. (2016). Rights in places: An analytic extension of the right to the city. Geoforum, *70*, 79–88. https://doi.org/10.1016/j.geoforum.2016.02.006
Puriri, A., & McIntosh, A. (2019). A cultural framework for Māori tourism: values and processes of a Whānau tourism business development. *Journal of the Royal Society of New Zealand*, *49*(sup1), 89–103. https://doi.org/10.1080/03036758.2019.1656260
Rawls, J. (1971). *A theory of justice*. Belknap Press of Harvard University Press.
Sachs, W. (2019). The Development Dictionary: A Guide to Knowledge as Power. Zed Book. https://www.zedbooks.net/shop/book/the-development-dictionary/
Sandel, M. (1998). *Liberalism and the limits of justice*. Cambridge University Press.
Scheyvens, R. (2002). *Tourism for development: Empowering communities*. Prentice Hall.
Scheyvens, R. (2011). Tourism and poverty. New York, NY: Routledge.
Scheyvens, R., & Biddulph, R. (2018). Inclusive tourism development. *Tourism Geographies*, *20*(4), 589–609. https://doi.org/10.1080/14616688.2017.1381985
Sen, A. (1999). *Development as freedom*. Oxford University Press.
Sharpley, R., & Telfer, D. (2016). *Tourism and development in the developing world* (2nd ed.). Routledge.
Smith, M., & Duffy, R. (2003). *The ethics of tourism development*. Routledge.
Stonich, S. (1998). Political ecology of tourism. *Annals of Tourism Research*, *25*(1), 25–54. https://doi.org/10.1016/S0160-7383(97)00037-6
Tuck, E., & McKenzie, M. (2015). *Place in research: Theory, methodology and methods*. New York and London.
UNWCED (1987). Towards sustainable development. In *Our common future: report of the world commission on environment and development* (pp. 1–16). United Nations World Commission on Environment and Development (UNWCED). http://www.un-documents.net/ocf-02.htm
UNWTO (1985). Tourism Bill of Rights and Tourist Code. http://www.aitr.org/wp-content/uploads/2014/04/omt_sofia1985.pdf
UNWTO (1999, October). Global code of ethics for tourism: For Responsible tourism. https://www.unwto.org/global-code-of-ethics-for-tourism.
UNWTO (United Nations World Tourism Organization) (1995). *Agenda 21 for the travel and tourism industry: Towards environmentally sustainable development*. UNWTO, WTTC and the Earth Council.
Wardana, A. (2018). Legal engineering over space in Bali. *Australian Journal of Asian Law*, *19*(1), 1–12.
Wardana, A. (2019). *Contemporary Bali: Contested space and governance*. Palgrave. https://doi.org/10.1007/978-981-13-2478-9
Whyte, K. P. (2010). An environmental justice framework for indigenous tourism. *Environmental Philosophy*, *7*(2), 75–92.
Young, I. M. (2000). *Inclusion and democracy*. Oxford University Press.
Young, I. M. (2006). Responsibility and global justice: A social connection model. *Social Philosophy and Policy*, *23*(01), 102–130. https://doi.org/10.1017/S0265052506060043
Young, I. M. (2011). *Justice and the politics of difference*. Princeton University Press.

Overtourism, place alienation and the right to the city: insights from the historic centre of Seville, Spain

Iban Diaz-Parra and Jaime Jover

ABSTRACT
Increasing international travel flows across the planet have drawn attention to socio-spatial justice concerning the impact of tourism and transnational gentrification in cities. We suggest that the growth of tourists and lifestyle migrants can potentially shrink or even deprive a social right to the city. The early work of philosopher and sociologist Henri Lefebvre, particularly *The Right to the City* (1968), offer valuable elements to understand the potential alienation of local residents from the city centre in the context of a tourism-led urban economy. Several questions arise in this context that are explored here: What injustices arise through the socio-spatial processes and transformation of social spaces due to rapid tourism growth and overtourism? What contribution does Lefebvre's theoretical work make to current studies on overtourism overall? We examine these questions through a case study in Seville, Andalusia (Spain). It illustrates how its residents lose the right to everyday life in the historic centre and become alienated from their local spaces. Mechanisms of commodification and gentrification contribute to socio-spatial injustices and also make the city "inauthentic" for visitors and residents. In this sense, the right to live and enjoy the city are lost to both groups.

Introduction: socio-spatial justice in times of overtourism

Political and social concern over tourism has grown in many southern European cities in the past few years, Athens, Barcelona, Lisbon, Florence, Madrid, and Rome, among others. Visitors have recently skyrocketed in these cultural tourism destinations triggering problematic socio-spatial consequences and leading to the conceptualization of "overtourism" in 2017 (see Peeters et al., 2018 for a brief overview of the origin of the term). Academic attention to the topic has grown since, including literature reviews (Capocchi, Vallone, Pierotti, & Amaduzzi, 2019 or Perkumiene & Pranskuniene, 2019) and special issues (Milano, Novelli, & Cheer, 2019). There is also growing political attention as the UN World Tourism Organization and the European Parliament have expressed their concern with tourism overcrowding (UNWTO, 2018; Peeters et al., 2018). Capocchi et. al. (2019) consider overtourism with respect to the proliferation of "tourismphobia" in Spanish tourists' destinations. Tourismphobia was first coined by Donaire in relation to Barcelona (2008) and refers to local aversion and antagonism between residents and visitors, driven by social concern and unrest due to tourism-related overcrowding and stress (see also Delgado, 2008). Like touristification (Jover & Diaz-Parra, 2019; Mansilla, 2019; Sequera & Nofre, 2018), overtourism describes a range of social-political issues resulting from mass tourism in urban cities. In European Mediterranean coastal destinations, for instance, ecological groups and scholars have denounced extreme urbanization following mass tourism and lifestyle migration (Blázquez & Murray, 2010; Fernández Durán, 2006; Naredo & Montiel, 2011). The problem has been exacerbated by factors such as low-cost airlines, cruise tourism and increasing short-term holiday rentals through the proliferation of online booking platforms such as Airbnb (Veiga, Santos, Águas, & Santos, 2018; Postma & Schmuecker, 2017; Jover, Berraquero-Díaz, Barrero-Rescalvo, & Jiménez, 2018). These have not only increased pressures on mature destinations but also expansion into new places, including urban areas that were neither designed for the activity in the first place nor probably prepared for it.

Overtourism is less about the number of tourists, and more about social injustices and stress on the destination and its residents (Butler, 2018), and is related to very specific socio-spatial and material processes, such as those of rising rents and displacement of locals (Capocchi et al., 2019). It has been characterised more holistically as an "excessive negative impact of tourism" (Capocchi et al., 2019, p. 2), influencing the "authenticity" of cities and neighbourhoods, affecting tourist experience and local quality of life, causing irritation and annoyance to residents (Goodwin, 2017, Milano et al., 2019). Among the socio-spatial processes that contribute to social injustices such as described above is gentrification, defined by Clark (2005: 25) as "a process involving a change in the population of land-users such that the new users are of a higher socio-economic status than the previous users, together with an associated change in the built environment through a reinvestment" (cf. Beauregard, 1986). Most scholars associate gentrification with different types of local population displacement, direct or indirect (Slater, 2009 following Marcuse, 1985) or even symbolic displacement (Janoschka & Sequera, 2016). The literature on tourism and gentrification has also grown (see, for instance, "tourism gentrification" by Gotham, 2005). Researchers have focused on gentrification and tourism as related processes in current capitalist urbanization on a global scale, especially studying historic centres in urban tourism destinations (García-Herrera, Smith, & Mejías Vera, 2013, Fuller & Michel, 2014, Cocola-Gant & Gago, 2019). Cocola-Gant (2018) understands tourism as a gentrifying process causing different forms of displacement, while Jover and Diaz-Parra (2019) deal with the articulation between gentrification related to the replacement of locals by new better-off, usually middle-class incomers, and their subsequent and overlapping replacement by temporary visitors. Debates also engage with the tensions and opportunities afforded by amenity migration and lifestyle migrants, regional or transnational newcomers arriving often with higher purchasing power (e.g., Sigler & Wachsmuth, 2016; Zaban, 2019).

However, while gentrification and lifestyle migration has been garnering research attention, a *critical* theoretical focus on social displacement and related social injustices within tourism studies has been slow to take hold. For instance, the notion of *alienation* has long been addressed by social theorists, who have moved the concept over time from its Marxist origin in the sphere of production to social reproduction, mainly following the work of French philosopher and sociologist Henri Lefebvre (1901-1991) (see Silver, 2018). Relevant concepts include "land alienation," related to of the loss of common or private lands by indigenous or local people (see, for example, Van Noorloos, 2011; Burlo, 1989). "Place alienation" characterizes the feeling of displacement or inability to develop a sense of belonging towards one's current community (Hummon, 1992; Relph, 1976). Gentrification, displacement and place alienation are intertwined socio-spatial processes that contribute to social injustice. All these elements are well informed by Lefebvre's theoretical and empirical work on urban space. The opposition between appropriation and place alienation as well as the relationship between resident displacement and tourist pressure on historic centres were articulated in his seminal work *The Right to the City* (Lefebvre, 1968). Yet little has been used of his immense oeuvre to address overtourism in tourism studies. Lefebvre's work offers a valuable opportunity to fill the current gap in critical research on the impacts of urban tourism and overtourism. Several questions arise in this context: What injustices arise through the socio-spatial processes and transformation of social spaces due to rapid tourism growth and overtourism? Are ban cultural tourism and the locals' right to the city mutually exclusive? What roles do Lefebvrian notions of *appropriation* and *place alienation* play in potential conflict that arise in this context? What contribution does Lefebvre's theoretical work make to current studies on overtourism overall?

We explore these questions through a case study of a lesser-known destination, the historic city of Seville, Spain. Research on overtourism has usually focused on large-scale global and touristic cities such as Hong Kong, Rio de Janeiro, Barcelona, Lisbon or Venice (Milano et al., 2019 or Capocchi et al., 2019). Far less attention has been paid to second-tier cities like Seville in increasingly touristic regions within southern Europe. We examine overtourism in Seville through empirical and theoretical study recovering the work of one of the most relevant writers on space and the city, Henri Lefebvre. Our aim is to confront the lack of critical and theoretical grounding of overtourism and examine socio-spatial injustice throughout a Lefebvrian framework, specifically the opposition between appropriation and alienation at the heart of Lefebvre's work on the right to the city.

The paper is organized as follows. We begin by discussing some key Lefebvrian theoretical concepts and their association with the problem under scrutiny. An empirical case study of Seville is then presented. Methodology includes both qualitative study (interviews) as well as quantitative data gathering and analysis; for instance, exploring datasets on urban tourism and population as well as analysing the City Council's specific urban renewal strategies oriented to attract tourism into the historic district. Results and discussion follow, which involve stressing the main contradictions generated by growing dependency on tourism growth, drawing on insights aided by Lefebvrian concepts. The paper closes with a short conclusion.

Theoretical framework: overtourism, alienation and the right to the city

City centres and overtourism in Lefebvrés work

Lefebvre's extensive work is omnipresent in liberal and radical approaches. Various authors have pointed out a tendency towards institutionalization and political moderation or reformism of his discourses by international organizations or frameworks like the UN-Habitat, as well as a sort of overinflation in using notions such as the right to the city (Merrifield, 2006; Purcell, 2014; see also https://www.un.org/ruleoflaw/un-and-the-rule-of-law/united-nations-human-settlements-programme/). However, Lefebvre remains the main critical (Marxist) philosopher whose writings on space are still indispensable in the analysis of urban processes. Among the wide range of topics Lefebvre reflected upon are historic cities and the effects of cultural tourism, which he addressed in relation to Mediterranean Europe (Spain, Italy, Greece and southern France).

Centrality is a key notion in studying these urban areas. The city is a concentration of resources, objective and symbolic, and cannot survive without a centre. On the one hand, historic city centres account for the historical process of urbanization, representing the deep transformation of the pre-modern and pre-capitalist city. On the other hand, they encompass an area where contradictions of modern urbanism emerge: exclusion, museumification and commodification. For Lefebvre, the western pre-capitalist city is mainly a concentration of use value and refers to the city as an *oeuvre* (broadly translated as "work" from French), where inhabitants produce and consciously shape their human space. The concentration of symbols and meanings associated with the monumentality of the built environment also plays a key role; Lefebvre states that the urban society is only conceivable with monuments, i.e., with a transfunctional and a transcultural city centre (Lefebvre, 1970, p. 103). He addresses the role of the urban and historical centre in works like *The Right to the City* (1968), *Space and Politics* (1972) and *The Production of Space* (1974). As he explains in the former, 'an urban reality cannot be forged [...] without the existence of a centre: without the ensemble of everything that is born and produced in space' (1972, p. 19). The city centre is a social and functional non-segregated space, allowing the greater virtue of the city: the possibility of encounters and the play of "difference" in local spaces, the way in which interaction and encounters of everyday life constructed sense of place and meaning. While power relations are dominant, for instance, he emphasizes the capacity of monuments to signify authority, he still recognized this latter meaning-making virtue of social spaces in the city centre: "who says 'monument' says oppression and domination, thus power, but also splendour and meaning" (1978, p. 264).

Cities existed before industrial capitalism, but this form of capitalism destroys them, argued Lefebvre. Use values are subordinated to exchange values through the institution of private property and the land market, inevitably entailing collective expropriation. These processes are so intense that they tend to destroy the city centre's use value. Here Lefebvre is describing the physical and symbolic decline of western city centres during the 20th century: depopulation, ruination or the emergence of slums, while land rents kept their potential value high (Smith, 1996). Most people abandoned the city centres and headed to new peripheral quarters, and subsequently a new polycentric and segregated urban space was produced. The city as an industrial product and a commodity substituted the urban space as an oeuvre (Lefebvre, 1968, 1970).

As value itself cannot exist without use value, the city centres had to house new functions: financial, commercial or touristic. The latter directed Lefebvre's attention towards cases in Mediterranean historic urban areas, which were being transformed into "a commodity for high standard consumption for foreigners, tourists, and suburban visitors. It survives thanks to this double function: place consumption and consumption of the place" (Lefebvre, 1968, pp. 27-28). Explaining further, Lefebvre (1971, p. 192) states that "tourists make the authenticity they seek in the historic city disappear. Venice, with 200,000 tourists, is not Venice anymore. The object vanishes with the activity by which it is used [...] tourists become their own spectacle." Discussing how tourists want to experience an elitist life in underdeveloped regions, and again referring to Mediterranean historic cities, he states: "Mass movement of tourist hordes escaping towards urban (Venice, Florence) or rural spaces, where their arrival is equal to its destruction, is one of the greatest spatial contradictions of modernity" (Lefebvre, 1974, p. 176).

Lefebvre mentions examples in Spain and Greece too. He links urban tourism with some kind of "urban neo-colonization" (Lefebvre, 1972, pp. 61 and 129). In southern European regions, lack of industrialization is substituted by the development of a non-productive circuit of accumulation sustained by tourism and urbanization. Lefebvre illustrated how overproduction in industrialized (northern European) countries threatens the functioning of capitalist economies and creates surplus capital flows to the non-productive circuit in southern Europe. Financial centres in the north control the fluxes of capital and tourists, generating a relation of subordination and dependency (*Ibíd*). The secondary circuit of accumulation that he traced was systematically theorized by David Harvey (1982, 2004). For Lefebvre (1968, 1972), this model creates an apparent wealth that conceals the underdevelopment which emerges during crisis periods.

Alienation, appropriation and the right to the city

Key to this theoretical development is how the processes described articulate with another Lefebvrian fundamental notion: *alienation*. Here Lefebvre departs from the Marxist notion of alienation that focuses on the production process, and how the object created is alien to the subject who produces it. In Marxist theory, this has been granted as a psychological attitude of workers towards the product of their labour and their productive activity, but it also has an objective content in the material pauperization of the working classes (Sanchez Vazquez, 2013, pp. 503-504). Lefebvre moves this alienation from the production of things to the production of space and social relations in capitalist systems. If private property ownership of means of production alienates the (waged) labour from its output, the private property ownership of urban land alienates inhabitants from the city, both subjectively and objectively. As Purcell (2014, p. 149) points out, Lefebvre understands the right to the city as a struggle to de-alienate urban space through the process of *appropriation*. Appropriation also entails making an oeuvre, i.e., inhabitants create and consciously transform the city, shaping and transmitting their own personality to it (Lefebvre, 1971). This oeuvre is the pre-capitalist city of unproductive consumption, characterized by the concentration of use value. However, the expansion of capitalist institutions, i.e., capitalist state, private property and the land market, dismantle this form of the city. The commodified product substitutes the oeuvre, the habitat does the same with inhabitancy, and alienation is imposed at the expense of appropriation (Lefebvre, 1968, p. 20). In this scenario, residents become alien to the city that they produce and transform, being at the same time objectively displaced from the oeuvre space par excellence: the historic city centre.

The historical process of commodification, displacement and gentrification of historic city centres is at the origin of modern place alienation. Lefebvre (1968) refers to Haussmann's Paris reforms (major public works project he undertook as Prefect of the Seine in the mid-1800s for Napolean III) as a violent eviction of the working classes from the inner city towards new peripheral neighbourhoods, as architecture design and planning configure a consumerist and alienated urban life (see also Harvey, 2003; Lefebvre, 1972, p. 167). In the new proletarian urban suburbs, Lefebvre (1968) identifies a similar double strategy of uprooting and disorganizing workers through the organization of social reproduction and continued inequities and alienation. He therefore posits a special meaning on the right to the city as *the right of the working classes to return to the city centre*—it is a right of reappropriation. Lefebvre also notices a resettlement process in 1970s Paris on previously abandoned historic quarters. New residents were middle-class professionals escaping from conservative and oppressive bourgeois suburbs with financial power to afford restoring run-down buildings, turning slums in inner Paris into picturesque quarters (Lefebvre, 1970, pp. 153-154). This movement towards the centre does not imply its collective appropriation; rather it describes further commodification and expansion of segregated habitats. The city centre's transformation into a place for middle-class hedonistic consumption and a touristic resource is the result of the commodification that drives a process of place alienation over time.

The notion of authenticity can be seen as complementary to Lefebvrian alienation in historic centres as the result of a historical process of capitalist modernization and commodification. Lefebvre did not give an explicit definition of authenticity in his work and only commented on it briefly. However, his use of the term is coherent with the notion of place authenticity used by humanistic and existentialist geographers, where discussion on place and placelessness relates authenticity closely to place alienation or place belonging. For instance, Relph (1976) describes the experience of place authenticity related to collective belonging stemming from collective identity and rooting. Sanguin (1981) describes inauthenticity as the result of the production of places for mass consumption, creating place alienation. Sociologists of tourism like Wang (1999) refer to notions of "existential authenticity", non-related to the authenticity of an object but to how people *experience* authenticity or loss of authenticity–in tourism activities as well as in daily life. It should be noted here that critics like Laguna (2006) state that authenticity is a modernist value in decay due to mass media and post-modern thought (see also Harvey, 1990). However, the consumption of authenticity is still a strong motivation to travel and continues to have a relevant role for urban tourism, such as in the renewal or revitalization of historic neighbourhoods as "authentic" spaces for tourist consumption (see Flores & Ramos, 2017; Fuller & Michel, 2014).

Going back to Lefebvre, it can be argued that the experience of inauthenticity either by residents or by visitors is a key result of place alienation due to the process of commodification of space. Such a place-based notion of authenticity involves the ideas of community and neighbourhood. According to Lefebvre, every community is grounded in some kind of collective

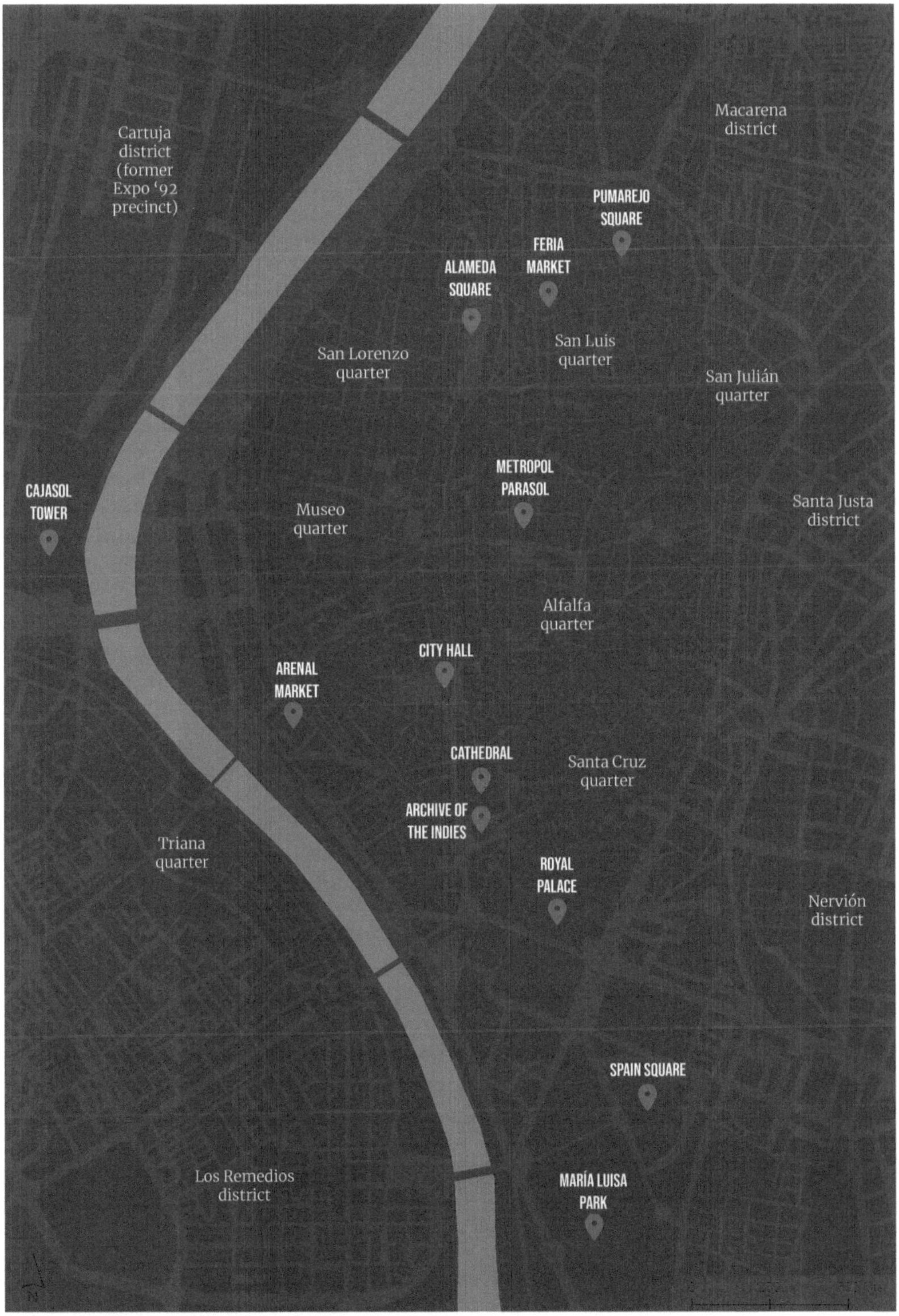

Figure 1. Seville's historic district: quarters and milestones. Source: authors.

property; urban communities as being discussed here depend on the collective appropriation of parts of their city by residents. A neighbourhood (*barrio*) is a specific community of residents appropriating a specific portion of urban space. Lefebvre describes neighbourhoods (*barrios*) as empirical units —more circumstantial than structural— that are necessary for the city's existence. Cities do not exist without the process of community appropriation, i.e., without neighbourhoods there is just urban space (Lefebvre, 1971), whereas these communities are the basis of lived experience and a rich authenticity of everyday life. To summarize, then, the Lefebvrian right to the city must be seen as a right of its residents to a non-alienated urban life and to collective *appropriation* by the locals that inevitably entails community and neighbourhood development, yielding authentic experiences for its inhabitants. The case study in Seville, Spain, below illustrates well Lefebvrian processes of place alienation, appropriation, and discourses of authenticity, and raise the critical question of the right to the city in the context of overtourism and gentrification.

Case study: Seville, Spain

Seville was a relevant centre of economic and political power during the Roman Empire's domination of the Iberian Peninsula as well as in the Islamic expansion during the Middle Ages. After the Christian conquest and the beginning of the colonization of America by the Spanish Empire, the city became one of the main commercial harbours in Western Europe. But by 1800 Seville had turned into an economic and demographic stagnated city, the capital of an agrarian and severe underdeveloped region. This situation extends along most of the 19th and 20th century when southern Spain remained disconnected from the European industrialization processes (Arenas Posadas, 2015). However, Seville succeeded in preserving a wide range of built heritage from different historical periods - it has one of the largest medieval city centres in Western Europe, partially still bound by its 12th-century Muslim walls. The UNESCO World Heritage Site declared in 1987 includes *The Royal Palace,* mixing Muslim and Catholic architecture, *the Cathedral*, built over the previous mosque but conserving its minaret and the ablution courtyard, and the *Archive of Indies* which contains the main archives on the process of colonization of America (see Figure 1). The tourist economy became highly significant as the 20th Century drew to a close and, in the aftermath of the 2008 economic crisis, tourism specialization grew even faster, so today Seville has turned in a fashionable destination. For the past three years, the city has broken the records in numbers of tourists, up to more than 3 million visitors in 2018 (see https://sevilla.abc.es/sevilla/sevi-nuevo-record-para-turismo-sevilla-supera-barrera-3-millones-visitantes-2018-201901312016_noticia.html). A detailed picture of the urban development and tourism-related transformations of Seville emerges in the research study undertaken below.

Research approach and methods

Within gentrification studies, finding reliable data on displacement tends to be a common obstacle. The same applies here; neither in Seville nor in Spain do we have statistics offering a clear picture of population displacement from the historic centre. This is due to the complex multidimensionality of the gentrification process as well as lack of attention of the state institutions to documenting the transformations and population flows. Our research of overtourism and gentrification in Seville examines its transformation not only with respect to physical displacement, but also in relation to its symbolic dimensions and intersections with tourism. As such, a mixed methods approach seemed appropriate, with quantitative data gathering to complement the qualitative study undertaken here.

We contextualize the case study with socio-demographic data coupled with secondary sources and information from Seville City Council's public urban and tourism policy documents. Quantitative data is provided by the municipal population census, consisting of basic demographic statistics disaggregated over time. Secondary sources include publicly available socio-demographic datasets provided by the Spanish Statistics Office, which have already been used in

Table 1. Table of participants.

Status	Nationality	Pseudonym	Profession	Age	Household size
Former residents	Spanish (Seville)	Ana	Architect	40	2
	Spanish (Madrid)	Jacinta	Waitress	36	3
	Spanish (Madrid)	Monica	Psychologist	38	2
	Spanish (Seville)	Mercedes	Artisan	43	4
	French	Angie	Translator	43	1
	Italian	Sara	Photographer	38	2
	Argentinian	Natividad	Carpenter	41	2
Current residents	Spanish (Seville)	María	Secondary school teacher	49	4
	Spanish (Seville)	José	Lawyer	65	2
	Spanish (Cordoba)	David	Architect	50	1
	Spanish (Seville)	Pepa	Geographer	55	2
	Spanish (Seville)	Manuel	Anthropologist, scholar	40	2
	Spanish (Cadiz)	Juan	Civil servant	36	1
	Spanish (Seville)	Silvia	Waitress	32	2
	Dutch	Peter	Retiree	66	1
	German	Karl	Language teacher	57	2
	Italian	Caroline	Hotel receptionist	32	1

previous research to test gentrification in Seville historic centre. The qualitative part of the study is based exclusively on in-depth interviews with residents in Seville. Interviews are used to identify what transformations are being perceived by inhabitants (if any) and how they are responding to changes, conflicts and contradictions regarding tourism specialization in Seville's city centre. We conducted five interviews with key informants (three of them current residents of the historic city centre): three planners working in the City Council's Urbanism Department and two social activists linked to the right to housing and anti-evictions movements. In addition, we interviewed participants belonging to households established in the last twenty years in the historic city centre. We explored here the differences between those who remain living in the area and those who moved out. We were also interested in differentiating between local residents, Spanish non-Sevillians and foreigners (potentially transnational gentrifiers) who settled in Seville. Most new incomers had first contact with the city as temporary visitors or tourists, so their perceptions of transformations around the historic centre since their first visit provided valuable insights. We conducted snowball sampling, looking for an equilibrium between displaced and non-displaced households, locals and foreigners. In total, we interviewed ten current residents and seven former residents of the historic district: Sevillian born (7), Spanish non-Sevillian (4) and foreigners (6) living in the city permanently. All foreigners are from Western Europe except for one Argentinian, who was among the group of former residents displaced (since 2014) from the city centre to which they had generally moved between the 1990s and the 2000s. The age range of participants was between 32 and 66. Family types include couples with children, childless couples, flat-mates without familiar bonds, three people living alone and one single mother. Most of them were skilled professionals, while four worked on unskilled jobs linked to the tourist sector in the city centre (only one participant had retired). We use pseudonyms for the participant described in Table 1.

Interview questions revolved around the participant's residential history and their spatial behaviours. We asked about their perception of the historic centre as well as their experiences of land use and built environment transformations, especially in relation to the intensification of tourists flows over the last decade and impacts on their daily spatial behaviours (paths to work, leisure, shopping), as well as relocation from the city centre (in the case of those displaced). Descriptive theme-based analysis of their response produced key themes around their perception of transformations of the historic centre (also related to authenticity or loss thereof), sense of belonging to their current or previous neighbourhood and their awareness of being physically or symbolically displaced by tourism.

Evolution and transformations of the historic city Centre of Seville

Seville's growth has oscillated between long periods of urban stagnation or uncontrolled expansion and two periods of planned restructuring, coinciding with international mega-events in 1929 and 1992 respectively, oriented to the attraction of public and private investment for tourism and commercial purposes.

Firstly, the 1929 Ibero-American Exhibition was used to create Seville's city-brand and to lay grounds for future tourist exploitation. It was the first well-planned urban enlargement, characterized by the construction of high-quality hotels and the reinvention of tourist attractions across and beyond the southern city centre (Diaz-Parra, 2016; Martín, 2009). The key operations took place in the surroundings of the current UNESCO World Heritage Site. Secondly, soon thereafter there was a long period of stagnation coinciding with the fascist coup d'état, the Civil War and the dictatorship. During the 20th century, Seville's historic city centre became polarized between a commercial and increasingly touristic south side that was sparsely populated by residents historically linked to the local high-class, and an overpopulated north side that had been downgraded for decades (González, 1975). The city centre was largely abandoned during the Francós dictatorship, where the urban focus then was on suburban expansion. Urban planning, especially in the historic district's north side, restructured medieval urban patterns to allow traffic and modern block of flats (Ruiz, 2005; Marín, 1980). As Spain transitioned to a democracy and progressive politicians took over the City Council, the 1980s saw a brake on large transformations and the start of a period of micro-scale interventions on the historic district. The city has stopped growing, and remains stable with around 700,000 inhabitants since the 1990s.

Shortly after these new ideas flowed into the City Council, a second moment of large-scale urban transformation occurred. The 1992 Universal Exposition meant urban restructuring ranging from the development of transport infrastructures –including improving accessibility to the historic city centre– to fostering urban renewal across previously downgraded quarters (Fernández Salinas, 1993). The 1987 Seville Master Plan gave legal support to the Exposition, pointing out some key issues in the historic centre such as the loss of historical buildings, attrition of residents, population aging and slums (Gerencia de Urbanismo, 1987, p.135). Editorials and opinion pieces complaining about the negative, downgraded image of parts of the city centre as well as need for security of tourists became commonplace pro-gentrification discourses by local intellectuals (León Vela, 2002). Urban planning discourses were also key in justifying neoliberal restructuring and the city centre's upgrade and rejuvenation around 1992. These policies blamed previous developmentalism plans for population decline due to displacement of low-income inhabitants (Gerencia de Urbanismo, 1995), but at the same time implemented renewal strategies that brought about real estate speculation and attracted gentrifiers that continued the trend of displacing rooted communities and neighborhoods in the 1990s and further (Diaz-Parra, 2014; García, 1998).

Urban renewal in the historic city centres northern quarters was possible thanks to the Urban European Initiative (*Ibidem*). Although European funds were initially aimed at social intervention, most of the investment was spent on repairing and rehabilitating various buildings and improving public infrastructure. This was accompanied by a strong campaign of security that included the eviction of squatted buildings, elimination of drug dealing hotspots and reduction of prostitution (Diaz-Parra, 2014). Private capital profited from the surplus created by these public interventions, restoring housing for new affluent residents. In a few years, the social character of the whole historic centre area had changed extensively, slums disappeared and many middle-class households substituted marginalized low-income groups who continued to be displaced to peripheral neighbourhoods. After further interventions at the beginning of the 21st century, tourist circuits expanded, the Alameda de Hercules being a

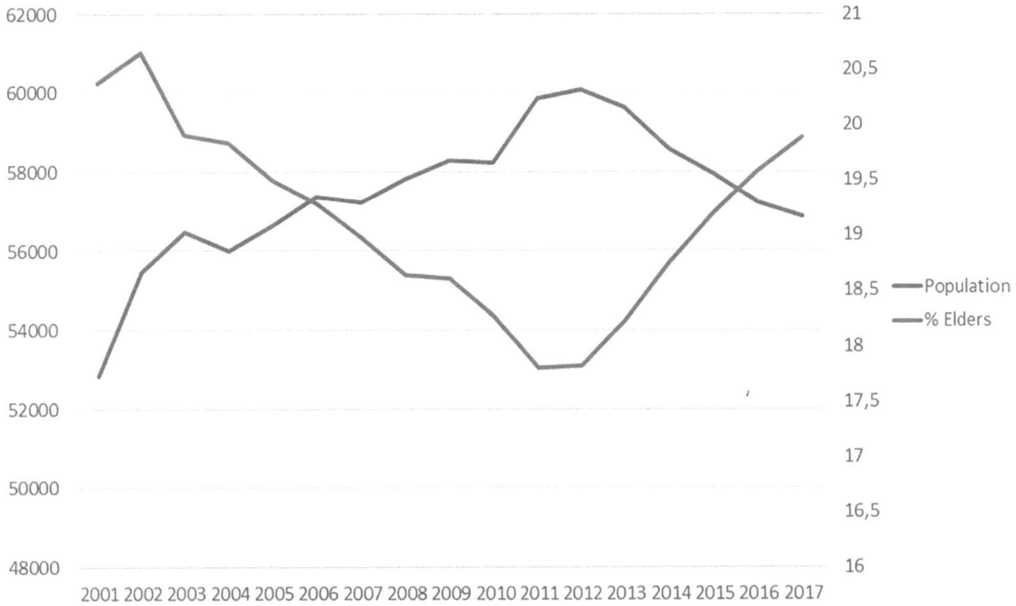

Figure 2. Seville's historic district: population and aging. Source: Municipal population census.

special case of this as chic restaurants and night pubs and clubs filled previously the side-lined area (Jover and Diaz-Parra, 2019).

Prior to the 2008 crisis, urban strategies also fostered two flagship projects in and around the historic district to attract visitors. The Metropol Parasol was a €120 million Council-financed project developed between 2006 and 2011 at the historic district's geographical heart. The idea was to restore an old marketplace by adding diverse facilities, new commercial spaces, an archaeological museum and a viewing platform. Between 2007 and 2015 the Council allowed the development of Torre Cajasol, the city's first skyscraper that also comprised a large commercial space in the vicinity of the historic centre (Figure 1). The regional government of Andalusia then listed the south-eastern commercial cluster of Seville's historic district as a Touristic Area in 2014. Businesses have since been free to choose their opening hours during periods of intensive tourist inflow and funding is available to modernize and revitalize the urban milieu. This has led to the loss of long-standing retail shops and the gourmetization of traditional markets in the area (Arenal, Encarnación and Feria), yet another step towards the commodification of urban space (Jover, 2019). Pubs and restaurants have mushroomed at a pace of 76% and small hotels at 36% between 2014 and 2017, while employment in such establishments has risen to make up 25% of total jobs in Seville (almost half of this 25% is located in the city centre). The City Council's goal is to expand the Touristic Area within and towards new areas (Ayuntamiento de Sevilla, 2018). The rapid increase of tourist flows goes hand-in-hand with the development of new tourist accommodation. Seville ranks fourth in terms of short-term holiday rentals in Spain, but is the second (only after Madrid) with respect to the percent of tourist accommodation compared to the total conventional dwelling in its historic centre. This type of lodging has not stopped growing since 2013 onwards, and in 2019 comprises up to 12,6% of the whole housing stock in the historic centre of Seville (data retrieved from http://datahippo.org). The increase coincides with a new U-turn in demographic tendencies. Between 2001 and 2014, Seville's historic centre increased to 6,000 inhabitants, but since 2014 it has lost more than 3,500 people, while aging resident increased in the historic core as most of them are landowners and cannot be easily displaced (see Figure 2).

Appropriation, displacement and place alienation in Seville

Appropriation

Despite the diverse origin of the participants, their reasons to settle in the city centre do not differ in general terms. Sevillian-born residents came mainly from middle-class neighbourhoods and moved into the centre with friends when they became self-sufficient and independent. The non-Sevillian born residents, both Spanish and foreigners, moved mostly directly to the centre when they first arrived in Seville. The main motivation for all participants to settle in the historic core was a previous network of friends. Many highlight a milieu of young people with multiple informal meeting points in squares, pubs and cultural or social venues. Based on their points of view, appropriation is related to the existence of a certain community and a strong feeling of belonging to (and identification with). a place. For example, when talking about Alameda, Natividad states: "the square was ours [...] to have a *barrio,* to control the space, made me feel good". This idea of a barrio also emerges in former residents' interviews when discussing which neighbourhood they chose to live in. For example Monica thought she would be "isolated" in a peripheral city neighbourhood. Sara had to move to the suburbs after 10 years of living in the city centre because the rising rents, making her feeling "lost and isolated." Most participants refer to the concept of barrio as attractive because it is "familiar", meaning that they would frequently come across people they are acquainted with on their streets and in the squares. This sense of familiarity and comfort is something that all participants realise has recently changed: they no longer know most of their neighbours and local shop owners. All former residents and some of the current ones who entered the area in the 1990s (Antonia, Silvia, Karl and Caroline) have particularly stressed they have noticed these transformations.

The barrio character ultimately refers to the use value of social relations. Lefebvre identified traditional festivities as the best example of this un-productive appropriation of space. Paradoxically, some of the foreign participants feel the opposite way. Peter, who bought a house in the historic city a few years ago, thinks that his quarter is very "authentic", not destroyed by tourism or "not too much", but then he recognises he still considers himself a tourist who is "accepted" on the streets. However, he adds: "I feel less integrated during the festivities... on a daily basis, when people work or go to the mall, I feel part of society, but when festivities arrive, I don't understand anything". By contrast, José talks about how the working classes that used to inhabit the city centre only come back to their former neighbourhoods when festivities such as the Holy Week take place.

There is also some degree of fetishization about the historical milieu and its aesthetics by non-Sevillian residents, both Spanish and foreigners. Monica speaks about residing in the city centre as living "in a fairy-tale, in a special place, as if it were a permanent holiday", a shared perception with Caroline. Authenticity of space appears throughout most interviews, especially in the cases of foreigners. Karl talks about defining the city and its historic quarters as containers of "love, enchantment, magic, fantasy", referring also to the local's culture of "marginality, freedom or rebellion" as very attractive. In Marx's Capital, fetishization is essentially part of the process of transforming use values into commodities, and it is part of the alienation of the objects from the producers and the social relations that create them (see Sanchez Vazquez, 2013). Consumers of the urban space describe the city centre as a "magical" place that seems to be independent or far away from the social processes going on (speculation, displacement) and from the role of visitors and inhabitants as consumers creating and transforming this space. In other words, the historical space in its current configuration appears to be perceived as outside of the social relations that are currently transforming them and that have historically shaped it—this is a fundamental part of the successful creation of a spatial commodity and the place alienation.

Displacement and place alienation

Just as alienation in Marxist terms is an objective process constituted by the exploitation of workers and separating from the means of production, place alienation can be described as an objective and physical process of displacement and/or deprivation from the city centre by those who have traditionally inhabited it. Former city centre residents interviewed offered key insights here. This group embodies middle-class households displaced towards neighbourhoods outside the city centre, and their movement is readily visible in the local demographic statistics, as the historic district has been losing young and adult households over the past five years. It is meaningful that these residents are moving to a place as close to the city centre as possible, because they wanted to keep living in the area. There is generally a feeling of forced displacement in all participants we interviewed, it is not exclusive to Spanish respondents. Sara and Caroline, both Italians, shared the same emotion once their landlords put an end to their long-term tenancy contracts to turn the flats into short-term vacation rentals. Sara states that evictions are currently widespread, and speaks of an "invisibilization". In Caroline's case, she had to move a few times within the city centre in the past three years, something that happened to her current boyfriend too. Abrupt termination of rent contracts has become commonplace as landlords decide to sell their properties to real estate businesses or to reform them to rent in the touristic short-term market. Other former residents have moved out of the centre because of rising rents and prices. For instance, Monica comments: "A lot of people who used to live around Alameda or San Luis are now moving here (towards a neighbourhood just outside the historic district). We are coming here not because we want to, but because we can't afford the cost of living in the city centre".

Long-term current residents such as Antonia or Juan who moved to the area at the beginning of the 1990s and bought relatively cheap properties, speak about how back then they witnessed policemen evicting entire buildings inhabited by low-income tenants or squatters. Moreover, the social activists interviewed (José and David), confirm a general displacement of poorer households in parallel to the area's renewal. Pepa, who worked on developing the EU Urban Plan between 1994 and 1999, recalls that the concentration of investment in a short period of time within the limited historic centre space brought about a first wave of speculation that displaced low-income families. As Ana explains, there are at least two differentiated processes of displacement: a classic gentrification process between the 1990s and the beginning of 21st century, and a more recent displacement of residents of diverse social backgrounds due to tourism pressures.

Alienation can also be viewed in subjective terms here. Former residents view the transformations of the city centre quite negatively (such perceptions are less evident in current residents' responses, especially those who have not been living in the area for a long time). Many older shops and businesses, especially pubs –very relevant in Seville's street culture–are perceived to have disappeared; some notice shops and pubs rapidly shifting from one year to another, and popular places where some participants used to hang out have shut down. Ana mentions that is no longer common to know the shop owner, who used to live in the same neighbourhood, nor the employees. There is also a perception of overcrowding in pubs which are targeting a different, higher-end consumer as prices have soared. Some participants talk about a sort of elitization in the retail fabric, on par with the disappearance of small and traditional establishments such as groceries or fisheries (Monica), while some remaining enterprises–for instance the revitalized Feria market–have become unaffordable (Angie or Ana). Antonia and Silvia regret the proliferation of modern aesthetics and different business models in the area, especially franchises, which are completely alien to the local culture.

Due to these transformations, participants experience a loss of familiarity with the neighbourhood, a sort of 'nobody knows anybody' feeling, to the extent that some do not wish to remain because there is no barrio anymore (Natividad). They are feeling the pressure of displacement as Marcuse (1985) would describe it. Sara and Angie left Spain for economic reasons during the 2008 economic crisis, and when they decided to come back, they chose not to live in the city centre again because the barrio that they knew had disappeared. Some foreign respondents currently living in the city recognize a similar loss of neighbourhood "authenticity" (Karl) and say "there is no barrio life anymore in the city centre" (Caroline). Likewise, Jacinta observes a depersonalization of the historic centre, driven by tourism and rising costs which only the elderly can sustain if they own their own homes:

> "Every six months it changes, businesses and people. It is difficult to create networks because there are many new people and many others are forced out because of rising rents. Tourism conditions all and nothing is defined. In five years, we have seen how a whole street was transformed, and only elderly people remain because they own their homes"

Perceptions of (over)tourism

Are these consequences on Seville's city centre strictly related to tourism? Ana and Sara think that the overcrowding of pubs and people in and around Alameda responds in part to tourism and partially to locals from other Sevillian neighbourhoods. Following Lefebvre, there should be a contradiction between collective appropriation of space –a community deeply rooted in a historical barrio, for instance–and the more recent access to the historic centre as a commodified place to visit and consume. Indeed, a few Spanish former residents like Mercedes and foreign current residents like Karl and Caroline avoiding touristic areas, and confess they only walk there at night when there are fewer tourists. In the tourism cluster in the southeast, there are queues to access the main monuments and it is increasingly difficult to walk through the busiest public spaces.

Respondents noticed, too, that tourism has spread across the whole city centre. Some were surprised when in the past few years they began to see groups of visitors in places that not so long ago would have been completely unnatural for them to be in (such as San Luis or the Feria market), or in places stigmatized as dangerous for outsiders (e.g., Pumarejo square and surroundings). The arrival of tourists has become a reason for residents to stop going to certain places. For example, Angie describes how she has been "scammed" in a pub in the city centre where she used to hang out because she is French, and the bartender thought she was a tourist. Mercedes thinks the transformation of the Feria market for tourism purposes has destroyed many traditional grocers, so she refuses to go shopping there anymore. She adds a spatial reflection: "before tourists were confined to hotels, but now they're all over the city centre." Similarly, Monica confesses that she hadn't usually noticed tourism because visitors did not go into residential areas as they do now. Ana and Jacinta no longer step foot in Alameda because it is overcrowded. Interestingly, other current residents, especially recent foreigners (Peter and Caroline) think these areas are still not excessively transformed by tourism.

It is worth mentioning that no participant blames tourism itself. Ana criticizes an excessive tourism specialization in certain areas, while Monica or Mercedes blame speculation and more specifically those doing business with tourists at the expense of residents. Others try to adapt to changes. Sara feels tourism took her space, but she makes the best of it and has thought about offering an alternative tour as part of her Airbnb offering, focused on contested memory and the story of social struggle in some inner-city quarters. This way, even a city walk becomes part of re-appropriating the city and countering the exacerbation of urban commodification by the digital economy's multinational companies.

Discussion

Examples abound in the literature about how investments in historic centres and urban tourism-led economies have turned deteriorating neighbourhoods into attractive places for local and foreign middle-class residents and visitors (see Cocola-Gant, 2018) city centre revitalization can also bring in more affluent consumers as the case of Seville illustrates. Herein lies the main contradiction that Lefebvre (and Harvey later) unveiled about tourism-oriented urban economies: successful commodification of the urban space destroys some of the elements (appropriation, community, authenticity) that initially attracted new residents and visitors. Both Spanish and foreign residents in Seville experience a chain of social injustices associated with the process, including physical household displacement driven by rising rents, as well as subjective alienation from their current or former neighbourhoods. Lefebvre's notion of *alienation* offers a critical way to understand overtourism's negative impact on everyday life and resident and community displacement due to gentrification and touristification. The never-ending transformation of urban space that is required for capital expansion translates into a non-appropriable place. Community building and meaning making become increasingly harder when public spaces and the built environment are constantly being commodified for capital gain.

The case also shows that place alienation and place authenticity are useful notions to consider jointly. Local and foreign residents' claims about the city centre becoming progressively inauthentic refer to the loss of neighbourhoods (barrios) and community, increasing lack of identification and familiarity with place as new faces and new businesses uproot traditional cultural institutions. As gentrification and increasing tourism turn historic districts into commodified and standardized places, they come to be non-appropriable and perceived as inauthentic by residents and visitors alike. Even though Lefebvre did not pay much attention to authenticity when dealing with place alienation, these ideas taken together cast a critical light on overtourism. The loss of authenticity and community can be interpreted from a Marxist-Lefebvrian perspective as part of place fetishization and commodification as described by Sanchez Vazquez (2013). Authenticity–or its appearance–can be manipulated to stimulate desire for consuming commodities, just as the experience or perception of authenticity (or loss thereof) can influence residents' capacity to appropriate space and neighbourhood. These notions can and should be integrated into a theoretical framework on place alienation and appropriation.

Seville's case also offers evidence that overtourism and gentrification can partially overlap, showing complementarity and paving the way to the historic centre's standardization. They share a common pattern: the commodification of space. Both processes are two sides of the same coin, working together to a certain extent, something that has already been pointed out (Jover & Diaz-Parra, 2019 …; Sequera & Nofre, 2018). When the city centre becomes a highly demanded commodity, market mechanisms begin to work and determine who has access to it and, more importantly, influence who owns it (and therefore makes a profit out of it), and who has the "right" to enter or leave. Displacement and segregation are some of the consequences since in the case study. However, at least in Seville's case, whereas early gentrification increased resident inflows and helped to rejuvenate the historic centre, current perceived overtourism is associated with depopulation (since 2013) and displacement of middle-class households because of the rising rents or the impossibility of buying a house.

Scholars of overtourism have also drawn attention to severe frictions between the right to tourism and the right to the city. Excessive tourism, linked to specific and material factors such as rising rents, denies the right to housing and to the historic centre for a variety of community residents. The case shows how the commodification of historic neighbourhoods not only displaces poorer households but also part of the middle-classes –especially renters–that moved into the historic centre over the last twenty years. Sevillian born residents that enjoy a special identity bond with the historic centre are uprooted, while displacement also spreads across non-Sevillians that were first attracted to live in the same area. Urban identity exists and the place is experienced as authentic because there is a neighbourhood community that uses and gives meaning to the area on a daily basis, i.e., that appropriate space. Communities are not structural to the city, but they need some stability, otherwise, they become an illusion, which is precisely what happens when the barrio turns un-appropriable due to growing commodification and commercialization: communities stop existing. In Lefebvrian terms, there is a conflict between collective appropriations and consuming urban space: a contradiction between the tourist hopes and desires, the sector's economic interests and the effects of both of these on places and their residents.

Conclusions

Lefebvre's theoretical framework on the right to the city offers a valuable critical approach to examine the emerging research on overtourism and gentrification. It enrichens conceptual building by linking tourism overcrowding to *place alienation* and unjust socio-spatial processes such as displacement or loss of neighbourhood (barrio), community and authenticity In Lefebvrian terms, the right to the city can be understood as a right to a non-alienated existence through urban *appropriation* that enables everyday life and dwelling in communities and neighborhoods (barrios). Yet appropriation is not a narrow concept. As the case study on Seville shows, it seems to be compatible with certain degrees of cultural tourism and is even a source of authenticity that attracts visitors and new residents. However, as economic commodification and tourist specialization accelerate, this appropriation becomes threatened and place alienation intensifies. The conceptualization of place alienation as discussed in this study is key to linking overtourism's consequences with socio-spatial justice and generalized processes of commodification in capitalist global metropolises.

Our current work has focused primarily on Lefebvre's early works on urbanization, particularly *The Right to The City* (1968). *The Urban Revolution (1970)* is also highly recommended to interested readers. Lefebvre developed his interest in ideology and alienation theories in these early works, building drawing on prior philosophical works. Rereading and reformulating Lefebvrian notions of alienation and appropriation today are essential to investigate current struggles over the right to the city in overtouristic urban destinies, and can help to open new urban political agendas and influence future research on tourism development and social contestation. If place alienation is part of a general process resulting from the commodification of social relations, to what extent is overtourism a paradigmatic and leading process of current capitalist commodification? Similarly, where could this process lead us in regards to spatial commodification processes going on in current global capitalism?

Instead of coupling the right to the city with a supposed right to tourism, we should perhaps advocate for a right to a non-alienated appropriation of historic centres for everyone, because they are key elements in identity production, generating a sense of belonging and rootedness in place. These lines of thought bring about more questions: Who really has the right to the city in a globalized world of mobilities? If we assume that communities are unavoidably created, dissolved and recreated, how are we to guarantee free access to historic urban centres without adversely impacting existing local communities? How can we ensure fair creation and re-creation of communities in spaces that are in high demanded by residents and visitors? What kind of policies are necessary to facilitate just (fair) access to diverse groups of residents and also conserve the monuments and other cultural heritage that constitute past, present and future of the historic city? And, most importantly, is it possible to manage what appear to be conflicting rights and divergent perceptions of tourism stress and place "authenticity" within the current context of capitalist space commodification?

Disclosure statement

No potential conflict of interest was reported by the author(s).

References

Arenas Posadas, C. (2015). *Poder y economía en el sur*. Seville: Fundación Centro de Estudios Andaluces.
Beauregard, R. A. (1986). The chaos and complexity of gentrification. In N. Smith and P. Williams (Eds.), *Gentrification of the city*. New York: Routledge.
Blázquez, M., & Murray, I. (2010). Una geohistoria de la turistizacion de las islas Baleares. *Periplo Sustentable, 18*, 69–118.
Burlo, C. (1989). Land alienation, land tenure and tourism in Vanuatu, a Melanesian Island Nation. *GeoJournal, 19*(3), 317–321.
Butler, R.W. (2018). Challenges and opportunities. *Worldwide Hospitality and Tourism Themes, 10*(6), 635–641. doi:10.1108/WHATT-07-2018-0042
Capocchi, A., Vallone, C., Pierotti, M., & Amaduzzi, A. (2019). Overtourism: A literature review to assess implications and future perspectives. *Sustainability, 11*: 3303. doi:10.3390/su11123303
Clark, E. (2005). The order and simplicity of gentrification. A political challenge. In: Atkinson R and Bridge G (Eds.), *Gentrification in a global context: The new urban colonialism*. Oxford: Routledge.
Cocola-Gant, A. (2018). Tourism gentrification. In Lees L and Phillips M (Eds.), *Handbook of Gentrification Studies*, Cheltenham and Northampton: Edward Elgar Publishing, 281–293
Cocola-Gant, A., & Gago, A. (2019). Airbnb, buy-to-let investment and tourism-driven displacement. *Environment and Planning A: Economy and Space*, Online first, 1–18. Retrieved from: https://journals.sagepub.com/doi/full/10.1177/0308518X19869012. doi:10.1177/0308518X19869012
de Sevilla, A. (2018). *Impacto ZGAT Sevilla 2014-2017*. Seville: Ayuntamiento de Sevilla.
Delgado, M. (2008). Turistofobia. El País, 12 de julio. Retrieved from: https://elpais.com/diario/2008/07/12/catalunya/1215824840_850215.html.
Diaz-Parra, I. (2014). *¿Gentrificación o barbarie? Disciplinamiento y transformación social del barrio de la Alameda de Sevilla*. Seville: Atrapasueños.
Diaz-Parra, I. (2016). Sevilla 1929-1992. La producción de una mercancía. In La Corrala (coord.), *Cartografía de la ciudad capitalista*. Madrid: Traficantes de Sueños, 195–201
Donaire, J. A. (2008). La efervescencia de la 'turismofobia'. Sobre turistas y turismo. *Barcelona Metrópolis*. Retrieved from: <https://ecoplann.wordpress.com/2010/12/02/barcelonametropolis-jose-antonio-donaire-la/>. [26/09/2019]
Fernández Durán, R. (2006). *El tsunami urbanizador español y mundial*. Barcelona: Virus editorial.
Fernández Salinas, V. (1993). Las grandes transformaciones urbanas de Sevilla durante los años previos a la exposición universal. *Estudios Geográficos, 54*(212), 387–407.
Flores, C. Y., & Ramos, R. (2017). Lo colonial como fuente de autenticidad de los centros históricos de las ciudades mexicanas. El doble colonialismo. *Scripta Nova, 21*(579), 1–15. Retrieved from: http://revistes.ub.edu/index.php/ScriptaNova/article/view/17771. doi:10.1344/sn2017.21.17771
Fuller, H., & Michel, B. (2014). Stop being a tourist! New dynamics of urban tourism in Berlin-Kreuzberg. *International Journal of Urban and Regional Research, 38*(4), 1304–1318. doi:10.1111/1468-2427.12124
García, P. (1998). Aplicaciones de la iniciativa comunitaria Urban. *Boletín de la Asociación de Geógrafos Españoles, 26*, 191–206. doi:10.21138/bage.1545

García-Herrera, L. M., Smith, N. y., & Mejías Vera, M. A. (2013). Gentrification, displacement and tourism in Santa Cruz de Tenerife. *Urban Geography, 28*(3), 276–298. doi:10.2747/0272-3638.28.3.276

Gerencia de Urbanismo. (1995). *El conjunto histórico de Sevilla. Avance del Plan Especial de Protección*. Seville: Ayuntamiento de Sevilla.

Gerencia de Urbanismo. (1987). *Memoria del Plan General de Ordenación Urbana 1987*. Seville: Ayuntamiento de Sevilla.

González, A. (1975). *Sevilla: Centralidad regional y organización interna de su espacio*. Madrid: Casa de la Moneda y Crédito.

Goodwin, H. (2017). The Challenge of Overtourism. *Responsible Tourism Partnership Working* Paper 4, October 2017. Retrieved from: https://haroldgoodwin.info/pubs/RTP'WP4Overtourism01'2017.pdf. [16/09/2019].

Gotham, K. F. (2005). Tourism Gentrification: The case of New Orleans' Vieux Carre. *Urban Studies, 42*(7), 1099–1121. doi:10.1080/00420980500120881

Harvey, D. (1982). *Limits to capital*. New York: Verso, 2006.

Harvey, D. (1990). *The condition of postmodernity. An enquiry into the origins of culture change*. Oxford: Basil Blackwell Ltd.

Harvey, D. (2003). *Paris, capital of modernity*. London: Routledge.

Harvey, D. (2004). *The new imperialism*. Oxford: Oxford University Press.

Hummon, D. M. (1992). Community attachment: Local sentiment and sense of place. In I. Altman and S. M. Low (Eds.) *Place Attachment*. New York: Plenum Press.

Janoschka, M., & Sequera, J. (2016). Gentrification in Latin America: Adressing the politics and geographies of displacement. *Urban Geography, 37*(8), 1175–1194. doi:10.1080/02723638.2015.1103995

Jover, J. (2019). Geografía comercial de los centros históricos: entre la gentifricación y la patrimonialización. El caso de Sevilla. *Boletín de la Asociación de Geógrafos Españoles, 82*, 1–33. doi:10.21138/bage.2788

Jover, J., & Diaz-Parra, I. (2019). Gentrification, transnational gentrification and touristification in Seville, Spain. *Urban Studies, Online first*. Retrieved from: https://journals.sagepub.com/doi/abs/10.1177/0042098019857585. doi:10.1177/0042098019857585

Jover, J., Berraquero-Díaz, L., Barrero-Rescalvo, M., & Jiménez, A. (2018). Turistización y movimientos urbanos de resistencia: experiencias desde Sevilla. In C. Milano & J. Mansilla (Eds.), *Ciudad de vacaciones*. Barcelona: Pol.len, 403–437

Laguna, D. (2006). El espacio del turismo. *Alteridades, 16*(31), 119–129.

Lefebvre, H. (1968). *El derecho a la ciudad*. Barcelona: Ediciones península, 1978.

Lefebvre, H. (1970). *La revolución urbana*. Barcelona: Ediciones península, 1972.

Lefebvre, H. (1971). *De lo rural a lo urbano*. Barcelona: Ediciones península.

Lefebvre, H. (1972). *Espacio y política. El derecho a la ciudad II*. Barcelona: Ediciones península, 1976.

Lefebvre, H. (1974). *La producción del espacio*. Madrid: Capitan Swing, 2013.

Lefebvre, H. (1978). *De l'Etat. Tome IV: Les contradictions de l'Etat moderne*. Paris: Union Générale d'Éditions.

León Vela, J. (2002). *La Alameda de Hércules y el Centro urbano de Sevilla: Historia de un reequilibrio del Casco Antiguo*. Seville: Universidad de Sevilla, Secretariado de Publicaciones.

Mansilla, J. (2019). Gentrificación, turistificación y clases sociales en las ciudades del Mediterráneo. In Cañada, E. (Ed.) *El turismo en la geopolítica del Mediterráneo*, Barcelona: Alba-Sud, 62–65

Marcuse, P. (1985). Gentrification, abandonment, and displacement: Connections, causes, and policy responses in New York City. *Journal of Urban and Contemporary Law, 28*, 195–240.

Marín, L. (1980). *Sevilla, centro urbano y barriadas*. Seville: Servicio de Publicaciones del Ayuntamiento de Sevilla

Martín, A. (2009). Primer proyecto español de marca-país: La Exposición Iberoamericana de Sevilla 1929. *I + Diseño, 1*(1), 7–20.

Merrifield, A. (2006). *Henri Lefebvre: A Critical Introduction*. London: Routledge.

Milano, C., Novelli, M., & Cheer, J. (2019). Overtourism and Tourismphobia: A journey through tour decades of tourism development, planning and local concerns. *Tourism Planning and Development, 16*(4), 353–357.

Naredo, J. M., & Montiel, A. (2011). *El modelo inmobiliario español y su culminación en el caso valenciano*. Barcelona: Icaria.

Peeters, P., Gössling, S., Klijs, J., Milano, C., Novelli, M., Dijkmans, C., & Postma, A. (2018). *Research for TRAN Committee - overtourism: Impact and possible policy responses*. Bruselas: European Parliament, Policy Department for Structural and Cohesion Policies.

Perkumiene, D., & Pranskuniene, R. (2019). Overtourism: Between the right to travel and resident's rights. *Sustainability*, 11, 2138.

Postma, A., & Schmuecker, D. (2017). Understanding and overcoming negative impacts of tourism in city destinations: Conceptual model and strategic framework. *Journal of Tourism Futures, 3* (2), 144–156. doi:10.1108/JTF-04-2017-0022

Purcell, M. (2014). Possible worlds: Henri Lefebvre and the Right to the City. *Journal of Urban Affairs, 36*(1), 141–154. doi:10.1111/juaf.12034

Relph, E. (1976). *Place and placelessness*. London: Pion.

Sanchez Vazquez, A. (2013). *Filosofía de la Práxis*. City of Mexico: Fondo de Cultura Económica.
Sanguin, A. L. (1981). La géographie humaine ou l'approche phénomenologique des lieux, des paysages et des espaces. *Annales de Geographie*, *501*, 568–584.
Sequera, J., & Nofre, J. (2018). Shaken, not stirred: New debates on touristification and the limits of gentrification. *City*, *22*(5-6), 843–855. doi:10.1080/13604813.2018.1548819
Sigler, T., & Wachsmuth, D. (2016). Transnational Gentrification: Globalisation and Neighbourhood Change in Panama's Casco Antiguo. *Urban Studies*, *53*(4), 705–722. doi:10.1177/0042098014568070
Silver, D. (2018). Alienation in a four factor world. *Journal for the Theory of Social Behaviour*, *49*(1), 84–105. doi:10.1111/jtsb.12192
Slater, T. (2009). Missing Marcuse. On gentrification and displacement. *City*, *13*(2-3), 292–311. doi:10.1080/13604810902982250
Smith, N. (1996). *The new urban frontier. Gentrification and the Revanchist City*. New York: Routledge.
Van Noorloos, F. (2011). Residential tourism causing land privatization and alienation: New pressures on Costa Ricás coasts. *Development*, *54*(1), 85–90. doi:10.1057/dev.2010.90
Veiga, C., Santos, M.C., Águas, P., & Santos, J.A.C. (2018). Sustainability as a key driver to address challenges. *Worldwide Hospitality and Tourism Themes*, *10*(6), 662–673. doi:10.1108/WHATT-08-2018-0054
Wang, N. (1999). Rethinking authenticity in tourism experience. *Annals of Tourism Research*, *10*(2), 349–370. doi:10.1016/S0160-7383(98)00103-0
Zaban, H. (2019). The real estate foothold in the Holy Land: Transnational gentrification in Jerusalem. *Urban Studies*, Online first. Retrieved from: https://journals.sagepub.com/doi/abs/10.1177/0042098019845614?journalCode=usja. doi:10.1177/0042098019845614

Who has the right to the rural? Place framing and negotiating the Dungog festival, New South Wales, Australia

Judith Mair and Michelle Duffy

ABSTRACT
Many small towns are using festivals and events as a way to boost tourism, by packaging and marketing existing festivals or by creating new events and festivals to attract tourists and the economic impact that they bring. However, increased mobility and migration, including amenity and lifestyle migrants, are changing the composition of many rural and regional centres, bringing both welcome diversity and potentially less welcome intrusion into established ways of life. So-called 'incomers' often bring higher levels of education and income and a different culture to small towns and villages, and, using their existing networks and connections, appear to be the driving force behind many cultural developments, including the creation or management of festivals and events. This paper examines the tensions arising as tourism and local festivals are being negotiated by various community stakeholders, and how place is framed through the rural idylls sought by amenity migrants. Drawing on Lefebvre's notion of the right to the city and the idea of rights in places, we investigate diverse interests, inclusions and exclusions through various place frames related to a local arts festival in Dungog, Australia, and question who has the right to the rural.

Introduction

For many small rural towns, community festivals and events are a way to boost tourism, either by packaging and marketing existing festivals to be more suitable for visitors, or by creating new events and festivals to attract tourists and the economic impact that they bring (Duffy & Mair, 2018). But festivals are not just economic spaces, they hold rich cultural and social meanings, depending on the kind of festival being enacted. For residents, they offer a way to engage in embodied, experiential, affective and emotional activities and practices that facilitate a sense of belonging to the local, and to lay claim to the sense of the community that arises through social interactions and engagement with the festival organization and enactment. They offer, in other words, ideas of community, belonging and place attachment–and notions of *rights* to the rural place they call home.

Tensions can then rise when newcomers arrive, bring diverse values, believes and customs, seeking the same rights to community and belonging as long-standing residents. Amenity and lifestyle migrants who move from urban centers to seek 'refuge' in attractive rural areas, for instance, can bring higher levels of education and income, new economic ideas and act as dynamic change makers in the community (Carson & Carson, 2018; Moss, 2006).Using their existing networks and connections, they may become the driving force behind new and existing social and cultural developments, including the creation or management of festivals and events (Mair & Duffy, 2018). Such reconceptualised or reinvigorated events are often supported by local authorities for the economic development and tourism opportunities that they offer, but may affect long-time residents in diverse ways (see below).

Our interest lies in how residents negotiate arts and cultural festivals in rural places as part of being and belonging to the local, and what kind of issues arise that may relate to infringement of social justice and community rights. These festivals are significant to issues of social justice because they represent a public performance of community related cultural goods and are also an important site for facilitating social relations. Important therefore to community identity, the festival is a political act, and we need to carefully consider how different understandings of place and culture are being negotiated, and ways in which possible notions of 'not belonging' are constructed in festival development. We explore these relationships in the context of a rural Australian community below, and seek to shed some preliminary light on some important questions that have received very little investigation, such as: What are the rights to the rural and how they are being negotiated or contested by diverse groups? Whose rights dominate, and why? Who has access, who is excluded and why?

We draw on Henri Lefebvre's (1991) notion of the right to the city as a starting point for considering how the assumed benefits of festivals may correspond to issues of social justice and rights. Lefebvre (1991) argued that lived space is not simply a passive stage on which social life unfolds, but represents an integral element of social life. Thus, the 'right to the city' refers to not just a right to physical space but more broadly to rights of access, participation, and, indeed, a right 'to habitat and to inhabit' (Lefebvre, 1996, p. 173). Lefebvre's focus on the *inhabitants* of the city is an argument that membership of a community is earned by living out the routines of everyday life in the spaces of that community (Purcell, 2002). Moreover, and relevant to our examination of the role of festivals in rural places, Lefebvre argues that art is an important political component of inhabiting place. He suggests that art is not about prettifying place through the addition of works of art; rather 'art can become *praxis* and *poiesis* on a social scale … art can create structures of enchantment' (Lefebvre, 1996, p. 173; italics in original). Fincher and Iveson (2008, p. 172, 173) point out that it is in just such ideas that Lefebvre 'imagined un-alienated urban life as a kind of permanent festival' and 'through participation in festive or carnivelesque moments, urban inhabitants could produce a more authentic form of being together'.

We suggest these ideas have important implications even for rural places, where increased mobilities and migrations are changing the composition of rural and regional centres, bringing change through performative activities in local spaces, such as that of community festivals. This paper reports on empirical research carried out in the small, rural town of Dungog, New South Wales, Australia and examines the tensions that arise in negotiating "community" and the place of festivals in the new rural idylls sought by amenity migrants. We seek to investigate how these practices, how community is enacted and produced, how the Dungog festival provides for equitable distribution of costs and benefits, how it facilitates connections between diverse groups, and what theoretical and conceptual approaches can help advance understandings of social justice and rights to the rural.

Festivals and social justice: rights to place, rights in places

Festivals 'have become an established part of the repertoire of contemporary planning' (Fincher & Iveson, 2008, p.176), and their benefits are seen to be not just economic but also social. Local government authorities facilitate them to bring about desired positive social outcomes such as increased social cohesion, community pride, sense of belonging and civic participation among others (Derrett, 2003; Duffy & Mair, 2018; Duffy & Waitt, 2011).It is surprising therefore to see how little very little research exists in this area to inform local government policy. For example, while increased community participation is an oft-stated benefit associated with holding a festival, local governments fail to incorporate clear objectives, strategies and evaluation that would provide measurable targets of enhanced feelings of inclusivity (Pugh & Wood, 2004). Instead, the emphasis has firmly been placed on how the economic benefits, job opportunities and local business investment (Whitford, 2009). Yet, festivals don't necessary provide fair or equal access to benefits and can act as spaces of exclusion through presentation of particular beliefs, ideologies or agendas of dominant interests involved. They function as both a form of social integration and cohesion as well as sites of subversion and protest—and these conflicting ways of framing the event may often occur concurrently (Mair & Duffy, 2014). As such they are political spaces. The different representations of community identity that circulate within the festival site are 'symbolic expressions of competitive social relations physically inscribed in space' (Jackson, 1988, p. 224). Greater attention to issues of diversity and *difference* in the festival spaces is also needed to better address injustices and inequalities that occur in the performance and enactment of community festivals. (Duffy & Mair, 2018).

Social justice, however, is a complex concept. It requires a geographical, sociological, political and cultural framework from which relations between individuals and groups can be understood, assessed, and characterized as just or unjust (UN, 2006, p. 12). Many countries around the world incorporate the principles of social justice into their governance structures; however, the nature of what is meant by social justice, and how it is conceptualised and operationalised, tend to differ. Three dimensions appear common to the way in which social justice is understood at a national level: equality of rights, equality of opportunities, and equity in living conditions. At the local level, it is often seen as relating to improving opportunities for participation in community life, improving decision-making, reducing economic disadvantage, and promoting respect and tolerance of diversity in all its various forms (Duffy & Mair, 2018). Fincher and Iveson (2008) argue for a socially just planning process so that participation of diverse groups and citizens in public spaces and events are facilitated. Such spaces are civic spaces reflecting democratic rights for 'when bodies assemble on the street, in the square, or in other forms of public space (including virtual ones) they are exercising a *plural* and performative right to appear' (Butler, 2015, p. 11; emphasis added). The ways in which our places are planned and developed are therefore important elements in improving everyday life and acting on issues such as enabling access, voice and equity in social spaces. The concept of the 'right to the city' developed by Lefebvre provides a useful approach to consider how performative social spaces like those of community festivals and events can facilitate (or inhibit) rights to the local.

The right to the city

The notion of the right to the city emerged from Lefebvre's thinking on how globalisation and neoliberalism were changing the way that our cities functioned. Lefebvre (1996) argued that the governance of our cities was being altered by the shifting political and economic landscape that began with industrialization and that as a result, urban inhabitants were increasingly marginalised and disenfranchised in relation to how cities were being planned, developed and used. As Purcell (2002) noted, what Lefebvre was suggesting was that the neoliberal agenda shifts policies and politics away from redistribution and towards competition, with an economic value placed on the urban spaces used by city inhabitants. This shift, as Harvey (2003) argues, means the interests of a small group of wealthy elites were dominating the way our cities were being shaped. In response to this, Lefebvre called for a radical rethinking of urban space, 'an expression of urban citizenship, understood not as membership in a polity … but as a practice of articulating, claiming and renewing group rights in and through the appropriation and creation of spaces in the city' (Lefebvre, 1996, p. 173). Lefebvre proposed the concept of the right to the city, a radical reimagining of how urban space should be produced and used. For Lefebvre, urban space was more than just the physical or material space within a city; rather, it related to how all aspects of city life are produced and reproduced. Further, only those who contribute to the lived experience of a city, by 'living out the routines of everyday life in the space of the city' should have the right to determine how urban space is used (Purcell, 2002, p. 102). Lefebvre (1996) identified two main elements to the right to the city—the right of participation (urban inhabitants should play a central role in decisions that affect the city) and the right of appropriation (not only should inhabitants have the right to access and use the urban space, they should have the right to produce urban space to meet their needs).

Lefebvre's interest in examining the production of, and rights to, urban spaces arose out of earlier studies on the urbanisation of mid 20th century rural landscapes (Elden & Morton, 2016). Drawing on a Marxist framework to examine the relations between land, the landowning class, rent and agriculture, his exploration of rural life and land reform led him to recognize the complicated processes that were transforming the local:

> Not only does the rural sociologist find themselves confronting structures originating from different historical epochs ... but they confront structures that are disintegrating, mixed with new forms and structures (Lefebvre, 2016, p. 67).

In the current landscape of local to global mobilities, amenity-led migration is one of the major processes that transform various property, social and material relations operating in rural places. New structures and processes are shaping representations of community identity as expressed in public places, social spaces and events like local arts and cultural festivals. Festivals constitute such a performative space in the community, not only in terms of who exercises authority over the form and activity of the festival, but in terms of democratic voice, diversity and difference, the inclusions and omissions of particular residents and participants in the festival's planning stages and its execution (Clarke & Jepson, 2011; Lefebvre, 1991). However, how well can Lefebvre's work address the mobilities and multiple performativities involved in the urbanizing of contemporary rural landscapes?

Rights in places

Scholars have debated the rights to the city concept over the years since it was first proposed. It has entered into common parlance as a 'political call to arms' (Pierce et al., 2016) and a 'rallying cry' (Attoh, 2011), and used in contexts as diverse as gentrification, immigration, housing and citizenship (Attoh, 2011). Purcell (2002) refers to it as equally exciting and disconcerting and notes that there is no guarantee that even if such rights were established in law or practice, that there would be a positive outcome for the city. Additionally, Purcell (2002) questions whether the right to the city treats all inhabitants equally, for as Marcuse (2009, p.189) points out, the problem is 'whose right, what right and what city'? Pierce et al. (2016, p.80) also emphasise that within the rights to the city framework, there are 'multiple, overlapping, fractious and contentious' rights, which require to be 'consistently (re)negotiated'. Lefebvre was clearly considering that those inhabitants who were being disenfranchised were the working class, but as both Purcell (2002) and Pierce et al. (2016) point out, there are many other marginalised groups who face struggles in terms of their ability to access and use urban space.

Drawing on Lefebvre's notion that social rights should be the result of the daily producing and reproducing of place, and also Massey's (2005) relational concept of place, Pierce et al. (2016) argue that a *rights in places* approach is a more useful way to understand how spaces and places are produced and consumed. For Pierce et al. (2016, p.80), social rights (unlike natural rights or civil rights) are earned 'through the constant informal political processes of daily life'.

Table 1. Specific stages for analysis for determining rights in places (after Pierce et al., 2016).

Rights in places claimants	Key identifiers
The relationships between different rights claimants	Who is making the rights claims? How are they differentiated in terms of power hierarchies?
The differentiated goals of those claimants	What are the claimants trying to achieve? To change the place-frame? To achieve a particular objective?
The rights claims themselves	What kind of rights are being sought? E.g. Property rights; Rights of access/use? Are the rights claims of different claimants mutually exclusive?
How those claims entail competing urban place-frames	Are the rights claimants trying to foreground a particular place frame e.g. community place-frame versus investment place-frame?
How the successes and failures of these different claimants in enrolling others shape the possibility for resisting oppressive hegemonies	How successful are the rights claimants at getting other people on board? Are they able to negotiate the persuasive place-frame/ narrative to overturn established hegemonies such as city councils, private investors etc.

They note that although these are not formal rights that are enshrined in law, nonetheless, such informal rights are socially enforced. Research shows that informal rights have a significant effect on daily lives (Blomley, 2005). For Pierce et al. (2016), rights are both differential (meaning that they can be specific to individuals and/or groups) and socially and politically negotiated. Thus rights in places as a concept acknowledges multiple and overlapping rights claims and claimants, and can help to illuminate the 'always ongoing' negotiations of different actors and processes (Pierce et al., 2016). Those claiming social rights in places need not necessarily be active or explicit in making their claims; simply performing their daily lives can be an articulation of their rights claim (Pierce et al., 2016).

Rights in places uses the idea of place-frames to understand the differing views and opinions of individuals and groups about a place. Place-frames relate to the various ways in which individuals and groups emphasise or de-emphasise aspects of a place, sometimes with the aim of furthering a particular political or social agenda or at other times simply as a result of different ways of understanding the place (Martin, 2003). Place-framing contributes to the way that narratives or discourses about particular places are shaped, and of course, competing place-frames can lead to contestation about place. As well as being a concept, rights in places can therefore also be used as an analytical process. Five specific stages to the process are identified by Pierce et al. (2016), as outlined in Table 1.

Although sharing some tenets with the notion of the right to the city, the rights in place approach may offer a better opportunity to consider how issues of rights, place and belonging are being negotiated in the rural context. Table 1 is particularly relevant when considering how mobilities and migrations have changed the face of the rural.

The changing face of the rural

Movement is a key characteristic of our twenty-first century lives. Theorists from a range of disciplines have argued that we need to recognise mobility as the central fact of modern (or post-modern) life (Sheller & Urry, 2006; Urry, 2008). In relation to rural places, one of the main elements of mobility is the increasing numbers of amenity migrants seeking to escape the city, or return 'back to the land' in search of more affordable homes, a better quality of life or just a change of scenery (Halfacree, 2007; Ruiz-Ballesteros & Cáceres-Feria, 2016). This phenomenon is known in Australia as sea-changing (moving to the coast) or tree-changing (moving to the countryside). Naturally, there is variation in the types of people who are moving out of the cities, but a notable number of these migrants are (early) retirees with greater affluence and mobility (Tonts & Greive, 2002). The movement of affluent in-migrants to rural areas seeking lifestyle changes and related amenities has led to significant changes in the make-up of a number of rural towns and villages and diverse interests and intentions towards these places. Differences in social and cultural capital between wealthy in-migrants and long-standing residents with a lower socio-economic statusm have invoked significant social tensions (Osbaldiston et al., 2015).

The notion of place-frames is useful here as a way to conceptualise the different perceptions and images of the place that can be created by differing groups of people in the changing mobilities of the rural community. A rural idyll, escape-to-the-country, bucolic place-frame may exist among the in-migrants, for instance, while a more historically-informed, community-based place-frame may be more evident among long-standing local residents. We therefore draw on Pierce et al. (2016) notions of rights in place and place-frames to consider how a raft of changes in the way that rural places are constituted in the case study of Dungog described below.

Case study: Dungog, New South Wales

Dungog is a small county town located in the Hunter Region of New South Wales, Australia. It is just over one hour's drive north of the city of Newcastle, and approximately three hours' drive north of Sydney. The Shire of Dungog covers an area of 2248 square kilometres and the main population centres are the towns of Dungog, Gresford, Paterson, Vacy and Clarence Town (Dungog Shire Council, 2019). Prior to European colonisation, this region was home to the Wonnarua people, part of the Indigenous population of Victoria and New South Wales collectively known as Koori. Their population is understood to have declined rapidly after European settlement, and their presence today is scattered through the Hunter Valley, an area that included Dungog (see https://australianmuseum.net.au/learn/cultures/atsi-collection/hunter-valley/the-wonnarua/). Surviving tribal groups and clans are actively exercising cultural identity and their rights and responsibilities associated with caring for Country (Wonnarua Nation Aboriginal Corporation 2014).The area around Dungog was colonised by Europeans in the early 1800s and has been an agricultural centre since this time, with a focus on dairy farms and timber production. The Shire has a current (2018) population of 9,346 people, of whom around 2,000 live in the town of Dungog making this the main settlement in the Shire. The 2016 Census report for Dungog Shire shows that 4.8% of residents were born overseas, and 5% of residents identified as Aboriginal and Torres Strait Islander Peoples. 91.6% of residents of Dungog were Australian citizens and between 2011 and 30 June 2016, 760 people had moved to the Shire from other parts of Australia (significant in a population of only 8,500 people) (Dungog Shire Council, 2019).

One of the main historic buildings in the town of Dungog is the James Theatre, known to be the oldest still operating purpose-built cinema in Australia and built in 1918 (Theatre, 2019). As a means to contribute towards the renovation and upkeep of this historic venue, a film festival was created in 2007. The Dungog Film Festival was very successful in terms of attracting famous names in Australian film, but after a period, the festival faced financial difficulties and ceased to operate in 2012. A new festival was inaugurated in 2016, called simply the 'Dungog Festival', and run primarily by volunteers. The 2018 Dungog Festival offered a wide range of elements, including a street parade, a long table dinner, a sculpture exhibition (Sculpture on the Farm), a range of films on show at the James Theatre, a country garden ramble (where several local gardens were opened to the public), a street party with local and national talent, a vintage and agricultural vehicle show, and daschund races amongst other things (Dungog Festival, 2018). The new incarnation of the festival appears to be gaining traction in terms of reputation and visitor numbers—numbers were estimated to be 5,000 in 2018 (Dungog Chronicle, 2018) an almost 5-fold increase from 2016 (Dungog Shire Council, 2019)—but there has arguably been a lack of clarity around the theme of the festival and its contents, which has caused some debate both within the festival organisation itself and in the community more broadly.

Methodology

This study employed qualitative research methodology and involved semi-structured individual interviews with stakeholders. Both authors also attended the event in the role of observer-as-participant, where the researcher is an observer who is not a member of the group under study (in this case part of the organising committee of the Dungog Festival, or a Dungog local resident) thus generating more complete understanding of the group's activities (Kawulich, 2005).Our sense-making of the data collected was informed by deep engagement with the community of Dungog over a period of one year (from around October 2017 to October 2018), including multiple visits, observations and discussions. In doing so, we attempt to take an emic perspective, described as a way of 'looking at things through the eyes of members of the culture being studied' (Willis, 2007, p.100).

Data gathering and analysis

Purposeful sampling was used to select individuals with particular knowledge of the research topic, which also helps to capture broad variations of a topic as well as commonalities (Patton, 2002).In this case, interviewees all had a good knowledge of the Dungog Festival and the Dungog community, either through being involved in the organisation and staging of the event, or through their professional capacities as business owners and local government officials. Additionally, all interviewees were local residents at the time of interviews. The use of semi-structured interviews—in which a set of key questions or themes are used to help guide the interview—allowed participants to discuss what is important to them (Brinkman, 2013) in the context of the Dungog Festival and its organisation. The open-ended questions also allowed participants to raise topics that may have not been considered when devising the interview schedule.

A total of 10 interviews were carried out between April and July 2018, including those who are involved in the festival organisation, staging and management, as well as those representing local businesses, tourism and the local council. Participants were invited to take part in the interviews to discuss their ideas about the role of festivals in the life of the community. Interviews were recorded with permission, transcribed, and then analysed using open and axial coding (Miles & Huberman, 1994). This involves the researcher breaking down the data into component parts to seek similarities and differences (Strauss & Corbin, 1990). Following this, a second round of coding is carried out using themes derived from the literature and themes that emerge from the data. In this case, both researchers coded the data individually, then met to discuss and synthesise the emergent themes in the second round of coding. The analysis and presentation of the interview material was informed by the Table 1 framework presented above.

Trustworthiness

In this research, trustworthiness was enhanced by a number of measures. Firstly, interviews were recorded with permission and transcribed. Member checking, where interview transcripts are made available to the interviewees for the purpose of checking their accuracy (Miles & Huberman, 1994) was carried out to ensure the credibility of the study. Triangulation of data, achieved by collecting data from multiple sources (interviews plus our extended engagement with the community) further helped to address the issue of credibility (Shenton, 2004). Full discussion of the context of the research, data collection and analysis methods, detailed presentation of data, help to address dependability. As with case studies in general, the results are transferable broadly to similar situations and sites elsewhere.

Findings and discussion

Competing place frames

Analysis of the data showed that each study participant had a specific place frame, albeit with some key links and connections with other place frames. For example, all interviewees spoke of the natural beauty of the town and its surroundings, calling it 'attractive'; 'pretty'; 'scenic'; 'visually aesthetic'; and 'beautiful'. Each of the place frames identified will now be presented as vignettes and discussed in turn. They show that although there are three main groups of rights claimants, they have a range of different ways of viewing the place. Underlining is used to emphasis specific points of relevance.

A town of Aussie battlers

This respondent, a long-term resident of the area, focused on Dungog as a place with a strong agricultural heritage, and where the famers are 'doing it tough' [a common expression in Australia denoting economic difficulties]. Dungog was framed as a town with tensions and divisions based on geography – 'Dungog Shire has two river valleys and one lot think the other lot don't care about them', and based on recent political decisions, including the realignment of the shire (local government authority) boundaries. The town was also framed as being home to some very "stubborn" locals, who, being rather set in their ways, get upset when the festival is being held because they 'can't sit in their favourite seat in the coffee shop'

It's a good life, here in Dungog

This participant focused on the amenities and practical and tangible benefits of living in Dungog – 'we have good schools and we are close to Newcastle and Sydney' and 'we live near a National Park, a State Forest and we have beautiful light for photography' - as well as the more intangible benefits – 'Dungog has a thriving arts and culture scene, including the festival'. The respondent had been born and brought up in Dungog, and welcomed inward migration, noting that it was 'helping to build the cultural and economic life of the community'. He viewed Dungog as an ideal tourist destination.

The perfect lifestyle, but…

This participant, an in-migrant, is a lifestyle migrant who viewed Dungog as having the perfect lifestyle after her previous hectic life in Sydney. She noted that Dungog is safe, accessible and had heritage value, and offers her the opportunity to run her own business (buying and selling antiques). However, she also noted that there had been a decline in retail in the town, which was leading to the town 'slowly dying', and that although the community was reasonably supportive of her business, 'running a small business in a rural town is not getting any easier' and 'in a small county town, local farmers are not always into the arts', so running a festival based on the arts might not be right for all local residents.

It's a dour town, but non-judgemental

This person had moved to Dungog in the past few years. He felt that the local community was quite hostile to change, and very 'ethnically homogenous', but at the same time was a very non-judgemental place to live. As well as offering fresh air, lots of space and no traffic (a huge bonus after his life in Sydney), the town was moving on from its historic reliance on agriculture and timber and towards a more tourism and arts based economy. He put much of this change down to the increasing numbers of in-migrants, but noted that it remains 'a community divided'.

A place that needs my help

This interviewee, also a migrant from Sydney and a previous second home owner in the area, felt that her calling is to 'contribute to communities in crisis'. She explained that she has a 'history' of being motivated to volunteer to help out various causes – churches, schools etc. She considered the place to be 'humble, non-pretentious'. She clearly viewed Dungog as another cause – but at the same time felt that she wanted to do something long term: 'not swan in and swan out, but [become] a member of a tight-knight community". She considers Dungog to be a place that needs her help and her network of contacts – 'I'm inviting all my friends from Sydney' - and wants Dungog to become famous for its public art. She was positive about the town, but noted that it needs 'more restaurants and a couple of good coffee shops'.

Stuck in the past

This interviewee enjoyed the sense of community and community involvement in Dungog, and as a former resident who moved away but has since returned, still has a close network of family and friends in the area. She described clear differences between long-term residents 'Old Dungog is generational and deeply embedded in the community' and more recent in-migrants - 'blow-ins' as they are known to long term residents. She said: 'Blow-ins are like, oh my God, this place is so beautiful, has such potential, let's make it really great' and suggested that these people are the ones driving change in the town, for example by getting involved in the running the festival. But she also pointed out that the majority of local residents still have very fixed ideas about how the town should be and don't always take kindly to change. She also highlighted the complete absence of Indigenous voices in any discussions in relation to the town and shire.

A happy little place to raise a family

This participant, the youngest, suggested that Dungog is a friendly place, with great amenities for young families and good accessibility to the larger cities of Newcastle and Sydney. He pointed out that he has a nice friendship group, his brother owns a local business and his own business is fairly successful. However, he noted that although it appears to be a friendly place, 'once you get in, you find that there are a lot of cliques'. He also stressed that as a small country town, it is 'white and aging' and that may not be ideal for the long term sustainability of the town.

Unfortunately, we're getting a roundabout

This is a long-term resident. She has lived in Dungog for a long time, and works in the tourist industry. Whilst being well aware of the attractions for visitors, such as the national park, the scenery and the walking trails, but she also had strong views on what makes Dungog a good place to live – 'we're a nice little country town, with good schools, we have no KFC, no Maccas [McDonalds], but unfortunately we are getting a roundabout'. She went on 'it's heaven, it's quiet, safe ... ' However, she did acknowledge that business was not as good as it once was, noting 'Dungog is not dying, but it is going downhill'. She put this down, in part, to what she called 'small town syndrome', where people are resistant to change and don't understand that the benefits of tourism accrue to the whole town, not just those explicitly in the tourism industry.

Dungog's not dying, it's alive and kicking

This respondent, who holds a senior position in local government, wanted to stress that although there had been some issues with a previous festival, and there are some economic difficulties in the town, the area was 'unique, beautiful, and pristine, with an outdoor rural lifestyle that is really important to me'. Acknowledging the aging population, she still noted that Dungog had positive population growth in recent years and that the strategic plan for the area involved focusing on building a sense of community as well as the wellbeing of the local children. The town had suffered 'divisions on a number of issues' (including the building of a dam) but that it was recognised that 'outside agriculture, tourism is the most important industry for the area'.

The identification of these place frames clearly shows that within the community, there are differing ideas of what the place could/should be and represent, and that the town is going through a process of negotiation and contestation about which of these place frames should be represented by the festival. The fact that the festival is organised by a committee including long term residents and new in-migrants is contributing to the lack of a focus/theme/topic for the festival, as individuals have very different ideas about the place and who and what the festival should represent. The '*place that needs my help*' place frame, where what is sought for the town is visitation, public art and improved restaurants and coffee shops, is always going to be at odds with the '*Aussie battler*' frame, where there is a sense that local, long-term residents (primarily farmers) are not going to benefit from tourism and art. At the same time, the '*Dungog is not dying, it's alive and kicking*' frame, while sharing the view that tourism is important, places the wellbeing of local children and building a sense of community at its heart; not altering the place to make it better for tourists.

The '*it's the perfect lifestyle but...*' frame shows the difficulties of making a living in the town as it currently stands, and contrasts with the '*happy little place to raise a family*' frame; yet both acknowledge that the town is divided, with cliques and tensions, a fact also highlighted by the interviewee frame '*it's a dour town, but non-judgemental*'. It might be a good place to live and to bring up a family, but economically, it is struggling.

Finally, the place frames of '*stuck in the past*' and '*unfortunately we are getting a roundabout*' focus strongly on the importance of the history and heritage of the town, the embeddedness of the local community and resistance to change. This is at odds with the *'place that needs my help'* framing of Dungog, which relies on in-migrants to know what will be best for the future success of the town and also with the *'it's a good life in Dungog'* frame, which focuses on the cultural and economic opportunities associated with increased tourism.

Rights claimants and their claims

The first element of a rights in places analysis requires the identification of the various rights claimants. There are several different groups which have rights claims in the place of Dungog. As noted in the literature review, rights claimants need not solely be those who are actively asserting claims to a place; simply performing one's everyday life and reflecting on this can be evidence of a rights claim. The place frames above help to identify key rights claimants. The first group is made up of long-term residents, and particularly those who have a focus on the agricultural history and heritage of the area, but also hold a view that the area remains an important farming centre. The strength of this claim is closely linked to the region's prosperity from the 1890s up until the 1980s that grew out of the dairy and timber industries, and the weakening and loss of these which has had a significant impact on this community. Their concerns relate more to the impact of drought and the struggles that local farmers have in maintaining their livelihoods. They have a very strong focus on what might be termed community spirit. Participant 1 describes this community spirit as follows: 'There is a real sense of community. I think this is very much a community program, the way we set the festival up and the festival board up and everything. It was very much about it being a community owned event'.

A second group is made up of the 'tree changers' or in-migrants, who have moved to the area within the past 2–5 years and represent very different levels of education and affluence to that of the long-term residents. Such in-migrants have very different ideas about rural place, in terms of its aesthetic value and the lifestyle that it offers. They appear to view the festival as an opportunity to (re)invigorate the local arts scene, and create a more vibrant place, based on arts and culture. For example, interviewee 2 feels that the festival offers 'an opportunity to showcase Dungog', while interviewee 7 explained that 'the time is right because the arts community are emerging from the bottom up'.

Finally, there are bodies charged with various responsibilities for economic development, including the local Council, the Chamber of Commerce and tourism promotion bodies. Their remit (amongst other things) is to create jobs and contribute to positive economic impacts, and they primarily look to tourism—including festivals and events—and the visitation that it brings, as a way to achieve this. This view is exemplified by interviewee 8, who explained, 'It's not a community festival. It is supposed to be Dungog bringing people in. So one of the aims of Destination New South Wales [a state destination marketing organisation] in their funding requirements … is to increase multiple overnight stays and year round visitation'.

Competing and sometimes contested views of a place are arguably a normal part of the fabric of any place or community. However, it can become an issue when the aims of different groups within a place seem to run counter to each other, and this may be especially challenging in locations where population and population density are low. Urban life is characterised by high density living and often markedly high social difference that produces sites of 'thrown togetherness' (Massey, 2005, p.11), which requires addressing a politics of coexistence as constituted 'through active living in the presence of others' (Kallio, 2016, p.2). Yet, rural life configures social life in different spatial configurations. Thus in the case of Dungog, it may be difficult to reconcile the three competing claims for rights in place outlined above—can one place concurrently be an agricultural community, a vibrant arts scene and a popular tourist destination? If so, how can these competing claims be managed? Will there always be a winner and therefore one (or more) losers? Examination of the competing place frames that are being used here may assist in unpacking some of these issues.

Festivals as sites for equitable distribution of the costs and benefits?

Many of those interviewed chose to be involved because of a strong sense of commitment to the Dungog community and support for its future. For many, this meant ensuring the economic viability of the town, given the weakening and loss of the region's prior agricultural industries. As one participant explained:

> I think this community needs help because it's still finding its mojo ... When those industries fell over, it's looking for a way of becoming economically sustainable (interview 6);

While another suggested that Dungog

> [needs] to build up local businesses and to give them a shot in the arm ... The businesses there battle, they struggle ... because people pile out of there into Newcastle and Maitland and Green Hills [larger cities and metro regions nearby], buying all their groceries and stuff (interview 5).

Thus, attracting new people into the town and region was viewed as an important means for ensuring that local businesses would remain and flourish, because 'like many regional communities there are businesses closing down' (interview 8). Some participants suggested that the tourism industry, and festivals in particular, offer an important opportunity to address this, that 'outside of agriculture, [tourism is] such an important part of the saving of Dungog Shire' (interview 1). Therefore, there was a very strong common theme arising from various members of the community interviewed as to the importance of a festival for Dungog beyond the scheduling of the event. Most of those interviewed were focused on ensuring the sustainability of Dungog as town *and* community, and this, participants suggested, would be brought about by establishing an economically viable town. Some festival scholars discuss this role in terms of a process of festivalisation, with festival activities not simply about renewing social bonds, but are a significant component of a community's socio-economic and cultural landscapes (Cremona, 2007; Roche, 2011). In this framework, festivals, rather than transcending the everyday, are intimately embedded within the public sphere as normative and at times transformative processes. Unsurprisingly, this connection between the festival and community viability was of particular importance for those attempting to establish new businesses in the area, who viewed the festival as a means to connect with a broader range of community members and promote business.

> Just showcasing the community and the businesses that we have here. There's a whole lot of really cool things happening that nobody knows about, or [they] aren't as well advertised as they should be. I think bring awareness to the district ... the festival is an opportunity to showcase Dungog (interview 2).

For a number of participants, while the festival is an important means to showcase Dungog, and inject economic benefits of repeat visitors, this needs to be part of a strategy that might encourage people to relocate to the area;

> I'd hope that people also visit the shire and come back again. Visit Dungog and the Shire and come back again sometime during the year ... I hope that people might visit Dungog and then take the opportunity to come and live here afterwards (interview 4).

Inherent in such responses is concern that Dungog, like many regional towns, has an ageing population. Yet, Dungog is one of a few regional towns in New South Wales that is not suffering population decline (Druce, 2018). Part of the reason for this is the town's location, as one participant noted;

> you've got a tree changer thing happening big time. It's still incredibly affordable ... you buy two houses her for the price of one in Newcastle, let alone Sydney. It is diversifying rapidly (interview 6).

The need for increased population to boost council revenue was noted by one participant, although caution was suggested given the potential impact this may have on lifestyle;

> There has to be some opportunity to grow the local area's government rate payer pool, because there's only 8,000 of them. Yeah, it's an interesting and delicate path, the potential of Dungog and how to grow it without destroying it (interview 8).

Therefore, while the festival may have a positive impact on the town's economy, there were concerns that its social impact has not been given adequate consideration. For some, the concern is a possible disruption to everyday activities, as one participant explained;

> The problem with festivals in country towns is that a proportion of the population loathe them because they can't get a park down the main street when they go down there ... or they can't sit in their favourite chair in the coffee shop on Saturday morning or the place runs out of bread or something (interview 5).

In this comment we hear reflected more than that of inconvenience; rather, the unsettling of one's capacity to live'out the routines of everyday life' (Purcell, 2002, p. 102) could be interpreted as impinging on the right to produce rural space to meet a person's needs. Such social rights may not be formally recognised, but as Pierce et al. (2016) suggest, challenges to these rights in place lie at the heart of negotiations about place and belonging. So how might rural communities navigate economic sustainability and social justice concerns?

The festival's role in facilitating connections between diverse groups

What most participants raised was the need to ensure that the community felt a sense of ownership of the festival. This was stated as being of particular importance given that a previous festival, the Dungog Film Festival, while successful, relied on the community for its success yet the festival organisers did not always ensure that the benefits flowed back into the Dungog community;

> The community just didn't feel they owned [the Dungog Film Festival] anymore. Once you lose the support of the community you just can't, it won't be a success. You can make it fabulous and just target outside people and hope that your market is your people coming from outside the town. But if you don't have the people willing to volunteer and help you do it, it's going to be a failure (interview 3).

The importance of a sense of community was a common reason given by participants when asked about their involvement with the festival, for example;

> To me, one of the greatest parts of the festival is the 140-odd volunteers that we get. And the experience that the volunteers have, and the pride that they have about the festival ... we all sit there the morning after the festival ... absolutely exhausted but having known we'd put on this wonderful event... there's many lasting friendships that have been forged by people volunteering together or meeting at the festival and you know that to me is a great outcome (interview 1).

The generation of such feelings plays a significant role in creating a sense of community, and for some participants the festival offered a way to help address feelings of disconnection and even antagonism between different parts of the Dungog community;

> I'm very much a community-minded person. I want to see the best for Dungog. I think that Dungog has gone through quite a number of hard issues – the Tillegra Dam was a huge issue that really divided the town. After that it was amalgamation, which again divided the town. So [the festival] was something to bring people into town (interview 1).

Nonetheless, nearly all of those interviewed voiced concern about the lack of an overarching theme for the festival, one that would encapsulate the community of Dungog, and the economic and social benefits that could be expected to follow on from this. Events promote social norms by showcasing a common social identity and facilitating a public expression of shared values and by being part of the shared life of a community, including its rituals and traditions (Mair & Duffy, 2018). Yet, the lack of a clear festival theme was of concern for our participants, and could arguably be indicative of a lack of common social identity in the town, and indeed may be reflective of the different place frames that are in play. This lack may also correspond to some form of class struggle, as competing notions of place unfold, and this is especially evident in responses to the role of art in the festival.

For some, the festival aligned with Dungog's art scene and that this potential could be strengthened. As one participant suggested, 'the time is right because the arts community are emerging from the bottom up' (interview 7). It is at this level of place framing that there is a clear divergence of views as to how best to represent Dungog. Rather than an image of Dungog as representative of a good life, the emphasis on art served to point to divisions between long-term farming families and more recent amenity-led migrants;

> Unfortunately a lot of people when they think of arts, they think arts as in painting and drawing ... I think a lot of people go, Oh, yes, that arts thing, they don't think they fit into that. That's for those arty-farty people, we're just farmers (interview 3).

Others argued that the festival required something radically different and unique that would not only pique the interest of potential visitors but could also contribute to the event's longevity;

> What do you do that's creative and different? Give a spin on that, that separates us and gives us something that we can build on. Parkes has got the Elvis Festival and an ABBA Festival ... anyway, something, a hook. The festival doesn't have a hook, is what my main concern is (interview 8).

Even so, a *few participants questioned how representative of the broader community the Dungog festival is*, and whether such themes are useful;

> I think there are too few people making the decisions and I don't think they're representative of – see, I could tell you the five people who are pivotal to this festival. But are they the right five people for the festival? (Interview 6).

One way to address concerns about broader community representation and engagement has been to incorporate events at a number of towns in Dungog Shire. Yet, as with strengthening networks beyond the immediate festival, the sustainability of this approach requires critical evaluation. As Reid (2007, p. 95) determined in her study of the social consequences of events on rural communities,

> The social divides or barriers that exist within rural communities are often deeply entrenched; in spite of this a neutral ground between these social groups may be achieved through organizing and planning rural events. Often this divide is based upon whether an individual is a "local" (born within the community) or a "nonlocal" (moved to the community).

When asked about the future of the Dungog festival, participants spoke about the hope that the festival would continue to grow and recognised as significant in the regional festival calendar; for example,

> I'd love to see [the Dungog festival] to be the same as the Elvis Festival or Tamworth Country Music Festival or some of the other festivals. ... I'm not talking about size but certainly that feeling that it's a great event (interview 4).

Some participants stated that the current framework for the event offers a means to contribute to this aspiration;

> I'm heartened that the festival will be successful this year in the model that we've got: the core program and the partner events. And I think it's brought people out of the woodwork ... that's great, if they can rally together extra people to just add to the celebration of community and that's what's wonderful (interview 7).

Others spoke of opportunities for further development, especially in terms of the sorts of new activities and approaches that could be incorporated into the program, as well as how the broader Shire community could be included;

> I think that unfortunately Dungog has been stuck in the past for too long in terms of everything from aesthetics to the way they do things ... I'd certainly like the festival to spread out into the community and

> for people to see opportunities where they can develop other businesses, tourism-related businesses (interview 8).

For some participants, such plans for future development requires greater motivation from members of the community, and that this will in turn be of benefit to the Dungog community;

> I like to see the community more motivated and supportive ... People get very motivated and they suddenly start being more positive. I'd like to see that ... I would like to see people say, wow, we're back on the map! We can put on an event, a festival, we can bring people to town. Let's all get involved and support it (interview 3).

Exclusion and "indifference"?

One key observation is the fact that Indigenous voices are almost completely absent from every place frame that was identified, and indeed Indigeneity was only briefly mentioned by one interviewee. This could arguably be a result of interview bias, but there was no mention of or reference to the traditional owners by any of the study respondents (except for one, see "Stuck in the past" above) and so the First Nations people do not appear to be involved in framing this place at all. This absence is also notable on the Dungog Shire Council website (www.dungog.nsw.gov.au/), which, unlike most Australian Council websites, lacks acknowledgement of Indigenous peoples and Country although some history is provided albeit anecdotally from Europeans moving into the region (www.dungog.nsw.gov.au/about-shire/our-history/history).

Even participants who represented the entrenched long-term residents, primarily farmers of European descent and part of the settler colonial history of the area, did not mention the need to include Indigenous Australian residents in Dungog town and the Shire in decision making on the festival, or in the distribution of the festival's potential economic and social benefits. Even the respondent noted in interview 6 above, who mentioned the need for broader representation, said nothing about including the traditional owners of the land either. The rights to the rural and the rights in places, at least in the case of Dungog and the arts festival, are limited to competing rights and place frames of dominant groups in the community, which comprise of generally affluent in-migrants (primary lifestyle migrants) and longer-term residents that range from well-settled local business owners (generally benefiting from tourism), plus long-established Australian farmers who were perceived to be resistant to change, including the growth of tourism (long established inter-generational farmers of European descent are part of the settler colonial history of the area). https://www.dungog.nsw.gov.au/about-shire/our-history/history).

Conclusions

Our study interest lay in how something like an arts festival can be a mechanism that may facilitate connections between the various and diverse members of a community in its geographical location. Data analysis, using Pierce et al. (2016) rights in places framework, uncovers diverse and contradictory imaginings of Dungog, competing interests and ideas of the community, of tourism and of the festival. Dungog, and its constituent parts of place and community, is strongly contested among dominant groups in the community. The various rights claimants and their place frames clearly demonstrate this. Such contestation is arguably a sign of democracy at work. Nonetheless, the new incomers (lifestyle migrants) with their desire to transform Dungog into a centre for the arts as well as a tourist attraction for wealthy visitors, appear to be getting the upper hand in terms of how the place of Dungog the town is being constituted. The conventional place frame with reliance on the agricultural heritage of the area, and long term residency as qualification for membership of the community, is potentially fading in the face of the economic importance of the tourism industry, although notions of the rural idyll underpin much of how this town is promoted.

It seems that the key reasons behind the strong voice and influence of the in-migrants are likely the high level of affluence plus knowledge, contacts and networks that they bring with them. These also provide these in-migrants with the confidence and skills to become involved in local politics, while the longer-term residents in our study appeared to see little need to be involved in local government. Examining the various place frames helps to identify the competing rights to shape Dungog, and the differing perspectives, influence and interests among the study participants and by extension, to various dominant groups in Dungog.

The rights in places approach also questions whether any of these rights claimants are able to overturn any established hegemonies in order to achieve their aims of reconstituting notions of Dungog as a particular place. In this case, the in-migrants are not overturning hegemonies, but actually are working with them (Council, local tourism bodies, etc.) in order to ensure that their place frames succeed. In order for the long-term residents to ensure that they are able to retain 'their' place, it is they who may need to overturn the Council hegemony. Yet, given the imperative to bring in visitors and their economic impact, it seems unlikely that they will be able to overturn the actions being implemented to make Dungog a successful tourist destination.

This study focused on interviews with key stakeholders associated with the festival, a fuller and more nuanced understanding of these differing processes requires interviews with a broader range of members of the Dungog community. To bring the discussion back full circle to Lefebvre requires a reminder of one of the key tenets of the rights to the city—that only those who 'live out the routines of everyday life in the space of the city' should have the right to determine how urban space is used (Purcell, 2002, p. 102). The participants in this research, those living out their lives in the space of the rural, demonstrated they felt they should have some say in determining the use of rural space. But what about the small, almost invisible, minority population of Indigenous Australians in the community and Shire? Diversity and *difference* was limited to gender equity, class difference, and differences in economic and social capital among a dominant population group who said little about the Indigenous People, their care for land, and how their perspective and voice mattered with respect to the festival and its intended benefits. Limited by time and resources, we were unable to meet with Indigenous representatives in Dungog and the Shire, but their exclusion and the inequity of democratic voice and participation in relation to festival planning and development suggests that greater effort needs to be made to ensure social equity, diversity and inclusiveness in festival planning and to include a decision making voice for Indigenous art and culture among the evolving festival themes should the traditional owners of the land wish to be involved.

In future, it would be advisable to incorporate decolonising methodologies into academic research on festivals and communities, and actively involving minority populations, diverse cultural and resident groups, in the research, planning and use of social spaces in the community. Local festivals and events not only affect local residents directly, they are a key portal to shaping the community through the interests of dominant local elites (Harvey, 2003) and the powerful influence of tourism.

Disclosure statement

No potential conflict of interest was reported by the author(s).

References

Attoh, K. (2011). What kind of right is the right to the city? *Progress in Human Geography, 35*(5), 669–685.

Australian Federal Government. (2010). *Australian law as applied to Aborigines.* Retrieved February 23, 2020, from www.alrc.gov.au/publication/recognition-of-aboriginal-customary-laws-alrc-report-31/4-aboriginal-customary-laws-and-anglo-australian-law-after-1788/australian-law-as-applied-to-aborigines/.

Blomley, N. (2005). Flowers in the bathtub: boundary crossings at the public-private divide. *Geoforum, 36*(3), 281–296.

Brinkman, S. (2013). *Qualitative interviewing: Understanding qualitative research.* Oxford University Press.

Butler, J. (2015). *Notes towards a performative theory of assembly.* Harvard University Press.

Carson, D., & Carson, D. (2018). International lifestyle immigrants and their contributions to rural tourism innovation: Experiences from Sweden's far north. *Journal of Rural Studies, 64,* 230–240.

Clarke, A., & Jepson, A. (2011). Power and hegemony within a community festival. *International Journal of Event and Festival Management, 2*(1), 7–19.

Cremona, V. (2007). Introduction—the festivalising process. In Hauptfleisch, T., Lev-Aladgem, S., Martin, J. Sauter, W. & Schoenmakers, H. (Eds.), *Festivalising! Theatrical events, politics and culture* (pp. 5–16). Rodopi.

Derrett, R. (2003). Making sense of how festivals demonstrate a community's sense of place. *Event Management, 8*(1), 49–58.

Druce, A. (2018). The 'unique' NSW regions defying the population drain. The Land. Retrieved June 3, 2019, from www.theland.com.au/story/5362642/unique-nsw-regions-defying-the-population-drain/.

Duffy, M., & Mair, J. (2018). *Festival encounters: Theoretical perspectives on festival events and social cohesion.* Routledge Advances in Events Research Book Series.

Duffy, M., & Waitt, G. (2011). Rural festivals and processes of belonging. In Gibson, C. & Connell, J. (Eds.), *Festival places: Revitalising rural Australia* (pp. 44–59). Channel View Press.

Dungog Chronicle. (2018, October 8). *Dungog festival attracts thousands.* Retrieved November 21, 2019, from https://www.dungogchronicle.com.au/story/5689658/festivals-new-look-format-a-winner/.

Dungog Festival. (2018). Retrieved March 12, 2020, from https://www.dungogfestival.com.au/.

Dungog Shire Council. (2019). *Dungog Shire Council Annual Report 2016–2017.* https://www.dungog.nsw.gov.au/annual-report-2016-2017.

Elden, S., & Morton, A. D. (2016). Thinking past Henri Lefebvre: Introducing "The theory of ground rent and rural sociology". *Antipode, 48*(1), 57–66.

Fincher, R., & Iveson, K. (2008). *Planning and* diversity in the city*: Redistribution*, recognition and encounter. Palgrave Macmillan.

Halfacree, K. (2007). Trial by space for a 'radical rural': Introducing alternative localities, representations and lives. *Journal of Rural Studies, 23*(2), 125–141.

Harvey, D. (2003). The right to the city. *International Journal of Urban and Regional Research, 27*(4), 939–941.

Jackson, P. (1988). Street life: The politics of carnival. *Environment and Planning D: Society and Space, 6*(2), 213–227.

Theatre, J. (2019). About the theatre. Retrieved March 12, 2020, from https://www.jamestheatre.com.au/about/.

Kallio, K. P. (2016). Youthful political presence: Right, reality and practice of the child. *Politics, citizenship and rights, 7*, 89–110.

Kawulich, B. (2005). Participant observation as a data collection method. *Forum Qualitative Sozialforschung/Forum: Qualitative Social Research, 6*(2), 1–28.

Lefebvre, H. (2016). The theory of ground rent and rural sociology. *Antipode, 48*(1), 67–73.

Lefebvre, H. (1991). *The* critique of everyday life (Vol. 1), John Moore Trans. Verso.

Lefebvre, H. (1996). *Writing on cities.* Blackwell.

Mair, J., & Duffy, M. (2014). Festivals and sense of community. In Jepson, A. & Clarke, A (Eds.), *Exploring community festivals and events.* Routledge.

Mair, J., & Duffy, M. (2018). The Role of festivals in strengthening social capital in rural communities. *Event Management, 22*(6), 875–889.

Marcuse, P. (2009). From critical urban theory to the right to the city. *City, 13*(2–3), 185–197.
Martin, D. (2003). 'Place-framing' as place-making: constituting a neighborhood for organizing and activism. *Annals of the Association of American Geographers, 93*(3), 730–750.
Massey, D. (2005). *For space*. Sage.
Miles, M. B., & Huberman, A. M. (1994). *Qualitative data analysis: An expanded sourcebook*. Sage.
Moss, L. (2006). *The amenity migrants: Seeking and sustaining mountains and their cultures*. CABI.
Osbaldiston, N., Picken, F., & Duffy, M. (2015). Characteristics and future intentions of second homeowners: A case study from Eastern Victoria, Australia. *Journal of Policy Research in Tourism, Leisure and Events Special Issue, 7*(1), 62–76.
Patton, M. Q. (2002). *Qualitative research and evaluation methods* (3rd ed.). Sage.
Pierce, J., Williams, O., & Martin, D. (2016). Rights in places: An analytic extension of the right to the city. *Geoforum, 70*, 79–88.
Pugh, C., & Wood, E. H. (2004). The strategic use of events within local government: A study of London borough councils. *Event Management, 9*(1), 61–71.
Purcell, M. (2002). Excavating Lefebvre: The Right to the city and its urban politics of the inhabitant. *GeoJournal, 58*(2/3), 99–108.
Reid, S. (2007). Identifying social consequences of rural events. *Event Management, 11*(1), 89–98.
Roche, M. (2011). Festivalisation, cosmopolitanism and European culture: On the sociocultural significance of mage-events. In Giorgi, L., Sassatelli, M. and Delanty, G. (Eds.), *Festivals and the cultural public sphere* (pp. 124–141). Routledge.
Ruiz-Ballesteros, E., & Cáceres-Feria, R. (2016). Community-building and amenity migration in community-based tourism development: An approach from southwest Spain. *Tourism Management, 54*, 513–523.
Sheller, M., & Urry, J. (2006). The new mobilities paradigm. *Environment and Planning A: Economy and Space, 38*(2), 207–226.
Shenton, A. (2004). Strategies for ensuring trustworthiness in qualitative research projects. *Education for Information, 22* (2), 63–75.
Strauss, A., & Corbin, J. (1990). *Basics of qualitative research: Grounded theory procedures and techniques*. Sage.
Tonts, M., & Greive, S. (2002). Commodification and creative destruction in the Australian rural landscape: The case of Bridgetown. *Australian Geographical Studies, 40*(1), 58–70.
United Nations (UN). (2006). *Social justice in an open world. The role of the United Nations. International forum for social development*. New York, United Nations. Retrieved September 29, 2019, from http://www.un.org/esa/soc-dev/documents/ifsd/SocialJustice.pdf.
Urry, J. (2008). Moving on the mobility turn. In W. Canzler, V. Kaufmann, & S. Kesselring (Eds.), *Tracing mobilities: Towards a cosmopolitan perspective* (pp. 13–23). Ashgate Publishing.
Whitford, M. (2009). A framework for the development of event public policy: Facilitating regional development. *Tourism Management, 30*(5), 674–682.
Willis, J. W. (2007). *Foundations of qualitative research: Interpretive and critical approaches*. Sage.

Locally situated rights and the 'doing' of responsibility for heritage conservation and tourism development at the cultural landscape of Honghe Hani Rice Terraces, China

Jun Gao, Hongxia Lin and Chaozhi Zhang

ABSTRACT

This study examines the understandings and practices of responsibility for heritage conservation and heritage tourism from a multiple-stakeholder perspective at the Cultural Landscape of Honghe Hani Rice Terraces, a unique World Heritage Site in China. Data were mainly collected via participant observations and semi-structured interviews. Local villagers comprised of ethnic minorities are conserving the heritage landscape in this protected heritage site through their traditional agricultural production and integrated religious and cultural practices. Conflict has arisen over the years due to inequities between the responsibilities they assume to maintain the cultural landscape and their ability to acquire compensation and share tourism's benefits. Other stakeholders (e.g., local government, the state-owned tourism enterprise, hostel operators) have benefited greatly from tourism to this area, but have not shared the responsibilities of actually maintaining the heritage landscape. The findings also show that an emphasis on the material heritage and technical approach towards the conservation of the heritage landscape by local authorities exacerbate the conflicts as the roles and needs of the local residents are not recognized. The study suggests that heritage responsibility discourses and practices justify and reinforce systematic unequal power relations between local community residents and the state at the Cultural Landscape of Honghe Hani Rice Terraces. The villagers play the greatest role in sustaining the cultural landscapes for current and future generations of Chinese citizens and visitors, yet experience numerous restrictions and impediments in the process. This study reveals complex relationships between the "doing" of responsibilities through local cultural practices of the Hani ethnic minority and their "rights," both of which are situated locally and socially constructed.

Introduction

Responsibility is often portrayed in various international conventions, charters, and policy documents as an abstract moral principle that can be followed with the help of a set of rules or guidelines. See, for example, the guiding principles listed in the Cape Town Declaration on Responsible Tourism.[1] Or see, for instance, the ever-increasing academic terms relating to responsibility (e.g., responsible tourism, destination social responsibility, heritage responsibility), which implicate that 'doing responsibilities' would help improve sustainability and tourist experiences alike (Zhang, Chen, & Wang, 2017). In practice, however, responsibility is enacted dynamically through the performance of responsibility, in the 'doing' of responsibility, as a process that is always situated and whose participants are always in a state of 'becoming' (Sin, 2014, p. 144). As such, responsibility is relational, complicated and unsettling (Sin, 2017). It is performatively embedded in discourse, conflict and power relations. Grimwood, Yudina, Muldoon, and Qiu (2015) find that touristic discourses of responsibility are socially produced and have normative power to silence minority Indigenous voices in the Thelon River area in northern Canada. Yet, while the notion of responsibility is being taken up in the context of tourism generally, research on responsibility in the context of cultural heritage conservation and heritage tourism is lacking, particularly with respect to the roles and relationships of various stakeholders in the destination (Zhang, 2014; Zhang et al., 2017). Furthermore, existing studies tend to be de-contextualized and lacks an emic perspective, that is, focusing on the situated, relational and contested nature of 'doing' responsibility on the ground, as Sin explains.

China currently is actively engaged in developing and promoting its cultural sites, and the customs and traditions that are part of their rich cultural heritage. For instance, local governments actively, even obsessively seek to inscribe sites on the World Heritage list, contributing to a World Heritage craze (Yan, 2018). Heritage tourism being promoted in these World Heritage destinations bring together tourist, corporation, state, locals, etc. and can serve as specific contexts to examine how responsibility for heritage conservation plays out on the ground. The question of responsibility is complicated by numerous factors in the Chinese context, such as the complexity of property rights (Li, Wang, & Chen, 2019). In this regard, securing rights, including land rights and usufruct rights (i.e., applications of property rights), is now seen as a precondition for the sustainability of conservation practices, giving local people long-term security and promoting long-lasting stewardship of the land (Oviedo & Puschkarsky, 2012). In China, however, the land essentially belongs to the state and is therefore not privately owned; the state can legally seize or expropriate the land with compensation to meet the 'public good'. Other rights are also at issue in the drive towards rapid heritage tourism development and promotion. For instance, Wang and Zheng (2015) point out that the rights of local community to participate in tourism development is not fully guaranteed by law in China. This article therefore sets to examine the understandings and practices of responsibility for heritage conservation and heritage tourism from a multi-stakeholder rights *and* responsibility perspective. A case study of the UNESCO protected Cultural Landscape of Honghe Hani Rice Terraces in China offers an opportunity to examine the performance, process and performativity of responsibility as it is enacted and practiced by the minority Hani residents whose productive work conserves the heritage landscapes of this World Heritage Site (WHS).[2] Religious and cultural practices interweave with their agricultural practices to enable crucial environmental and heritage conservation of these unique cultural landscapes. Complex relationships between rights and responsibilities arise and play out among multiple stakeholders, including the local and state governments, tourism interests and the Hani communities that have carved and tended the irrigated terraces that support their rice paddy fields for 1300 years in southern Yunnan province.

Responsibility and rights-based approaches to world heritage

Conservation interventions often exert profound impacts on the livelihoods and living conditions of local communities in or near protected areas. Oviedo and Puschkarsky (2012) assert that it can be very challenging to protect heritage sites without compromising the rights of local communities; examples may include forced evictions and violence, development restrictions that entrench poverty, destruction of means of subsistence, disruption of cultural practices, etc. Such problematic link between heritage conservation and rights has engendered many negative social impacts, hampering the sustainability of heritage conservation. Indeed, with the increasing recognition of rights at stake in and around World Heritage Sites, there has been an emergence of interest in the adoption of rights-based approaches (RBAs) to heritage conservation (Kraak, 2017). In terms of cultural heritage, RBAs 'oblige one to go beyond preserving and safeguarding an object or a manifestation in itself to take into account the rights of individuals and groups in relation to such object or manifestation and to connect cultural heritage with its source of production' (Bennoune, 2016, p. 15). This echoes the global trend in heritage conservation towards being 'people-centered' (Chitty, 2016; ICCROM, 2015) and the call to engage local communities to move towards more inclusive and socially just conservation in the wider conservation literature (Oviedo & Puschkarsky, 2012; Shultis & Heffner, 2016). Thus it challenges the traditional heritage conservation approach underpinned by authorized heritage discourse, featuring state and expert-led mechanisms that tend to prioritize materiality with technical conservation matters and exclude local communities (Smith, 2006; Zhang & Lenzer, 2019).

However, in terms of RBAs to heritage, there is inconsistency and ambiguity in the understanding of 'rights'; orientations towards 'rights' range from strictly legal to more discursive (Kraak & Aykan, 2018). Regarding a legal understanding of rights, international human rights laws are first referred to, whereas domestic legislations are usually also taken into consideration. In such circumstances, the focus is on legal obligations owed to local communities by virtue of their rights under international and national laws (e.g., civil and political rights, economic, social, and cultural rights; Jodoin, 2014). Such an understanding is often adopted in theoretical discussions about RBAs and heritage conservation, and in some cases, the term 'human rights', instead of 'rights', is more generally used (e.g., Blake, 2011; Ekern, Logan, Sauge, & Sinding-Larsen, 2012; Vrdoljak, 2018).

On the other hand, in the empirical investigation of heritage practices on the ground, a more discursive approach—which may take the legal side into account to a greater or lesser extent—is generally adopted (Coombe & Weiss, 2015; Kraak & Aykan, 2018). In this context, rights refer to the fundamental normative rules about what is allowed of people or owed to people in relation to heritage, according to laws, regulations, or moralities, as embedded in what people say and do (Escallón, 2018). Compared with a legal perspective on rights, it is thus a much broader understanding, and the meaning of rights can be highly contingent and contextual (Kraak, 2018). Therefore, 'the debate over the meanings and uses of rights cannot be considered in the abstract but, instead, must be seen as reflecting how rights are contextual, malleable, and strategic' (Escallón, 2018, p. 78). Even so, regarding RBAs to heritage, there seems to be an emerging focus on the right of local communities to participate effectively in decision-making, which requires relevant parties to obtain free, prior, and informed consent from local communities for heritage identification and governance (Kraak & Aykan, 2018; Vrdoljak, 2018).

Rights should go hand in hand with responsibilities. Oviedo and Puschkarsky (2012), for instance, comment on the conservation paradigm shift:

> Efforts to empower local communities so that they can affirm their rights and act on their responsibilities, and to integrate natural and cultural values at sites overlapping with traditional lands, are increasingly seen as fundamental elements of conservation approaches (p. 285).

Likewise, Ekern et al. (2012) point out that the application of RBAs to heritage involves clarifying the relationships between all the stakeholders in a project or a policy initiative in terms of their rights and responsibilities, and seeking ways to overcome the power differentials that might block the realization of rights and assumption of responsibilities (see also Kraak & Aykan, 2018). Responsibilities are seen here to be enacted as a performative process in a pluralistic domain in which multiple people, states and organizations make active and situated choices with the help of practical reasoning (Clarke, Barnett, Cloke, & Malpass, 2007; Sin, 2014).

Stakeholder responsibilities and rights in the heritage destination

Tourism development brings new groups and subgroups, varies and complicates conflict nature and forms in tourism destinations (Yang, Ryan, & Zhang, 2013), especially in the context of heritage tourism. Zhang, Fyall, and Zheng (2015) point out that conflict is often an inevitable outcome of heritage and tourism interaction as the two are always in a challenging relationship. Their study identifies six types of conflicts in heritage tourism development, including resource-use conflicts, commercialization-authenticity/modernity-tradition conflicts, interest/collaboration conflicts, cultural/ethnic conflicts, conceptual conflicts, and human rights-World Heritage conflicts. Among these conflicts, the interest conflict between stakeholder groups has become the major contradiction in heritage tourism (Zhang, 2014). What lies behind this type of conflict is often the disparity in power across stakeholders, specifically, the change of power balance with regard to the stewardship of heritage is often the source of contradictions and conflicts (Porter & Salazar, 2005). For example, regarding tourism development in traditional villages in China, Gao and Wu (2017) note that national and local authorities would usually intervene and take over development rights, designating villages as tourism attractions and collecting tickets. This, in turn, often leads to the marginalization of local villagers in benefit-sharing as well as decision-making power, which leaves space for potential conflicts in further developments. This is especially so in China's ethnic minority areas where local authorities, often hand in hand with external Han Chinese entrepreneurs, dominate the tourism industry, leaving limited space for ethnic minorities to participate in and control tourism development (Yang, Wall, & Smith, 2008; Zhu, Jin, & Graburn, 2017), though a few recent studies indicate the growth of ethnic minorities' capacity to participate in local socio-economic development (Cornet, 2015; Gao, Ryan, Cave, & Zhang, 2019). Yang et al. (2008), for instance, note that the tourism sector in Xishuangbanna of Yunnan is dominated by the majority Han Chinese population and local ethnic minorities are frequently marginalized as they are not able to compete with experienced Han.

Empowering local communities to effectively participate in tourism decision making and be able to share equitably in the benefits of tourism development is a key principle of sustainable tourism (Cole, 2006; Strzelecka & Wicks, 2015; Zuo & Bao, 2008). And Western writings often assume that participation in decision making precedes the acquisition of benefits (Su & Wall, 2015). In China, however, studies have suggested that local communities can receive benefits from tourism despite their minimal participation in tourism decision making (Su & Wall, 2015). In fact, it is often benefit sharing that is more valued by local communities than the right to participate in tourism decision making (Su & Wall, 2014, 2015). Possibly due to China's long tradition of being a centralized state with powerful governments, not all community members are interested in decision making, which is often seen as the responsibility of the government and their managers (Su & Wall, 2015). Not surprisingly, it is benefit distribution mechanisms that have typically drawn much academic attention in China (Zhang et al., 2019). For example, partly to guarantee the right of local communities to share the benefits of tourism development, Chinese scholars have proposed the adoption of a new type of property rights in law, that is, 'tourist attraction property right'—the exclusive right to acquire differential earnings out of the tourist attraction value of land and its appendages (Bao & Zuo, 2012; Zuo & Bao, 2016).

Case study: the cultural landscape of Honghe Hani Rice Terraces

The Cultural Landscape of Honghe Hani Rice Terraces, designated as a UNESCO WHS in 2013, is a farming civilization spectacle mainly created by Hani people, an ethnic minority in China. It is also a Globally Important Agricultural Heritage Systems site designated by the Food and Agriculture Organization of the UN in 2010. These terraces, located in Yuanyang County, Honghe Prefecture, Yunnan Province in southwest China (Figure 1), are over 1300 years old. As a cultural landscape, the Outstanding Universal Value (OUV) of the site relates to the harmonious

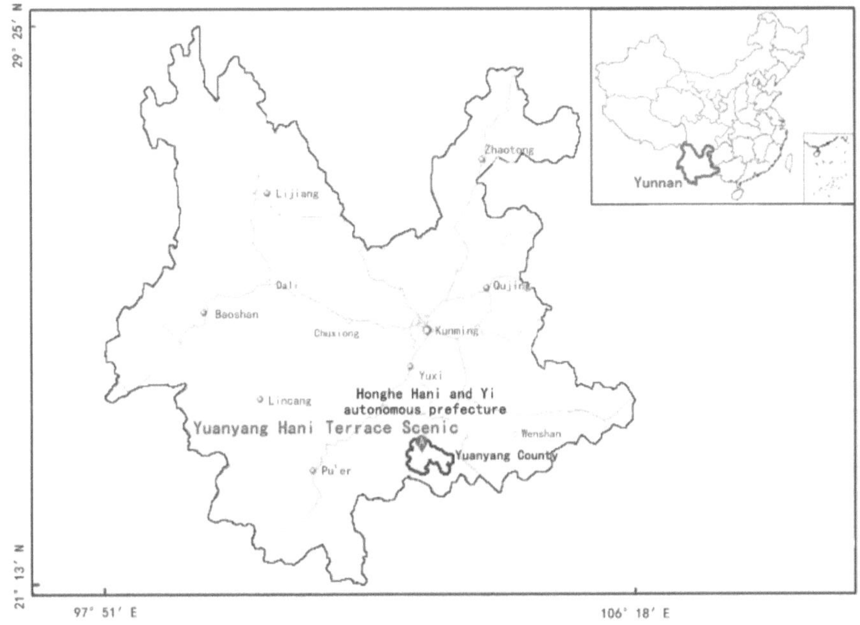

Figure 1. Location of the WHS.

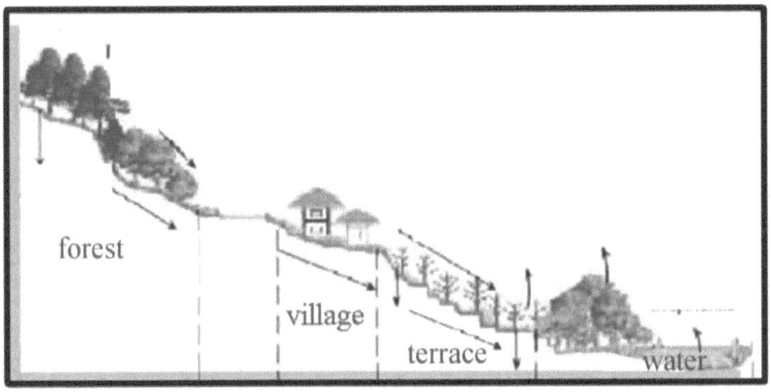

Figure 2. Schematic map of the forest-village-terrace-water system.

interaction of human beings and nature, based on exceptional and long-standing social and religious structures, resulting in an elaborate and finely tuned agricultural, forestry and water distribution system.[3] Specifically, the system consists of four core elements, including forest, village, terrace, and water, which are all integrated vertically in order to take advantage of the environment of high mountainous terrains, tropical climates, and heavy rainfall (Figure 2; Li, 2017). The forest plays the role of a water reservoir, supplying ever-running streams via a complex web of natural and man-made drainage channels and sluice gates distributing water throughout the terraces via a democratic social system with an elected irrigation headman to manage and apportion water resources (Adachi, 2007). The village is the central focus of the landscape, providing active labor, livestock, and technology, and is featured by mushroom like traditional dwellings—commonly two-and-a-half storeys with a thatched roof. The maintenance of the terraces landscape to date is attributed to the collective work of the Hani communities, and villagers have a strong sense of community. The terraces produce agricultural products that sustain life, and

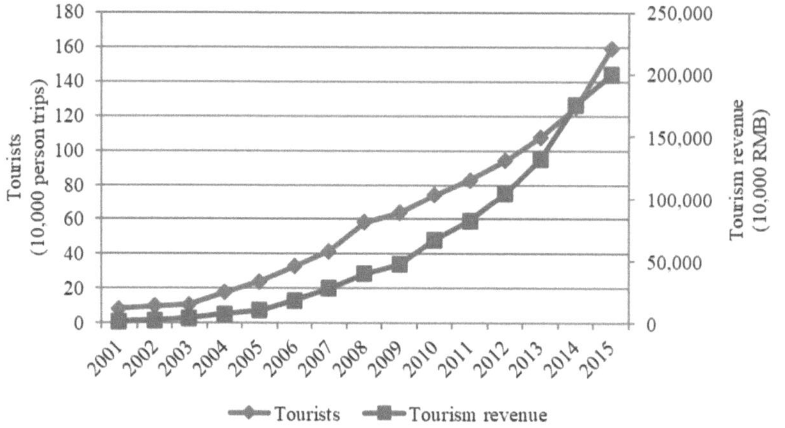

Figure 3. Tourist arrival and tourism revenue of Yuanyang County (2001–2015). Source: Yuanyang County Tourism Administration. *Note*: The Rice Terraces is the core attraction and must-see of the county, thereby the tourism development of the county reflecting the trend of tourism development at the site.

provide a backdrop containing the other elements (Jiao, 2003). The river, at the bottom, principally function as a climate regulator.

While a few 'pioneers' (mainly photographers) had started to visit the area in the early 1980s, domestic mass tourism there only began in the early 2000s. Since then, its tourism development has been in a steady increase, with the successful designation of Hani terraces as a WHS providing a further impetus for tourism development in 2013 (Figure 3). Even so, tourism is still at an early stage of development in the area, compared with that of well-known destinations in Yunnan, such as Xishuangbanna, Dali and Lijiang. The nominated property covers an area of 166 square kilometers, where 82 relatively small villages with between 50 and 100 households are constructed above the terraces just below the mountain top forests. Yet, most of the villages are currently off the tourist trails, and *Qingkou*, *Dayutang*, and *Pugaolaozhai* are the three main villages that are involved in tourism. These three villages all lie in the protected zone of the WHS, hence face the same conservation laws and regulations. Due to the early stage of tourism development, formal tourism businesses at the villages were limited to hostels that offered accommodation and dining for tourists. To date, poverty is still a big problem facing the indigenous people there. While the study involves all the three tourist villages, *Pugaolaozhai* is the focus of this study due to the intensity of tourism business development there combined with the intensity of interactions among villagers, migrant entrepreneurs, local government and a state-owned tourism enterprise (i.e., Expo Company). Even so, when our field work was conducted in 2014 and 2015, the annual individual income of the village was only RMB 2000–3000 ($284–$426), and most of the young villagers went to cities around as migrant workers.

Methodology

The current study takes a discursive approach to responsibilities and rights in relation to heritage, assuming that they are intersubjectively and socially constructed. An interpretive approach and qualitative data collection and analysis was undertaken. Two fieldtrips to the site were conducted from 24 August to 6 September 2014 and from 2 to 17 February 2015, respectively, reaching up to 30 days. First-hand data were collected via participant observation and semi-structure interview. For the observation, photos and notes were taken for further analysis. A snowball sampling approach was employed for interviewing 51 respondents (see Appendix Table A1), among whom 41 were residents of the three tourist villages, including 22 ordinary villagers who are not much involved in tourism (Ov1–Ov22), 13 local hostel operators (Lo1–Lo13), 6 outside

Table 1. Major events in the World Heritage designation process.

Year	Major events
2000	World Cultural Heritage Nomination Committee established at the prefecture level and county level
2001	The Interim Measures for the Administration of Hani Terraces in Honghe Hani and Yi Autonomous Prefecture made by the prefecture government
2002	Heritage nomination planning made by a group of experts organized by World Heritage Administration Office at the provincial level
2007	Hani Terraces Administration Office established at the prefectural level
2008	Hani Terraces Administration Office established at the county level
2009	Hani terraces protection planning and heritage nomination materials made by experts organized by Hani Terraces Administration Office established at the prefectural level
2012	Hani Terraces Protection Regulations of Yuan Yang County, Honghe Hani and Yi Autonomous Prefecture, Yunnan Province made at the prefecture level Village Residential House Protection and Management Regulation for Honghe Hani Terrace Cultural Landscape (Trial Implementation) made at the county level The terraces landscape approved by State Council as China's World Culture Heritage nomination site for 2013
2013	Successful nomination of the terraces landscape as a WHS

Han Chinese hostel operators (Oo1–Oo6). The rest were seven government officials, six at the county level and one at the prefecture level (Go1–Go7), and three staff from the state-owned tourism enterprise (Te1–Te3) that operates tourism at the site.

The semi-structured face-to-face interviews, each lasting between 15 and 121 min, were all recorded with consent from the respondents, and later transcribed. For government officials, questions about history of heritage nomination and tourism development, governmental involvement in tourism development, heritage protection, conflicts and problems at the site, were asked. For the Expo Company respondents, the questions focused on tourism operation, conflicts and problems at the site, heritage protection, corporate social responsibility visions and practices. For community residents, questions related to livelihood, conflicts and problems at the site, attitudes towards local government, the enterprise, tourism development and terraces, were asked. Moreover, relevant secondary sources of data to supplement and triangulate the first-hand data, including laws, regulations, planning, etc. were obtained from local government and the Expo Company. It should be noted that although the fieldwork data were from 2014 to 2015 (several years ago), it was right after the WHS designation of Honghe Hani Rice Terraces, and provides vital documentation of this period. Also relevant laws and regulations have not changed since 2014–2015, so the findings continue to be highly applicable today.

Consistent with our discursive approach to responsibilities and corresponding rights in relation to heritage, discourse analysis was adopted to analyze the data. Discourse analysis 'examines how discourses are constituted and circulated within texts and representations, which in turn function to produce a particular understanding or knowledge about the world that is accepted as "truth"' (Waitt, 2005, p. 168). Embedded within a social constructivist paradigm, this approach holds that broader assumptions, structures and/or meanings underpin what is actually articulated in the data (Braun & Clarke, 2006). It is worth pointing out that understanding and elaboration of discourse involves not only the text but also the practices based on it, hence for heritage studies, 'a particular discourse is unfolded both in its language (understanding of the heritage) and practice (conservation, identification and management of the heritage)' (Yan, 2015, p. 68). Our analysis began with familiarizing ourselves with the different sources of data via repeated reading. Then we identified themes in relation to the semantic content of the data. Lastly, we related to the literature on heritage administration and tourism development in China and determined the underlying ideas, assumptions, and conceptualizations (i.e., latent themes) that may shape the semantic themes identified previously. It should be noted that the analysis was not a linear process as we constantly moved back and forward between the entire data set, the coded extracts of data, and the themes identified (Braun & Clarke, 2006).

Table 2. Protective titles and corresponding laws/regulations.

Year	Protective titles	Corresponding laws/regulations
2002	Preparatory Directory for World Cultural Heritages of China	Convention Concerning the Protection of the World Cultural and Natural Heritage Operational Guidelines for the Implementation of the World Heritage Convention
2006	Prime Cropland Preservation Area of Yuanyang County	Regulations of the China on the Protection of Basic Farmland
2007	National Wetland Park	Measures for the Administration of National Wetland Park
2008 2012 2014	County-level cultural relics protection units Province-level cultural relics protection units Major Historical and Cultural Site Protected at the National Level	Act of the Preservation of Cultural Relics Promulgated by China Regulations for the Implementation of the Act of the Preservation of Cultural Relics Promulgated by China
2013	Important Agricultural Cultural Heritage of China	Regulations for the Administration of Important Agricultural Cultural Heritage of China
2013	World Cultural Heritage Site	Measures for Conservation and Management of World Cultural Heritage Sites

Source: Administration Office for Hani Terraces of Yuanyang County.

Findings

Multi-stakeholders' perspectives on responsibilities in relation to heritage

In this section, the understandings of responsibilities in relation to heritage, with reference to rights, are examined from the perspectives of local government, the state-owned tourism enterprise, local villagers, and outside hostel operators (Han new-comers).

The local government

In 1999, a culture anthropologist at Yunnan Provincial Academy of Social Sciences put forward the application of Hani terraces as a WHS. The idea was soon recognized by the prefecture government, ushering the heritage making process. As shown in Table 1, local government established corresponding governmental bodies for heritage nomination and administration, organized experts to prepare nomination materials, made specific planning, regulations and laws to manage heritage. Such a heritage making approach is a typical reflection of the decentralized administrative system for heritage in China, that is, the local government (county and prefecture level) is responsible for heritage administration (Bodolec, 2012). Also, the process is usually supported by experts, as stated in article six of Measures for Conservation and Management of World Cultural Heritage Sites issued by the Ministry of Culture of China:

> The state shall adopt the expert consulting system for the important matters relating to the protection of World Cultural Heritages.

While configured with international heritage discourse (mainly represented by UNESCO) and the heritage administration structure of the state, the heritage making process is largely dominated by local government, in alliance with heritage experts, dismissing the voices of local communities. Such is illustrated in the prefecture governor's summary of the heritage nomination process:

> The precondition of heritage nomination is that it must have OUV. Though Honghe Hani terrace culture landscape has abundant heritage value, [one] **must accord with the newest format and request for World Culture Heritage nomination**, unearth the most outstanding and unique heritage value, meet the standard of heritage nomination. Therefore, **Honghe prefecture government organized expert groups to make elaborate planning**... finally identified two main value as breaking point... Centering on such a value approach, we made heritage nomination document...

The heritage making also endows local government authority over the forest, villages, terraces and water system that prior to the heritage nomination the government rarely paid any attention to or controlled. On the one hand, with the heritage making and tourism development, the terraces become increasingly valuable. Local government applied for a variety of protective titles for the terraces, which correspond to the specific laws and regulations that bestow it the authority to administrate the heritage landscape (Table 2). On the other hand, in order to promote heritage nomination and protection and tourism development, local government established specific bodies and made specific laws, regulations and planning to manage the heritage, which greatly reinforces its control over the heritage site. For example, article 14 of Hani Terraces Protection Regulations of Yuanyang County, Honghe Hani and Yi Autonomous Prefecture, Yunnan Province (hereafter Terraces Regulation) states:

> The construction of house, road and other buildings and business activities within the key protected areas of Hani Terraces should conform to the protective management and planning regulations. Also, in the approval process, the relevant departments shall solicit opinions of the Hani Terraces Administration at the county level.

Nevertheless, with the right to 'make' and manage heritage, comes responsibilities. To protect the heritage is frequently mentioned as governments' primary responsibility, be it in local heritage laws, regulations and planning, or in official inspection of the heritage site, or in government officials' depiction of governmental responsibility for heritage. For instance, Article 1 of Village Residential Houses Protection and Management Regulation for Honghe Hani Terrace Cultural Landscape (Trial Implementation) (hereafter Residential House Regulation) justifies the making of the regulation as follows:

> To practically strengthen the protection and management of village residential houses of Honghe Hani Terraces Cultural Landscape, maintain the location feature, layout structure and building style of the traditional village, highlight traditional culture landscape elements...

On top of the responsibility for heritage protection, to improve the well-being of local communities is also identified as governmental responsibility, as reflected in local regulations/laws as well as local officials' account. For example, article 12 of the Terraces Regulation requests:

> The governments of the prefecture, Yuanyang County, Honghe County, Lvchun County and Jinping County should make industry policies that facilitate the sustainable development of Hani terraces, arrange **preferential** agricultural fund, support, lead and help villagers in the key protected zone **develop economy, increase their income**.

Notably, while the identified World Heritage value relates to the interconnectedness of the local population and the rice terraces, on-the-ground reality is that the intangible cultural aspect of the heritage landscape is less recognized, as one local government official commented (Go3):

> Living subsidies are too limited to support their [community cultural elites'] livelihood. Those villagers, who are talented and willing to learn their culture [Hani culture], are not recognized by the government, not gaining governmental attention and support.

Content analysis of the specific laws and regulations in regards to the Hani Terraces also confirms such a view. Often, while these laws and regulations would specify how to conserve the tangible aspects of the heritage landscape (e.g., the terraces, the traditional 'mushroom' houses), they either fail to refer to the intangible cultural aspects underpinning the tangible heritage landscape or simply mention the need to protect indigenous culture in a general way, without any specifics. Article 22 of the Terraces Regulations, for example, reads:

> The county governments of Yuanyang, Honghe, Lvchun, and Jinping should take measures to protect the tangible cultural heritage relating to Hani Terraces, including relevant cultural relics, traditional festival venues, and the like, as well as intangible cultural heritage of local ethnic minorities, such as traditional craftsmanship, festivals, dancing and singing.

Yet, throughout the Terraces Regulations, it neither relates the importance of local intangible cultural aspects to the conservation of the terraces, nor does it specify how to protect such intangible aspects of the heritage landscape. Such neglect appears to reflect the authorized heritage discourse in China which tends to focus on the materiality of heritage, notably reflected by its official adoption of the traditional 'object-based' authenticity—the authenticity of the original (usually based on expert evaluations or certified documents) (Su, 2019; Zhu, 2016).

The state-owned tourism enterprise

Mass tourism did not start until the twenty-first century when the county government stepped in local tourism development, such as making a series of tourism planning, building tourism facilities, and marketing the terraces. A tourism committee made up of county governmental officials, town governmental officials, and villagers was established. Yet, in practice, the county government developed the resource, managed the resource and operated tourism development, village heads did not care and understand the process, villagers knew nothing about it (Liu, 2008). Nevertheless, the development scale was still limited, as one official at Yuanyang County Tourism Administration (Go6) said:

> Tourism development, to be exact, started in 2008 when the Expo Company came. Before that it's not really formal tourism development.

At the end of 2008, with the introduction of Yunnan Expo Group, a provincial government holding enterprise, the Expo Company was established as a cooperation of the Group and Yuanyang county government. Yunnan Expo Group owns 66.67% of the share, and the remaining 33.33% of the share goes to Yuanyang County State Assets Management Company. The latter is a *de facto* representative of the county government as the government cannot be any shareholder of a company according to the law. In this aspect, the official explained:

> To develop the scenic area by government, in terms of management, is against the policy. You know, it involves entrance ticket income. Moreover, **with regard to its development and management, government is no doubt not as experienced as companies**, and there is trouble coordinating different governmental bodies.

To that end, the heritage tourism development and operation shifted to the Expo Company, while the government focused on infrastructure construction, such as road building. The company built visitor center, and three terrace observation decks on land occupied by local residents. In terms of land acquisition, the land technically belongs to the Chinese state. The state can legally seize or expropriate land with compensation out of public funds, which is also stressed in the Village Regulation and Non-governmental Agreement of Xinjie Town. In order to guarantee the land acquisition smoothly, the government negotiated with local residents, on behalf of the Expo Company. On the ground, the village cadres completed this process with little compensation via a combination of their *guanxi*[4]and tough measures. The company has started to charge tourists entrance fee since early 2009, 60% of which goes to the company and the government enjoys 30%. The rest of the entrance income nominally goes to local villagers, but is, in practice, gained by the government as infrastructure construction fund. Besides the entrance ticket income, the enterprise also makes money from other channels, such as hotels, travel agencies, and tourist buses.

Therefore, in alliance with local government, the Expo Company has largely obtained the right to use heritage for tourism development. Correspondingly, the company recognizes its responsibility for heritage protection, and the responsibility to share tourism benefits with local community, as one manager (Te1) commented:

> Since we came here, we haven't done anything to seek quick success and instant benefits. We completely abide by the request of World Cultural Heritage nomination. A deputy manager at our group company who is in charge of us puts forward idea of "three advantages principles" and "three good": To benefit the construction of world-renowned agriculture tourism destination, **to benefit villagers to acquire wealth**, to benefit the construction of the scenic area with international reputation. Meanwhile, since it's a World Cultural Heritage, we need to inherit it. **Firstly to protect the heritage well, then inherit it well, then use it well**, you know, the "three good"... Our scenic area planning, and facility construction, including building appearances, are consistent with heritage nomination... Our buildings are grey, so are the lookouts and other facilities. **Our idea is ecological landscape, that is, the built facilities match the ecosystem**...

Table 3. Examples of villagers' responsibilities to heritage as defined in regulations/laws.

Responsibilities	Regulations/laws
To terraces	Article 25 The following behaviors are prohibited in key protected areas of Hani terraces: Abandon or damage the terraces (Retrieved from Terrace Regulation)
To forest	Article 18 The development scale of timber forest and commercial forest, the selection of tree species should be coordinated with the original ecological system and landscape style of Hani terraces in the key protected zone. (Retrieved from Terrace Regulation)
	Article 10 (17) The inhabitants of the village in the heritage area must not excavate sand and stones, or deforest at large scale in the heritage area. If villagers really need the construction materials to repair residence, the county government delimit certain zones on the condition that it doesn't affect heritage landscape, and give some economic compensation. (Retrieved from Residential House Regulation)
To village	Article 6 The protection and management of villages of Cultural Landscape of Honghe Hani Rice Terraces sticks to **the principle of heritage authenticity and integrity.** Article 10 The residents have the responsibility to maintain and repair the traditional residences, and also, they are not allowed to tear down, reconstruct, expand the existing buildings, or change the construction of the thatched roof and the color of the wall. (Retrieved from Residential House Regulation)
To water system	Article 10 (7) According to villager regulations, residents should consciously maintain and repair the pool, water grind, water mill, worship room, mill autumn field and other public service facilities, and effectively make the most of their functions. (Retrieved from Residential House Regulation)

The indigenous community

Before heritage making and tourism development, the local community was isolated and marginalized, to which local government hardly paid any attention, not to mention outside enterprise. Villagers farmed the rice terraces from generation to generation for over a millennium, which was their means of livelihood. The terraces, however, are not only about material living, but Hani culture, as reflected by various festivals related to them (Wang & Jue, 2013). For the villagers, cultivating terraces and protecting forests as well as water have become the established behavioral norms over a long time, and internalized in deep-seated values. Depopulation in the area has started since 1990s, gained momentum since 2000s when links with the outside world improve (e.g., transport, tourist flow), as young generation go to the nearby cities for a better living, leaving the terraces to the old. In this regard, one village official (Lo1) noted:

> Nowadays, more than 90% of the young adults in *Pugaolaozhai* are migrant workers. **Some of them come back in times of peak labor needs**, while others hire labor to do that for them.

In the heritage making process, due to the specific laws and regulations mentioned previously, the villagers are imposed the responsibility to protect and conserve the heritage in the way imbedded in authorized heritage discourse, such as the idea of heritage authenticity and integrity (Table 3). Regarding the restrictions on their daily lives out of heritage conservation, they demand for corresponding compensation. For example, one villager (Lo9) commented on the restriction on housing site:

> The government has lots of money. They don't let us build houses where want, then they can plan a site for us to build houses. Then the confrontation will be solved, but [they don't] ...

It should be noted that for villagers, the responsibility to conserve the terrace system is mostly justified by living need and culture norm, as some stated:

> We can't damage the terrace, but need to protect it well, otherwise how can we survive (Ov3)?

> For these villagers who don't go out as migrant workers, only by cultivating terrace can they make a living (Ov8).

> The rice terrace is passed to us by our ancestors, we need to pass it to our next generations (Ov4).

However, such motivation to maintain terraces is weakening. On the one hand, alternative living from tourism and working in the cities means that the terrace is no longer the only source of living. For instance, some local hostel operators, and villagers who go to the cities for a living just hire labors to cultivate the terrace. On the other hand, with the intrusion of modernization, there appears to exist a decreasing attachment to the terrace and agriculture, especially among the young generation who are unwilling, or just do not know how to cultivate the land. Indeed, while most people plan to continue farming and support landscape conservation, they do not want the next generation to farm (Zhang et al., 2017).

The heritage tourism development has resulted in sub-groups among the community. Seeing the tourist flow, villagers with good political tie, social network and financial capital start to actively do tourism businesses, such as opening family inns like the outsiders, running car rental business. A majority of villagers, nevertheless, are seldomly or not involved in tourism, but cultivate terrace and/or go out to work to make a living. But they have realized the (potential) benefits of tourism, 'the road was fixed, and the village is much cleaner than before' (Ov9). They hold the idea that as the caretaker of terraces, they should benefit from heritage tourism. In this aspect, one villager complaint (Ov22), 'They charge entrance fee, do business, giving us nothing. We cultivate terraces, they make money.[In this case], we don't want to cultivate the land!'

The outside hostel operators

As tourism has been booming in recent years, some outsiders, mainly Han Chinese, have moved to the communities and rented locals' houses to run family inns. By the end of 2014, there were 29 family inns at *Pugaolaozhai*, of which 20 were ran by outsiders. For the Han new comers, they are usually there for a lifestyle business; they enjoy the slow pace of life and the beautiful environment, which are seen as the opposite of what they have in the cities. Compared with the Hani minority, they usually have advantages in terms of capital, education, language, work experience, etc., hence do a better business than local hostel operators. Local authorities recognize their importance in developing tourism, and let them operate with preferential policies. For example, commercial usage of water and electricity (e.g., golf courts, hotels) is usually charged much higher fees than ordinary household usage in China; yet, the usage of electricity by the hostels in the studied villages were charged the same as that of ordinary village household, while water usage were not charged at all. The newcomers do not own the land or other elements of the heritage landscape, nor do they know much about cultivating the rice terraces. Also they see their hostel business as not directly related to heritage. Emotionally, they regard themselves as outsiders in the indigenous communities. 'Here I do my own things with my own pace, If I can't, I will leave this place' (Oo2). Hence, they perceive little responsibility of their own in relation to the conservation of the heritage landscape, and correspondingly, they do not demand any specific rights in relation to the heritage.

Interestingly, other stakeholders' responsibilities and rights in relation to the heritage landscape quite often emerged out of their narratives. They held that local government and the tourism company should share tourism benefits with villagers. In this regard, one operator (Oo5) noted: 'The starting point of local authorities is wrong. Tourism development here should help improve locals' wellbeing and income. Now local authority cooperates with the Expo Company, and the company has obtained most of the money. It should be the other way round, villagers having the most of the tourism income while the company obtaining a small portion'. They pointed out that local government should help solve villagers' housing problems, thus echoing villagers' demand for compensation for the restrictions posed on their daily lives (e.g., not to build houses as they want, but in a way that is consistent with 'heritage authenticity and integrity').

They also talked about the responsibility of local authorities and villagers to protect the physical environment of the heritage landscape. However, for villagers, they suggested that such responsibility must be accompanied by the acquisition of practical benefits to make it more solid. Hostel operator Oo3, for example, argued:

Table 4. Responsibilities and rights in relation to heritage imbedded in stakeholders' respective discourses.

Stakeholders	Perceived responsibilities	Perceived rights
Local government	To conserve the (tangible) heritage To improve the well-being of the indigenous community	To 'make' heritage Heritage stewardship
Indigenous community	To conserve the (tangible) heritage	Benefit compensation Tourism benefit sharing
State-owned tourism enterprise	To conserve the (tangible) heritage Tourism benefit sharing with the indigenous community	Use heritage for development
The outside hostel operators (Han new comers)	Local government and the tourism company to share tourism benefits with villagers Local government to compensate villagers for the protection restrictions on their lives Local government and villagers to conserve the (tangible) heritage	Villagers to benefit from tourism and have compensation for the protection restrictions

> **When talking about responsibility, one must first solve the problem of food and clothing.** Local villagers have little income, all the money goes to local government and the tourism company. Then they demand villagers plow the rice fields, transplant rice seedlings, nobody is happy with that. Why should villagers not throw rubbish away? **No matter how well they behave, it doesn't change their income or life.**

Hence to sum up, the responsibilities and associated rights in regards to heritage embedded in the four major stakeholders' discourse are listed in Table 4. It is of interest to note that while the site is a cultural landscape where culture-nature interconnectedness is a much valued aspect of the identified World Heritage value, responsibilities for the conservation of the site are focused almost exclusively on the tangible aspects of the site. Also, no stakeholder groups, even the indigenous community themselves, identify the rights of local people to participate in decision-making in relation to heritage making, stewardship or use. This is in contrast to an emphasis on decision-making power in relation to community participation in heritage practices in many recent international documents as well as heritage studies (Oviedo & Puschkarsky, 2012).

Stakeholder conflicts viewed through the lens of doing responsibility

In the previous section, we have elaborated on what stakeholders say about their respective responsibilities. In this section, we turn to what stakeholders actually do in relation to their perceived responsibilities and its consequences via unraveling a few emergent conflicts among the stakeholder groups.

The demolition of newly built residential houses

According to the heritage laws and regulations, villagers have the responsibility to protect and conserve the traditional village landscape. One typical illustration is:

> Any new building, rebuilding or building extension must be approved by the county government ... each household can only have one housing site ... after villagers sell or rent residential house out, they cannot apply for housing site (Retrieved from Residential House Regulation)

These regulations significantly reduce villagers' possibilities for action, as some villagers complaint:

> In the past, we could build it wherever we want, but now we can't (Lo13).

> If their [governmental] demand for building 'mushroom house' [Figure 4] is not met, governmental subsistence allowances will be gone, all governmental compensation will be gone (Lo4).

Such regulations not only bring inconvenience for living, such as the need to build new house for family division among brothers (which is a local tradition), but also limit the possibility

Figure 4. Traditional 'mushroom' house (picture by author, August 2014).

to gain benefits from tourism, such as renting houses to outsiders or using houses to run hostel business. In this regard, local authorities committed to compensating with a piece of designated land that would not 'spoil' heritage landscape, with infrastructures and facilities built, to meet villagers' housing demand. Yet, this commitment lasted for years and never came into practice, leading to growing discontent among villagers.

What's worse, those villagers who have good *guanxi* with government managed to build houses as they wished, as noted by some villagers:

> Nowadays it's not allowed to build new houses, but my son is a labor contractor, he knows people in the government, so he can get the approval (Lo10).

> Government officials here are unjust, nothing is just. If you have money, have good *guanxi*, you can build houses, otherwise no (Ov22).

Then in the first half of 2014, several local residents of *Pugaolaozhai* started to build new houses without governmental approval, and were asked to stop. However, these residents insisted that they could do it like those who had already done so successfully. So they ignored the government's ruling, and continued with their construction. A few days later, their houses were demolished by armed police, albeit villagers' tough resistance (Figure 5).

The terrace cultivation strike

In March 2014, when it was the season for ploughing and transplanting rice, villagers of *Pugaolaozhai* and *Huangcaoling* near the terrace observation deck collectively had a terrace cultivation strike to express their anger towards the distribution of entrance ticket income as mentioned before. The strike, however, lasted only two months, for the government threatened to take back the terraces "if villagers failed to take the responsibility to cultivate them". In this regard, one villager recalled:

> The officials came to us, saying if you don't plant rice there, the terraces are not **beautiful**. If you don't plant (rice), we will take them back. We then planted rice, we had no option, we need to eat. Also, **old villagers** said the terraces must be cultivated (Ov19).

Figure 5. Remnant of the removed concrete house (picture by author, February 2015).

The violent collection of water bill

Water for daily use in the villages as well as terrace cultivation comes down from the mountains. In recent years, there has been a remarkable increase of water usage due to the family inns, leading to a somewhat water shortage. One villager complaint:

> The more family inns, the larger amount of water usage, the less water for terraces. This year there is no water for tens *mu* (1 *mu* = 0.067 ha) of terraces, thus no rice shoots were planted (Ov3).

Likewise, another villager argued:

> The family inns are using **our water**, they should pay some water fee (Ov21).

In consequence, villagers not operating the business got increasingly unsatisfied. In May 2014, these villagers at *Pugaolaozhai*, armed with 'weapons' such as knives, sticks and hoes, gathered together to force every hostel operator to pay a water fee. Local operator were asked to pay several hundred RMB as water fee, while outsider operators were requested to pay several thousand. The situation did not calm down until government intervened, detaining several villagers who led the incident.

To conclude, the responsibilities for the conservation of a 'living cultural heritage' like the Hani Terraces, translate into many restrictions on local villagers in their daily lives, figuratively freezing the landscape in time with World Heritage status and performance for tourists. Nonetheless, for many local villagers, their 'doing' of responsibilities does not go hand in hand with the realization of desired rights and benefits. This has become a direct source of conflicts between ordinary villagers and those stakeholders who have benefited greatly from heritage tourism development, yet have not shared the responsibilities of actually maintaining the heritage landscape (e.g., local government, the state-owned tourism enterprise, hostel operators).

Discussion

This study investigates the understandings and practices of responsibilities for heritage from a multi-stakeholder rights *and* responsibility perspective at Cultural Landscape of Honghe Hani

Rice Terraces in the context of heritage tourism. Local government, supported by the knowledge of heritage experts, exerts influence over the heritage site through heritage making. Accordingly, the right to manage the increasingly valuable heritage landscape has largely shifted from local communities to local government. In this regard, the government also sees its responsibility for heritage protection. Also, as the local heritage becomes World Heritage with good economic value, notably for tourism development in the Chinese context (Gao, Zhang, & Liu, 2019; Yan, 2018), to bring benefits to the local community is seen as another stream of governmental obligation. For the state-owned tourism enterprise, in alliance with local government, it has largely gained the right to use the heritage landscape for tourism. Delimited by the authorized heritage discourse as well as the value of the heritage as core tourist attraction, the responsibility to protect heritage and share tourism benefits with the local community is imbedded in the discourse of the company. For local residents, they realize their responsibility for heritage protection largely out of living needs and tradition, but increasingly feel the imposed responsibilities from heritage nomination and tourism development. Correspondingly, they demand the right to acquire benefit compensation and tourism benefits accorded to them as the creator as well as the caretaker of the heritage. For the Han new-comers, they do not relate themselves to the heritage landscape, hence perceive little responsibility or rights in relation to heritage. Nevertheless, they echo locals' view that villagers as the protectors of the heritage landscape should acquire tourism benefit sharing and benefit compensation. Hence, all suggests a balance of rights and responsibilities in relation to heritage is desired.

A noteworthy point is that the nature of perceived responsibilities for heritage conservation is unanimously consistent among the four stakeholder groups, as all focus on the material aspects of the heritage. Such can be attributed to the authorized heritage discourse in China that prioritizes the materiality of heritage and values technical expertise towards heritage (Su, 2019; Zhu, 2016). In this regard, Jamal and Camargo (2018) notes, 'Intangibles such as cultural values can be easily omitted in planning and decision making, particularly where evaluation is driven by scientific, measurable, managerial tools' (p. 206). This can undermine the sustainability of the cultural landscape in the long term, for it is the interconnectedness of the local population and the rice terraces that has helped sustained the landscape for over a millennium (Chan, Iankova, Zhang, McDonald, & Qi, 2016). Hence intangible aspects such as traditional farming technology, management system (e.g., water distribution system), belief system (e.g., the belief in 'village god's forest'), and popular folk festivals warrant more attention to sustain the cultural landscape. For instance, local authorities might enhance their support for traditional Hani cultural elites, and offer incentives for the learning of traditional culture among the younger generations. Regarding tourism, ethnic tourism products and activities (which are currently focused on the rice terraces) can be enhanced, which may help the indigenous people participate more in tourism, and improve their cultural confidence.

The absence of indigenous people in decision making, and the wane of their traditional culture along with the intrusion of modernization, accelerated by tourism, would likely threaten the future of the 'living' cultural landscape. However, surprisingly, no stakeholder groups, even the indigenous people themselves, identify the rights of local people to participate in decision making regarding heritage making, stewardship or use, but all emphasize on practical benefits, be it benefit compensation or sharing. One may relate this to local context as well as the broad context of China. On the one hand, one should note that Yunnan is a poverty-centered province in Southwestern China, and that Yuanyang and its neighboring counties in Honghe are all deeply poverty-stricken areas. To alleviate poverty in Yunnan, and especially its border areas like Honghe where social stability is of great concern to the Chinese state, has become a prevalent discourse since the twenty-first century, notably when president Xi jinping came into power (Gao et al., 2019). In fact, tourism has been envisaged as a tool for socio-economic development and poverty reduction at China's borders (State Council of China, 2016). On the ground, 'to live a better life' is thus the primary aspiration of local population, which may orient their attention to more practical benefit acquisition instead of decision making. On the other hand, the long tradition of China's being a centralized state, where the government generally plays a dominant role in social life and lacks Western democracy, may account for low interest in decision making among ordinary Chinese (Su & Wall, 2015). In this aspect of heritage administration in China, the authorized heritage discourse justifies the important role of government, with the support of expert knowledge, excluding local communities (Zhang & Lenzer Jr, 2019). Indeed, the situation that villagers demand extra compensation for being the protector of the heritage landscape, seems similar to the compensation model already employed in China's environmental protection policy (e.g., industries and cities in the downstream will compensate for the upstream protection by villagers) (Zheng & Zhang, 2006).

When it comes to the 'doing' of responsibility, the indigenous Hani are bearing burden and inconveniences to maintain the cultural landscape, and have not achieved the right to acquire compensation for the restrictions this poses on their daily lives or the right to share the benefits of local heritage tourism development. This becomes a direct source of conflict between local people and these stakeholders who have benefited greatly from heritage tourism development (local government, or the state-owned tourism enterprise, hostel operators, etc.). A power disparity among stakeholders is clearly at play. The heritage administration system of China endows local government dominant power in heritage management and the authority of defining what the responsibilities are and who has the responsibilities. The state-owned enterprise, an ally of the local government in tourism development, and the Han newcomers (viewed as more capable in development by local government), have a strong influence over the use of heritage in tourism development. It could be argued based on this case that heritage responsibility discourses and practices are embedded in and reinforce the systematic unequal power relations between local community and the state. The governments, with the support of expert knowledge, decide what the responsibilities for heritage are and who should bear the responsibilities, and institutionalize such via corresponding laws, regulations and planning. The power of local government over the heritage site extends to the removal of 'illegal' buildings, and they can threaten to take back the terraces if the Hani do not exercise their responsibility to conserve them and the practices that shape these cultural landscapes. Such responsibility discourse has the normative power to structure and shape social realities (Grimwood et al., 2015; Shultis & Heffner, 2016; Sin, 2017). Whether it is to cultivate the terraces and grow rice, or maintain the traditional villagescape, or construct appropriate 'ecological' tourism facilities that match the cultural landscape, both everyday traditional practices and tourism development can be used to justify that local villages must 'do' their responsibilities to protect the authenticity and integrity of their cultural heritage without fair compensation or benefits.

Conclusions

Viewing through the lens of multi-stakeholder responsibility in relation to heritage conservation and use for tourism development, this study sheds some light on an equitable sharing of the costs and benefits of heritage conservation. Procedural rights, in particular the rights of local communities to participate in decision-making regarding heritage identification and governance, are increasingly evident in international heritage documents (Oviedo & Puschkarsky, 2012). But these are not recognized by major stakeholders of the Rice Terraces Cultural Landscape, and have a complicated relationship with the local people themselves. In the case of the Hani and in the Chinese context, participatory rights are emergent, socially constructed and context specific—they have to be examined in relation to issues such as equitable compensation and benefit sharing for heritage practices by local residents 'doing responsibilities', the ways in which the meanings and practices of responsibility are constructed as they engage in their cultural practices and agricultural traditions. In short, the rights of community participation in heritage practices in China are more about benefit acquisition instead of decision making. Particular attentions should also be paid to the power relations and performativity of heritage discourses involving ethnic minority groups such as the Hani.

This study also has some implications for future studies. The current study is a place-specific research and the insights may be useful in similar settings, but they are not generalizable. Further studies can examine the current results in different types of heritage tourism destinations in China, such as natural heritage sites, and heritage destinations with a mature tourism industry. Further studies would benefit by taking a longitudinal approach to tracking the evolution of such understandings and practices in heritage tourism destinations in China. The study of the Cultural Landscapes of the Honghe Hani Rice Terraces shows that rights and responsibilities are performed through situated practices that involve 'doing' responsibilities over time in a particular place, and conflicts can emerge over time during the process.

Notes

1. https://responsibletourismpartnership.org/cape-town-declaration-on-responsible-tourism/.
2. https://whc.unesco.org/en/list/1111/.
3. https://whc.unesco.org/en/list/1111/.
4. *Guanxi* refers to Chinese-style social connections and reciprocal relations (Siriphon, 2015).

Acknowledgments

We would like to thank the anonymous reviews for their constructive feedback on earlier drafts of this paper, and Prof.Tazim Jamal for her very helpful and impressive editing.

Disclosure statement

No potential conflict of interest was reported by the author(s).

Funding

The authors would like to express their gratitude for the National Science Foundation of China (project serial number: 41471122; 41901162) and China Postdoctoral Science Foundation (project serial number: 2019M653260) for supporting the research .

ORCID

Chaozhi Zhang http://orcid.org/0000-0003-2588-8838

References

Adachi, S. (2007). Agricultural technologies of terraced rice cultivation in the Ailao Mountains, Yunnan, China. *Asian and African Area Studies*, *6*(2), 173–196.

Bao, J., & Zuo, B. (2012). Legislating for tourist attractions rights. *Tourism Tribune*, *27*(7), 11–18.

Bennoune, K. (2016). *Report of the special rapporteur in the field of cultural rights*. New York: UN Human Rights Council.
Blake, J. (2011). Taking a human rights approach to cultural heritage protection. *Heritage & Society, 4*(2), 199–238. doi:10.1179/hso.2011.4.2.199
Bodolec, C. (2012). The Chinese paper-cut: From local inventories to the UNESCO Representative List of the Intangible Cultural Heritage of Humanity. In R. F. Bendix, A. Eggert, & A. Peselmann (Eds.), *Heritage regimes and the state* (Vol. 6, pp. 249–264). Gottingen: Universitätsverlag Göttingen.
Braun, V., & Clarke, V. (2006). Using thematic analysis in psychology. *Qualitative Research in Psychology, 3*(2), 77–101. doi:10.1191/1478088706qp063oa
Chan, J. H., Iankova, K., Zhang, Y., McDonald, T., & Qi, X. (2016). The role of self-gentrification in sustainable tourism: Indigenous entrepreneurship at Honghe Hani Rice Terraces World Heritage Site. *Journal of Sustainable Tourism, 24*(8-9), 1262–1279.
Chitty, G. (2016). *Heritage, conservation and communities: Engagement, participation and capacity building*. New York: Routledge.
Clarke, N., Barnett, C., Cloke, P., & Malpass, A. (2007). Globalising the consumer: Doing politics in an ethical register. *Political Geography, 26*(3), 231–249. doi:10.1016/j.polgeo.2006.10.009
Cole, S. (2006). Information and empowerment: The keys to achieving sustainable tourism. *Journal of Sustainable Tourism, 14*(6), 629–644. doi:10.2167/jost607.0
Coombe, R. J., & Weiss, L. M. (2015). Neoliberalism, regimes, and cultural rights. In. L. Meskell (Ed.), *Global heritage: A reader* (pp. 43–69). Chichester: Wiley.
Cornet, C. (2015). Tourism development and resistance in China. *Annals of Tourism Research, 52*, 29–43. doi:10.1016/j.annals.2015.02.002
Ekern, S., Logan, W., Sauge, B., & Sinding-Larsen, A. (2012). Human rights and World Heritage: Preserving our common dignity through rights-based approaches to site management. *International Journal of Heritage Studies, 18*(3), 213–225. doi:10.1080/13527258.2012.656253
Escallón, M. F. (2018). Heritage, land, labor, and competing claims for Afro-Colombian rights. *International Journal of Cultural Property, 25*(1), 59–83. doi:10.1017/S0940739118000061
Gao, J., Ryan, C., Cave, J., & Zhang, C. (2019). Tourism border-making: A political economy of China's border tourism. *Annals of Tourism Research, 76*, 1–13. doi:10.1016/j.annals.2019.02.010
Gao, J., & Wu, B. (2017). Revitalizing traditional villages through rural tourism: A case study of Yuanjia Village, Shaanxi Province, China. *Tourism Management, 63*, 223–233. doi:10.1016/j.tourman.2017.04.003
Gao, J., Zhang, C., & Liu, L. (2019). Communicating the outstanding universal value of World Heritage in China? The tour guides' perspective. *Asia Pacific Journal of Tourism Research*. doi:10.1080/10941665.2018.1564340
Grimwood, B. S., Yudina, O., Muldoon, M., & Qiu, J. (2015). Responsibility in tourism: A discursive analysis. *Annals of Tourism Research, 50*, 22–38. doi:10.1016/j.annals.2014.10.006
ICCROM. (2015). *People-centred approaches to the conservation of cultural heritage: Living heritage*. Via di San Michele: ICCROM. Retrieved from https://www.iccrom.org/.
Jamal, T., & Camargo, B. A. (2018). Tourism governance and policy: Whither justice?. *Tourism Management Perspectives, 25*, 205–208. doi:10.1016/j.tmp.2017.11.009
Jiao, Y. (2003). The feature of Hani cultural region–the Cultural Landscape of Hani Rice Terraces. *Yunnan Geographic Environment Research, 21*(6), 734–743.
Jodoin, S. (2014). Can rights-based approaches enhance levels of legitimacy and cooperation in conservation? A relational account. *Human Rights Review, 15*(3), 283–303. doi:10.1007/s12142-014-0312-8
Kraak, A. L. (2017). Impediments to a human rights-based approach to heritage conservation: The case of Bagan, Myanmar. *International Journal of Cultural Policy, 23*(4), 433–445. doi:10.1080/10286632.2015.1066782
Kraak, A.-L. (2018). Human rights-based approaches to World Heritage conservation in Bagan, Myanmar: Conceptual, political, and practical considerations. *International Journal of Cultural Property, 25*(1), 111–133. doi:10.1017/S0940739118000012
Kraak, A.-L., & Aykan, B. (2018). The possibilities and limitations of rights-based approaches to heritage practice. *International Journal of Cultural Property, 25*(1), 1–10. doi:10.1017/S0940739118000073
Li, Y. (2017). A brief analysis of the conservation and management of the Honghe Hani Rice Terraces in China. *Journal of World Heritage Studies*, (Special issue), 6–10.
Li, Y., Wang, X., & Chen, P. (2019). Development and testing of heritage responsibility scale: Based on perspectives of tourists and residents. *Tourism Tribune, 34*(10), 60–75.
Liu, D. (2008). *Tourism gaze: Chinese local study*. Tianjin: Tianjin University Press.
Oviedo, G., & Puschkarsky, T. (2012). World Heritage and rights-based approaches to nature conservation. *International Journal of Heritage Studies, 18*(3), 285–296. doi:10.1080/13527258.2012.652146
Porter, B. W., & Salazar, N. B. (2005). Heritage tourism, conflict, and the public interest: An introduction. *International Journal of Heritage Studies, 11*(5), 361–370. doi:10.1080/13527250500337397
Shultis, J., & Heffner, S. (2016). Hegemonic and emerging concepts of conservation: A critical examination of barriers to incorporating Indigenous perspectives in protected area conservation policies and practice. *Journal of Sustainable Tourism, 24*(8-9), 1227–1242.

Sin, H. L. (2014). Realities of doing responsibilities: Performances and practices in tourism. *Geografiska Annaler: Series B, Human Geography, 96*(2), 141–157. doi:10.1111/geob.12042
Sin, H. L. (2017). Selling ethics: Discourses of responsibility in tourism. *Annals of the American Association of Geographers, 107*(1), 218–234. doi:10.1080/24694452.2016.1218266
Siriphon, A. (2015). Xinyimin, new Chinese migrants, and the influence of the PRC and Taiwan on the northern Thai border. In Y. Santasombat (Ed.), *Impact of China's rise on the Mekong Region* (pp. 147–166). New York: Palgrave Macmillan.
Smith, L. (2006). *Uses of heritage*. London: Routledge.
State Council of China. (2016). *Opinions of the State Council on supporting the development and opening of key areas along the border*. Beijing: Author.
Strzelecka, M., & Wicks, B. E. (2015). Community participation and empowerment in rural post-communist societies: Lessons from the leader approach in Pomerania, Poland. *Tourism Planning & Development, 12*(4), 381–397. doi:10.1080/21568316.2015.1013564
Su, J. (2019). Managing intangible cultural heritage in the context of tourism: Chinese officials' perspectives. *Journal of Tourism and Cultural Change*. . doi:10.1080/14766825.2019.1604720
Su, M., & Wall, G. (2014). Community participation in tourism at a world heritage site: Mutianyu Great Wall, Beijing, China. *International Journal of Tourism Research, 16*(2), 146–156. doi:10.1002/jtr.1909
Su, M., & Wall, G. (2015). Community involvement at Great Wall World Heritage sites, Beijing, China. *Current Issues in Tourism, 18*(2), 137–157. doi:10.1080/13683500.2013.823917
Vrdoljak, A. F. (2018). Indigenous peoples, world heritage, and human rights. *International Journal of Cultural Property, 25*(3), 245–281. doi:10.1017/S0940739118000218
Waitt, G. R. (2005). Doing discourse analysis. In I. Hay (Ed.), *Qualitative research methods in human geography* (pp. 163–191). London: Oxford University Press.
Wang, D., & Jue, Y. (2013). Study on the diversity of festival culture at villages of Hani Terrace Culture Landscape. *Yunnan Geographic Environment Research, 25*(3), 100–103.
Wang, H., & Zheng, Y. (2015). Where are the rights of community on participation in tourism? A content analysis of Chinese tourism laws and regulations. *Tourism Tribune, 30*(5), 74–84.
Yan, H. (2015). World Heritage as discourse: Knowledge, discipline and dissonance in Fujian Tulou sites. *International Journal of Heritage Studies, 21*(1), 65–80. doi:10.1080/13527258.2014.894930
Yan, H. (2018). *World heritage craze in China: Universal discourse, national culture, and local memory*. New York: Berghahn Books.
Yang, J., Ryan, C., & Zhang, L. (2013). Social conflict in communities impacted by tourism. *Tourism Management, 35*, 82–93. doi:10.1016/j.tourman.2012.06.002
Yang, L., Wall, G., & Smith, S. L. (2008). Ethnic tourism development: Chinese government perspectives. *Annals of Tourism Research, 35*(3), 751–771. doi:10.1016/j.annals.2008.06.005
Zhang, C. (2014). Heritage responsibility: Conception, characteristics and research issues. *Tourism Tribune, 29*(11), 45–51.
Zhang, C., Chen, P., & Wang, X. (2017). Conceptualizing heritage responsibility in world heritage sites: Insights from Levinas' ethics of responsibility. *Asian Journal of Tourism Research, 2*(3), 80–101. doi:10.12982/AJTR.2017.0018
Zhang, C., Fyall, A., & Zheng, Y. (2015). Heritage and tourism conflict within world heritage sites in China: A longitudinal study. *Current Issues in Tourism, 18*(2), 110–136. doi:10.1080/13683500.2014.912204
Zhang, M., & Lenzer, J. H. Jr. (2019). Mismatched canal conservation and the authorized heritage discourse in urban China: A case of the Hangzhou Section of the Grand Canal. *International Journal of Heritage Studies, 26*, 105–119. doi:10.1080/13527258.2019.1608458
Zhang, Y., He, L., Li, X., Zhang, C., Qian, C., Li, J., & Zhang, A. (2019). Why are the Longji Terraces in Southwest China maintained well? A conservation mechanism for agricultural landscapes based on agricultural multi-functions developed by multi-stakeholders. *Land Use Policy, 85*, 42–51. doi:10.1016/j.landusepol.2019.03.046
Zhang, Y., Min, Q., Zhang, C., He, L., Zhang, S., Yang, L., ... Xiong, Y. (2017). Traditional culture as an important power for maintaining agricultural landscapes in cultural heritage sites: A case study of the Hani terraces. *Journal of Cultural Heritage, 25*, 170–179. doi:10.1016/j.culher.2016.12.002
Zheng, H., & Zhang, L. (2006). *Chinese practices of ecological compensation and payments for ecological and environmental services and its policies in river basins*. Washington: World Bank.
Zhu, Y. (2016). Authenticity and heritage conservation in China: Translation, interpretation, practices. In K. Weiler & N. Gutschow (Eds.), *Authenticity in architectural heritage conservation* (pp. 187–200). Heidelberg: Springer.
Zhu, Y., Jin, L., & Graburn, N. (2017). Domesticating tourism anthropology in China. *American Anthropologist, 119*(4), 730–735. doi:10.1111/aman.12956
Zuo, B., & Bao, J. (2008). From community participation to community empowerment: Review on theoretical study of "tourism empowerment" in Western countries. *Tourism Tribune, 23*(4), 58–63.
Zuo, B., & Bao, J. (2016). Revisiting tourist attractions rights. *Tourism Tribune, 31*(7), 14–23.

Appendix

Table A1. The profile of the respondents.

No.	Gender	Age	Ethnicity	Interview time & location
Ov1	male	40+	Hani	Aug 2014 *Qingkou*
Ov2	male	50+	Hani	Aug 2014 *Qingkou*
Ov3	male	40+	Hani	Aug 2014 *Pugaolaozhai*
Ov4	female	15	Hani	Aug 2014 *Pugaolaozhai*
Ov5	female	80+	Hani	Aug 2014 *Pugaolaozhai*
Ov6	female	38	Hani	Aug 2014 & Feb 2015 *Pugaolaozhai*
Ov7	male	50+	Hani	Aug 2014 *Pugaolaozhai*
Ov8	female	23	Hani	Aug 2014 *Pugaolaozhai*
Ov9	male	30	Hani	Aug 2014 *Pugaolaozhai*
Ov10	male	50+	Hani	Aug 2014 *Pugaolaozhai*
Ov11	male	30	Hani	Aug 2014 *Pugaolaozhai*
Ov12	male	70+	Hani	Aug 2014 *Pugaolaozhai*
Ov13	male	80+	Hani	Aug 2014 *Pugaolaozhai*
Ov14	female	30	Hani	Aug 2014 *Pugaolaozhai*
Ov15	female	25	Hani	Aug 2014 *Dayutang*
Ov16	male	40	Hani	Feb 2015 *Pugaolaozhai*
Ov17	female	40+	Hani	Feb 2015 *Pugaolaozhai*
Ov18	male	50+	Hani	Feb 2015 *Pugaolaozhai*
Ov19	male	40+	Hani	Feb 2015 *Pugaolaozhai*
Ov20	male	40+	Hani	Feb 2015 *Pugaolaozhai*
Ov21	male	60	Hani	Feb 2015 *Pugaolaozhai*
Ov22	male	40+	Hani	Feb 2015 *Pugaolaozhai*
Lo1	male	40+	Hani	Sep 2014 & Feb 2015 *Pugaolaozhai*
Lo2	female	50	Hani	Sep 2014 *Pugaolaozhai*
Lo3	female	28	Hani	Sep 2014 *Pugaolaozhai*
Lo4	male	54	Hani	Sep 2014 & Feb 2015 *Pugaolaozhai*
Lo5	female	24	Hani	Sep 2014 *Pugaolaozhai*
Lo6	female	22	Hani	Sep 2014 *Pugaolaozhai*
Lo7	male	20	Hani	Sep 2014 *Pugaolaozhai*
Lo8	female	25	Hani	Sep 2014 *Pugaolaozhai*
Lo9	female	40	Hani	Sep 2014 *Pugaolaozhai*
Lo10	male	50+	Hani	Feb 2015 *Pugaolaozhai*
Lo11	female	24	Hani	Feb 2015 *Pugaolaozhai*
Lo12	male	18	Hani	Feb 2015 *Pugaolaozhai*
Lo13	male	24	Hani	Feb 2015 *Pugaolaozhai*
Oo1	female	60+	Han	Sep 2014 *Pugaolaozhai*
Oo2	male	40	Han	Sep 2014 *Pugaolaozhai*
Oo3	male	30	Han	Sep 2014 & Feb 2015 *Pugaolaozhai*
Oo4	male	30+	Han	Feb 2015 *Pugaolaozhai*
Oo5	male	30+	Han	Feb 2015 *Pugaolaozhai*
Oo6	male	30+	Han	Feb 2015 *Pugaolaozhai*
Go1	male	30+	Hani	Aug 2014 & Feb 2015 the county government
Go2	female	30	Hani	Aug 2014 the county government
Go3	male	40	Hani	Aug 2014 the county government
Go4	male	40+	Han	Aug 2014 the county government
Go5	female	50+	Hani	Feb 2015 the prefecture government
Go6	male	40+	Hani	Feb 2015 the prefecture government
Go7	male	30+	Han	Feb 2015 the prefecture government
Te1	male	40+	Han	Feb 2015 the Expo Company
Te2	female	26	Hani	Feb 2015 *Bada* viewing platform
Te3	female	30+	Hani	Feb 2015 the Expo Company

Indigenous tourism and cultural justice in a Tz'utujil Maya community, Guatemala

Lucy C. Harbor and Carter A. Hunt

ABSTRACT

Despite explicit links to justice issues inherent in indigenous rights movements, little research has been undertaken to understand Indigenous Tourism from a justice perspective. This study employs ethnographic interviewing, participant observation, and archival data to study tourism in the Tz'utujil Maya community of Santiago Atitlán, Guatemala. Findings emphasize emic views of local community members that offer valuable insights for understanding justice as it relates to Tz'utujil culture. We argue that the loss of Indigenous culture, ways of living, and ways of knowing would almost certainly be hastened if tourism and the associated cultural valuation were not present. Furthermore, direct participation in negotiation on tourism related matters is a key principle to facilitate autonomy, agency, fairness and equity in cultural justice. Authenticity, similarly, is a negotiated concept, requiring direct participation to facilitate fairness and equity in cultural tourism, as seen being practiced by the Tz'utujil people. The cultural justice framing here makes a valuable contribution to recent writing in tourism studies on indigenous environmental justice.

Introduction

Nowhere is there more at stake regarding the consumption of cultural assets through tourism, or greater need for performative resistance to globalization, than in Indigenous communities. Indeed, a large body of writing has unfolded since the 1970s to analyze tourism and Indigenous peoples with an emphasis on cultural impacts, cultural commodification, and Indigenous control—or lack of—in tourism development (for an overview, see Whitford & Ruhanen, 2016), yet the study of justice in Indigenous Tourism has seen little momentum. In the 2016 special issue of *Journal of Sustainable Tourism* on "Sustainable Tourism and Indigenous Peoples," the term "justice" itself was mentioned only in passing in the otherwise outstanding contributions. A few studies in the context of Indigenous tourism can be identified, such as Schellhorn (2010) on social justice, Whyte (2010) on environmental justice, and Jamal and Camargo (2014) on justice related in issues among the Maya in Quitana Roo, Mexico. Given a lengthy history of marginalization and endangerment of Indigenous communities through irresponsible forms of tourism (Coria & Calfucura, 2012; Johnston, 2006; van den Berghe, 1994), and the ongoing loss of cultural diversity underway worldwide (Gorenflo et al., 2012), much greater research attention is needed to understands how Indigenous peoples negotiate the tourism in ways that facilitate just and fair outcomes for them.

This paper undertakes this task through ethnographic study of a geographically constrained Indigenous group in highland Guatemala. Composed of just 60,000 members, the Tz'utujil Maya are restricted to the town of Santiago Atitlán and the two smaller neighboring villages of San Pedro La Laguna and San Juan la Laguna. These communities are located on the shores of Lake Atitlán, one of the Guatemala's primary attractions for international tourists. We take a situated approach to our ethnographic study in Santiago Atitlán, attempting as much as possible to ensure our approach was guided by decolonizing methodologies (Smith, 2012, 2nd edition). While familiar with western notions such as of procedural and distributive justice, our participatory, facilitative approach strove to understand the emic views of key participants in cultural tourism in this community. Our findings, while mixed in some instances, collectively provide strong indications that tourism offers an important vehicle for the Tz'utujil people to negotiate their cultural identity and well-being within the changing context of broader Guatemalan society and forces of globalization. We argue that tourism provides a valuable tool to the Tz'utujil people for ensuring their agency, autonomy, and rights to shape their cultural tourism offering and their cultural heritage. These are performative endeavors facilitated, too, by direct participation in controlling and managing cultural tourism. In doing so, we provide a much-needed contribution to the literature to facilitate thinking on *cultural justice* in the context of tourism and Indigenous People.

Indigenous tourism and justice research

Challenges to indigenous survival in a globalized world

Historical forms of colonial and neocolonial challenges have greatly endangered cultural survival of native populations, and globalization and tourism mobilities continue to exacerbate the problem. Many find themselves struggling to maintain or reconcile their indigenous identity. A strong critique of tourism arises here as facilitating dependency and neo-colonialism through tourism (see Sharpley & Telfer, 2015).[1] As they negotiate external forces of globalization that perpetuate colonial and neocolonial power structures, native populations find themselves struggling to maintain or reconcile their Indigenous identity when tourism arrives to subject them to the tourist gaze (Hollinshead, 1992). Both material culture, such as language and clothing (Whitney-Squire, 2016), as well as immaterial culture (Ruhanen & Whitford, 2019), become salient yet susceptible expressions of Indigenous identity. The demonstration effect introduces and reifies the perception that non-Western modes of consumption (e.g., Indigenous dress, food, language, etc.) are a proxy for poverty (de Kadt, 1979). Indigenous people typically serve as the basis for the *imaginaries* used to promote tourism (Salazar, 2012). Through visual imagery that includes, but is not limited to, archeological ruins and colorful ethnic clothing, Indigenous tourism encourages travelers from around the world to seek out windows into the way the "exotic other" lives (Butler & Hinch, 2007; Johnston, 2006; van den Berghe, 1994; Zeppel, 2006).

In the context of Indigenous tourism, many have been excluded from fair and equitable economic benefits and may suffer appropriation of their traditional knowledge, traditional medicines and healing practices to serve touristic interests (e.g., Johnston, 2006; Salazar, 2012). Examples abound in tourism studies of appropriation of land and exclusionary practices, such as the exclusion of nomadic tribes from newly formed protected areas (Charnley, 2005) or disregard and disrespect of sacred lands (Figueroa & Waitt, 2010; White et al., 2013). In such instances, historical and economic exploitation and exclusionary practices are exacerbated by tourism in a way that can contribute cumulatively to cultural disintegration, political disenfranchisement, discontent, increased violence, and conflict in Indigenous communities (Johnston, 2006).

Challenges related to indigenous tourism

The UN Declaration on the Rights of Indigenous Peoples make an explicit link to justice that had always been inherent, if less explicitly addressed, in the writings on tourism in Indigenous communities (e.g., Greenwood, 1977; Ritchie & Zins, 1978; Swain, 1977; Whitford & Ruhanen, 2016). Rapid loss of cultural diversity at the global scale (Gorenflo et al., 2012) makes research into the effects of tourism development on Indigenous communities, and their ability to obtain just outcomes, particularly urgent. Regrettably, the scholarly record—too large to fully cite here—is clear that tourism often causes disruption, exacerbation of inequalities, exploitation, and conflict in Indigenous communities (e.g., Coria & Calfucura, 2012; Johnston, 2006; Stronza, 2008; Zeppel, 2006). Colonial histories often leave complex *power dynamics* in place within local and regional tourism industries (Nepal & Saarinen, 2016; Pereiro, 2016). Social inequities can flare up at the local level, leading to local competition and rivalry for access to tourism resources (Belsky, 2009; Carlsen, 1993). Wealth redistribution associated with rapid and often inequitable economic change can result in social destabilization as dominant local elites work with foreign interests to accumulate capital and resources at the expense of local populations and disadvantaged, minority groups (Belsky, 2009; Chan et al., 2016). Unequal power relations affect the ability of indigenous peoples to exert their *agency* in management decisions related to tourism; thus, despite acknowledged potential for alternative forms of tourism, such as ecotourism, to improve indigenous livelihood possibilities in certain circumstances (e.g., (Stronza, 2007)), in practice the distribution of economic benefits are often uneven and tend to favor community elites (Coria & Calfucura, 2012).

A key challenge that arises here is the weakening of communality and *a sense of community* and the corresponding changes in informal cultural institutions. Communities are founded in social settings that extend beyond one's own life, resulting in a form of social control reinforced by ongoing relationships and the need to take responsibility for one's actions (Goodsell et al., 2014). In traditional rural communities, these social settings were established and reinforced by spatial proximity; however, changes in global connectivity and communication technology have "'melted down' the various institutions that provided stable contexts and frames of reference" as well as those that have helped to maintain social equity (Goodsell et al., 2014, p. 633). Macro-level processes such as urbanization, industrialization, and centralization of bureaucratic power can weaken the autonomy of once-isolated rural and indigenous communities, aligning them more with the needs and decisions of mass society. This "institutionalized individualism" means that people, especially members of indigenous groups, are forced to adapt to changing life conditions without the support formerly supplied by family, religion, class, and culture, bringing in new elements of risk and destabilizing community trust (Goodsell et al., 2014).

Despite the negative socio-cultural processes and outcomes resulting from tourism development identified in the preceding section, scholars have identified other situations where tourism has provided empowering opportunities for the preservation—and even the revitalization—of Indigenous identity, knowledge, and tradition (e.g. Butler & Hinch, 2007; Chan et al., 2016; Higgins-Desbiolles, 2016; Holmes et al., 2016; Medina, 2003; Stronza, 2008; Zeppel, 2006). As is true with the conservation value of tourism (Stronza et al., 2019), the positive effects of tourism on culture are likely be come into sharper relief when we consider how Indigenous communities would have otherwise faired in tourism's absence. In any case, individuals are often differentially exposed to processes of cultural reinforcement and revitalization as a function of differential access to capital, markets, training, and/or other capabilities. Bunten (2010) contends that, with careful planning, tourism can be a powerful tool for peace and prosperity within Indigenous communities—not only for supporting preservation of knowledge and cultural values, but also for bridging cultural boundaries. She points out that value judgments made by (expert) researchers such as "cultural commodification" have to be carefully scrutinized. They arise in relation to particular, situated practices in place, and assessing what they mean requires understanding *local perspectives,* how for both individuals as well as collectively within the place and space of the community (Bunten, 2010).

The promise of tourism and "justice" through tourism

As seen above, an increasingly interconnected global environment creates numerous threats to the cultural, linguistic, and traditional ecological heritage of indigenous peoples (Gorenflo et al., 2012). International tourism development has led to improved economic conditions (Wunder, 2000; Stronza, 2007) and strengthened cultural identity (Stronza, 2007; Zeppel, 2006) for indigenous communities, but has also resulted in numerous adverse impacts such as commodification of culture and dispossession from traditional lands (Coria & Calfucura, 2012; Johnston, 2006). Tourism is frequently promoted as a viable strategy for addressing some of the challenges noted above, and achieving some or all of the Sustainable Development Goals (SDGs), in which issues of inclusion, justice, and strong institutions (e.g., SDG16) feature prominently (UNWTO, 2017). Sustainability must consider not only the promotion and perpetuation the local biophysical environments, but also conservation of their local commons and maintenance of their cultural relationships to those environments. Costs and benefits (e.g. environmental, social, financial, and cultural) ought to be *distributed fairly* such that environments, culture, and relationships can be maintained over time and across generations, avoiding "colonization" and dependencies. It must facilitate agency and autonomy, and as well as respect for those involved in developing and selling their environmental and social-cultural goods (Hollinshead, 1992; Whyte, 2010). A key principle to ensure such equitable and fair outcomes is that the products and experiences being developed for tourist consumption must be *negotiated by those whose ecological and social-cultural goods and relationships, and their identity and well-being, stand to be most impacted* by the commodification of their ecological and cultural goods. Discussions of justice have been slow to arise in tourism and Indigenous Tourism contexts, but this important norm is being addressed in discussions of environmental justice (e.g., Mohai et al., 2009) and in the context of environmental justice as related to environmental tourism in Indigenous communities (Whyte, 2010).

Theoretical notions of justice span a range of approaches including *distributive justice, procedural justice, recognition justice,* and *environmental justice* (see, for example, Johnston, 2006; Rawls, 1997). Scholars of tourism are beginning to explore these from a destination perspective (e.g., Jamal & Camargo, 2014), while others examine diverse ethical perspectives that offer justice insights (e.g., Hunt & Stronza, 2011, Smith & Duffy, 2003). Forms of *justice tourism* at the destination level have arisen (e.g., Scheyvens, 2002) with related discussions of the influence of globalization (Higgins-Desbiolles, 2008). The broad questions these raise, such as of fairness and equity in the use and distribution of societal goods, as well as fair participation for due legal recourse, recognition of political status as a cultural group, offer a valuable guide to approaching the study of justice in tourism (see Whyte, 2010). So far, however, very little has been done to specifically examine notions of *cultural justice* in tourism from an emic perspective, i.e., from a local, place-based perspective of cultural participants in developing cultural tourism. How they negotiate sharing what they wish to of their cultural practices and cultural good to provide what they perceive to fair and equitable outcomes for themselves and their community? How do tourism researchers ensure that their approaches are fair and attentive to the well-being of the cultural group, their cultural heritage, and their intangible as well as tangible cultural goods? These are among the questions taken up in the present study. Given that it is impossible to separate ourselves from our theoretical influences means that we approached this study with care and thoughtfulness, attentive to approaches to justice widely and in the Indigenous context (see below), implementing decolonizing methodologies and situated, place-based approaches in Indigenous Tourism that facilitate local, emic views to participate fairly in dialogue and discussion (e.g., Smith, 2012; Tuck & McKenzie, 2015).

Study setting: the Tz'utujil people of Santiago Atitlán, Guatemala

In the five centuries since their conquest by the Spanish in the 1500s, the Indigenous population of highland Guatemala "has been repeatedly victimized" (Verrillo & Earle, 1993, p. 226). More recently, the nation's brutal 36-year Civil War bore an immense toll, resulting in the genocide of over 200,000 people, 83% of whom were Indigenous descendants of the Maya (de Onis & Yates, 2011; Devine, 2016). Fueled by structural racism, this violent conflict resulted in the militarization of rural areas (Carlsen, 1993) and the appropriation of Indigenous land (de Onis & Yates, 2011; Carlsen, 1993). More than two decades after the United Nations-facilitated Peace Accords of 1996, Indigenous Guatemalans continue to negotiate injustices relating to identity, land tenure, and livelihoods (Devine, 2016).

The multicultural discourse used by the government in tourism promotion today comes in sharp contrast to decades of discrimination and state-sponsored acts of genocide against Guatemala's Indigenous population. Since the late 1990s, Guatemala has joined Honduras, El Salvador, Belize, and Mexico in the *Mundo Maya* project, a large-scale promotion supported by the World Tourism Organization (Brown, 1999; Devine, 2016; Magnoni et al., 2007). This effort recharacterizes Guatemalan national identity in an attempt to capitalize on the "Maya culture" as a source of competitive advantage on the global tourism stage (Brown, 1999; Devine, 2016; Little, 2004). International campaigns that promote Guatemala as a cultural tourism destination also present the state's Indigenous communities with opportunities for economic and cultural revitalization.

Around Lake Atitlán in particular, over 95% of the population self-identifies as belonging to one of three Indigenous ethnic groups of Maya descent: the Tz'utujil, Quiche & Kaqchikel peoples (INE, 2002). This geographic concentration of indigenous population is a relic of the Spanish conquest of the 1500s, during which indigenous people of Guatemala were pushed from their central and coastal lowland homes into the highlands (Carlsen, 2011). Coupled with the highly aesthetic landscape, the modern culture of the local Kakchiquel, Ki'che', and Tz'utujil people is a primary pull factor bringing over 300,000 international and domestic visitors to the Lake Atitlán region each year (Centro de Salud Santiago Atitlán, 2011).

As the capital of the Tz'utujil people, Santiago Atitlán is the largest of three villages where the Tz'utujil reside and is a pueblo rich with the "living culture" that attracts visitors to the lake region. The municipality of the same name encompasses 136 square kilometers, comprised of the urban center and small satellite settlements located within expansive rural surroundings that are a mix of forest and agricultural production (Carlsen, 2011). The town is growing at a notable rate, close to 3% annually, and projections put the 2018 population of the urban center alone over 37,000 (Centro de Salud Santiago Atitlán, 2011). Much of the economy is based on farming and the production of artisan goods. Individual families manage small plots of farmland, or milpas, on the volcanoes' slopes, where they grow coffee, avocados, corn, beans, and other produce for subsistence use, sale in the local market, and – when yield permits – widespread distribution through large corporations (Carlsen, 2011).

The town is known for its quality handmade woven fabrics, clothing, and glass beads, which are sold to locals, national and international tourists, and Indigenous peoples from other regions of Guatemala. Santiago also holds a wide variety of lesser-known attractions of the type that tourists travel the world to seek, including both religious and nature-based attractions. Mirroring steady growth on an international scale—from 25 million tourist arrivals in 1950 to over 1.3 billion in 2017 (UNWTO, 2018), tourism to Guatemala, and thus tourism to the Lake Atitlán basin, has grown dramatically in recent decades. There is a body of anthropological writing on distinct Maya ethno-linguistic groups engaging in tourism in other regions of Guatemala, (e.g., Little, 2004, and numerous writings by Walter Little about tourism in the colonial city of Antigua), yet there remains limited fine-grain data on tourism-related outcomes within the Tz'utujil or the two

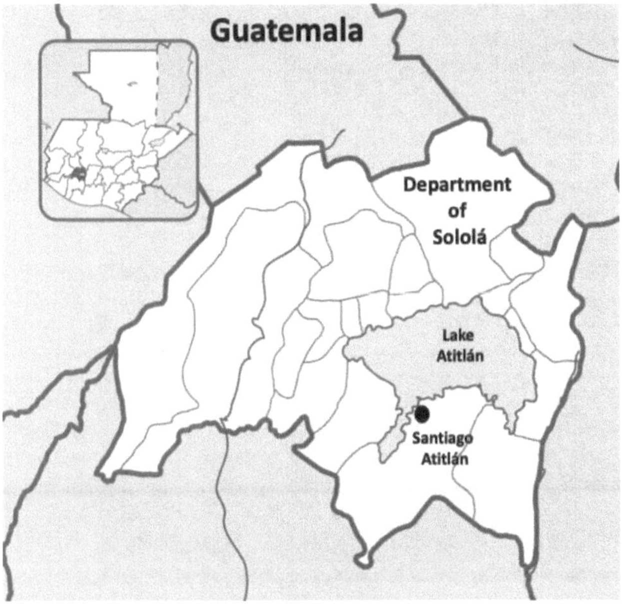

Figure 1. Map of Guatemala, with Sololá & Atitlán Inset.

other Maya Indigenous groups (*Quiche* & *Kaqchikel*) residing on Atitlán's shores, despite the lake being among Guatemala's most visited tourist destinations (Figures 1 and 2).

Methodology

This study undertook qualitative work grounded in the emic perspectives of those involved—and those to whom the cultures and environments belong. Ethnographic methods and an exploratory and inductive approach provide a platform for participants to share their emic views (Bernard, 2017; Harris, 1976). Given that 98% of Atitecos self-identity as Tz'utujil (Centro de Salud Santiago Atitlán, 2011), an ethnic group whose involvement in tourism is unique and relatively undocumented, this emic perspective is not one that can be easily found within existing literature or by using quantitative, deductive approaches (Creswell, 2013). While the history of virtually all social science methodologies is subject to much warranted critique, the ethnographic approach taken here is situated at the more humanist and interpretivist end of the social science spectrum, providing more in the way a "decolonizing methodology" than more positivist, survey-based approaches that characterize much of tourism research. The ethnography conducted here sought to facilitate fair and equitable participation of residents and communities in developing and enacting the research prosses, discussing the findings, and participating in how the results were to be shared, and how they could contribute towards communal well-being and culturally just tourism.

Both authors are white North Americans who speak Spanish fluently as a second language. The second author engaged in ethnographic work in ethnographic work in this region in 2003, returning briefly in 2005. Relationships with Santiago community members established during the 2003 fieldwork, as well as with anthropologists with longer-term presence in this region, facilitated re-immersion into Santiago daily life and access to community leaders (e.g., Santiago's Tz'utujil mayor), facilitating collaborative engagement and research design on the present research being undertaken in Santiago. Both authors collaborated with local participants on the current research design. The first author resided in the community of Santiago Atitlán to oversee new data collection during a period of 10 weeks in 2017.

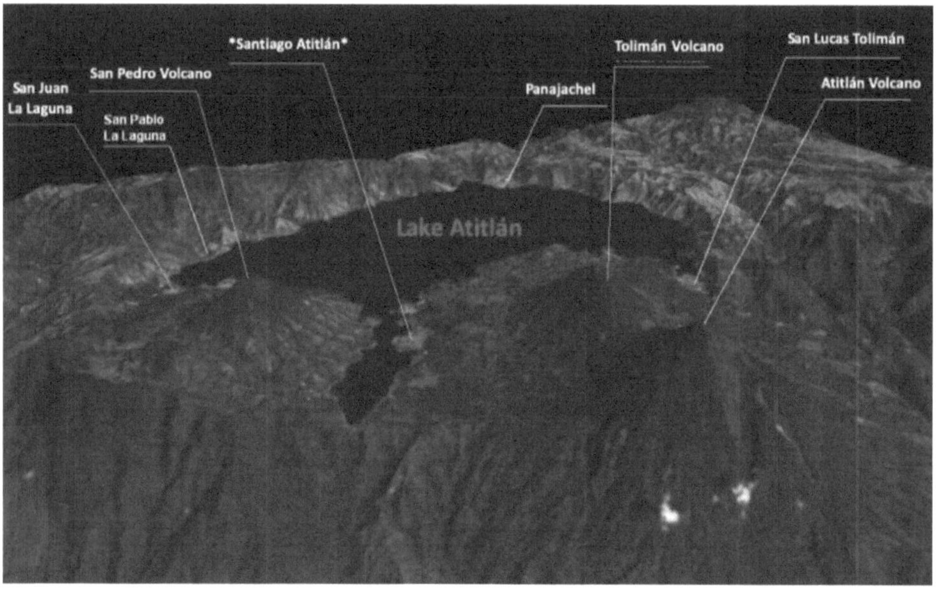

Figure 2. Lake Atitlán map with relief.

Data collection and analysis

Ethnographic methods that facilitated storytelling and rich, joint discussions were employed to capture emic views of tourism and its influence on justice. Data collection activities fall into two categories. Existing archival data, in the form of relevant scientific and academic research, local government planning documents, external reports, and social media posts of those involved in tourism were sought and compiled from various local and online sources. Ethnographic data were generated through (i) daily participant observations (ii) structured interviews based on typical case sampling of tourism-related business owners or employees ($n = 34$) in the shops, cafes, restaurants, and stores along the *Calle Real* (primary tourism street) in Santiago, and (iii) semi-structured key informant interviews ($n = 15$) that were conducted with individuals purposively sampled for expertise in tourism. In an effort to capture a diverse range of perspectives from community members involved in tourism, these key informants were sought within the following strata: local tourism business owners, tour guides, hotel owners or employees, Municipal Government employees, members of Santiago's original Tourism Self-Management Committee (CAT) formed in 2004, members of Santiago's recently-appointed CAT formed in 2016, and members of other local committees. Some interview subjects belong to two or more categories. Of these 15 key informants, 6 self-identified entirely with the indigenous Tz'utujil race. One informant identified as a mix of Tz'utujil and Ki'che', while another informant identified as Kakchiquel. As is common within the tourism industry, one informant – a hotel owner – was an expatriate American citizen.

Both structured and semi-structured interview themes explored how participants experienced and negotiated tourism development and their interactions with tourists, Tz'utujil identity and heritage, the impacts of tourism in Santiago, factors influencing quality of life for local residents, and a census of local institutions related to tourism. These efforts resulted in interviews ranging from 25 to 92 minutes in length. Informal interviews addressed similarly themes, though also diverged into other topics depending on the expertise of the informants encountered. With one exception, all interviews except were conducted in Spanish, a second language for the researchers as well as the informants (Tz'utujil being the primary language). Repetition and summary feedback were frequently employed as they were particularly useful to facilitate further discussion and storytelling, expanding on insights as they arose (Musante & DeWalt, 2011).

All data sources described above were converted into a text format prior to analysis. Audio-recorded key informant interviews conducted in Spanish were transcribed directly into English by the first author, prioritizing preservation of the interpreted meaning of the conversation over a strict word-for-word conversion. The transcriptions from the recorded interviews, along with field notes documenting participant observation and archival data constituted the knowledge base from which we engaged with further analysis. We were able to further understand the emergent trends by thinking through how they might inform notions of justice, but our focus here was not to peg them to specific etic terms or concepts like distributive or procedural or environmental justice. Rather, we sought to identify broad ethical and moral principles of justice that might comport with the emergent themes, e.g., fairness, equity, as well as key values that were important for Indigenous well-being (e.g., agency, autonomy, reciprocity). While not consistent with a traditional definition of "grounded theory" that indicates the development of new theory, the authors' process nonetheless adheres to more liberal and interpretivist approaches to this method (e.g. Bernard et al., 2016).

Study findings of indigenous cultural tourism in Santiago Atitlán

The sale of culturally-based commodities and services is central to tourist encounters in Santiago. This section presents examples of the complex ways in which Tz'utujil culture is commodified in the name of tourism. The data presented here show how the influence of tourism—both positive and negative—on social processes indeed leads to the commodification of non-material (e.g., language, knowledge) and material (e.g., clothing) aspects of culture. Yet, the study participants also see tourism as invoking a corresponding re-arrangement and re-negotiation of a complex Indigenous identity in ways that preserve threatened Tz'utujil culture.

Market opportunities for cultural flourishing

Indigeneity across Guatemala is characterized by place-based identities, something that is particularly notable around Lake Atitlán (Carlsen, 2011). Through the use of specific colors, design patterns, and objects depicted, clothing is used as a signal not just of one's ethnicity, but of one's village (Little, 2004). While many Atitecos, especially women, continue to dress in *traje (traditional Mayan vestments)*, globalization has resulted in a blurring of this visual signal of indigeneity. As Carlsen (2011) writes, tourism demand influences the amount of social control Tz'utujil weavers have to experiment with wider color palates and creative new designs. At the same time, access to wider markets mean that more people are buying clothing based on other characteristics—like preference, quality, and price (van den Berghe, 1994).

Santiago Atitlán's reputation for high quality handwoven and hand embroidered clothing, in combination with relaxed social norms around *traje (Figure 3)*, has enabled Atitecos to expand their market to Indigenous people from all over Guatemala, as demonstrated in the following participant observation vignette:

> *June 11, 2017: I approached a stall hung with row upon row of huipiles (traditional embroidered blouses worn by women), stretching across the 3-meter-wide stall from the ground to the roof. The vendor, Rosario, was a young woman in traje, perhaps about my age. Her stall was on the outskirts of the market, and it was still early in the day—around her, other stalls were still just setting up. She wasn't busy, and happily obliged to my questioning.*
>
> *When I asked where the huipiles came from, she said here in Santiago Atitlán—pointing to the top two rows. The huipiles in the top row were adorned with colorful birds of various sizes; the next row was covered with flowers. The designs in these two rows were embroidered onto the vertically striped fabric that is typical of the Atiteco traje. Sometimes the patterns were concentrated just around the collar area, with large areas of open cloth. On others, the vertical stripes had been crossed with embroidered horizontal stripes, turning the cloth into rows of boxes, each of which contained a different bird. They were very detailed, and complex.*

Figure 3. Tz'ujujil traje, typical indigenous apparel.

Author 1: [pointing to the top two rows] Who made them? You?

Rosario: Yes.

Author 1: By hand, right?

Rosario: Yes, always by hand, these ones, from Santiago Atitlán.

Author 1: Wow, they're beautiful. How long does it take you to make them?

Rosario: Two months.

Below the top two rows of birds and flowers, there were several rows of huipiles that were completely different. They were entirely covered with colorful geometric patterns, with none of the base cloth visible. The embroidery was much more uniform, and the stitches had a cross-stitch like pattern (a bunch of Xs) instead of the handmade ones, whose patterns were made from longer, straight parallel stitches and no Xs. Below these, there were several more rows of huipiles that were more similar to the upper rows (more exposed cloth, distinct objects depicted, like flowers, instead of geometric patterns) but had a more "modern" look. They were adorned with sequins, sparkles, and 3D designs of coiled material.

Rosario indicated that these bottom two sections were all machine made, and not from Santiago. She told me that these ones came from other places, like Xela[2]*. While she had originally answered in the affirmative when I asked her if she made them, she now spoke of it more as a network of people working together. The hand-embroidered, Atiteco-style huipiles in the top two rows were all made by hand here in Santiago Atitlán, by various local women. The other ones that were machine made were purchased from Xela, and brought back here to be sold.*

Author 1: So the ones that come from Xela, where do you buy them? From the market?

Rosario: Yes, in the market. We go buy them there, and bring them back here to sell.

Author 1: And do people from Xela also come to Santiago Atitlán and buy your huipiles, to take back and sell them there?

Rosario: Yes, it's an exchange of business. We buy from each other to sell in our markets.

Author 1: But in Xela, they're Kaqchiquel, right?

Rosario: Yes, they're Kaqchiquel and here we're Tz'utujil.

Author 1: But you buy their things and sell them here?

Rosario: Yes, these ones [indicating top row] are our traditional *traje*. These are the style of Xela [indicating the middle section described above]. But people like to choose.

As indicated here, weavers diversify their offerings in order to target several niche markets simultaneously. Far from passive reaction to an imposed market, this reflects active negotiation of market-related opportunities, a type of involvement in tourism that Bunten (2010) refers to as "Indigenous capitalism."

In another conversation at the market, a young man selling bolts of the woven cloth that is used by to make *faldas*, the long skirt that is part of the female *traje*, confirms these arrangements:

Author 1: And who buys your fabric? Only locals? Or do tourists buy them too?

Misael: Some tourists. Mostly Atiteca women, but also women come from other places, like Xela and Panajachel. They come here to Santiago to buy their skirts.

Author 1: Why?

Misael: The quality. They know the quality of the weaving here.

Author 1: But in Xela and Panajachel they're Kaqchiquel, right?

Misael: Yes, they speak Kaqchiquel.

Author 1: They wear different *traje*, right?

Misael: Yes, but they can choose what they want to wear.

Again, this informant indicates how globalization in the form of tourism leads to relaxation of rigid norms, yet this relaxation opens up additional opportunities for skilled local artisans to gain recognition in the national market. Maria, an artisan who earns money hosting foreigners in her family home and who is an artisan skilled in hand embroidery, expressed pride in the ability to share her life and culture with guests. "If you bought a *huipil* that I made," she said, "It would make me happy because it shows your appreciation for my unique skill and hard work." In this regard, the vending of handicrafts in tourism is a source of individual as well as collective pride and reflection on Tz'utujil culture (Little, 2004; Stronza, 2008). Among the many homogenizing forces of globalization, work in tourism is seen as holding promise for conserving the Tz'utujil culture for Maria. Whether tourism is indeed seen more broadly as facilitating *cultural justice* subsequent the colonial-era suppression of longstanding Tz'utujil traditions, knowledge, and practices, is an important question that we pursue below .

Resisting colonization of Tz'utujil language

In an increasingly globalized world, loss of Indigenous language is a prevalent concern (e.g., Gorenflo et al., 2012). Language is key part of Tz'utujil identity, and something of which Tz'utujil informants indicate that they are most proud. Unfortunately, tourism development in the community also creates pressure to reject Indigenous language. When asked about criteria for being Indigenous, Pedro – a Santiago tour guide who picks up business informally by waiting for tourists disembarking at the boat dock – explains:

> The negative side is that nowadays, some Indigenous people, when a child is born—the first language is Spanish. Only Spanish. But also there are—for example, [pointing to his young son, who he brought with him]—he speaks very little Spanish. He only speaks the Maya language ... because my primary interest is the Maya language. Spanish, he's going to learn in school. (Pedro, Tour Guide)

Atitecos who only speak Tz'utujil have little opportunity to engage in the tourism industry, with the exception of artisans who sell their work through middlemen (van den Berghe, 1994), or the few women who get by selling souvenirs with a basic vocabulary. While tourism incentivizes empowering skill development for certain languages, like Spanish and English, this can come at the expense of a devaluation of the Tz'utujil language, as Pedro notes here. The process he notes is consistent with the findings about local language loss in other Indigenous tourism contexts (Coria & Calfucura, 2012; Johnston, 2006; Whitney-Squire, 2016). Yet, as seen below, there are other spaces and opportunities in which the language facilitates both economic opportunities and cultural well-being

Negotiating cultural exploitation and discrimination for mutual benefits

The use of culture to earn money is not inherently problematic, as numerous tourism scholars have noted (e.g., Bunten, 2010; Medina, 2003). Indeed, when asked if their capacity to earn money was connected to their Indigenous identity, many informants who work in the tourism industry answered affirmatively. When asked if he has noticed any misuses of Tz'utujil culture in Santiago Atitlán, tour guide Pedro indicates that culture is definitely misused by some in order to earn money, explaining:

> So—this happens more touristically. Sometimes a tourist comes and says, I'm looking for a shaman to talk to with, to do a ceremony. And I could say, "Ah, I'm one!" But I'm not one. (Pedro, Tour Guide)

Here tourism creates a monetary temptation to stage Indigenous authenticity for visitors in Santiago Atitlán (MacCannell, 1976). However, as Miguel, another tour guide notes, staging is just as likely to occur on the part of non-Indigenous guides. He feels it makes sense for him to call upon his genuine Indigenous identity for mutual benefit (his and for visitors):

> I have friends that aren't Indigenous, and although they want to identify as Indigenous ... first of all, they don't speak a Maya language. And visually, or physically, it's very noticeable. You can say "I'm Indigenous," but if your physical appearance doesn't show it ... So then, when people see me—dark-skinned, short stature—they say "Ah, you're Indigenous!" and they start to ask me things. And people value that a lot. They ask how I live, how I have overcome all of those social problems, especially discrimination. So then, in the end, they appreciate it, and they can value you for a certain amount of money—they can give you tips.

While the leveraging of culture to earn income within the tourism industry might lead some scholars to frame these manifestations as exploitive, the way the Tz'utujil use their culture suggests active and conscious appropriation to generate benefits, overcome discrimination (see quote above), and "perform" their culture for mutual benefit–appreciation by visitors and some income for the Indigenous tour guide.

Miguel later elaborates on the features identified above as generating pride and sense of well-being and worth—physical appearance and language. When asked how his interactions with tourists make him feel about his identity, Miguel continues, "they make me feel very proud, very happy. And them too. When I teach everything that I am, and that I feel—they also appreciate that. They realize that I am proud of my culture, and of my work, what I'm doing. And it's a combination—because they leave satisfied, seeing someone so proud." Tourism here is offering opportunities to counter entrenched colonial injustices and to facilitate Tz'utujil cultural survival, but it is also contributing immensely to individual well-being (Miguel's, for instance) and to a good experience for visitors.

Reciprocity and equity in interactions

Overwhelmingly, participants interviewed in this study agree with Miguel, indicating that their interactions with tourists, who travel from across the globe to experience authentic Maya culture, create feelings of immense pride in Tz'utujil cultural heritage, offering a new postcolonial "mirror" through which to see themselves (Stronza, 2008). Informants say that their interactions with tourists, in addition to bringing income for individuals, help to reinforce and sustain a collective pride, a finding that has been noted in other Indigenous tourism contexts (Bunten, 2010; Coria & Calfucura, 2012). As such, tourism not only offers a valuable economic supplement to current livelihood alternatives in agriculture, fishing, and other sectors, but also crucial opportunities to enact and "perform" their culture for individual and communal identity and well-being. But there is also a sense of reciprocity and equity that weaves through discussions such as with Miguel above.

Numerous examples in the case show how tourism allows local artisans to leverage cultural heritage to generate increased income and support continuity of their cultural traditions. In Santiago, Atitecos do not feel that the artisanal markets necessarily result in less authentic products or the loss of cultural practices, nor has it led *away* from traditional materials and methods of production in favor of advanced technology, as scholars have feared in other contexts (e.g., Coria & Calfucura, 2012; Johnston, 2014; van den Berghe, 1994; Zeppel, 2006). In fact, the view conveyed by informants is that the opposite outcome has occurred, as tour guide Daniel notes, "Because, for example, an artisan, its more—[tourists] like their work more. Because what the tourists want is more so the craftsmanship. And so, more is made by hand, rather than embroidered by machine." Local artisans thus also experience increased valuation of their traditional cultural skills developed through learning from interactions with visitors. There is a sense of *equity* and *reciprocity* in this relationship, where both sides benefit mutually, neither feel exploited.

Learning related reciprocities

Some study participants emphasize the influence of tourism on improved environmental conditions within Santiago. The following two exchanges demonstrate this. When asked what he sees as the advantages to working in tourism, a local tour guide responds:

> **Guide:** The advantage is, let's say... one can learn other cultures. Other cultures, and also, to share—right? To share with people, because cultures are different. Here in Guatemala, the reality is, we have a culture, let's say... For example, here the majority of the people—their trash? They throw it in the street. But a tourist, even if it's the smallest piece of plastic, they always put it in the trash bag. So then, that's a culture that we have to copy, because we need it. To conserve our lake, our volcanoes, and... many things. To develop, to treat people well...
>
> **Author 1**: It's not only to share your culture, but also to learn the cultures of the visitors.
>
> **Guide:** Exactly, yes. That's it.

Similarly, David, a weaver and Santiago municipal government employee, discusses the influence of tourism on environmental learning among the Atitecos. Again, reciprocal benefits and relationships are observed here:

> **David:** Yes, yes. The tourist is an example for us, of education. A tourist comes asking where he can put this trash, and the people watch him—this is how you do it. So [the tourists] come, they [Atitecos] learn, and they leave their trash there. This is the help, the support, which tourism gives us. ... Tourism is helping us in an education.

In another compelling quote, Felipe explains how working as a guide has encouraged him to remain engaged with, and to continue to learn from, local elders:

> Ok, when I started working as a tour guide, it interested me a lot, because it's a very special subject for me. Because within the area of tourism, as a guide, one works in subjects like archaeology, anthropology, ecology, the Maya worldview, Maya spirituality... all those subjects. And I take part in the Cofradias a lot, with the elders. For me, they're great people, they're the best teachers. I use a lot of the customs, the oral traditions from our ancestors, our grandparents, our parents—and today, well, I manage quite a lot of information, and I continue learning. Because when a tourist arrives in Santiago, he asks about everything, he wants to know about everything, right? And as a guide, you have to be well prepared, to give good service. (Felipe, Tour Guide and Artisan)

The spirit in which Felipe conveys these views makes it clear that this engagement with local elders, and in particular the perpetuation of Tz'utujil knowledge and practices, would not occur without his involvement in tourism. The new valuation of Tz'utujil knowledge is extended from younger generations with direct links to tourism to members of older generations without direct interaction with tourists. Felipe's account thus conveys outcomes of tourism where different social strata within Santiago's Tz'utujil population benefit from the complex negotiation between culture, tourism, commodification, authenticity, and justice.

Control over tourism development and equitable access to tourists

A number of social stressors are evident in the accounts presented by several of Santiago's local tourism stakeholders. One key issue that informants identified is the lack of a central tourism office to help direct tourism to the community. An older guide and artisan, who has been operating in Santiago Atitlán since 1996, is at the center of this drama. In his words:

> We had the beginning of a guide office. Before, there wasn't an association, there were just guides. We had the office, but the office... tourists didn't get over to it, because the commission agents, the unauthorized guides—they go down directly onto the dock. So, they divert the tourists, they misinform the tourists, and the office—although it was open, although we had a good display, the tourists didn't come there. Because of the disorder.

Access to visitors here is viewed much like a common-pool resource that is being overexploited by unregulated tour guides, just as described in a recent study of Indigenous involvement in ecotourism in Ecuador (Marcinek & Hunt, 2019). This appears to be consistent with Felipe's account, and this is corroborated in an additional account for why the office and guide group fell apart. As another guide Diego describes:

> There isn't activity. They left the group... Felipe is a very... He doesn't work transparently. That's why they fall apart, the associations, because if something comes for the group, only one person does it. And he, as president—every group that comes, he's guiding it. And there isn't anything for the others. They get angry. Only one person benefiting, while the others contribute their time. He is everywhere. I don't like how he works. (Diego, Local Committee Member)

Several informants indicate that Felipe took more than his fair share of the guiding business, while Felipe writes this off as jealousy over the increased attention he received after he had the good fortune to be featured in a *Lonely Planet* travel book. In this case, an external publication governed Felipe's ability to pursue what he feels are fair outcomes of tourism for himself, though for others this situation created an impediment for other tour guide informants. Lack of control and processes for fair and equitable access to tourists is creating "winners" and "losers" – has served to destabilize the sense of community among these guides (Belsky, 2009; Carlsen, 1993). Yet another informant, Juan, describes how Felipe destroyed his business more than twenty years ago, resulting in Juan leaving tourism altogether and starting a new career. Juan describes how he started an office and tour guide cooperative down by the dock but, because he is illiterate, he relied on Felipe, a blood relative, to be his secretary. By Juan's account, Felipe became jealous and angry, claiming that Juan was taking more than his share of the tours. Felipe extorted 1000 quetzales (about 140 USD) from Juan, and then sold Juan's office space to the next-door vendor, who now uses the space to sell tourist trinkets. Inequities are clearly evident here at the micro-level for fair and equitable access to the benefits of tourism.

Environmental conservation through ecological tourism

Atitecos view their natural environment as an asset that can be leveraged to proactively influence how tourism manifests in their community, and thus to optimize tourism outcomes. Recognition of a growing demand for ecotourism and nature-based tourist activities, such as bird watching and climbing volcanoes, is prompting the creation of new products that incorporate the natural environment. However, they are thoughtful to the importance of intrinsic Tz'utujil cultural value of the environment being preserved in the process. For example, when asked the reason for building a trail and viewpoints at a popular birdwatching spot, an employee for the municipality's Commission for Economic Development, Tourism, Environment, and Natural Resources explains the economic argument towards the end of conserving vital local habitats:

> Because in Santiago Atitlán, there is lots of biodiversity. So, we want to implement this... it was implemented, most, to conserve it. Because right now, well, many people go there, but they don't respect the biodiversity. So, we want to implement this to attract more tourists—foreigners, national and international—in order to generate the economic gains needed to conserve the forest.

But do all these changes mean that Santiago, too, will undergo commodification of its cultural and now ecological goods, to the point of unsustainability of the biophysical environment and the well-being of local residents?

Tourism has dramatically changed the appearance and cultural character of other adjacent towns on the Lake Atitlán shoreline, including Santiago's Tz'utujil neighbor San Pedro La Laguna, where one can find a plethora of businesses adorned with colorful signs, written in both Spanish and English, announcing their goods and services. While the imagery present in San Pedro would not seem unfamiliar to travelers who have visited touristy, tropical locations, tourism has yet to lead to such manifestations in Santiago Atitlán. Tourism stakeholders in Santiago do recognize that changes are needed to improve access to tourism's benefits. But they are taking control in establishing processes and strategies for better outcomes after regarding the types of changes in built environments that they feel have undermined the cultural identity of San Pedro. As one experienced local guide elaborates:

> What I'm trying to implement right now more than anything is bird tourism. Bird tourism, because Santiago Atitlán is known as the House of the Birds. If Santiago Atitlán is known throughout the whole country as the House of the Birds - it's because there is a great concentration of birds here. And it's one of the things that I like to do, because, apart from being a way of life for me, it's something quite interesting for many people from outside. And what I want now basically, is to have the most suitable places to take them, because some of the birders from other places, they're older people. So, this the reason I say that more than a job for me, it's a passion. This is what I want - to convey all my knowledge to people.

In sharp contrast to the neighboring villages, tourism-related changes to the ecological and cultural environment are navigated with care in Santiago—both within and from outside the community—in order to preserve it as the primary bastion of Tz'utujil Maya identity and cultural integrity.

Discussion and directions for future research

Much research has been conducted that shows tourism can improve economic and social-cultural conditions for Indigenous communities (Stronza, 2007; Wunder, 2000; Zeppel, 2006; Carr et al., 2016) and that it can also result in adverse impacts commodification of culture and dispossession from traditional lands (Coria & Calfucura, 2012; Johnston, 2006). A plethora of analyses on tourism "impacts" and principles for responsible, sustainable and community-based tourism and ecotourism complement such studies (the list would be too long to cite here). Our study above shows that it is important to investigate not just the net effect of tourism development on the community as a whole, but *how* these costs and benefits are distributed and negotiated among individual local people and institutions, and what norms guide good tourism development that facilitates communal well-being and justice. Given a lengthy history of marginalization and endangerment of indigenous communities through irresponsible forms of tourism (Coria & Calfucura, 2012; Johnston, 2006, 2nd edition, 2014; van den Berghe, 1994), research that determines how Indigenous peoples can negotiate the arrival of international tourism in ways that promote social, cultural, and environmental justice is especially important (Carr et al., 2016). Yet, despite a large body of scholarship on sustainable tourism and Indigenous peoples, and the centrality of justice within the Indigenous rights movement, very little scholarly effort has been dedicated to the study of justice in the context of Indigenous Tourism.

Our aim here was to attempt to explore how the Tz'utujil Maya were engaging with tourism, what they felt were fair and good outcomes of tourism development, how they were negotiating these and the resulting changes. We were especially concerned to approach the study in a way that did not colonize the meanings that arose by imposing a "western" frame of justice to guide the study, but sought instead to explore in rich detail *how* tourism development was being enacted and negotiated, and principles we could identify that facilitated fair and just outcomes as they perceived and experienced it. It is clear from our case that tourism development as being currently enacted by Tz'utujil Maya facilitate critical Indigenous principles like self-determination, autonomy, reciprocity, and perhaps even mutual respect (less clear but implicit) (de Lima & King, 2017; Escobar, 2018; Hollinshead, 1992; Whyte, 2010).

Emic views of Tz'utujil Maya community members indicate that tourism is slowing the erosion of threatened culture heritage in Santiago Atitlán, just as it is often slowing the erosion of threatened natural heritage in other contexts (Stronza et al., 2019). While there remains evidence of the tourism industry's appropriation of knowledge, intellectual property, and other Tz'utujil resources in some cases, by and large the Tz'utujil people involved in tourism in Santiago Atitlán are negotiating tourism—and even the practice of commodification itself—in ways that resist historically entrenched colonial influences to situate their own ways of being and knowing. Their language is currently flourishing, their visitors are curious to learn more about their culture and heritage, and there is evidence of reciprocity in learning and sharing knowledge and insights. Rather than attempt to "nail" down specific theoretical perspectives that may do little justice to the emerging form of cultural tourism in Santiago Atitlán, we draw on Whyte's (2010) cogent argument for norms guiding environmental justice for Indigenous Tourism, which includes close consideration of environmental heritage among other environmental relationships. His careful analysis identifies *compensation justice* (related to distributive justice, and involves fair exchange of goods, bads, and risks), *participative justice* (opportunity to give informed consent by all who stand to be benefit or be harmed) and *direct participation* (related to *recognition justice*, and involves being able to integrate one's own cultural terms and social circumstances in decision making) as key dimensions of environmental justice in Indigenous Tourism.

The Tz'utujil people are developing cultural tourism and also getting into environmental tourism. The close relationship between 'culture' and 'environment' for Indigenous people (Hollinshead, 1992; Tuck & McKenzie, 2015), including the Tz'utujil Maya (Carlsen, 2011), suggest that Whyte's (2010) framework may a helpful one to build on.[3] Based on our study, we draw on the general principles forwarded by Whyte to suggest a preliminary framework to guide future research on cultural justice in Indigenous Tourism:

1. *Compensation justice*: A number of examples in the case note challenges with distributing the economic benefits of tourism fairly. There were also some emerging debates within the community regarding the procedures needed for managing tourism more sustainably in the future. However, tourism is in early stages of development, and while conflict can be identified (e.g., among the tour guides seeking access to tourists), processes and structures (e.g., local tourist office) will need to be developed to organize and manage the tourist flows for more *fair and equitable distribution* of economic costs and benefits. However, in the context of *cultural justice*, much greater attention will need to be paid to addressing the intangible as well as tangible impacts of cultural tourism development.
2. *Recognition justice*: Despite the history of persecution, racism, and cultural devaluation that Indigenous peoples across all of Guatemala face, the Tz'utujil participants in this study expressed immense pride when discussing their cultural identity and practices, and the opportunity that tourism presents to be *recognized* and share it with visitors. However, a long history of colonization, discrimination and racism experienced by the Tz'utujil Atitecos (Carlsen, 2011) suggest that mechanisms for are needed to ensure that they are accorded recognition status and due recourse if they experience discrimination and unfair treatment. This is closely related to procedural and participative justice, as well as direct participation as argued by Whyte (2010).
3. *Procedural justice/Participative Justice and Direct Participation*[4]: The Tz'utujil Atitecos are exercising *agency* to capitalize on the opportunities provided by tourism (see also Bunten, 2010). They are currently actively engaged to preserving the Tz'utujil culture in Santiago, and highly aware of how they don't want it to develop watching other communities in the area. As numerous approaches to community-based tourism and ecotourism advocate (Belsky, 2009; Schellhorn, 2010; Scheyvens, 2002; Stronza, 2008), the Tz'utujil people will need to continue to be directly involved in tourism development, as well as in the management and marketing of their tourism offerings. Tourism was playing a significant role to facilitate economic well-being as well as cultural well-being, facilitating not just agency but also *autonomy* to develop cultural tourism as they wished, through *direct engagement and participation in the process*. When asked how he learned the job skills he needed to be a guide, one respondent replied simply, "It's my life. I lived it."

Directions for future research and practice

It cannot be denied that there will be factors in every particular cultural setting that make it unique. However, here the use of rich, thick description by the authors in presenting these findings—including direct quotations of emic perspectives and sufficient detail of the environment in which conversations and observations occurred—allows the reader to determine how and where the empirical findings are transferable to other settings or themes. The drivers of development and globalization—including tourism—have many complex and overlapping impacts on Indigenous communities that can only meaningfully be assessed by capturing local emic views with ethnographic sensitivity. Their perceptions of what constitutes fair and equitable outcomes of tourism emerges through the meshing of their economic needs, and social and cultural norms, values and beliefs with the enactment and negotiation of tourism development. In this sense, it could be argued that insights on norms that guide fairness and justice are arising performatively in the community, guided by communal social and cultural norms. However, institutional structures for the regulation of tourism, mechanisms for procedural justice as well as participatory planning to ensure that community members have voice and *control* over tourism development, may be needed as tourism continues to develop. Direct participation is a crucial principle, as Whyte (2010) argues.

Based on our joint experience, the following suggestions are offered for future research and practice, and to help build a robust framework for cultural justice in Indigenous Tourism:

- Use decolonizing methodologies and develop research topics and research design collaboratively with the Indigenous community; involve them directly in the data gathering, analysis and sharing of research that emerges (e.g., Smith, 2012). Also, partner with researchers familiar with the language and cultural context when undertaking research in Indigenous communities. Anthropologists, in particular, can serve a valuable role as "cultural brokers" to facilitate communication of diverse interests between Indigenous peoples, state institutions, and visiting tourists.
- Establish and make accessible for joint discussion during tourism development the cultural laws of the community (which are generally intangible), and how these would influence tourism related practices, cultural tourism 'products' and experiences to ensure cultural justice (e.g., Holmes et al., 2016). This might include establishing off-limits areas of the community to prevent tourist intrusion and appropriation of cultural practices that the Indigenous people do not wish to share.
- Encourage hospitality, service quality, and language training that emphasizes cultural heritage interpretation through direct participation of local Indigenous residents (as has been done with environmental interpretation & hospitality training to expand the opportunities in the ecotourism sector for rural youth in Costa Rica, for example; see http://inogo.stanford.edu/selal%3Den?language=en)
- Campaign for greater links with INGUAT and other national agencies involved with tourism planning and promotion in the region. Demand greater participation and representation in tourism-related decision making for local residents, drawing on principles of inclusivity and communality. This may help unify rival local guide groups described above in an effort to promote their collective interests.

Notes

1. Commonly known as Quetzaltenango, but often referred to in this part of Guatemala as Xela (pronounced SHAY-lah), a shortened version of its Indigenous name
2. Such inequities will only continue to be exacerbated as biodiversity loss, species extinction, global warming and its related impacts continue through the Anthropocene (Steffen et al., 2007).
3. Note that Jamal et al. (2010) also attempted to dissolve the artificial nature-culture divide by grappling with the notion of eco-cultural justice, drawing strongly on principles of environmental justice.
4. As Whyte (2010, p. xx) explains, "[t]he norm of direct participation requires accountability on the part of tourism operators and tourists in ways that are not covered by the norm of participative justice."

Acknowledgements

The authors thank The Pennsylvania State University's Interinstitutional Center for Indigenous Knowledge, which supported this work with an M. G. Whiting Graduate Student Research Award, and The Penn State College of Health and Human Development, which supported this work with a Small Projects Grant. Additional thanks to many friends, informants, and colleagues in Santiago Atitlán and to the NC State University Ethnographic Field School.

Disclosure statement

No potential conflict of interest was reported by the author(s).

References

Belsky, J. M. (1999). Misrepresenting communities: The politics of community-based rural ecotourism in Gales Point Manatee. *Rural Sociology, 64*(4), 641–666.

Bernard, H. R. (2017). *Research methods in anthropology: Qualitative and quantitative approaches* (6th ed.). Rowan & Littlefield.

Bernard, H. R., Wutich, A., & Ryan, G. W. (2016). *Analyzing qualitative data: Systematic approaches* (2nd ed.). SAGE Publications.

Brown, D. F. (1999). Mayas and tourists in the Maya world. *Human Organization, 58*(3), 295–304.

Bunten, A. C. (2010). More like ourselves: Indigenous capitalism through tourism. *American Indian Quarterly, 34*(3), 285–311.

Butler, R., & Hinch, T. (Eds.). (2007). *Tourism and indigenous peoples: Issues and implications*. Butterworth-Heinemann.

Carlsen, R. S. (1993). Discontinuous warps: Textile production and ethnicity in contemporary highland Guatemala. In J. Nash (Ed.), *Crafts in the world market: The impact of global exchange on Middle American artisans* (pp. 199–222). State University of New York Press.

Carlsen, R. S. (2011). *The war for the heart and soul of a highland Maya town*. University of Texas Press.

Carr, A., Ruhanen, L., & Whitford, M. (2016). Indigenous peoples and tourism: The challenges and opportunities for sustainable tourism. *Journal of Sustainable Tourism, 24*(8–9), 1067–1079.

Centro de Salud Santiago Atitlán. (2011). *Memoria de estadística vitales y vigilancia epidemiológica*. Centro de Salud Santiago Atitlán.

Chan, J. H., Iankova, K., Zhang, Y., McDonald, T., & Qi, X. (2016). The role of self-gentrification in sustainable tourism: Indigenous entrepreneurship at Honghe Hani Rice Terraces World Heritage Site. *Journal* of Sustainable Tourism, *24*(8–9), 1262–1279.

Charnley, S. (2005). From nature tourism to ecotourism? The case of the Ngorongoro Conservation Area. *Human* Organization, *64*(1), 75–88.

Coria, J., & Calfucura, E. (2012). Ecotourism and the development of Indigenous communities: The good, the bad, and the ugly. *Ecological Economics, 73*, 47–55.

Creswell, J. W. (2013). *Qualitative inquiry & research design: Choosing among five approaches* (3rd ed.). Sage Publications.

de Kadt, E. (1979). *Tourism—passport to development: Perspectives on the social and cultural effects of tourism in developing countries*. Oxford University Press.

de Lima, I. B., & King, V. T. (2017). *Tourism and ethnodevelopment: Inclusion, empowerment and self-determination*. Routledge.

de Onis, P., (Producer), & Yates, P. (Director). (2011, January 25). *Granito: How to nail a dictator.* [Motion picture]. Skylight Pictures.

Devine, J. A. (2016). Politics of post-war tourism in Guatemala: Contested identities, histories, and futures. *Journal of Political Geography and Geopolitics, 28,*1–22. https://journals.openedition.org/espacepolitique/3723

Escobar, A. (2018). *Designs for the pluriverse: Radical interdependence, autonomy, and the making of worlds*. Duke University Press.

Figueroa, R. M., & Waitt, G. (2010). Climb: Restorative Justice, Environmental Heritage, and the Moral Terrains of Ulur\bar u-Kata Tjuṯa National Park. *Environmental Philosophy, 7*(2), 135–164.

Goodsell, T. L., Flaherty, J., & Brown, R. B. (2014). Community as moral proximity: Theorizing community in a global economy. In C. Bailey, L. Jensen, & E. Ransom (Eds.), *Rural America in a globalizing world: Problems and prospects for the 2010s* (pp. 626–641). West Virginia University Press.

Gorenflo, L. J., Romaine, S., Mittermeier, R. A., & Walker-Painemilla, K. (2012). Co-occurrence of linguistic and biological diversity in biodiversity hotspots and high biodiversity wilderness areas. *Proceedings of the National Academy of Sciences, 109*(21), 8032–8037.
Greenwood, D. J. (1977/1989). Culture by the pound: An anthropological perspective on tourism as cultural commoditization. In V. Smith (Ed), *Hosts and guests: The anthropology of tourism* (pp. 171–185). University of Pennsylvania Press.
Harris, M. (1976). History and significance of the emic/etic distinction. *Annual Review of Anthropology, 5*(1), 329–350.
Higgins-Desbiolles, F. (2008). Justice tourism and alternative globalisation. *Journal of Sustainable Tourism, 16*(3), 345–364.
Higgins-Desbiolles, F. (2016). Sustaining spirit: A review and analysis of an urban Indigenous Australian cultural festival. *Journal of Sustainable Tourism, 24*(8–9), 1280–1297.
Hollinshead, K. (1992). White'gaze,'red'people—Shadow visions: The disidentification of 'Indians' in cultural tourism. *Leisure Studies, 11*(1), 43–64.
Holmes, A. P., Grimwood, B. S., King, L. J., L. & K'E Dene First Nation, (2016). Creating an Indigenized visitor code of conduct: The development of Denesoline self-determination for sustainable tourism. *Journal of Sustainable Tourism, 24*(8–9), 1177–1193.
Hunt, C. A., & Stronza, A. L. (2011). Missing the forest for the trees?: Incongruous local perspectives on ecotourism in Nicaragua converge on ethical issues. *Human Organization, 70*(4), 376–386.
Instituto Nacional de Estadística (INE). (2002). *XI Censo nacional de poblacion y VI de habitacion*. Instituto Nacional de Estadística.
Jamal, T. (2019). *Justice and ethics in tourism*. Earthscan.
Jamal, T., & Camargo, B. A. (2014). Sustainable tourism, justice and an ethic of care: Toward the just destination. *Journal of Sustainable Tourism, 22*(1), 11–30.
Jamal, T., Camargo, B., Sandlin, J., & Segrado, R. (2010). Tourism and cultural sustainability: Towards an eco-cultural justice for place and people. *Tourism Recreation Research, 35*(3), 269–279.
Johnston, A. M. (2006). *Is the sacred for sale?: Tourism and Indigenous peoples*. Earthscan.
Little, W. E. (2004). *Mayas in the marketplace: Tourism, globalization, and cultural identity*. University of Texas Press.
MacCannell, D. (1976). *The tourist: A new theory of the leisure class*. Schocken Books.
Magnoni, A., Ardren, T., & Hutson, S. (2007). Tourism in the Mundo Maya: Inventions and (mis)representations of Maya identities and heritage. *Archaeologies, 3*(3), 353–383.
Marcinek, A. A., & Hunt, C. A. (2019). Tourism and cultural commons in the Ecuadorian Amazon. *Journal of Tourism and Cultural Change, 17*(4), 449–466.
Medina, L. K. (2003). Commoditizing culture: Tourism and Maya identity. *Annals of Tourism Research, 30*(2), 353–368.
Mohai, P., Pellow, D., & Roberts, J. T. (2009). Environmental justice. *Annual Review of Environment and Resources, 34*(1), 405–430.
Musante, D. K. M., & DeWalt, B. R. (2011). *Participant observation: A guide for fieldworkers*. AltaMira Press.
Nepal, S., & Saarinen, J. (Eds.). (2016). *Political ecology and tourism*. Routledge.
Pereiro, X. (2016). A review of Indigenous tourism in Latin America: Reflections on an anthropological study of Guna tourism (Panama). *Journal of Sustainable Tourism, 24*(8–9), 1121–1138.
Rawls, J. (1997/2009). *A theory of justice*. Harvard University Press.
Ritchie, J. B., & Zins, M. (1978). Culture as determinant of the attractiveness of a tourism region. *Annals of Tourism Research, 5*(2), 252–267.
Ruhanen, L., & Whitford, M. (2019). Cultural heritage and Indigenous tourism. *Journal of Heritage Tourism, 14*(3), 179–191.
Salazar, N. B. (2012). Tourism imaginaries: A conceptual approach. *Annals of Tourism Research, 39*(2), 863–882.
Schellhorn, M. (2010). Development for whom? Social justice and the business of ecotourism. *Journal of Sustainable Tourism, 18*(1), 115–135.
Scheyvens, R. (2002). Justice tourism. In *Tourism for development: Empowering communities* (pp. 102–119). Prentice-Hall.
Sharpley, R., & Telfer, D. (2015). *Tourism and Development: Concepts and Issues*. Channel View.
Smith, L. T. (2012). *Decolonizing methodologies: Research and indigenous peoples* (2nd ed.). Zed Books Ltd.
Smith, M., & Duffy, R. (2003). *The ethics of tourism development*. Routledge.
Steffen, W., Crutzen, P. J., & McNeill, J. R. (2007). The Anthropocene: Are humans now overwhelming the great forces of nature. *Ambio, 36*(8), 614–621.2.0.co;2] [18240674]
Stronza, A. (2007). The Economic Promise of Ecotourism for Conservation. *Journal of Ecotourism, 6*(3), 210–230. https://doi.org/10.2167/joe177.0
Stronza, A. (2008). Through a new mirror: Reflections on tourism and identity in the Amazon. *Human Organization, 67*(3), 244–257.
Stronza, A. L., Hunt, C. A., & Fitzgerald, L. A. (2019). Ecotourism for Conservation? *Annual Review of Environment and Resources, 44*(1), 229–253.

Swain, M. B. (1977/1989). Gender roles in indigenous tourism: Kuna Mola, Kuna Yola, and cultural survival. In V. Smith (Ed), *Hosts and guests: The anthropology of tourism* (pp. 83–104). University of Pennsylvania Press.

Tuck, E., & McKenzie, M. (2015). *Place in research: Theory, methodology, and methods*. Routledge.

van den Berghe, P. L. (1994). *The quest for the other: Ethnic tourism in San Cristobal.*, University of Washington Press.

Verrillo, E., & Earle, D. M. (1993). The Guatemalan refugee crafts project: Artisan production in times of crisis. In J. Nash (Ed.), *Crafts in the world market: The impact of global exchange on Middle American artisans* (pp. 225–245).State University of New York Press.

White, N. E., Buultjens, J., & Shoebridge, A. (2013). Complex interrelationships between ecotourism and indigenous peoples. In R. Ballantyne & J. Packer (Eds.), *International handbook on ecotourism* (pp.78–94). Edward Elgar Publishing.

Whitford, M., & Ruhanen, L. (2016). Indigenous tourism research, past and present: Where to from here? *Journal of Sustainable Tourism*, *24*(8–9), 1080–1099.

Whitney-Squire, K. (2016). Sustaining local language relationships through indigenous community-based tourism initiatives. *Journal of Sustainable Tourism*, *24*(8–9), 1156–1176.

Whyte, K. P. (2010). An environmental justice framework for indigenous tourism. *Environmental Philosophy*, *7*(2), 75–92.

World Tourism Organization. (2017). *Tourism and the sustainable development goals – Journey to 2030*. United Nations World Tourism Organization [UNWTO].

World Tourism Organization. (2018). *UNWTO annual report 2017*. https://www.e-unwto.org/doi/pdf/10.18111/9789284419807

Wunder, S. (2000). Ecotourism and economic incentives — an empirical approach. *Ecological Economics*, *32*(3), 465–479. https://doi.org/10.1016/S0921-8009(99)00119-6

Zeppel, H. D. (2006). *Indigenous ecotourism: Sustainable development and management*. CABI.

Becoming common plantain: metaphor, settler responsibility, and decolonizing tourism

Michela J. Stinson, Bryan S. R. Grimwood and Kellee Caton

ABSTRACT

As tourism scholars have turned to matters of reflexivity, epistemology, and ethics in research and practice, questions have been raised about how those in positions of privilege ought to situate their knowledge/power and take responsibility for enacting justice. In this article, we convey and engage the merits of *becoming common plantain* (i.e., *Plantago major*)—a familiar, low-lying plant species that has become "naturalized" to North America—as a metaphor that positions Settlers as constructive participants in decolonizing tourism and tourism research. By working through experiential, imaginative, and narrative moments associated with our tourism research on Indigenous-Settler relations in Canada, we illuminate how *becoming common plantain* works to foster Settler accountability for colonization and colonial complicity; place Settlers in relation (e.g., to land, identity, Indigeneity); and augment conceptualizations of justice as healing. The article contributes to theoretical and methodological discussions on the power of metaphor in sustainable tourism worldmaking and the relationships between tourism, justice, and Settler (de)colonization.

Introduction

How might one live? This is the question that Todd May (2005) uses to invite readers into his deciphering of Deleuze's philosophical project. May, a philosopher himself, argues that Deleuze's extensive, influential, and yet often dense body of scholarship simmers down to this provocative question. On the surface, *how might one live?*, may seem rather prosaic. But, as May asserts, the question cuts to the heart of recurrent concerns circulating within continental philosophy and of Deleuze's incessant task of articulating "how we might think about things in ways that would open up new regions for living" (p. 3). It's a question inviting the possibility that no matter how we are living, we might think and learn to live otherwise (May, 2005).

This article is not about Deleuze, nor is it about infusing erudite philosophical discourse into the study of sustainability and justice in tourism. This paper is, however, concerned with opening up meanings and possibilities for how we might live well together in this world: a place where the devastating effects of imperial and capitalist expansion are increasingly felt and witnessed, intensifying the urgency to shift into styles of collective living that might just give cultures, lands, waters, plants, and animals a fighting chance to survive. Living well together on this finite planet is a considerable but necessary moral weight that we carry, and one that is no doubt tethered to the fact that the three of us are doing quite all right in terms of relative social privilege and economic opportunity. Still, that weight exists. And it swells when we read the news; when we listen to the voices of those living and resisting various breeds of oppression; and when we contemplate relationships between justice, ethics, and tourism (Fennell, 2018; Jamal, 2019), our responsibilities as tourism academics to serving the public (Caton, 2012), and the worldmaking effects and potentialities of our social and scholarly endeavours (Hollinshead, 2007).

We can of course only take on so much in a single paper. Our aspirations and imaginings for how we might live well together are approached in this paper by exploring and composing a pathway for taking responsibility in the decolonization of tourism and tourism research. Taking responsibility in academia is often distilled into adhering to ethical standards and protocols governed by university research ethics boards (Jazeel & McFarlane, 2010). To be responsible translates into such things as ensuring the benefits of research outweigh the risks; securing informed consent of research participants and their communities; following approved templates for participant recruitment, data collection, and sharing feedback; and managing data in ways that safeguard confidentiality and anonymity. What constitutes responsibility as tourism researchers, however, cannot be fully captured in policy documents and compliance with protocol. The practice of research, after all, entails lived and negotiated circumstances that elude guarantees of ethical relationships even when a researcher receives ethics clearance from their host university (Grimwood, Doubleday, Ljubicic, Donaldson, & Blangy, 2012). Moreover, given what we have learned over the years about the neocolonial power relations marshalled by institutions like universities (Jazeel & McFarlane, 2010), we must be ever vigilant about how structures and systems establish and enforce norms that tend to suppress multiple and emergent understandings. What we need then, we argue, is a pluralistic orientation where responsibilities are enabled and enacted in relation to the particular contexts of peoples and places we encounter and collaborate with (Jamal, 2019; Sundberg, 2014).

Appealing to the power and pervasiveness of metaphor in constituting lives and worlds, our specific aim is to convey the merits of *becoming common plantain* (i.e., *Plantago major*) - a metaphor that we argue can usefully position White Settlers[1] as constructive participants in ongoing struggles of decolonization. In the wake of European colonization, Indigenous peoples in North America observed common plantain growing in the paths of Settlers. This invasive, low-lying plant species was named "White Man's Footstep" by several Indigenous communities and became appreciated for its medicinal and nutritional benefits. Potawatomi scholar Robin Kimmerer (2013) advises that common plantain offers important teachings for Settlers striving "to become naturalized to place" (p. 214). She explains that while common plantain is now omnipresent in her homelands, it is both useful and discreet.

> When it became clear that White Man's Footstep would be staying on Turtle Island, [Indigenous peoples] began to learn about its gifts. In spring it makes a good pot of greens, before summer heat turns the leaves tough. The people became glad for its constant presence when they learned that the leaves, when they are rolled or chewed to a poultice, make a fine first aid for cuts, burns, and especially insect bites. Every part of the plant is useful. Those tiny seeds are good medicine for digestion. The leaves can halt bleeding right away and heal wounds without infection. (Kimmerer, p. 214)

Unlike other immigrant plant teachers that model "how *not* to make themselves welcome on a new continent," common plantain's "strategy was to be useful, to fit into small places, to coexist with others around the dooryard, to heal wounds" (Kimmerer, 2013, p. 214). For Kimmerer, common plantain teaches Settlers that being naturalized to place involves giving your gifts, meeting your responsibilities, and taking care of the land in ways that honour ancestors and safeguard future generations.

Taking our cue from Kimmerer, we aim to show in this article how *becoming common plantain* opens meanings and possibilities for living tourism and tourism research in ways that are simultaneously accountable to and tuned away from colonization. In effect, *becoming common plantain* moves us towards thinking about, relating to, being with, and enacting a pluralistic approach to decolonial justice through our situated tourism endeavours, a reverberation of Jamal's (2019) deliberations on the need for flexible, context-sensitive approaches to justice and ethics in tourism. Drawing on—or better, working through—experiential, imaginative, and narrative moments of tourism research on Indigenous-Settler relations primarily in Canada, we illuminate how *becoming common plantain* works to (1) foster Settler accountability for colonization and colonial complicity; (2) place Settlers in relation (e.g., to land, identity, Indigeneity); and (3) augment conceptualizations of justice as healing. Our work here is bolstered by an intention to scrutinize our own practices: a careful looking-back that may diffuse into similar habits among students, peers, and colleagues (while attempting to normalize such meditation in ourselves). In turn, the paper complements the rising tide of tourism scholarship attentive to matters of reflexivity, epistemology, and ethics in research and practice (e.g., Belhassen & Caton, 2009; Grimwood, 2015; Jamal & Hollinshead, 2001; Wilson & Hollinshead, 2015), and to the more specific issue of how those in positions of relative privilege ought to situate their knowledge/power when endeavoring to take responsibility for enacting justice in particular contexts of Indigenous-Settler relations (Regan, 2010). We also seek to ground—in both a figurative and literal sense—theoretical and methodological discussions of sustainable tourism worldmaking and relationships between tourism, justice, and Settler (de)colonization.

Worldmaking and metaphors

Scholars working in tourism are increasingly drawing on the notion of "worldmaking". This idea originated in the antifoundational theory of Goodman (1978), who argued that "in order to understand how people exist, live, and create, it is more useful to map the particular 'worlds' they inhabit"—indeed, build around themselves—than it is to make claims about some sort of universalized world (Hollinshead & Caton, 2017, p. 15). If such a universalized world exists, we could never access it in a way unmediated by symbolic systems like language and culture. As Nietzsche contended, knowledge production cannot escape such systems; knowledge claims are simply metaphorical redescriptions of our objects of study (Babette, 1999). Thus, it makes more sense to work at the level of social construction than to plumb for ontological foundations. In particular, Goodman (1978) emphasizes the provisional nature of people's constructed "worlds". People are at least somewhat flexible in their ability to take on new understandings, priorities, values, and so forth (House & Howe, 1999), and so worlds are not made once and for all but are constantly in the process of revision—they *become*, rather than resting steady.

The ontological perspective of critical realism holds that social reality is composed of two distinct layers: the transitive domain and the intransitive domain. The transitive domain consists of human concepts, theories, and understandings about phenomena, whereas the intransitive domain consists of the structures and properties of phenomena which are largely enduring (but not immutable) and which exert constraints on human agency (Platenkamp & Botterill, 2013). An example of an object from the intransitive domain is a national border—a social construction, yes, but one which goes beyond the will of any individual to define or change, and which has an inertia, due to the layers of collective belief that underpin it: that citizenship derives from affiliation with a nation-state, that nation-states have sovereignty over particular geographical areas and the mobilities that can occur within them and at their edges, and so forth. This perspective helps us see how symbolic worlds like Goodman's, and metaphorical descriptions of those worlds like Nietzsche's, shade into tangible worlds that have material consequences for humans and other species. Thus, tourism scholars have tended to view worldmaking as a holistic concept, involving,

> on one hand, collectively or disjointedly produced ideologies, or mythologies, or imaginaries of spaces and cultures (which can be held by groups about others, by groups about themselves, or both), and on the other hand, material outcomes, such as changes to the built environment, changes in human behaviors and practices, changes in resource levels and distributions, and changes in life chances in terms of what real people are actually able to do and to be. (Caton, 2013, p. 342)

That imagined worlds can have consequences for how we function together materially is foundational to those who study metaphor as a window into human values and actions. Metaphors are ubiquitous. For many of us, they are one of the earliest linguistic devices we learn, and one we use most frequently. According to the Oxford Learner's Dictionaries (2019), metaphor is defined as a "word or phrase used to describe someone or something else, in a way that is different from its normal use, in order to show that the two things have the same qualities and to make the description more powerful" (n.p.). Put differently, a metaphor takes the basic form "x *is* y" and has the primary function of enhancing understanding by conceiving one thing in terms of another. For example, with the metaphor *time is money*, we understand time as a currency with particular value that can be exchanged or misused. While metaphors allow us to step away from the literalism of experience into a space of imaginative potential—one where we can process what is going on around us through rich frameworks of comparison and contrast—they also obscure other potential comparisons that could illuminate different aspects of reality (Lakoff & Johnson, 1980). In this way, metaphors are a highly powerful.

While some metaphors are unique individual interpretations, many metaphors are pervasive and familiar within a culture or community. How common are references to *love is a rose, the world is a stage, that classroom was a zoo*, or *nature is wild*? In addition, some metaphors imply relationships between things. When we say, for instance, *your research gets to the root of the problem* or *my research stems from hers*, we make an implied comparison between the activities of research and structural features of plants. Recurrent metaphors in a culture tell us about how that culture's members imagine their world and their place in it. In doing so, metaphors also say something about cultural values and how these get circulated. Importantly, metaphors don't simply reflect worlds; they actively shape worlds, socializing people into particular understandings of and expectations about reality (Lakoff & Johnson, 1980), in which particular values and courses of action make sense (Hollinshead & Caton, 2017).

Tourism scholars have developed a rich literature examining the use and utility of metaphor in tourism. Back in 2002, Dann and other sociologists published a volume centred on theorizing the tourist as a metaphor for the social world (Dann, 2002). In his introductory chapter, Dann observes that in much tourism theory the tourist is cast metaphorically as sightseer, stranger, pilgrim, or performer, while the institution of tourism itself is described in terms of international social fact, play, or imperialism. Major focal points in tourism analysis are also metaphorical. The study of motivation, for instance, is often framed in terms of "push" and "pull" factors (Uysal, Li, & Sirakaya-Turk, 2008), while tourism development is classically described to follow the trajectory of a life cycle (Butler, 1980) or processes of consumption and production that equate tourism as a performance (Doorne & Ateljevic, 2005). Metaphors are also used to prompt debate concerning conventional conceptions and approaches in tourism studies. McCabe's (2002) ethnomethodological analysis triggered important shifts from understanding tourism as an escape to understanding tourism as the everyday. Urry (2000), meanwhile, equated tourism with a fluid-like substance, insofar as it increasingly flows across borders and through various networks, changing and being changed as it goes. Like Bauman's (2006) "liquid modernity", in which the contemporary era is defined by its intense rate of change, with social forms and norms shifting too quickly for people to use them as a dependable frame of life-reference, contemporary tourism also does not stay stable long enough to solidify.

The prevalence of metaphors in tourism theory and conceptualization is complemented by their use in tourism practice. For instance, recent empirical research demonstrates that metaphors have a crucial selling function in tourism development (Adu-Ampong, 2016). One notable study by Jaworska (2017) reveals the significant use of metaphor in tropical destination promotion and marketing. Jaworska illustrates how metaphors are used to appeal to multiple sensory registers (e.g., vision, taste, imagination), thus amplifying tourists' desires to consume places and assisting in the ideological work and power relations that characterize tourism's colonial undercurrents. In the context of tourism education, Belhassen, Caton, and Vahaba (in press) considered metaphors as used by undergraduate students when reflecting on their internships, finding that these metaphors articulated the complexities of their experiences with industry and shed light on why many chose ultimately not to pursue careers in tourism. More immediate to our cause here, research has also engaged metaphor to illuminate systems of value that jar normalized understandings. As a case in point, Grimwood (2015) interprets relational values of *emplacement*, *wayfaring*, and *gathering* within the practices of Arctic riverscape tourists and Indigenous inhabitants. These metaphors refuse the sort of universal and divisive expressions of value encoded within perceptions and representations of Arctic tourism "nature". Echoing insights of Belhassen and Caton (2009) and Prince (2019), Grimwood demonstrates how metaphors surfacing through the creation of values-based tourism knowledge can advance our epistemological understandings and responsibilities in tourism research (see also Chambers & Buzinde, 2015; Grimwood, Stinson, & King, 2019; Tribe & Liburd, 2016). In short, whether they are used for capital production and soft power (Nye, 2004), to promote and stage destinations in ways that create particular imaginaries for target audiences (Salazar, 2010), or whether they are used pedagogically, to challenge dominant tourism discourses and practices and bring about social change, metaphors are the stuff of worldmaking in tourism.

Decolonization and enacting responsibilities

Tourism scholars are increasingly alert to multiple meanings and practices of decolonization (Chambers & Buzinde, 2015; Lee, 2017). As we've elaborated elsewhere (e.g., Grimwood et al., 2019; Stevens, Grimwood, & Caton, 2019), our relationship to decolonization is structured by settler colonialism as a particular form of colonial domination. Following Tuck and Yang (2012), we approach decolonization as a tangible, lived, and material process that entails returning autonomy over Indigenous lands to Indigenous peoples[2] and, consequently, the restoration of Indigenous self-determined lifeways. Such Indigenous futurity requires dismantling and rebuilding institutions, structures, systems, identities, and narratives of settler colonialism. Decolonization is, therefore, an "ethical imperative" (de Leeuw & Hunt, 2018, p. 4) that must be mobilized into concerted efforts to change tourism, ourselves as Settlers, the stories we tell, and, indeed, the research practices and language systems we deploy. These efforts must be ongoing. As Wilson and Hollinshead (2015) note, tourism scholars wishing to be actors in decolonization must be ever vigilant of both the shifting mundane governances of their setting and the "multiplex and perpetually recalibrating" (p. 34) identities and aspirations of those whose lives their academic efforts seek to touch—while also being always reflexive about themselves as authors of what comes to be viewed as knowledge. They refer to this mode of engagement as "critical multilogicality" (after Kincheloe & Steinberg, 2008).

How do we position ourselves today as Settlers in relation to the projects of decolonization? How do we responsibly situate ourselves to do decolonizing work in tourism and tourism research? Recognizing that both research and tourism involve a set of ethical relationships—perhaps most visibly when cross-cultural encounters and understandings are involved—tourism scholars have toiled with questions about positionality and reflexivity (Hall, 2011; Pritchard, Morgan, & Ateljevic, 2011; Wilson & Hollinshead, 2015). Discussions of positionality are regularly framed around an "insider/outsider" dynamic, the basic premise being that we negotiate multiple identities and degrees of proximity and distance from those places and peoples being studied and/or visited. Full insider status amounts to the dubious notion of "going native", while full outsider status resonates with desires for unblemished objectivity. Of course, given what we have learned over the years from feminist and poststructural scholars (e.g., England, 1994; Haraway, 1991), neither extreme is desirable or fully realized. Being reflexive—that is, the methodological "process of critical self-reflection on one's biases, theoretical predispositions, preferences, and so forth" (Schwandt, 2015, p. 268)—is often deemed a necessary attribute of researchers committed to responsibly negotiating these and other dynamics. Indeed, reflexivity is said to enable the analysis of how personal, intersubjective, and social processes influence an investigation and ultimately rests on one's awareness of self (Killion & Fisher, 2018). It has also recently been applied to identify with the agencies and capacities of tourists as they grapple with the ethical implications of their travel decisions and behaviours (Mkono, 2016).

When it comes to decolonizing tourism, however, we are unconvinced that our current repertoire of concepts and practices offers us the tools we need. After all, it has been said that decolonization requires us to consistently imagine relationships differently (Lowman & Barker, 2015). Our objective, therefore, is to engage *becoming common plantain* as an ontological metaphor that opens new meanings and possibilities for Settler responsibility in tourism and tourism research. Ontological metaphors are those that enable us to understand some aspect of our experience as an object or substance (Lakoff & Johnson, 1980). As one illustration, Lakoff and Johnson point to the way ideas are often understood in terms of plants:

> His ideas have finally come to *fruition*. That idea *died on the vine*. That's a *budding* theory. It will take years for that idea to *come to full flower*. He views chemistry as a mere *offshoot* of physics. Mathematics has many *branches*. The *seeds* of his great ideas were *planted* in his youth. She has a *fertile* imagination. Here's an idea that I'd like to *plant* in your mind. He has a *barren* mind. (p. 47)

Lakoff and Johnson (1980) explain that it is our lived experiences with physical objects that provides the basis for viewing events, activities, emotions, and such as entities and substances. Accordingly, we want to illustrate how *becoming common plantain* makes visible and organizes certain qualities in the experiences of tourism and tourism research that we feel our conventional conceptual system does not make available or actively obscures. More specifically, *becoming common plantain* might just enable Settler scholars and tourism actors to better interrogate their subjectivity, reflexivity, and responsibilities such that we can figure out how to finally stop colonizing (Grimwood et al., 2019; Lowman & Barker, 2015).

The following sections of this article are an experiment, but also an enactment (Stewart, 2007). They are full of potencies, pieces of possibility, a move to both wonder about and wander through a cascade of potentialities contributing to work on the power of metaphor in tourism, and its subsequent worldmaking capacities (Belhassen & Caton, 2009; Grimwood, 2015; Hollinshead et al., 2007). Descriptively speaking, these sections function as a recollection of ecological encounter with common plantain, as well as a process of *becoming*—a play at the possibility of new, creative existences born from a constant vigilance tuned toward the unexpected (Deleuze & Guattari, 1987). They draw on literature, moments from tourism research (some of our own, some of others, some only hypothetical), and aspects of common plantain's ecology to suggest alternative and potent (yet inevitably partial) ways of engaging our work in tourism responsibly. To these ends, we echo Prince's (2019) suggestion that "new discursive anchors to speak of tourism are important to enable researchers and practitioners to retain a conceptual ability to imagine and apply new worldviews that would make local experiences matter" (p. 793).

Below, the narrator both *is* Plantain and *engages* plantain as she—we have already met her (Grimwood et al., 2019)—reworks herself by considering accountability, relationality, and justice-as-healing. In this sense, both Plantain (the narrator) and plantain (the plant) are denoted with the pronoun "she". By bearing the name of Plantain, the narrator seeks to inhabit the qualities of her namesake as she moves through spaces of encounter, adaptation, and doubt. We are unsure if this narrative *works,* but it is our intention to show the practice of its *working—the being*—a rough rendering of what is "already happening" (Morton, 2018), and what may or may not "continue happening" in becoming common plantain. Stylistically, our presentation stitches italicized sections recounting moments of collision with common plantain and non-italicized sections fused with additional literature and reflective commentary. Our intention here is not to present a series of story/evidence couplings in which prose is subsequently "backed up" with literature and data—this type of representational choice would ultimately reinforce a false binary between being/knowing and research/practice (Berbary, 2015). Instead, our storying is kin to creative analytic practices that disrupt traditional research conceptualizations of truth, validity, reliability, and authority (Berbary, 2015), while simultaneously attempting to be generative.

As we offer various glimpses into the many fertile offshoots that can come from a single walk around the block, we urge readers to consider how everyday encounters may allow them to rethink collisions of tourism, research, and story in novel and unconventional ways. With this in mind, the following narrative is honest, awkward, and uncertain: its tentative vagueness is a functional choice to leave room for the narrative to surprise us, linger, and work on its own. As Terry Tafoya explains, "the closer you get to defining or explaining an idea, the more it loses its context" (as cited by Wilson, 2008, p. 99).

Fostering settler accountability

Being common

I am out walking. It's Tuesday, maybe Wednesday. There is a loose, liminal quality to recent time for me, as I spend many hours cooped up with a computer. Despite my technological interface, I feel disconnected—from an ecological self, from any responsible reconciliatory work. I feel strange and uprooted, restless and reflective. Why am I trying to find plantain online? I know her shape and form, her broadleaf disposition, her histories hiding in plain sight. I will stumble upon her, because she is everywhere.

Common plantain is low-lying and perennial, with wide, lush leaves interspersed with tensile veins. She comes back. Spurring from a rosette, a cluster, with intermittent stamens and stems and central stalk of seeds. You would know her if you saw her. Despite common plantain's frequency and capacity to live almost-anywhere, we hardly ever recall any of her many names. So, there is this: becoming common plantain means recognizing and taking ownership of our identity—a humble accounting for who and what we are. Some of this is a practice of reflexive and reflective naming, realizing that we may assume a Settler identity while also being friend-parent-student (or whatever diversity of descriptors), that these identities are partial but important, that they are flexible and invaluable parts of what hold us up.

Kimmerer (2013) explains that practices of naming are a first responsibility, but warns that designating a name does not discharge the responsibility of learning more about that which is named (including ourselves). Although common plantain exists with various nomenclatures—*Plantago major,* White Man's Footstep—there is a certain frequency in her designation as *common,* one which is useful to the responsibility of normalizing Settler identity. Becoming common plantain assumes we have a responsibility to our naming—a relationship built *through* our naming (Kimmerer, 2013)—and positions this as an active and ongoing choice to both explore and commit to an identity that accounts for its violent, harmful, and complex roots. This opposes a "positional piety", a declarative positionally of moral authority often used as a substitute for engagement (Lowman & Barker, 2015).

Our insistence on normalizing Settler identity in tourism returns to the persistent need to interrogate work that positions Indigenous nations, knowledges, and being as *other* (Regan, 2010). In other words, it means tourism studies must decidedly interrogate *the Settler problem* (Grimwood et al., 2019). Taking responsibility for decolonization must become a common, *daily practice* in tourism, such that efforts to disrupt or dismantle colonial systems are no longer seen as radical or fringe, but rather commonplace and consistent. The undertaking here is not about

Figure 1. Common plantain.

empty-gesture territorial acknowledgements or pithy collective action (Whyte, 2018), but genuine and active consideration of how we might build resistance and healing into our communities, families, friendships, and relationships with the place(s) we live and visit. As *daily action*, becoming common plantain means we are resolved in doing our *homework* (Spivak, 1990)—situating self in contexts of where we are and what we know (Sundberg, 2014). We must own ourselves as Settlers.

Interrogating roots

Here she is! Common plantain (Figure 1). Walking out, I didn't have to go far—I found her clustered in the cracks of the path down the street from my home. I crouch down and note how green everything is. It's spring. I experiment. I lean down to pull plantain from the ground and immediately find resistance. Despite plantain's feathery roots she is difficult to isolate. I pry her carefully from a concrete seam and shake dust off her matted roots, bundled, innervated.

Common plantain's roots are fibrous, with a smaller, central taproot. They hold fast to the top few inches of soil. Once emplaced, common plantain stretches and bursts its way through assorted surfaces, showing up unharmed in driveways, lawns, parking lots, parks, and fields. Common plantain, with a tenacious grip through tenuous fibres. Common plantain, with a small and silent shaking-up. In becoming naturalized, common plantain has worked to interrupt structures and foundations—you'll see her reaching basal rosettes out from fractures in city sidewalks and driveway asphalt. Common plantain is subtly disruptive to settlement at large, while simultaneously active in holding together rich, fertile soil underneath; enriching endemic ecosystems while being unobtrusive. This dual practice—a destabilizing of colonial institutions, a solidarity with and fostering of Indigenous livelihoods—is imperative for decolonial work (Tuck & Yang, 2012).

The practice of locating ourselves in relation to decolonization can stem from interrogating and utilizing Settler memories and stories—our roots—however tentative or shallow they may be. As common plantain teaches, even shallow roots are valuable to grounding us, especially when they subsequently work to dismantle dominant and oppressive structures, forces, and ideologies. Taking responsibility in tourism means acknowledging the dispersion of our roots—what they have interfered with and have possibly run through. In our research practice, this means both contextualizing our complicity in the historical, political, and material violences of settler colonialism (Chambers & Buzinde, 2015), but also attending to our own personal narratives of owning/rejecting Settler identity and their recurrently harmful effects.

Roots are also memories, familiarities, relationships, identities, feelings, and stories. Some are stronger than others, but they all feed us, anchor us, and allow us to grow. As we engage responsibly in tourism, altering our rooted stories—refusing dominant narratives—has productive, disruptive potentialities (Grimwood et al., 2019). Perhaps, instead of attending to rootedness as an independent, arboreal quality, we can encourage possible alternative connections, assemblages, and fertilities of common plantain. The decolonizing settler story published elsewhere (Grimwood et al., 2019) is designed to cultivate growth of new, different roots—perhaps rhizomes. Rhizomes avoid singular stagnancy, connecting multiply and necessarily in strange new ways. Indeed, even while common plantain relies on roots, its holistic function always shapes, and is shaped, by rhizomatic relations with something else outside: wind, water, air, soils, animals, us (Deleuze & Guattari, 1987). What might flourish when roots are undone? As an illustration from our own practice, we are participating in a collective memory work study designed to unsettle personal stories of travel, particularly those associated with meanings of "Canadian" and meanings of "Indigeneity" (Grimwood & Johnson, 2019). The process of recalling, writing, sharing, and restorying these is structured to locate and rework harmful narratives and tropes like *terra nullius*, settler entitlement to Indigenous knowledge, and representations of Canada as harbinger of peace. Like common plantain, our roots are tools used to dismantle colonial systems and structures and to fertilize futures outside of colonial control.

Placing settlers in relation

Sharing soils

It's strange how close plantain is, how everywhere she is. To me, to the ground, to this soil—thin, dusty, flecked with flint. And here—rich, riparian. Rambunctious! It makes me wonder where else she might have been hiding, or who else may be close by. I plunge my fingers in to search the soils around her, knowing certain stands of plants tend to be proximal—neighbours share tendencies to grow in specific soils. I remember learning that where I find yarrow I will indeed find other plants of the same nature. Who else grows here? Can I grow here?

Becoming common plantain means that Settlers place themselves (and their stories, systems, and institutions) *in relation* to Indigenous peoples. As common plantain was never a species endemic to North America, it is by definition a trace of colonization. Kimmerer (2013) explains that despite common plantain's colonial roots, it has lived for generations as a good neighbour, coexisting with native plants such that it is now integrated and welcomed as fully "naturalized"—plantain and other plants grow from the same dirt, the same waste lots, the same damp loam. This shared-soil doesn't mean that common plantain "becomes Indigenous", as seeking to do is an altogether parasitic position (Whyte, 2018) (something Kimmerer (2013) associates with the vine *kudzu* and its propensity to choke out native plants). Instead of becoming-*like* those Indigenous to many spaces and lands, common plantain teaches us to focus on becoming-*with* our surrounding plants, animals, lands, stories, and peoples. Becoming naturalized means sharing soil: being grounded, couched, present, close-by, and resilient. It means being influential, respectful, working in tandem, joining in fertilizing and propagating a vibrant and vital ecosystem of multi-species, multimodal inter-workings.

In tourism research and practice, becoming common plantain is about being relational ontologically, epistemologically, methodologically, and axiologically (Wilson, 2008). All of this is intertwined—as relational practices are likened to be—as practices of being and researching (and doing so responsibly) are decidedly inseparable. Relational accountability, in these ways, must remain intuitive, ongoing, immersive, and active—relationships do not solely shape reality, relationships *are* reality (Wilson, 2008). This relational being, again, does not mean blindly assuming Indigenous ontologies and ways of knowing, but contextualizing, owning, and recognizing our own knowings and beings as Settlers (Roisek, Synder, & Pratt, 2020).

Smith (2012) explains that methodologically, decolonizing efforts need not outright reject Western theory or methods, but rather attempt to place Indigenous knowledges in relation to Western knowledges, without ascribing false "ownership", temporality, or supremacy. We may now share the same intellectual soils—and they may be enough to sustain us together—but we must not gloss over, simplify, or be hasty to understand complex, long-lived processes of being or knowing (Tuck & Yang, 2012). Like plantain living alongside milkweed or vervian, we should take the time to discover for ourselves, in our own context, which theories and ideas feed us, can coexist with ours (Wilson, 2008). Indeed, Western theorists like Latour, Barad, and Bennett, among others, often speak of relational, connective agency, which has marshalled concepts and methodologies increasingly mobilized within tourism (see e.g., van der Duim, Ren, & Jóhannesson, 2017). Roisek et al. (2020) remind us that these ideas, theories, and lifeways are not without their owing a debt to contemporary Indigenous thinkers and the rich and robust history of Indigenous thought concerning relational knowing and being (Todd, 2016). In sharing soils, we enable cross-fertilizations, like Bell, Instone, and Kathleen (2018) pulling from the Indigenous Australian concept of Dadirri to articulate engaged witnessing, or Sundberg's (2014) building on Zapatista practices of performative social transformative walking. In sharing soils, we can learn from one another.

Proximity and practices

I have a memory of plantain, clustered and close, lining the driveway of my new home after a move, the very first time we met. The summer heat that year was unrelenting—mirages of water populated my vision, strange shimmering patches appearing on the baking asphalt as grassy front yards lay parched and browning. Between cracked slabs of this molten ground-cover I first encountered plantain, a persistently vibrant intrusion alongside dandelion, goldenrod, and thistle in an expanse of concrete. So many weeds, I frowned, piling moving boxes on my front lawn. I hated moving. I hated feeling lonely. Later that week, a friend came by to help me tidy the yard. "There's plantain", she said, bending over to pull off a leaf. She offered it to me to chew and I hesitated before crushing the bitter leaves in my mouth. The sun seemed to recede; I smelled dirt and loam as cars tested their brakes at the corner. Sitting on the retaining wall by my driveway, we stopped our work and felt sunburn cross our cheekbones, nodded when a passing cyclist waved. "Met your neighbours yet?" my friend asked. I shook my head, then felt the acrid reminder of plantain on my tongue. But close by: St. John's Wort, summer's blood, yellow-petaled and buoyant; skunky catnip that lingered, dusty on the air; wild carrot, rampant and rough, too-close to poison hemlock for me to trust; garlic mustard, invasive and frustrating, a problematic pesto.

Common plantain is literally grounded: she is trampled, stepped-upon, dragged in the footprints and paths of colonizers. Her spatiality renders a dwelling, an emplacement, a taking-of-space wrangled and rendered by fitting into cracks and corners, by poking around fences, by creeping close across the ground. Knowing her, following her puddles of possibility, requires that we get low to the ground. While common plantain can subsist in many places, her constitution and fortitude are grown in togetherness with trash, seed-pods, weed-killer, glass, and other noxious substances.

Engaging responsible tourism practices means recognizing ethical practice is an aspect inherent to physical spaces that emerges relationally, or proximally (Raghuram, Madge, & Noxolo, 2009). Or, as Wilson (2008) explains, "we cannot remove ourselves from the world in order to examine it" (p. 14). Returning to Kimmerer's (2013) concept of "becoming naturalized", we must live with and within the land responsibly, care-fully, and deeply, while recognizing the diversity and persistence of the active participation in tourism *by the land itself* (Waitt, Figueroa, & McGee, 2007). We must be close-to, nudged-by, together-with, and responsible-to the many moral terrains of tourism, research, and their situated spaces (Grimwood, 2015; Waitt et al., 2007). As Marker (2018) explains, "recognizing and acknowledging the ways places contain spiritual and mythic energy is a methodological position that opens possibilities for both expanding consciousness and building relationships with Indigenous people" (p. 460).

Drawing from research across many modalities of tourism, we suggest that becoming common plantain engages us in localized, proximal, and material becomings that recognize the ongoing vibrancy of present, responsive agencies (Barad, 2007). These "new materialisms" allow collisions with other plants, strata, things, and dirt—they embed us and our tourism relationships in the very messy matters of earth and land. There are many examples of research that weave together new materialist, relational ontologies: ideas of proximity and dwelling (Prince, 2019; Rantala, 2019), being ecological (Morton, 2018), and ecological vibrancy (Evers, 2019). Grimwood is currently exploring a multi-sensory, multi-modal tracing of cranberries through both an oral history and an Actor-Network theory inspired inquiry into the histories and legacies of the Iroquois Cranberry Growers marsh on Haudenosaunee reserve lands in Ontario's Muskoka area. Tracing cranberries might mean engaging different or surprising ways of understanding tourism spaces, or learning and caring about land relations through a wandering of methodologies, tourism experiences, or research practices (Kuntz, 2015). While Smith (2012) outlines twenty-five projects of Indigenous research efforts that are transformative and productive, we too might consider what alternative tourism projects may come from proximal, dwelt, or grounded efforts. Local, ethical, or relationally responsible tourisms could mean theorizing encounters with tourism objects, dreams about tourism journeys, engaging with pipeline and rail blockades and direct action, planting native species, or just staying home. What tourism initiatives might we find in our own backyards?

Justice as healing

Sowing seeds

It's strange, my preoccupation with locating her flowers, as if a bloom is the only visual signifier of growth. Flowers are easy, but they are the result of much hidden work. I run my finger and thumb along plantain's centre stamen, pulling them away to release a chestnut-colour smudge of seeds. Strange. Sticky. There are no bugs to pollinate her, and yet…

Common plantain does not show off her flowers. In fact, until close inspection, it is unclear if common plantain even *has* flowers, let alone petals of such delicate and humble constitution. Alongside her strangely evasive flowers, common plantain carries capsules of minute, bitter seeds along central stalks, numbering roughly 20,000 per plant (Samuelsen, 2000). Common plantain reproduces in two ways: her seeds dispersed almost-invisibly by the wind; or, in humid conditions, her sticky seeds attach themselves to animals and people on the move (Samuelsen, 2000). Common plantain's reproduction is a rather concealed process.

Affects—a dynamic relationality between bodies—move similarly, with seemingly "dangerous causative flickering: magic" (Morton, 2018, p. 157). A productive invisibility, affects are circulative, transmissive, and contagious (Pile, 2010) and, indeed, "sticky" (Ahmed, 2010). It is hard to determine the dispersion of affects or consider how they move through and around human and non-human bodies; sometimes, encountering affects feels disconnected, as potent disruptions diffusing upwards with no apparent material or physical trace. Yet affect—if we follow Thrift (2004)—can be "harnessed" to create differing political and ethical effects, especially in tourism spaces; they can *do* and *become* many things (Singh, 2018). Affects bump into, bounce off, and are generated by a wild range of bodies—they mingle about in the world like common plantain's seeds, latching onto people and things, planting possibilities, being carried around and lodged into even the smallest cracks or openings. Much like common plantain makes public its germination habits, attending to affects in tourism "challenges assumptions that embodied feelings are a private matter, or that they simply begin within—and belong only to—individuals" (Buda, d'Hauteserre, & Johnston, 2014, p. 103).

And like common plantain's regenerative abilities, affects are productive, despite the invisibility of their work. As much as their processes are more-than-representational, you can *feel* their force (like sunlight, like wind) as they stick to you—the resulting change is generative, nurturing, and transformative, even with uncomfortable feelings like fear, anger, and shame. How might we be similarly productive in decolonizing tourism research and practice for enhanced justice? Drawing on sensory ethnographic research from Darwin, Australia, Lobo (2014) explores visceral histories of white occupation, and the affective resistance of Indigenous presence at a public beach. Lobo (2014) explains that affective energies, as opposed to intentional ethical acts, interrupt "force fields of whiteness", especially those relating to land. In tourism spaces, this means allowing for empathy, emplacement, and transformative change in ways that are vulnerable, open, and welcome, both *as researchers* and *as tourists* (Singh, 2018). Common plantain, too, is insistent on growth and change, sending out tens of thousands of hints at germination, hoping that maybe, just maybe, one will stick, planting within us seeds of care that perhaps could—unexpectedly—nurture and grow richer, just worlds (Lobo, 2014). To echo Wilson (2008), research (and living!) is done right when it changes you as a person.

Researching care-fully

Reaching into my pocket to retrieve my phone, I pull the back of my hand over the rivet of my jeans. It's sudden—blood. I bring hand to mouth, then spit to the ground. Like some synchronic thing, I pause in disbelief at the opportunity. I bend to the ground and, slowly, I pull a leaf from plantain and rub it between my fingers, a slippery slime. She bleeds too. Soothing. I press the leaves to my skin and pass the paste toward my bleeding. Stop, sweet. Staunch. It is cooling. Calm.

A wealth of literature speaks to the healing capacities of common plantain—we have learned about her soothing mucilage and cooling temperament from friends, nature schools, long-forgotten books on dusty shelves, academic writing, and folk and Indigenous healing traditions. The Swedish name for common plantain is *groblad,* or "healing leaves"—common plantain's healing capacities are built into many of her designations (Samuelsen, 2000). Brill and Dean (1994) and Kimmerer (2013) detail her medicinal and healing properties, which range from aiding digestion to halting infection to stopping a bleed. Plantain is antibacterial, anti-inflammatory, analgesic, antioxidant, and a delicious source of sustenance (Samuelsen, 2000).

Justice, like healing, is tricky. There is no single path or localized definition toward either, and both are often fumbled and dropped, marked by sabbaticals of further harm, interrupted by the reopening of old wounds. Healing is never linear—it is a holistic event both erratic and plain, and means different things to different people at different times. Becoming common plantain demands that we resist doing research, doing tourism, in ways that produce realities which are harmful to Indigenous nations and individuals, to the land, even to ourselves.

Like common plantain, our practice of research and tourism must be flexible, comprehensive, and holistic. When we offer up to the process every part of ourselves, we must be hardy and adaptable. We must be discreet if needed (e.g., those of us doing participatory research), but we must also show up. We must be responsive to calls to practice justice and care in a variety of sometimes-unclear means, but consistently offer a wealth of healing tools and modalities—space, time, money, resources, humility, presence, absence, and care. This offering is a contextual practice and a continual process of trying to reduce harm in whatever ways necessary—by not seeking stories of trauma from Indigenous peoples in the pursuit of Settler agendas and desires, by not speaking to Indigenous nations if research protocol is in any way tokenizing, by *not going* to certain tourism spaces, or *not doing* certain types of research.

Promoting care and healing means enabling certain types of research practices or protocols that decentre Settler desire and needs. This may further mean not publishing, or even not doing research in recognizable ways (Smith, 2012). As an alternative to traditional "drive by" ethnography, Neilsen and Wilson (2012) advocate for "Indigenous driven" research priorities such as Indigenous self-determination, control, co-authorship, ontologies, and methodology, with the important corollary that tenets of this work can also be engaged by non-Indigenous researchers. An attempt at employing such a decolonizing methodology is outlined in Grimwood et al.'s (2016) chapter, describing a community-based research partnership and planning processes with the Lutsel K'e Dene First Nation in the Northwest Territories, Canada. This process resulted in a formalized research agreement, place-based research protocols, and community-focused priorities and outcomes.

In becoming common plantain, entanglements of justice and care also mean encouraging healing by staying out of the way, or being discrete. In our own research, this has meant renegotiating both the intention and the execution of a project on Anishinabeg, Huron-Wendat, and Algonquin First Nations land, in the provincially regulated Algonquin Provincial Park. The project considers the raising of a totem pole carved by Whitney, Ontario area resident and Algonquin man Dan Bowers, which was gifted to the park in the spirit of peace and reconciliation in 2015. In this context, we have been tracing the material histories of this totem, its connection to Eastern white pine political ecology, and its capacity to affect reconciliatory change in Settler Canadians. Initially, we had intentions of talking to the carver, as well as ill-defined Indigenous stakeholders—intentions that were ultimately crafted by ourselves in order to further specific, individual research agendas. While drafting a clearance form to conduct research in Algonquin Park, we realized our efforts up until that point were very possibly harmful, and contributing to discourses that positioned knowledge as something we were entitled to obtain from local Indigenous nations (Jones & Jenkins, 2008). This not only places an additional burden on Indigenous peoples to inform and educate Settlers, but also perpetuates a history of viewing Indigenous knowledges, stories, and feelings as commodities to be extracted for Settler research agendas (Latulippe & Klenk, 2020). In our case, not only did we not engage in any consultation with the Algonquins of Ontario about possible projects or nation research focuses, we also misconstrued the importance of apprehending Indigenous voices, time, and energy for a project that considered *Settler responsibility*. This project is being renegotiated to instead position ourselves and our Settler identities as the objects of interrogation, with a contextualized reliance on the *already available* wealth of Indigenous voices speaking about Algonquin Park and reconciliatory processes. Becoming plantain means knowing when to step back and away.

I drop to the ground again and find myself on my knees. It's Wednesday—this time I am sure—and I find myself sitting on the sidewalk three blocks from home, tending to a tendency to stray. I am uprooted; I want to remain grounded. I am parched; I would like to be filled. The sun drifts with a lazy lull as clouds rush to cover me. The creases of my hands are filled with traces of dirt, and they embolden the lines that may tell my future. I am finding that being disconnected might be an impossibility. Instead: a connection, a responsibility. I am finding that plantain is (here with) me, with and without notice—now that I have met her once will I ever be able to turn away?

Conclusion

When we contemplate justice as contemporary tourism scholars working on colonized territory, we find little help from the sorts of theories of justice typically drawn on in works on ethics in the tourism literature. These theories tend to descend from a white, European, masculinist tradition that centers the individual and their entitlements and responsibilities within the context of a political entity, such as a nation-state (e.g., Rawls, Miller, Nussbaum). Such liberal democratic theories of justice may work well within the context of a self-determined state, but they offer little guidance to help us navigate across political entities, much less when we are working as citizens in an entity whose very existence is founded on the dispossession and cultural destruction of others. Other prominent theories of justice make moral rather than political demands (e.g., Singer's utilitarianism, Levinas' immediacy of responsibility), but they tend to be universalistic. To ask what we owe other human beings categorically, simply by virtue of their being human, yields a very different answer than to ask what we owe in particular contexts, such as settler-colonial citizenship. To ask how we should be toward and with one another generally does not help us to know how we should be toward particular others with particular histories, traumas, present experiences, and dreams. As such, to consider how we might help to enact justice in the places we live, we have turned not to abstract theories but to the direct ecology in which we are embedded, using imagination rather than analysis to explore, metaphorically, how we might be, under the guidance of a teacher from the more-than-human world.

Indeed, this paper has aimed to infuse *becoming common plantain* into the language of tourism worldmaking, illuminating along the way its promise for guiding Settlers in taking responsibility for settler colonialism and enacting decolonial justice. In this paper, *becoming common plantain* emerges as both a productive and disruptive force of change, renewal, and relational responsibility that asks us to consider new life ways branching and stemming from those in which we already dwell. Returning to May's (2005) question *how might one live?*, *becoming common plantain* teaches us that emergent possibilities for change are abundant, and that potentialities for living otherwise can be found by attending carefully to the places in which we already find ourselves.

Attending to the landscapes, plants, and ecologies with which we co-constitute our lives, research, and tourisms reminds us that it is not only human beings that have agency—many matters are agential, including plants, and they have stories and histories that we can learn from (Caton, 2018). While scholars in tourism are taking up new materialisms, posthumanisms, and more-than-human wanderings, acknowledgement of the relational and responsible teachings of plants in tourism remains limited, though Cohen and Fennell (2019) have recently written on our ethical responsibilities towards plants in tourism and Caton (2018) has welcomed trees to teach us about what a moral gaze in tourism might look like. We can move things further if, by following Oberndorfer, Winters, Gear, Ljubici, and Lundholm (2017), we do more to recognize plants as vital facilitators of cultural practices, values, and community well-being, and as storytellers that convey different kinds of relationships to cultivate.

In learning from common plantain, we have awakened an ontological metaphor for decolonizing tourism that we find is akin to what Morton (2018) describes as *living* ecological knowledge. Morton (2018) argues that,

> It seems to be not enough just to know stuff. In fact, it seems like "just knowing stuff" is never just knowing stuff ... [but rather] a way of living things too. And knowing that there is a way of living things implies there could be other ways too. (p. 11)

Living ecological knowledge shies away from conventional environmental discourses that place us in a traumatized paralysis of irresponsibility—our state of being with and in our environment is often positioned as insurmountable crisis wherein we try to cope with our fear by trying to *do something*, anything. Instead, living ecological knowledge requires situating ourselves in the middle, in continuous relation with our environment (and our communities, ecologies, politics, selves ...) such that our many identities, practices, and actions are already part of the systems we are trying to responsibly respond to. But truly living ecological knowledge is difficult; like *becoming common plantain,* it requires that we don't immediately—and maybe shouldn't immediately—know exactly what to do (Morton, 2018). In this way, ecological awareness is awareness of unintended consequences and their freeing, flirtatious futurism, ridding us of our burdensome want to control what we cannot, and instead allowing us to responsibly relate to what we can. *Becoming common plantain* is thus an experiment, an attempt towards ecological awareness: towards the art of living ecological knowledge. Making messy movements from the middle—being what we already are *but more so.*

Bringing this once again into discussions of justice, *becoming common plantain* orients us towards the sort of relational and contextualized approach called for by Jamal (2019). Emphasized above in our narrative moments are understandings situated in relation to place and historical, political, and social contexts. Common plantain teaches us about respecting difference, making space for others to thrive, and offering our gifts to aid and heal, all the while learning to take account of our past and emergent futures. These are tenets in Jamal's (2019) vision for a pluralistic and flexible approach to justice in tourism, a vision that seems really quite necessary for decolonial projects where end points are uncertain and always in process of being created (Grimwood et al., 2019).

One final aspect to the metaphor of *becoming common plantain* warrants comment, and that is the ontological concept of *becoming*—an ethical position of radical, responsible subjectivity (Braidotti, 2013). While we maintain that this paper is not about Deleuze, there is an ironic circling back to this consideration of *becoming,* and how it relates to our committed relationship with metaphor. In the ongoing practice of *becoming common plantain* the emphasis rests on encouraging a radical, ethical subjectivity positioned as part of the inseparability of natures and cultures (Braidotti, 2013). This is a responsible subjectivity which is emergent and productive; which is generative and unavoidably material; which welcomes new ways of *being*. But indeed, *becoming* is never-complete, never-fixed, and always up-for-grabs—the newness of *being* must be made constantly-new, and remains forever-negotiated in relation as something entangled, emergent, and in flux. What Métis scholar Sebastian De Line (2016) considers *Niw_hk_m_kanak*, or the Cree/Métis ceremonial acknowledgement and philosophical proposition *all my/our relations*. Haraway might call it *diffraction*: "marking differences from within and as part of an entangled state" (Barad, 2007, p. 90). This is a changeable, workable subjectivity, responsible to living ecologically, and held in relation. This sentiment, too, is a working metaphor—Barad (2007) pulls the notion of diffraction from attention to phenomena of light and its behaviour as rippling, bending, and multiplying of waves depending on the interference of objects and other waves. Through diffraction, our practices *matter,* and our visions for justice *matter*: we can both take responsibility for, and make a difference through, our *being* and *becoming* in the world.

Taken alongside *common plantain,* the modification of *becoming* situates tourism research as a process of inquiry with an embedded ethic of responsibility to the human and more-than-human relations within which we are entangled (Roisek & Snyder, 2018). In this way, the metaphorical nature of *becoming common plantain* has certain inherent risks—while the power of metaphor lies in its ability to encourage certain abundant futures, understanding "one aspect of a concept in terms of another… will necessarily hide other aspects of the concept" (Lakoff & Johnson, 1980, p. 10). In attending to only certain potentialities of *becoming common plantain*, we ultimately shape the understandings that can arise from such open-ended concepts as bountiful, vital, transformative, and responsible. Of course, this *becoming* could also result in the reinstatement of paralysis, in experiences of alienation, in fear, in further harm at the hands of Settlers. We could fail at our attempts to be accountable, to share fertile soil, to engender change in tourism and in ourselves. Metaphors—much like stories and research—provide us only a partial understanding of responsibility and relationality. Our aim is that our practice herein of "reimagining the possibilities within experience" (Roisek & Snyder, 2018, p. 8) will result in Settlers taking responsibility for worldmaking practices in tourism. And perhaps, just maybe, if *becoming common plantain* or other similar metaphors resonate within tourism and tourism research, our worlds and ways of being might just shift as well.

Notes

1. Lowman and Barker (2015) position Settler identity as a situated, place-based, and processed-based identity that connects a common group of people and practices. Identifying as Settlers, they argue, is a choice to "foreground issues of agency, responsibility, and accountability with respect to Indigenous nations" through a relational, embodied analysis (Lowman & Barker, p. 14). Although Whiteness is not a universal aspect of Settler identity, we understand it as a crucial system of value that consistently intersects with settler colonialism through eliminatory, extractive, and genocidal means (Yusoff, 2018).
2. Indigenous peoples in Canada include diverse First Nations, Inuit, and Métis peoples. Recognizing identities as fluid and hybrid, our aim is to respect those individuals and communities that self-identify as "Indigenous", especially in and where Indigenous identities mean "being oppositional to colonization" (Lowman & Barker, 2015, p. 14).

Disclosure statement

No potential conflict of interest was reported by the authors.

Funding

This work was supported by Ontario Ministry of Research, Innovation and Science [Early Researcher Award, project number ER15-11-182].

ORCID

Michela J. Stinson http://orcid.org/0000-0002-8449-4049
Bryan S. R. Grimwood http://orcid.org/0000-0003-1555-7541

References

Adu-Ampong, E. A. (2016). A metaphor analysis research agenda for tourism studies. *Annals of Tourism Research, 57*, 248–250.

Ahmed, S. (2010). Happy objects. In M. Gregg, & J. G. Seigworth (Eds.), *The affect theory reader* (pp. 29–51). Durham, NC: Duke University Press

Babette, B. (1999). Nietzsche's critical theory. In B. Babette & R. S. Cohen (Eds.), *Nietzsche, theories of knowledge and critical theory* (pp. 1–14). Dordrecht: Kluwer.

Barad, K. (2007). *Meeting the universe halfway: Quantum physics and the entanglement of matter and meaning.* Durham, NC: Duke University Press.

Bauman, Z. (2006). *Liquid times: Living in an age of uncertainty.* Cambridge, UK: Polity.

Belhassen, Y., & Caton, K. (2009). Advancing understandings: A linguistic approach to tourism epistemology. *Annals of Tourism Research*, *36*(2), 335–352.
Belhassen, Y., Caton, K., & Vahaba, C. (In Press). Boot camps, bugs, and dreams: Metaphor analysis of internship experiences in the hospitality industry. *Journal of Hospitality, Leisure, Sport, and Tourism Education*. 1–11. doi:10.1016/j.jhlste.2019.100228.
Bell, S. J., Instone, L., & Kathleen, J. M. (2018). Engaged witnessing: Researching with the more-than-human. *Area*, *50*(1), 136–144.
Berbary, L. A. (2015). Creative analytic practices: Onto-epistemological and theoretical attachments, uses, and constructions within humanist qualitative leisure research. *International Literature Review*, *2*(4), 27–55.
Braidotti, R. (2013). *The Posthuman*. Cambridge, UK: Polity Press.
Brill, S., & Dean, E. (1994). *Identifying and harvesting edible and medicinal plants in wile (and not so wild) places*. New York, NW: Harper.
Buda, D. M., d'Hauteserre, A.-M., & Johnston, L. (2014). Feeling and tourism studies. *Annals of Tourism Research*, *46*(1), 102–114.
Butler, R. W. (1980). The concept of a tourist area cycle of evolution: Implications for management of resources. *The Canadian Geographer/Le Géographe Canadien*, *24*(1), 5–12.
Caton, K. (2012). Taking the moral turn in tourism studies. *Annals of Tourism Research*, *39*(4), 1906–1928. doi:10.1016/j.annals.2012.05.021
Caton, K. (2013). The risky business of understanding: Philosophical hermeneutics and worldmaking in tourism. *Tourism Analysis*, *18*(3), 341–351.
Caton, K. (2018). Conclusion: In the forest. In B. S. R. Grimwood, K. Caton, & L. Cooke (Eds.), *New moral natures in tourism* (pp. 194–205). London, UK: Routledge.
Chambers, D., & Buzinde, C. (2015). Tourism and decolonisation: Locating research and self. *Annals of Tourism Research*, *51*(1), 1–16.
Cohen, E., & Fennell, D. (2019). Plants and tourism: Not seeing the forest [n]or the trees. *Tourist Studies*, *19*(4), 585–606.
Dann, G. M. S. (2002). *The tourist as a metaphor of the social world*. New York, NY: CABI.
de Leeuw, S., & Hunt, S. (2018). Unsettling decolonizing geographies. *Geography Compass*, *12*(7), e12376. 1-14.
Deleuze, G., & Guattari, F. (1987). *A thousand plateaus: Capitalism and schizophrenia* (B. Massumi, Trans.). Minneapolis, Minnesota: University of Minnesota Press. doi:10.1086/ahr/57.2.471
De Line, S. (2016). All my/our relations: Can posthumanism be decolonized? *Open! Platform for art, culture & the public domain*. Retrieved from https://onlineopen.org/all-my-our-relations.
Doorne, S., & Ateljevic, I. (2005). Tourism performance as metaphor: Enacting backpacker travel in the Fiji Islands. In A. Jaworski & A. Pritchard (Eds.), *Discourse, communication, and tourism* (pp. 173–198). Bristol, UK: Channel View.
England, K. (1994). Getting personal: Reflexivity, positionality and feminist research. *The Professional Geographer*, *46*(1), 80–89.
Evers, C. W. (2019). Polluted leisure. *Leisure Sciences*, *41*(5), 335–423. doi:10.1080/01490400.2019.1627963
Fennell, D. A. (2018). *Tourism ethics* (2nd ed.). Bristol, UK: Channel View.
Goodman, N. (1978). *Ways of worldmaking*. Hassocks, UK: Harvester.
Grimwood, B. S. R. (2015). Advancing tourism's moral morphology: Relational metaphors for just and sustainable Arctic tourism. *Tourist Studies*, *15*(1), 3–26.
Grimwood, B. S. R., Doubleday, N. C., Ljubicic, G. J., Donaldson, S. G., & Blangy, S. (2012). Engaged acclimatization: Towards responsible community-based participatory research in Nunavut. *The Canadian Geographer/Le Géographe Canadien*, *56*(2), 211–230.
Grimwood, B. S. R., King, L., & Holmes, A, & the Lutsel K'e Dene First Nation. (2017). Decolonizing tourism mobilities? Planning research within a First Nations community in northern Canada. In J. Rickly, K. Hannam, & M. Mostafanezhad (Eds.), *Tourism and leisure mobilities: Politics, work, and play* (pp. 232–247). New York, NY: Routledge.
Grimwood, B. S. R., & Johnson, C. W. (2019). Collective memory work as an unsettling methodology in tourism. *Tourism Geographies*, 1–22. doi:10.1080/14616688.2019.1619823
Grimwood, B. S. R., Stinson, M. J., & King, L. (2019). A decolonizing settler story. *Annals of Tourism Research*, *79*, 102763.
Hall, C. M. (2011). *Fieldwork in tourism: Methods, issues and reflections*. New York, NY: Routledge.
Haraway, D. (1991). *Simian, cyborgs, and women: The reinvention of nature*. New York, NY: Routledge.
Hollinshead, K., (2007). Worldmaking" and the transformation of place and culture. In I. Ateljevic, A., Pritchard, N. & Morgan (Eds.), *The critical turn in tourism studies: Innovative research methodologies* (pp. 165–193). Oxford, UK: Elsevier.
Hollinshead, K., & Caton, K. (2017). Worldmaking and the representation of peoples and places in/through tourism. In L. Lowry (Ed.), *The SAGE international encyclopedia of travel and tourism* (pp. 1486–1490). Thousand Oaks, CA: Sage.
House, E., & Howe, K. (1999). *Values in evaluation and social research*. Thousand Oaks, CA: Sage.

Jamal, T. (2019). *Justice and ethics in tourism*. New York, NY: Routledge.
Jamal, T., & Hollinshead, K. (2001). Tourism and the forbidden zone. The underserved power of qualitative inquiry. *Tourism Management*, *22*(1), 63–82.
Jaworska, S. (2017). Metaphors we travel by: A corpus-assisted study of metaphors in promotional tourism discourse. *Metaphor and Symbol*, *32*(3), 161–177.
Jazeel, T., & McFarlane, C. (2010). The limits of responsibility: A postcolonial politics of academic knowledge production. *Transactions of the Institute of British Geographers*, *35*(1), 109–124.
Jones, A., & Jenkins, K. (2008). Rethinking collaboration: Working the Indigene-Colonizer hyphen. In N. K. Denzin, Y. S. Lincoln, & L. T. Smith (Eds.), *Handbook of critical indigenous methodologies* (pp. 471–486). Los Angeles, CA: SAGE.
Killion, L., & Fisher, R. (2018). Ontology, epistemology: Paradigms and parameters for qualitative approaches to tourism research. In W. Hillman & K. Radel (Eds.), *Qualitative methods in tourism research* (pp. 1–28). Bristol, UK: Channel View.
Kimmerer, R. W. (2013). *Braiding sweetgrass: Indigenous wisdom, scientific knowledge and the teachings of plants*. Minneapolis, MN: Milkweed Editions.
Kincheloe, J., & Steinberg, R. (2008). Indigenous knowledge in education. In N. Denzin, Y. Lincoln, & L. Smith (Eds.), *Handbook of critical indigenous methodologies* (pp. 135–136). Los Angeles: Sage.
Kuntz, A. (2015). *The responsible methodologist: Inquiry, truth-telling, and social justice*. New York: Routledge.
Lakoff, G., & Johnson, M. (1980). *Metaphors we live by*. Chicago, IL: University of Chicago Press.
Latulippe, N., & Klenk, N. (2020). Making room and moving over: Knowledge co-production, Indigenous knowledge sovereignty and the politics of global environmental change decision-making. *Current Opinion in Environmental Sustainability*, *42*, 7–14. (1),
Lee, E. (2017). Performing colonisation: The manufacture of Black female bodies in tourism research. *Annals of Tourism Research*, *66*, 95–104.
Lobo, M. (2014). Affective energies: Sensory bodies on the beach in Darwin, Australia. *Emotion, Space, and Society*, *12*(1), 101–109.
Lowman, E. B., & Barker, A. J. (2015). *Settler: Identity and colonialism in 21st century Canada*. Black Point, NS: Fernwood.
May, T. (2005). *Gilles Deleuze: An introduction*. Cambridge, UK: Cambridge University Press.
Marker, M. (2018). There is no *place of nature;* there is only the *nature of place:* Animate landscapes as methodology for inquiry in the Coast Salish territory. *International Journal of Qualitative Studies in Education*, *31*(6), 453–464.
McCabe, S. (2002). The tourist experience and everyday life. In G. M. S. Dann (Ed.), *The tourist as a metaphor of the social world* (pp. 61–75). New York, NY: CABI.
Mkono, M. (2016). The reflexive tourist. *Annals of Tourism Research*, *57*, 206–219. doi:10.1016/j.annals.2016.01.004
Morton, T. (2018). *Being ecological*. London, England: Pelican.
Neilsen, N., & Wilson, E. (2012). From invisible to indigenous-driven: A critical typology of research in Indigenous-tourism. *Journal of Hospitality and Tourism Management*, *19*(1-9), e5.
Nye, J. (2004). *Soft power: The means to success in world politics*. Cambridge, MA: Perseus.
Oberndorfer, E., Winters, N., Gear, C., Ljubici, G. J., & Lundholm, J. (2017). Plants in a "sea of relationships": Networks of plants and fishing in Makkovik, Nunatsiavut (Labrador, Canada). *Journal of Ethnobiology*, *37*(3), 458–477.
Oxford Learner's Dictionaries. (2019). Retrieved from https://www.oxfordlearnersdictionaries.com/definition/american_english/metaphor.
Pile, S. (2010). Emotions and affect in recent human geography. *Transactions of the Institute of British Geographers*, *35*(1), 5–20.
Platenkamp, V., & Botterill, D. (2013). Critical realism, rationality and tourism knowledge. *Annals of Tourism Research*, *41*, 110–129.
Prince, S. (2019). Dwelling and tourism: Embracing the non-representational in the tourist landscape. *Landscape Research*, *44*(6), 731–742.
Pritchard, A., Morgan, N., & Ateljevic, I. (2011). Hopeful tourism: A new transformative perspective. *Annals of Tourism Research*, *38*(3), 941–963.
Raghuram, P., Madge, C., & Noxolo, P. (2009). Rethinking responsibility and care for a postcolonial world. *Geoforum*, *40*(1), 5–13.
Rantala, O. (2019). With the rhythm of nature: Reordering everyday life through holiday living. In L. James, C. Ren, & H. Halkier (Eds.), *Theories of practice in tourism* (pp. 58–76). New York, NY: Routledge.
Regan, P. (2010). *Unsettling the settler within: Indian residential schools, truth telling, and reconciliation in Canada*. Vancouver: UBC Press.
Roisek, J. L., & Snyder, J. (2018). Narrative inquiry and new materialism: Stories as (not necessarily benign) agents. *Qualitative Inquiry*, 1–12. doi:10.1177/1077800418784326
Roisek, J. L., Synder, J., & Pratt, S. L. (2020). The new materialisms and Indigenous theories of Non-human agency: Making the case for respectful anti-colonial engagement. *Qualitative Inquiry*, 1–16.
Salazar, N. (2010). *Envisioning Eden: Mobilizing imaginaries in tourism and beyond*. New York, NY: Berghahn.

Samuelsen, A. B. (2000). The traditional uses, chemical constituents and biological activities of Plantago major L: A review. *Journal of Ethnopharmacology*, *71*(1–2), 1–21.

Schwandt, T. (2015). *The Sage dictionary of qualitative inquiry* (4th ed.). Thousand Oaks, CA: Sage.

Singh, N. M. (2018). Introduction: Affective ecologies and conservation. *Conservation and Society*, *16*(1), 1–7.

Smith, L. T. (2012). *Decolonizing methodologies: Research and Indigenous peoples* (2nd ed.). New York, NY: Zed Books.

Spivak, G. (1990). *The post-colonial critic: Interviews, strategies, dialogues*. New York, NY: Routledge.

Stewart, K. (2007). Ordinary Affects. Durham, NC: Duke University Press.

Stevens, Z., Grimwood, B. S. R., & Caton, K. (2019). Story and moral development in tourism education. *Journal of Teaching in Travel and Tourism*, *19*(1), 22–38.

Sundberg, J. (2014). Decolonizing posthumanist geographies. *Cultural Geographies*, *21*(1), 33–47.

Thrift, N. (2004). Intensities of feeling: Towards a spatial politics of affect. *Geografiska Annaler*, *86*(1), 57–78. (B),

Todd, Z. (2016). An Indigenous feminist's take on the ontological turn: "Ontology" is just another word for colonialism. *Journal of Historical Sociology*, *29*(1), 4–22.

Tribe, J., & Liburd, J. J. (2016). The tourism knowledge system. *Annals of Tourism Research*, *57*, 44–61.

Tuck, E., & Yang, W. K. (2012). Decolonization is not a metaphor. *Decolonization: Indigeneity, Education & Society*, *1*(1), 1–40.

Urry, J. (2000). *Sociology beyond societies: Mobilities for the twenty-first century*. New York, NY: Routledge.

Uysal, M., Li, X., & Sirakaya-Turk, E. (2008). Push-pull dynamics in travel decisions. In H. Oh (Ed.), *Handbook of hospitality management* (pp. 412–439). New York, NY: Elsevier.

van der Duim, R., Ren, C., & Jóhannesson, G. T. (2017). A decade of interfering with tourism. *Annals of Tourism Research*, *64*(1), 139–149.

Waitt, G., Figueroa, R., & McGee, L. (2007). Fissures in the rock: Rethinking pride and shame in the moral terrains of Uluru. *Transactions of the Institute of British Geographers*, *32*(2), 248–263.

Whyte, K. P. (2018). On resilient parasitisms, or why I'm sceptical of Indigenous/settler reconciliation. *Journal of Global Ethics*, *14*(2), 277–289.

Wilson, E., & Hollinshead, K. (2015). Qualitative tourism research: Opportunities in the emergent soft sciences. *Annals of Tourism Research*, *54*, 30–47. 10.1016/j.annals.2015.06.001

Wilson, S. (2008). *Research is ceremony: Indigenous research methods*. Winnipeg, MB: Fernwood.

Yusoff, K. (2018). *A Billion Black Anthropocenes or None*. Minneapolis, Minnesota: University of Minnesota Press.

Heritage justice, conservation, and tourism in the Greater Caribbean

Brent R. Fortenberry

ABSTRACT

Heritage Justice in the Greater Caribbean is a vital component of the region's sustainable development. Its Early Modern heritage is inextricably linked to the brutal regimes of European colonial expansion that enslaved African and Indigenous populations. Equitable and inclusive narratives, set within a framework of heritage justice, sit uneasily within the tourism landscape and are often erased or hidden in favor of histories that celebrate White colonial achievement. This article examines the port towns of St. George's Bermuda, and Falmouth, Jamaica, to explore how policies and priorities of heritage tourism and conservation practices, are used to exclude, erase, and diminish the contributions of communities of African descent, their material histories, and disconnect communities from tourism economies. The discussion argues for a more inclusive and equitable approach to heritage tourism that champions heritage justice for under-represented communities. Ultimately, this discussion calls for a transformation of heritage, preservation policy, and tourism in the Greater Caribbean.

Introduction

Cultural heritage conservation is vital for the development of healthy, sustainable communities, providing a sense of individual and community identity through the conservation of intangible cultural rituals and traditions, as well as a tangible connection to the past through the conservation of built heritage, tied to place. It is also pragmatically important for tourism and economic development opportunities for communities in historic landscapes at local, regional, and national levels (see Hampton, 2005). However, heritage is a product of the present and is appropriated through practices related to its conservation, production, and consumption (Graham et al., 2000). Heritage tourism is a performative concept in that it involves *appropriating*, interpreting, and communicating aspects of the past to visitors. The conservation of heritage for tourism is similarly performative and involves processes and relationships of power and agency (or lack thereof) in the production and consumption of heritage for tourism.

In the Greater Caribbean, cultural heritage often connects with the colonial past, particularly the brutal regimes of the slavery system, to narratives of Emancipation and, later, independence, with a shared sense of ownership of place in the 21st-century.

The residuals of these colonial regimes and their inequalities of power still permeate the region. The interplay of these forces dictates that the production of cultural heritage, both

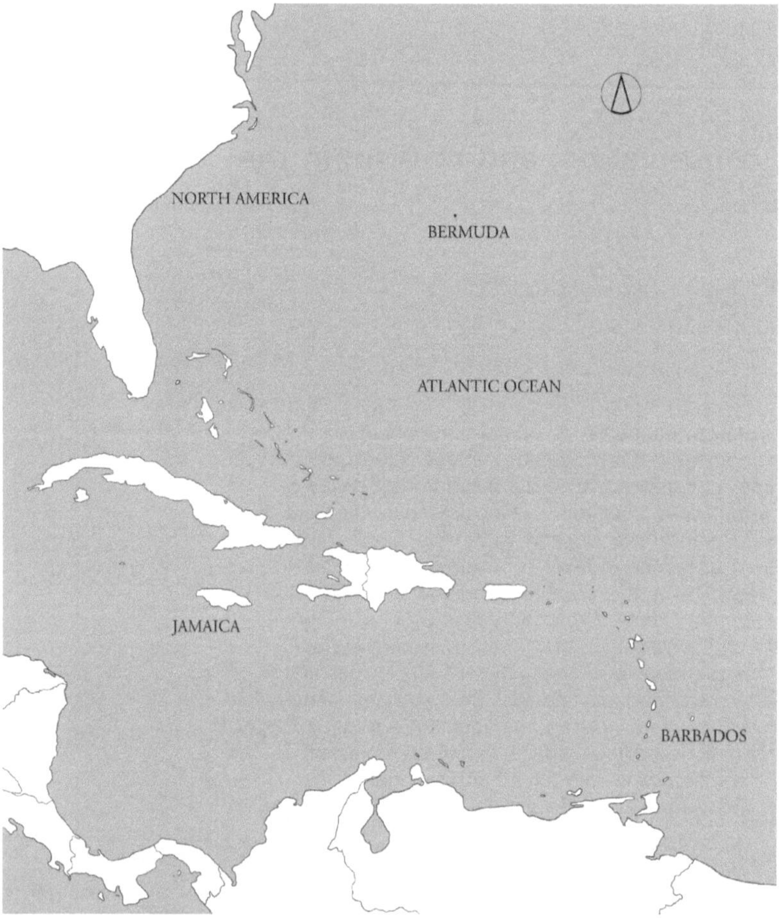

Figure 1. Map of the Greater Caribbean showing the regional location of Bermuda and Jamaica (Image: Brent R. Fortenberry).

independently and as a part of tourism programming, is directly tied to contemporary issues of equity, representation, and social justice for Caribbean communities (see Palmer, 1994 for an example from the Bahamas).

The purpose of this article is to demonstrate how identities are represented, repressed, hidden, and walled off as a part of the cultural heritage conservation and tourism economies in the Greater Caribbean. It illustrates the ways that heritage conservation and programming decisions have a direct impact on the heritage narratives presented to citizens and tourists, and it points to the wider challenges of the region in the 21st century. The two case studies presented below on two destinations, St. George's, Bermuda, and Falmouth, Jamaica, demonstrate how, though both occupy varying degrees of "post-colonial" conditions, in practice, they are bound up in a much more complicated web of relationships among their former colonizers, and wider neo-colonial processes which take varying political, social and economic forms in the tourism sector (see Figure 1).

The discussion draws linkages among structural heritage planning practices, heritage programming, and transnational tourism practices to demonstrate the disconnect between these and *heritage justice* in the Greater Caribbean. Through a comparative approach of the two destinations, this article serves as a departure point to explore the structural inequities and exclusionary tactics involved in the conservation and production of knowledge as a part of heritage tourism programming in the Greater Caribbean. It highlights foundational reforms that need to be undertaken to provide a just and equitable approach to heritage conservation and sustainable tourism development in the Caribbean. *Fair representation* and *inclusive participation* of diverse resident groups, especially those whose heritage is being appropriated for tourism, are crucial elements of *heritage justice* related to cultural heritage conservation and tourism in the Greater Caribbean (Jordan & Duval, 2009).

Caribbean heritage, tourism and heritage justice

Heritage justice as it relates to cultural heritage tourism in the Greater Caribbean is bound up in a complex web of relationships among heritage groups, including individuals and communities, stakeholder bodies such as National Trusts, heritage advocacy groups, heritage planning policies, and government entities, and transnational tourism companies. In this section, I trace these networks by first discussing cultural heritage tourism in the region, followed by a more specific context discussion of port towns as venues for heritage tourism before turning to how these problems are connected to wider issues of heritage justice and the well-being of historically oppressed communities.

Tourism is a vital part of the Greater Caribbean economy (see Honey, 2019 for a recent discussion). But the relationships among tourism industry transnational actors, local economies, and stakeholder communities is complicated. Many argue these transnational actors are merely forces of neo-colonialism, with Western-based companies taking advantage of developing island nations that were once former colonies. In the case studies below, Caribbean tourism functions as a form of neo-colonialism. As I am deploying it here, neo-colonialism refers to the political and economic power that developed nations (in this case the United States and Western Europe) exercise over former and existing (e.g., Bermuda) colonial holdings (after Hall & Tucker, 2004). Moreover, following Hall and Tucker, "Students of Neo-colonialism are therefore interested in the spatial and temporal dimensions of cultural production and social formations of the colony and post colony and the ongoing construction and representation of specific spaces and experiences" (Hall & Tucker, 2004, p. 2). Heritage tourism is one segment of the tourism economy, where we can see how the colonial past influences the neo-colonial present.

Cultural heritage tourism is an under-developed component of the region's economy (Brown, 2011; Harrigan, 1974; Wilkinson, 2004). This is not for lack of cultural sites of local, national, and international significance. Indeed, the region boasts eight World Heritage Sites with over 1,000 contributing properties; there are also nineteen regional sites on the World Heritage tentative list (Jordan & Jolliffe, 2013, p. 3). On the national and local levels, National Trust organizations and other heritage groups are working to create comprehensive heritage registers and partner with governmental and non-governmental agencies to conserve heritage places (Jordan & Jolliffe, 2013, p. 4). One of the region's greatest challenge is in the translation of these places into sustainable heritage tourism sites.

Programming and sustaining a heritage site is a difficult task. In a region with a Modern history of enslavement, racism, and colonialism, the ways that stakeholders and sites represent that complicated history is one of the most significant underlying challenges to regional heritage tourism, for example, how sugar plantation sites in the English Caribbean are interpreted as landscapes of violence (see discussions in Best & Phulgence, 2013; Inniss & Jolliffe, 2013). Beyond individual sites, as shown below, port towns are the optimal contexts to explore the challenges of heritage tourism and social justice because they can provide heterogeneous sites that illustrate the diversity of society. Still, too, these urban landscapes provide diverse experiences for tourists, in particular (as will be discussed below, this is not always the case as illustrated by case study 2 in Falmouth, Jamaica).

Historically, port towns served as hubs for society as spaces of interaction, trade, and sociability and were the driving forces of the early modern Greater Caribbean (Knight & Liss, 1991,

p. 5). Today, they are often the hubs for the Caribbean tourism economies (Martha, 2019). As "living landscapes," port towns must balance the needs of the tourism economy with those of its inhabitants. From a policy perspective, urban landscapes are the most heavily legislated, and thus cultural heritage conservation policies and politics have a direct influence on how the past is made more durable. These policies and politics modify how the past is presented to tourism audiences. To date, much of the heritage tourism scholarship surrounding port towns have centered on the Mediterranean, with the recent focus on Venice and Barcelona which have begun to push back on the tourism economy, placing heavy restrictions on cruise ship arrivals (Martha, 2019, pp. 52–54). But Caribbean port towns face many of the same challenges of economic development and predatory practices of tourism entities as their Mediterranean counterparts. Unlike Western Europe, the legacies of colonial institutions pervade tourism landscapes here; still too, under-represented and disenfranchised communities are often cut out of the benefits of tourism-based economic development. Understanding the dimensions of these heritage tourism inequities will ultimately help chart course correction towards fairer policies and process for sustainable tourism development and heritage conservation.

Heritage tourism and heritage justice in the Greater Caribbean

Beyond the site-specific challenges, Greater Caribbean heritage tourism suffers from a lack of education frameworks, resources (financial and structural), and local expertise to effectively manage, conserve, and package heritage sites for tourism purposes (for a national example see Jordan, 2013). The explosion of the tourism industry means that these heritage stakeholders must find a balance among local, equitable heritage representation, set within ever-increasingly complicated landscape of tourism priorities. This balance (or lack thereof) in heritage conservation and related tourism development is deeply linked to issues of social justice and community well-being: in the context of heritage, I refer to it as *heritage justice.*

Cultural heritage is a crucial component of community well-being. It provides a tangible and intangible connection to the past, group, and collective identity, and crucially, an important source of economic development, chiefly through tourism. Thus, it is imperative for heritage stakeholders to be represented as a part of the heritage planning process, be represented in the heritage tourism narratives, and have equitable access to the tourism industry to capitalize on these economic opportunities. Power structures among governmental, heritage and tourism entities dictate the complex relationships of heritage tourism that directly contribute to social justice and local well-being.

Notions of justice as they relate to heritage studies have been widely discussed by practitioners and academics alike (Baird, 2014; Ireland & Schofield, 2015; Johnston & Marwood, 2017). Key defining areas including the representation and acknowledgement of under-represented, subaltern communities, most notably, indigenous and African descendent populations (Smith, 2010; Waterton & Smith, 2010). These and related ideas have been situated within sustainability concerns by tourism scholars (e.g., Jamal & Camargo, 2014). I argue following Miller (2017) that heritage justice in the context of heritage tourism is related to how individual and group representation impact heritage knowledge production *and* the ability of these groups to access the benefits of the tourism economy through narrative production. It is important to recognize here that the situated, interrelated, historical context of tourism (Jamal, 2019) is inseparable from the domain of heritage conservation. What are the underlying structural conditions and institutional (policy-oriented) power relations that have resulted in the exclusion and under-representation of the now majority African descendent population in cultural heritage conservation and tourism development in the Greater Caribbean?

Research approach and methodology

Understanding heritage tourism as it relates to heritage justice requires a multi-scalar and heterogenous approach from the policy level to the "on-the-ground" conditions. The arguments and interpretations presented below are a product of my work as a heritage professional with longstanding relationships in both Bermuda and Jamaica. My perspective and interpretation of the two case studies below (St. George's, Bermuda, and Falmouth, Jamaica) are derived from extensive research of heritage planning regulations, a survey of the major historic elements of heritage production, and first-person participant observation of heritage sites, events, and policies. The methodological approach used here incorporates my experience and other published data. Together, they facilitate *crystallization*, i.e., creative, analytical thinking and the creation of new insights (Ellingson, 2009).

The research methodology was congruent in both Bermuda and Jamaica:

i. As a researcher and collaborator with Bermudian and Jamaican heritage stakeholders, I have been a participant both in the study, recording, of heritage knowledge and programming, as well as a participant in the process of generating narratives as a part of heritage tourism programming.[1] With this integrated research and technical approach, I have also analyzed the heritage planning policies and procedures, and been a first-person observer of the outcomes of these policies as iterated in St. George's and Bermuda as a whole, over the last thirteen years. In Bermuda, these heritage stakeholders (in addition to others), directly work with the Bermuda Tourism Authority (BTA, https://www.gotobermuda.com/), a non-governmental agency tasked with the island's holistic tourism plan, as well as the recording, reporting, and interpreting of the island's tourism statistics. BTA data forms the foundation for how heritage programming is linked to the wider Bermuda tourism landscape.
ii. I have worked in Jamaica since 2014 as a researcher and collaborator.[2] Working with scholars from the United States and architects and engineers, I have engaged with the Jamaica National Heritage Trust (JNHT), the University of Technology (UTech) located in Kingston and the Caribbean School of Architecture to conserve built heritage sites. I have also served as a consultant for conservation strategies at several major heritage tourism sites, . I similarly have analyzed both the heritage planning and policies of the Jamaican government, as well as the national tourism policy, and been on the ground in Falmouth observing and contributing to the creation of the heritage tourism landscape.

Case study 1–St. George's Bermuda

The heritage tourism landscape of St. George's, Bermuda, illustrates the ways that structural, neo-colonial heritage planning policy, and the challenges of material heritage durability and survival serve to under-represent the Afro-Bermudian past. These issues of inequity are troubling given the town's UNESCO World Heritage status.

Background and setting

Bermuda and its early capital of St. George's is a key case study in understanding the impacts of cultural heritage tourism on heritage representation in Greater Caribbean port towns. St. George's has been occupied for over 400 years, making it the longest British occupied urban landscape in the Western Hemisphere. Surprisingly, the extended occupation of St. George's has not resulted in a degradation of its built heritage. The town retains over 200 buildings that predate 1900–a major contributing factor to its World Heritage status (UNESCO, 2000).

Bermuda's and St. George's history are intertwined. Located roughly 600 miles east of Cape Hatteras, North Carolina, the island is slight, only 12 square miles and comprises over 100 small

Figure 2. Aerial image of St George's, 1997 (Image: Bermuda Government).

islands connected by a series of bridges. Bermuda's permanent European occupation began after 1609 with the accidental arrival of the *Sea Venture,* which shipwrecked on the island on its way to Virginia; the island was hosted an indigenous population due to its distance from North America. St. George's became the capital in 1612 and took on a predominantly administrative role for the colony. Few buildings survive of this early landscape, save the administrative buildings, and excavations have yet to uncover much of the material culture of this early period (Lowry, 2011, p. 8). (Figure 2).

The endeavored tobacco economy ultimately failed by the third quarter of the 17th century, and the Bermuda Company folded. Bermuda was then integrated into the growing portfolio of Crown-managed colonies. The turn to the Crown had the added impact of allowing Bermuda to trade with other colonies. This new economic freedom resulted in Bermudians "turning to the sea" and constructing a fleet of ships to shuttle goods across the Atlantic marketplace (Jarvis, 2010; Wilkinson, 1950).

St. George's was rapidly transformed from a relatively empty landscape to a mercantile hub. A change in the urban built environment accompanied this shift in colonial administration—merchants rapidly erected a series of warehouses and residences, constructed in local limestone. St. George's, as it survives today, is a direct product of this administrative and economic transition. Indeed, the World Heritage Status (discussed further below) connects the town's outstanding universal value with the integrity of the 18th-century buildings (UNESCO, 2000).

With the loss of the American colonies, particularly the northern ports of Boston and New York, Bermuda rapidly became a strategic British naval hub in the North Atlantic. A garrison was brought to the island, initially stationed in St. George's and then later relocated to the extreme West End at Ireland Island and the Royal Naval Dockyard (White, 1999, pp. 106–123). This slow shift in administrative focus from St. George's to points west on the island would continue throughout the 19th-century. By 1815, the capital had been moved to Hamilton in Pembroke Parish (White, 2015, pp. 1–9). The departure of the capital ultimately had the secondary benefit of lessening development (Jarvis, 1998, pp. 119–136). As such, the architectural heritage of the town was conserved (Fortenberry, 2016b, p. 603). In the 20th century, Bermuda transitioned to a tourism-based economy, with holidaymakers coming primarily from the UK and the east coast of North America (see McDowall, 1999 for a history of Bermuda tourism).

The Afro-Bermudian past

Enslaved Africans were first brought to Bermuda in the 1620s for their skills as agricultural cultivators of tobacco and as pearl divers (Chappell, 2010, p. 67; Jarvis, 2010, pp. 26–27) They formally entered the laboring population on the island by the middle of the 17th century, but excess labor prompted a ban on the importation of slaves by 1674. With the rise of Bermuda's maritime economy, Black Bermudians were primarily crewmen of the island's growing mercantile fleet and domestic servants and (Dawson, 2013; Jarvis, 2002). Blacks became a majority population by 1790, and without a large-scale plantation system, bondsmen and -women worked in close proximity to the White Bermudian population and lived in cellars and attics (Chappell, 2010, pp. 68–69). Emancipation in 1834 shifted the demographics and power relations of the island. But Black Bermudians continued to work as domestic servants–freedom was now attained, but equality remained elusive.

Segregation and disenfranchisement were pervasive in Bermuda throughout the late 19th and early 20th centuries. The Black majority held little political power, and by the middle of the 20th century, growing frustrations among Black Bermudians started to manifest through theater boycotts and growing public demonstrations (Jones, 2004, p. 208). In 1963, suffrage was granted to all Bermudians, and the Progressive Labor Party (PLP), the island's first political party, was formed. The 1968 riots, however, brought into sharp relief the tensions among White elites and the Black majority (Jones, 2004, p. 204). The riots were a small measure compared to the assassination of Police Commissioner George Dukett in 1972, followed by the assassinations of Governor Richard Sharples and his aid in 1973 by Erskine Burrows. Burrows sought, as noted in his confession, to illustrate to the island's Black population and their colonial oppression at the hands of the White elite (Jones, 2004, p. 216). It was not until 1998 that the first PLP government was formed, the first Black majority government in the island's history (Jones, 2004, p. 218).

Bermuda, geo-politics, and tourism today

Bermuda today comprises roughly 71,000 residents. Relative to the rest of the Caribbean, Bermuda has a smaller percentage of the population of African descent, just 53% with Whites comprising 31%. It is per capita the wealthiest island in the Greater Caribbean due to its financial sector accounting for 85% of GDP. Tourism accounts for just 5%, but conversely is the largest employer of the population ('Bermuda', 2019).

Geo-politically, Bermuda continues to sit awkwardly in the Atlantic world. Unlike many of the Caribbean island nations who declared independence throughout the middle of the 20th century, Bermuda has remained a loyal part of Britain's colonial portfolio, in modern political terms, an "overseas dependent territory" (Royle, 2010). Bermuda sits at the most independent end of the sliding scale of Overseas territory autonomy. It is effectively self-governing for domestic affairs with an elected Parliament with an upper and lower house. The Crown still manages defense and foreign affairs, but with the Bermuda Assembly driving the priority of arrangements (Kim & Rodriguez, 2006). There is regular talk of Bermuda going independent by leaders of political parties, but this is unlikely in the short-term. The current arrangement provides extreme social, political, and economic stability for the island. Their in-between status also allows

Bermuda to operate an exempt company economy, particularly the off-shore banking and re-insurance sector, boosting the island's GDP and quality of life (Butkiewicz & Gordon, 2013). Ultimately, Bermuda needs to maintain inks to the United Kingdom to maintain a safe, stable and welcoming economy, particularly in the areas of international trade (for the exempt companies and the reinsurance business) and international tourism (Royle, 2010, p. 207).

The most recent tourism figures for Bermuda come from 2018, where the island welcomed roughly 770,700 visitors. Of that 63% (or roughly 484,000) arrived via cruise ships and 36% (roughly 282,000) arrived via air; the final 1% arrived via yacht. Of note is that 2018 marked an 11% rise in air arrivals, the highest since 2007. The tourism market skews heavily towards North America and the northeastern United States, with the most significant percentage of arrivals originating in Boston, New York, Philadelphia, and Washington D.C. (Bermuda Tourism Authority, 2016, p. 6).

Findings and discussion

Within these geo-political and socio-economic realities, St. George's primarily relies on its architectural history for cultural heritage tourism. The town's primary heritage sites include St. Peter's church (c. 1612), the World Heritage Center in a 19th-century warehouse, the President Henry Tucker (c. 1762) and Globe Hotels (c. 1691) (both operated as museums by the Bermuda National Trust), as well as other historic house museums run by smaller heritage groups such as Mitchell House Museum (c. 1640) operated by the St. George's Historical Society, the Bermuda Heritage Museum. The Government also operates a museum at Fort St. Catherine's on the eastern extreme of the island. These sites are connected by a dense urban landscape consisting of 18th and 19th-century domestic structures.

St. George's is one of the most studied urban landscapes in the region. "The Old House Survey," spearheaded by the Bermuda National Trust in the 1980s, created a baseline database for the historic buildings on the island. That initial survey became the basis for the Trust's Architectural Heritage Series, a Parish by Parish publication series; the first, focusing on St. George's, was published in 1998 (Jarvis, 1998). The archaeological study of St. George's has been similarly intensive, excavations have taken place throughout the World Heritage Site since the 1980s (Figure 3).

Thus, the challenges of equitable representation are not due to a lack of knowledge, infrastructure, or architectural integrity. Instead, they relate to policy, programmatic foci, the challenges of tourism clientele, all a product of Bermuda's unique place within the 21st-century Atlantic world.

Heritage planning biases

In planning policy, biases in the elements of conservation ordinances and the World Heritage inscription only further de-emphasize the subaltern narratives of Black Bermudian's heritage. Bermuda has some of the strongest built-? and archaeological heritage protections in the region. Built on the framework of Great Britain's cultural heritage laws, with a listing system, these protections are even more robust in St. George's which received special status in the 1930s, due to its architectural integrity (Bermuda Government Department of Planning, 2008, 2010, 2012, 2014; Bermuda National Trust and Bermuda Government Department of Planning, 2002). Strengthening of these provisions in the 1950s, and the implementation of the listing system in the 1980s and World Heritage status in 2000 means that the cultural heritage in St. George's is well protected.[3]

The conservation policy of St. George's is modeled on the foundations of the UK's listing system. This system prioritizes architectural integrity, the material remains of the past, meaning that, in many cases, buildings are protected *only* if they retain original fabric. But enslaved

Figure 3. Architectural character of St. George's (Image: Brent R. Fortenberry).

Bermudians often did not inhabit stand-alone buildings. Instead, they occupied cellars and attic spaces living in close proximity to their owners. Still too, after Emancipation, the continued occupation of buildings through successive owners often dilutes the meaning of space. In a related vein, archaeological excavations often can shed light on the ephemeral remains of enslaved African lifeways on Bermuda (e.g., Fortenberry, 2013). But the current heritage planning guidelines do not afford protection to archaeological sites; they only suggested guidance (Bermuda Government Department of Planning, 2008). These factors combined, much of the Afro-Bermudian past comprises intangible, celebrated through practice-based performances such as the Bermuda Gombeys (see Gibson, 2019) which are closely linked to the Caribbean's wider rituals of Jonkonnu (see Smalligan, 2011 and Mullin, 2014) and oral histories, for example St. David's Islander's connections to the indigenous cultures of North America (see Gaieski et al., 2011), and not physical spaces, elements of the past that current conservation policy is ill-equipped to protect and celebrate.

Eurocentric world heritage status

While the conservation policy has an underlying bias against the representation of Black Bermudian heritage, the World Heritage nomination is more overt in its neglect of recognizing the contributions of enslaved Africans and Black Bermudians to the town's universal outstanding value. St. George's and its related fortification were inscribed on the World Heritage List under criterion iv: as outstanding examples of a continuously occupied, fortified, colonial town dating from the early 17th century, and the oldest English town in the New World" (UNESCO, 2000). This short description is telling of the structural issues of World Heritage Status in St. George's, invoking a Eurocentric view with the use of the "New World."

Furthermore, the UNESCO status description almost exclusively engaged with the White colonial history of the island and its broader connections with the fortification traditions of the Atlantic world. There is little discussion of the enslaved African arrival on Bermuda as a part of the transatlantic slave trade, other than marking the date of arrival on the island after 1620. More troubling is that while the World Heritage Listing champions the integrity of historic building stock in the World Heritage Area (roughly 65% of the surviving buildings pre-date 1900) (UNESCO, 2000), there is little discussion of the fact that without enslaved labor none of these buildings or fortifications would have been constructed. While discussions have taken place to add to the World Heritage listing with more contributing properties, there has yet to be a discussion of ways to make the listing more inclusive and more equitable of Bermuda's Black heritage. These policy challenges directly contribute to the narratives currently being deployed by stakeholders in St. George's.

Thus, both the heritage planning guidelines and the World Heritage status language favor White colonial heritage on Bermuda. This structural framework means that the built heritage elements that are designated for conservation and protection are inequitable to Bermuda's Black majority. In heritage planning explicitly, linking higher levels of protection with the integrity of standing heritage only further skews surviving built fabric to historically White colonial buildings. Reform of these guidelines to include societal values will more equitably represent the Afro-Bermudian past.

Programming challenges of studying the Afro-Bermudian past

A further complicating factor is the foreign audiences that these historic house museums need to attract–predominantly white, North American tourists (Bermuda Tourism Authority, 2016, p. 7). This demographic has historically been interested in exploring the historical connections between Bermuda and the east coast of the United States; it's these stories that have motivated them to visit heritage sites rather than simply head for the beaches.

Nowhere is this tension more evident than at the Globe Hotel, an early 18th-century building owned by the Bermuda National Trust and operated as a historic house museum. Beyond the gift shop on the ground floor, the upper floor museum space presents St. George's role in the American Civil War, particularly the safe harbor St. Georgians gave Confederate blockade runners in an exhibition called *Rogues and Runners* (the museum was formerly called the Confederate Museum). This exhibit is the most widely visited museum space in the World Heritage Site and is anecdotally most popular among White, southern American visitors to the town. The exhibit's popularity demonstrates the uneasy reality of heritage tourism in the Greater Caribbean. It is one thing to ignore, hide, or obfuscate the heritage narratives of groups of African or Indigenous descent; it is another to have a museum exhibit that celebrates the roles that White Bermudians played in a foreign war primarily predicated on the retention of the American south's chattel slavery system (Fortenberry, 2016b). Thankfully, the Trust has acknowledged that this needs to change, and they will be replacing *Rogues and Runners* with a rotating exhibit that focuses on St. George's architectural heritage (Figure 4).

This thread of connecting 18th-century Bermudians to the broader narratives of U.S. history is continued at the President Henry Tucker House Museum, also operated by the Bermuda National Trust as a historic house museum. The museum relays the familial connections between the Tucker Family and Henry Tucker's son, St. George Tucker, who traveled to Virginia to study law at the College of William and Mary in Williamsburg. But unlike the Globe Hotel, the heritage programming at the Tucker House also begins to interpret the archaeological signature of enslaved Bermudians at the site, with reference to the archaeological work completed by the Colonial Williamsburg Foundation during the 1980s and 90s (Pickett & Brown, 1998).

The present administration of the Trust is not to blame for this programming. The creation of these North American-centric exhibits was directly related to catering to foreign audiences. As discussed above, with the flood of North American visitors to the islands (mainly from the coastal ports on the east coast), the Trust sought to capitalize on the shared heritage of their predominantly white American visitors. So it was only natural to design their exhibits for specific

Figure 4. Globe Hotel (Image: Brent R. Fortenberry).

demographics. While having this focus, equitable heritage representations were secondary, if included at all, from the outward-facing cultural heritage programming. Changing these foci is an underlying challenge for all Caribbean heritage stakeholders. Local expertise is scarce; for example, the Bermuda National Trust employs a single curator with limited research and financial resources. They must rely on volunteers and external researchers to push forward and amend the existing museum installations. This lack of local capacity is an underlying, long-term challenge to the further development of the island's (and region's) cultural heritage programming.

While museum programming in St. George's has intensely focused on the town's connection to American history, the live-action events in the town, too, celebrate the relationship between the American Revolution and St. George's. Live-action performances commemorate the events of the Gunpowder Plot, in which a group of American sympathizers stole British gunpowder in St. George's and shipped it to rebelling Americans in strongholds on the eastern seaboard (Kerr, 1995, pp. 49–59). Such dramatizations are designed to provide heritage tourists with a deeper connection to the past as understood through historic spaces. Nevertheless, the focus of these events again lies in their outward programmatic gaze that celebrates the contributions Bermudians made to another country's independence efforts. At the same time, ironically, the island remains a colony.

Engaging tourism and tourists. A major challenge of engaging with Black Bermudian heritage, as alluded to above, relates its ephemeral material signature. The recovery of the ephemeral archaeological and architectural remains of these lifeways is challenging (Fortenberry, 2016a, pp. 610–611). It is only on rare occasions that researchers can isolate the material signatures of these individuals and groups in the form of undisturbed cellars and attic spaces (Fortenberry, 2013, 2015).

This is not to say that there is a complete lack of heritage programming with respect to Black Bermudian Heritage. The African Diaspora Trail provides a passive tour of important sites of pre-

and post-Emancipatory Black Bermudian history (Fortenberry, 2016a, p. 611). Still too, the Bermuda heritage museum on the eastern extreme of St. George's provides dedicated museum programming to historic Black St. Georgian's lifeways. The challenge for this museum is that it sits on the periphery of St. George's, well off the typical tourist pathways in the town. Moreover, at the time of writing, the Bermuda Community and Cultural Heritage Board had recently introduced a new live-action dramatization associated with the life of Mary Prince on Bermuda for tourists in both St. George's and Hamilton (Bell, 2019). These efforts signal a rising interest from the public, tourism programmers, and stakeholders to better represent Black Bermudian past in the heritage narratives for both islanders and tourists. Instead of lower tourism interest, the hope is that these new narratives will draw international tourists in the diverse and vibrant components of Bermuda's cultural heritage, and ultimately provide representation as a form of social justice for the island's majority Black population.

These are isolated efforts in the wider approach to cultural heritage programming in St. George's. The systemic stakeholder fragmentation is the largest obstacle to a sustainable cultural heritage tourism landscape in St. George's. Over ten stakeholder groups are present in the town, each with varying priorities, resources, and histories. Advisory bodies such as the St. George's Preservation Authority and the World Heritage Site Management Committee are designed to provide a forum for collaboration among the stakeholders. Still, these committees are merely venues for the airing of grievances and arguments among the stakeholders. They, in effect, have a negligible impact on the St. George's community.

In 2011, the Bermuda Tourism plan was overhauled. Instead of discrete sites, the plan established five hubs that would be the center of tourism activity. St. George's was Hub 1 and was organized as the cultural heritage-focused aspect of the Tourism plan. The idea was to draw visitors into an integrated cultural landscape where the town would offer a set of site visits to museums, historic buildings, and built spaces to convey the diversity of Bermuda's cultural heritage past (Bermuda Tourism Authority, 2012). Funding was allocated to hubs to create integrated experiences that would best represent the diverse heritage of St. George's. This unified narratives never materialized amid the continued isolationist attitudes of stakeholders and a lack of centralized organization. A recent shift in tourism policy changed the strategy from "hubs" to an "experience"-based framework, ultimately de-emphasizing the shared cultural heritage spaces. Such an approach only further fragments the cultural heritage stakeholders (Bermuda Tourism Authority, 2019). By organizing sites as individuals and not as connected spaces within the landscape, this approach will only further delay any opportunities for sustainable heritage programming on a town-wide scale.

Programmatic development opportunities

Despite the challenges of programming, there are economic resources available to transform the cultural heritage tourism narratives in St. George's. In 2013, the Bermuda Government passed the UNESCO World Heritage Fund and Levy Act, "for the purpose of maintaining, developing, and promoting the UNESCO World Heritage site' (Corporation of St. George's (World Heritage Fund and Levy) Act, 2013). A tax of 0.25 cents per liter is levied on all petroleum imported at the St. George's oil docks. In 2016, over 1.2 million Bermudian dollars were distributed for a wide variety of heritage conservation-related projects. Nearly $50,000BMD was used to promote UNESCO projects, including historical reenactments, new World Heritage signage, the direct support of heritage sites, UNESCO due, and local and international research projects. The remainder was used for public works projects, such as roof replacements, sewer infrastructure, and dock repairs (Bernews, 2017). The UNESCO World Heritage Fund is a sustainable means of supporting a small number of heritage initiatives while at the same time providing more robust infrastructure support for the World Heritage area. But it has yet to aid a majority of St. George's stakeholders directly. This is due to the fact that many of them do not regularly apply to the fund due to lack of time, expertise, and resources. Most of these small heritage groups are staffed by part-time volunteers, and so the task of annually applying to the World Heritage Levy Fund is not practical.

The heritage tourism landscape of St. George's remains tied to colonial narratives. However, these structures are slowly being dismantled in favor of more inclusive programming due to the willingness on the part of stakeholders to recognize the under-representation of Black Bermudian narratives. By engaging with a broader array of stakeholders and working to better program these sites with more comprehensive and equitable heritage narratives, Bermuda and its visitors will better understand the diverse cultural heritage of this Atlantic island. More critically, the heritage planning frameworks need to be more inclusive of non-tangible elements of the past.

Case study 2: Falmouth, Jamaica

Further south in the heart of the Caribbean, the town of Falmouth sits centrally on Jamaica's north coast. Here, challenges to equitable heritage tourism landscapes are brought about by the exclusionary tactics of the cruise ship industry, and the relative isolation of the town from larger urban centers. These factors, coupled with a lack of conservation enforcement, cripple cultural heritage tourism efforts that directly benefit the residents of Falmouth. Here the neo-colonial nature of tourism in historic port towns is on full display.

Background and setting

Jamaica was Britain's largest colonial holding in the Caribbean (4,244 square miles). The island was initially colonized by the Spanish in the 16th century. Still, it was never a significant holding within the Spanish Caribbean relative to Florida and Cuba, which were more central to their colonization strategy. Little architectural evidence of Spanish presence survives on the island.

The British were desperate for space to capitalize on the 18th-century sugar boom, which first took off in the 17th century in Barbados to the south. With only a small band of invaders, the British rapidly populated the island and established a new capital at Port Royal (and later Kingston after the earthquake). Sugar cultivation quickly occupied all arable land on the island (Robertson, 2008).

Enslaved Africans arrived in Jamaica as early as 1510 as part of the early Spanish presence on the island. Spain's colonial agenda was primarily focused on recovering mineral resources (e.g., gold) and not agriculture. As a result, they put up little resistance when the English captured the island in 1655 (Robertson, 2008). The English brought enslaved Africans to the island on mass as they transformed significant portions of the island into sugar plantations, which required the mass labor for cultivation (Nelson, 2016, pp. 97–132). Emancipation in 1834 initiated a wholescale change in Jamaican society. With a labor force that now needed wages, the social landscape of the island slowly began the shift. Four years of "apprenticeship" between 1834 and 1838 would transition Jamaican society out of the chattel slavery system. Widespread poverty characterized the period after Emancipation for Black Jamaicans. With frustrations growing, the Morant Bay rebellion called into sharp relief the racial and economic tensions of the island. The White authorities moved quickly to end the uprising, ultimately resulting in the execution of 500 Black Jamaicans (Chivallon & Howard, 2017). Social and economic development occurred in the years following Morant Bay with improving infrastructure and public works projects for the island. The concluding decades of the 19th century also witnessed the decline of the sugar industry in Jamaica (Cumper, 1954). The 20th century ushered in more power to the Black Jamaican Majority, mainly through the union activism of Marcus Garvey, who founded the island's first political party in 1929 (Lawler & Davenport, 2009). Labor unrest in the 1930s ultimately led to broader political participation buoyed by Norman Manley and the formation of the People's National Party (Levi, 1992). By the 1940s, Britain ceded more copious amounts of autonomy to Jamaica under the Crown colonial system in 1944. A desire for self-government only grew through the middle of the century.

Jamaica declared independence in 1962 as a part of the colonial independence movements in the Caribbean during the middle of the 20th century (Kelly, 1963). The island re-oriented itself as a major hub of the English-speaking Caribbean, agricultural export, and a growing tourism economy. Today Jamaica is a nation of 28 million people, 92.1% of whom are Black, and 6.1% are of mixed descent. ('Jamaica', 2019). It is the most impoverished English-speaking nation in the Caribbean, and second overall in poverty behind Haiti.

The economy is heavily reliant on tourism, making up over 20% of the economy, and sustainable and safe tourism is vital to the Jamaican economy ('Jamaica', 2019). A rise in violence in Kingston, and more recently on the north coast in Montego Bay, have resulted in a heightened level of tourism trepidation. St. James Parish has been under a curfew with regular military checks associated with gang violence since 2018. These factors complicate tourism (inclusive of cultural heritage tourism) efforts outside of the bubbles of cruise ships (Hall, 2012). While often the threat of crime has been isolated to Kingston, Spanish Town, and the south coast, gang violence in the tourism hub of Montego Bay caused great pause for the north coast–the driving engine of the island's tourism economy. At the same time, the rise of violence, even within the all-inclusive resorts, has further pushed tourists towards cruise ships as safer and cheaper options to visit Jamaica (McKee & Chase, 2003). The move towards a more significant percentage of cruise ship tourists lessens the positive tourism impact on the Jamaican economy–fewer dollars directly reach communities via cruise ship tourism (Kerswill & Mair, 2015).

Tourism as part and parcel of the service industry comprises 70% of Jamaica's economy ('Jamaica', 2019) In 2016, the island had 2.1 million air visitors and 1.6 million cruise ship passengers for a total of over 3.8 million visitors, with a combined total annual expenditure of 2.6 million USD (Jamaica Tourist Board, 2016, p. 7). Since 2000, Jamaica has seen a steady growth of both air and cruise ship arrivals at 5.5% and 10% respectively. Like much of the Caribbean, Jamaica draws heavily from the United States for its visitors (68%), with Europe a distant second at 17% (Jamaica Tourist Board, 2016, p. 5). Ironically, Jamaica's Master Tourism Plan calls for the island's heritage to become a significant component of tourism as a key to its future success (Jamaica Tourist Board, 2016, p. 10) In 2016 the Tourism Master Plan went so far as to call for a heritage fund to jump-start heritage tourism on the island (Jamaica Tourist Board, 2016, p. 96). Urban landscapes are crucial components of this tourism strategy with Port Royal, Spanish Town, Falmouth, and Seville/St. Ann's Bay, identified as critical locales of development (Jamaica Tourist Board, 2016, pp. 104–112). Three of the four designated towns are also historic ports that cater to cruise ship tourists.

Findings and discussion–Falmouth challenges of place

Falmouth is a major cruise ship hub for the Jamaica tourism economy. Located 18 miles east of Montego Bay and the seat of Trelawny Parish, the town sits on a deep harbor, ideally situated as an export hub for the burgeoning sugar plantations in the hills to the south (Nelson, 2016, pp. 156–186). Today, Falmouth and its environs are a relatively small town in Jamaica with 8,188 (this in comparison to Kingston that has a population of 937,000, and neighboring Montego Bay with 110,000) ('Jamaica', 2019).

Similar to St George's, the architectural heritage of Falmouth is unrivaled in Jamaica and much of the Greater Caribbean. While other port towns such as Kingston and Port Royal boast a more extended history, Falmouth, because of its rapid rise and fall in concert with the sugar industry, retains much of its 19th-century architectural heritage. Its stock of historic buildings

Figure 5. Before and after demolition of 7 Market Street (Image: Peregrine Bryant).

have been studied in detail by architectural historians, Louis Nelson and Edward Chappell, as a part of annual architectural history field schools, and published in monographs and GIS websites (Nelson & Chappell, 2014).

Heritage conservation unchecked

Successful cultural heritage tourism representation requires some level of material integrity of the historic built environment. While Falmouth retains a high degree of integrity, its survival is fragile at best. At the local level, non-existent, non-enforced heritage planning regulations mean that development projects can proceed with little oversight and consultation often to the detriment of cultural heritage assets. No local authority checks over demolition permits and demolition by neglect is rampant. One needs to only look to the destruction of urban store-houses on Market Street (the town's main thoroughfare) to understand the level change to the historic urban landscapes.

The national heritage body, the JNHT, often boasts of its strong protections over the built heritage of Falmouth. However, they lack enforcement. Demolition permits are often granted if they suit perceived development needs, particularly with the backing of foreign money. In Falmouth, the practice of foreign ownership has led to the loss of many outstanding examples of Jamaican buildings. 14 Market Street, once an exceptional example of a merchant's "store house" residence, was razed, and now a concrete "TESCO Family Dollar Superstore" has taken its place (Bryant, 2019). (Figure 5).

If the protection of historic structures hangs on the whim of new concrete grocery stores and not an enforced heritage conservation ethic, then much of Falmouth's historic structures are destined for demolition. At the same time, demolition by neglect is evident everywhere. For example, poor maintenance of the electrical system at 7 Market Street, the Molton Barrett House, resulted in a fire four years ago that all but destroyed the urban fabric of the upper floors (Chappell et al., 2014) (Figure 6).

Little heritage conservation oversight puts the tangible elements of Falmouth's heritage at risk, but local, highly-skilled, craftspeople are as vital to retaining historic architectural integrity. For fifteen years, Falmouth had such a force. In 2000, Falmouth Heritage Renewal (FHR) was founded by international advocates in cooperation with local Jamaican stakeholders. Its mission was simple: build local heritage conservation capacity, to better care for the historic buildings in Falmouth (Parrent, 2005). International experts trained local artisans in historic conservation techniques and ethics, and some were sent to the UK to hone their skills. FHR took on restoration and rehabilitation projects in Falmouth and sought to revitalize the town's historic urban environment. FHR also hosted international architectural history field schools where university

Figure 6. Molton Barrett House after fire (Image: Brent R. Fortenberry).

students recorded and conserved historic buildings alongside Jamaican experts. FHR's business model, however, was not sustainable, and it required constant fundraising to maintain. With few local dollars to provide financial feasibility, FHR ultimately wound down its operations in 2016. The local economy could not support the relatively high costs of expert heritage conservation work in contrast to the cheaper, less-skilled labor of 21st-century Jamaican construction.

FHR played an essential role as a stopgap and watchdog for Falmouth's historic buildings. They were in effect the local conservation enforcement. Their loss means that now there is no local entity to ensure development went unchecked. FHR is an optimal model for heritage conservation; it just lacked the financial viability. Government grant programs provided some funding, but not enough to maintain the non-profit.

Tourism enhancement fund as economic infrastructure

A small positive in this landscape is the presence of government grants for heritage conservation projects. The Jamaican Tourism Enhancement Fund (TEF) was intended to provide a grant source for the betterment of tourism-relative products (Jamaica Tourist Board, 2016). The TEF is funded by arriving tourists through a tax, as a part of purchased tickets. Each year, tourism-related entities could apply to the TEF to fund new initiatives. The supported projects vary widely; thankfully, heritage conservation stakeholders have been successful in being granted these funds. But the TEF is not a perfect system, they have since been de-ear-marked for strictly tourism activities, and moved into a general government fund. Still too, like the World Heritage Levy fund in St.

Figure 7. "Historic Falmouth" Arrival Terminal (Image: Brent R. Fortenberry).

George's, local expertise and resources to write grants are often lacking, meaning that smaller, less well-equipped heritage stakeholders might not even have the capacity to submit a grant through the system.

The lack of influential local heritage stakeholders has resulted in a vacuum for heritage tourism in Falmouth. The building blocks for comprehensive protection and programming exist, but without local capacity, heritage tourism remains untapped. Coupled with the rise of cruise ship tourism in Falmouth, there were optimal conditions for the creation of a neo-colonial heritage tourism landscape.

Tourism challenges and "historic Falmouth"

Falmouth's size and relative distance from the larger northern ports of Montego Bay and Ocho Rios mean that it functions as primarily a day trip destination, lessening the potential impact of tourism on Falmouth. The town does not boast a major hotel, despite several efforts in recent years. Ironically, it is this lack of development that has provided the heritage conservation opportunities for Falmouth. Cruise ships similarly visit Falmouth for a single day, leaving ports of Montego Bay and Ocho Rios in the evening and arriving in Falmouth in the morning and then departing in the late afternoon. This creates a tight window to capitalize on tourism revenue, and one only hampered by the design of the tourism experience in Falmouth, most notably the cruise terminal "Historic Falmouth" (Hori, 2016).

The Historic Falmouth Cruise Ship Port, located on the northern coast of the town, is the largest purpose-built cruise ship port of call in the Caribbean (Figure 7). "Historic Falmouth" is confusingly named. It creates an exclusion zone on the town's harbor front, a form of enclave tourism (McFarlane-Morris, 2019). This walled cruise ship compound creates a barrier between tourists and Jamaica's built heritage, and a recent study has found that Falmouth residents feel a deep sense of marginalization (McFarlane-Morris, 2019, pp. 11–12). This is not surprising given the nature of the exclusionary zone. Ironically, to leave or enter the "Historic Falmouth" one needs to go through a passport control procedure, and Jamaicans (and visiting scholars) need to surrender their passports when entering the compound. Within the walls, cruise ship companies endeavor to create a tourist's tropical paradise, complete with high walls, faux Georgian buildings, and American restaurants, such as hotdog stands, sandwich shops, and casual dining. The only surviving historic building in the compound is Tharp House, which was the grandest Town House in Falmouth (Chappell & Fortenberry, 2016). Tharp House is intended to be the primary historical programming within Historic Falmouth, but its rehabilitation has taken years to complete with little local expertise left with the loss of FHR.

Few (and now almost no) local vendors are present within the exclusion zone, and instead, the "local market" has been moved outside of the compound on the other side of passport control, with Falmouth merchants eager to sell to tourists. But, they seldom make it due to the complex control of movement in and out of the compound. In truth, the area that comprises the compound is not technically Jamaican territory. Instead, the compound operates effectively as an international liminal zone between the cruise ship and Jamaica. It has its own private security, utilities, and control of space. It is a tourist third space created on the North Coast of Jamaica with a fantasized historic built environment.

This exclusion zone bifurcates the heritage tourism landscape of Falmouth. The cruise ship industry has created an island, similar to their "private islands" that one can visit for the day before leaving again. Still too, it encourages cruise ship visitors to remain within the walled "Historic Falmouth" compound (Bray & Redfearn, 2014, pp. 228–235). There is little encouragement to leave the exclusion zone, though two main options are promoted: a "Historic Falmouth Train" tour organized through the cruise ship companies and the Historic Falmouth Port Authority, or directed excursions to other leisure experiences in Trelawny Parish.

The Historic Falmouth Train experience takes cruise ship visitors from the compound and provides them a tour of some of the historic sites of Falmouth (the author has directly taken this heritage experience). During the tour, passengers never leave the train, and sites are only visited from the public right of way. As is the case with many similar tours, much of the information provided to visitors is inaccurate (Figure 8). By offering a tour that is devoid of interaction with the people of Falmouth and its historic sites, the Historic Falmouth Train experience further deepens the bifurcation of the tourist spaces and only exacerbates the neocolonial aspects of cultural heritage tourism. Small walking tour companies do exist; however, they often cannot compete with the cruise ship industry and cater more directly to the arriving land-based passengers. What is the benefit to heritage stakeholders if tourists do not get to directly engage with sites? Multinational corporations, not the people of Falmouth, are capitalizing on the integrity of the town's material past.

In Falmouth, Jamaica, equitable representation in heritage tourism does not exist due to the lack of heritage programming infrastructure and the exclusionary and enclave practices of the cruise ship industry. The ultimate threat to Falmouth was recognized on a global scale when it was placed on the World Monuments Fund Watch List. While this did much to raise awareness about the threats to the town's historic built environment, in the long term, this listing has done little to change the crushing impacts of the cruise ship tourism industry on the town. It is not until a local heritage capacity is built, a rigorous heritage conservation ethic is practiced, and the predatory, neo-colonial practices of the cruise ship industry stop, that equitable representation of Jamaicans in their heritage tourism landscape will occur.

Discussion

The two case studies presented above represent two distinct yet interconnected challenges for cultural heritage conservation and sustainable tourism development. Bermuda is per-capita, the

Figure 8. Image of Falmouth with docked cruise ship in background (Image: Peregrine Bryant).

wealthiest island in the Greater Caribbean. Jamaica, in contrast, is the poorest of the English-speaking islands. Still too, Bermuda is an overseas dependent territory with a good deal of autonomy, while Jamaica went independent in the middle of the 20th century. Despite their differences, the cultural heritage tourism landscapes of both St. George's and Falmouth are united by the non-inclusive, non-equitable nature of their cultural heritage landscapes. Both suffer from colonialist and neo-colonial pressures that marginalize local, Black, and minority heritage narratives in favor of white colonialist history. The irony of the 21st-century Caribbean is that while independence allowed islands to administratively break free from colonial power structures, these places are still reliant on former colonial and neo-colonial powers for their economic viability—most notably, in the form of tourism and with respect to heritage conservation, as shown above.

The tourism economy is a driving force for these challenges in both case studies in different ways. In Bermuda, the majority White, North American tourist demographic are pushing stakeholders to engage in colonial heritage out of a perceived financial necessity. In Jamaica, the enclave tourism model of "Historic Falmouth" has led to the disenfranchisement of the Jamaican people in the production of their heritage tourism narratives, in favor of a whitewashed history inside the walls of the cruise ship compound. Outside of this fantasy landscape, negligent heritage conservation policies have led to the destruction of the historically significant buildings that only degrade the integrity of Falmouth's heritage landscape.

The irony of these contexts is that they illustrate two ways in which historically oppressed, enslaved, and subaltern communities that now occupy majority political and social positions are still being under-represented and ignored in the cultural heritage tourism sector in distinct ways: in Bermuda, by structural disenfranchisement on the part of heritage planning policy and underdeveloped subaltern heritage programming and, in Jamaica, by the neo-colonial enclave tourism practices of the cruise ship industry, and negligent heritage conservation oversight. These disconnects are detrimental to the well-being of the local residents and occur at the cost of cultural heritage programming that could serve to build shared senses of identity and connections to the past.

As the cases show, cultural heritage and its conservation can be understood as a major component of neo-colonial tourism forces through the manifold ways that the tangible past is commodified, packaged, and performed for tourism audiences. This presented past is ultimately asymmetrical. Inequitable sets of narratives and relationships are skewed towards a tourism product of comfort and ease for predominantly North American and Western European audiences. The inequities and lack of representation evident in the cases are a product of four connected processes: 1) a lack of cohesion among heritage stakeholder groups; 2) a lack of enforcement of conservation ordinances and guidelines, despite codification by national planning offices; 3) a lack of local capacity and resources to create more inclusive heritage programming; 4) a lack of heritage programming that is inclusive focused, directly engaging visitors while also serving local populations. Heritage justice entails addressing these four processes to facilitate equity, fair representation and inclusiveness as discussed below.

Facilitating heritage justice

Within this context, I offer several structural changes that should be addressed by planning offices, heritage conservation professionals, and stakeholders in Bermuda and Jamaica. Above all, there needs to be a re-orientation of the heritage planning structure in several key areas: 1). The built heritage protection guidelines need to be altered to make them less focused on European, colonial built heritage, and these policies should be enforced. 2.) The heritage planning priorities need to include aspects of intangible heritage, with an equitable emphasis on oral histories and traditions that are vital to the cultural heritage of the Afro-Bermudians. 3.) Incorporating a deeper civic engagement in the planning process will provide multiple, diverse voices for the protection of tangible and intangible heritage.

Changes are also needed to address the structurally and institutionally engrained marginalization of residents (Afro-Bermudians and local Jamaicans) in order to facilitate just and equitable heritage conservation and tourism development. Al heritage stakeholders must directly address their lack of inclusion in the heritage tourism narratives that acknowledge the complicated Caribbean past. Governmental and heritage stakeholders must also directly address the exclusionary, predatory, and neocolonial aspects of the tourism industry, such as the enclave tourism in Falmouth. In Bermuda, the structural, colonial practices of heritage planning exclude Afro-Bermudians entirely from the heritage process. This can only be accomplished through proactive stakeholder and community engagement through public forums and workshops that foster authentic dialogue among the parties. The wider, global heritage conservation community can and should have a key role in facilitating these discussions.

The greatest need in every case is local capacity building. While external expertise is needed, relying on off-island researchers for capacity perpetuates the neo-colonial aspects of heritage tourism. While Bermuda has been challenged to foster local cultural heritage researchers, Jamaica has two major universities (University of the West Indies, Mona, and the University of Technology) training future cultural heritage specialists. Unfortunately, in both contexts, there are few well-paying jobs for cultural heritage practitioners. Opportunities for the capacity building would provide more sustainable solutions for inclusive cultural heritage narratives on both islands.

Conclusion

St. George's and Falmouth offer differing iterations of regional challenges to inclusive and equitable cultural heritage programming. It was argued that the conservation and commodification of heritage landscapes of St. George's and Falmouth must be decolonized from historic colonial narratives and 21st-century neo-colonial forces in order to facilitate *heritage*. In both cases, building local expertise and heritage conservation policies that benefit local populations are paramount for heritage justice, and crucial to providing economic and heritage return to Bermudians and Jamaicans, not just to international tourism companies. Also crucial is collaborative planning that includes local stakeholders, governments, transnational tourism companies, and the broader global heritage conservation community as key stakeholders. Here, including local residents and those diverse groups that stand to be most affected by cultural heritage conservation and tourism development in the Caribbean is a key principle of *heritage justice*. They must be included in heritage planning and policymaking, not just to have their voice heard but also in decision making on the representation and use of their cultural heritage for tourism purposes.

Directions for future research include examining resident and visitor perceptions of heritage tourism and heritage conservation, particularly with respect to colonial history, heritage representation, and interpretation. More considerable research attention is needed on collaborative planning and equity planning processes that can lead to developing an *inclusive and equitable framework for heritage justice* in the context of heritage conservation and tourism development (Metzger, 1996; Reardon et al., 2008; Zapata & Bates, 2015).

Notes

1. In Bermuda, since 2007 I have undertaken intensive research investigation and heritage collaborations through archaeological excavations (Fortenberry, 2008, 2009, 2015); architectural investigation and recordation (Fortenberry, 2017, 2018, 2020). Through this engagement, I have analyzed and been a participant in the national planning process and UNESCO World Heritage governance and procedures. Still, too, I have been an advisor to the Bermuda National Trust, National Museum of Bermuda, the St. George's Historical Society, the St. George's UNESCO World Heritage Planning Committee, and the National Planning Office.
2. I have worked with Falmouth Heritage Renewal (FHR), a heritage non-profit, as both a collaborator and a board member (Chappell and Fortenberry, 2017; Fortenberry, 2019).
3. The construction of a hotel on Gates Bay has caused widespread panic among heritage stakeholders in St. George's. The site sits within the buffer zone of the World Heritage area and is just meters from one of the major fortification contributing properties, Fort St. Catherine's. The hotel's impact on St. George's World Heritage status is still uncertain (Fortenberry, 2016a).

Disclosure statement

No potential conflict of interest was reported by the author.

ORCID

Brent R. Fortenberry http://orcid.org/0000-0001-9544-9633

References

Baird, M. F. (2014). Heritage, human rights, and social justice. *Heritage & Society*, *7*(2), 139–155. https://doi.org/10.1179/2159032X14Z.00000000031

Bell, J. (2019, August 1). Paper unveils new prince timeline, *The Royal Gazette*.

'Bermuda' (2019). *CIA factbook*. Central Intelligence Agency. https://www.cia.gov/library/publications/the-world-factbook/geos/bd.html.

Bermuda Government Department of Planning. (2008). *The Bermuda plan, 2008*. Hamilton.
Bermuda Government Department of Planning. (2010). *Development in the town of St. George*. Bermuda Government, Department of Plannin.
Bermuda Government Department of Planning. (2012). *Alterations or additions to listed buildings and/or buildings located within historic areas GN203*. Bermuda Government.
Bermuda Government Department of Planning. (2014). *Historic town of St. George and related fortifications: World heritage site management plan*.
Bermuda National Trust and Bermuda Government Department of Planning. (2002). *The traditional building guide: Advice for preserving Bermuda's architectural heritage*. Island Press.
Bermuda Tourism Authority. (2012). *Bermuda national tourism master plan*.
Bermuda Tourism Authority. (2016). *2016 visitor arrivals report*. Hamilton.
Bermuda Tourism Authority. (2019). *Bermuda national tourism master plan*.
Bernews. (2017, February 27). *Minister: Fund helping to improve St. George's*, Bernews. http://bernews.com/2017/02/minister-fund-helping-to-improve-st-georges/.
Best, M. N., & Phulgence, W. F. (2013). Interpretation of contested heritage at an attraction in St. Lucia. *Journal of Heritage Tourism*, *8*(1), 21–35. https://doi.org/10.1080/1743873X.2013.765744
Bray, K., & Redfearn, C. H. (2014). Poverty in paradise. In Nelson, L. P. and Chappell, E. A. (Eds.), *Falmouth Jamaica: Architecture as history* (pp. 228–249). University of the West Indies Press.
Brown, H. (2011). *More to be done to strengthen heritage tourism says CARICOM cultural manager*. CARICOM Heritage Experts Meeting.
Bryant, P. (2019). Preservation challenges in Jamaica. In *Culture, heritage, technology seminar 2019*. University of Technology, Jamaica.
Butkiewicz, J. L., & Gordon, L.-R C. (2013). The Economic growth effect of offshore banking in host territories: Evidence from the Caribbean. *World Development*, *44*, 165–179. https://doi.org/10.1016/j.worlddev.2012.09.017
Chappell, E. A. (2010). Accommodating slavery in Bermuda. In Ellis, C. and Ginsburg, R. (Eds.), *Cabin, quarter, plantation: Architecture and landscapes of North American slavery* (pp. 67–98). Yale University Press.
Chappell, E. A., & Fortenberry, B. R. (2016). *Architectural investigations at Tharp's Town House, Falmouth, Jamaica*. Falmouth Heritage Renewal.
Chappell, E. A., & Fortenberry, B. R. (2017). *Architectural investigations at Cinnamon Hill, St. James Jamaica*. Falmouth Heritage Renewal.
Chappell, E. A., Fortenberry, B. R., & Harding, K. V. (2014). *Architectural investigations at the Molton Barrett House*. Falmouth.
Chivallon, C., & Howard, D. (2017). Colonial violence and civilising Utopias in the French and British empires: the Morant Bay Rebellion (1865) and the insurrection of the South (1870). *Slavery & Abolition*, *38*(3), 534–558. https://doi.org/10.1080/0144039X.2016.1267549
Corporation of St. George's (World Heritage Fund and Levy) Act. (2013).
Cumper, G. E. (1954). Labour demand and supply in the Jamaican sugar industry, 1830–1950. *Social and Economic Studies*, *2*(4), 37–86.
Dawson, K. (2013). Enslaved ship pilots in the age of revolutions: Challenging notions of race and slavery between the boundaries of land and sea. *Journal of Social History*, *47*(1), 71–100. https://doi.org/10.1093/jsh/sht052
Ellingson, L. (2009). *Engaging in crystallization in qualitative research: An introduction*. Sage.
Fortenberry, B. R. (2008). Recent excavations at St Peter's Church, St George's Bermuda. *Church Archaeology*, 67–72.
Fortenberry, B. R. (2009). *Archaeological investigations at Ming House*. Bermuda National Trust.
Fortenberry, B. (2013). Preliminary Archaeological Investigations at Verdmont, Smith's Parish. *Bermuda Journal of Archaeology and Maritime History*, *20*, 90–106.
Fortenberry, B. R. (2015). *Archaeological investigations at the Southampton Cottage*. Bermuda National Trust.
Fortenberry, B. R. (2016a). *Archaeological investigations at the St. George's Golf Club, St. George's Bermuda*. Clemson University, National Museum of Bermuda.
Fortenberry, B. R. (2016b). Life among ruins, Bermuda and Britain's colonial heritage. *International Journal of Historical Archaeology*, *20*(3), 601–613. https://doi.org/10.1007/s10761-016-0365-2
Fortenberry, B. R. (2017). *Architectural investigations at Bridge House, St. George's Bermuda*. Bermuda National Trust.
Fortenberry, B. R. (2018). *Architectural investigations at Mitchell House, St. George's Bermuda*. Bermuda National Trust.
Fortenberry, B. R. (2019). *Architectural investigations at the Spanish town barracks, Spanish Town, Jamaica*. Falmouth Heritage Renewal.
Fortenberry, B. R. (2020). *Architectural investigations at St. Peter's Church*. Bermuda National Trust.
Gaieski, J. B., Owings, A. C., Vilar, M. G., Dulik, M. C., Gaieski, D. F., Gittelman, R. M., Lindo, J., Gau, L., & Schurr, T. G. and Genographic Consortium (2011). Genetic Ancestry and indigenous heritage in a native American descendant community in Bermuda. *American Journal of Physical Anthropology*, *146*(3), 392–405. https://doi.org/10.1002/ajpa.21588

Gibson, S. (2019). The same but not quite: An exploration of the mythology and mimicry of the Bermudian Gombey costume. University of Toronto Art Journals, 7. https://utaj.library.utoronto.ca/index.php/utaj/article/view/29944.
Graham, B., Ashworth, G. J., & Tunbridge, J. E. (2000). *A geography of heritage: Power, culture and economy*. Arnold.
Hall, A. (2012). The response of the tourism industry in Jamaica to crime and the threat of terrorism. *Worldwide Hospitality and Tourism Themes*, *4*(1), 59–72. https://doi.org/10.1108/17554211211198598
Hall, C. M., & Tucker, H. (2004). Tourism and postcolonialism: An introduction. In Hall, C. M. and Tucker, H. (Eds.), *Tourism and postcolonialism: Contested discourses, identities, and representations* (pp. 1–24). Routledge.
Hall, M. C., & Tucker, H. (2004). *Tourism and postcolonialism: Contested discourses, identities and representations*. Routledge.
Hampton, M. P. (2005). Heritage, local communities and economic development. *Annals of Tourism Research*, *32*(3), 735–759. https://doi.org/10.1016/j.annals.2004.10.010
Harrigan, N. (1974). The legacy of Caribbean history and tourism. *Annals of Tourism Research*, *2*(1), 13–25. https://doi.org/10.1016/0160-7383(74)90053-X
Honey, M. (2019). (Ed.) *Cruise tourism in the Caribbean: Selling sunshine*. Routledge.
Hori, J. M. (2016). Berthing violent nostalgia: Restored slave ports and the Royal Caribbean historic Falmouth Cruise Terminal. *American Quarterly*, *68*(3), 669–694. https://doi.org/10.1353/aq.2016.0058
Inniss, T., & Jolliffe, L. (2013). The Industrial heritage of sugar at world heritage sites in the Caribbean. In Jolliffe, L. (Ed.), *Sugar heritage and tourism in transition* (pp. 89–109). Channel View Publications.
Ireland, T., & Schofield, J. (2015). The ethics of cultural heritage. In Ireland, T. and Schofield, J. (Eds.), *The ethics of cultural heritage* (pp. 1–10). Springer.
'Jamaica'. (2019). *CIA Factbook*. Central Intelligence Agency. https://www.cia.gov/library/publications/the-world-factbook/geos/jm.html.
Jamaica Tourist Board. (2016). *Jamaica master tourism plan*.
Jamal, T. (2019). *Justice and ethics in tourism*. Routledge.
Jamal, T., & Camargo, B. A. (2014). Sustainable tourism, justice and an ethic of care: Toward the just destination. *Journal of Sustainable Tourism*, *22*(1), 11–30. https://doi.org/10.1080/09669582.2013.786084
Jamal, T., & Hales, R. (2016). Performative justice: New directions in environmental and social justice. *Geoforum*, *76*, 176–180. https://doi.org/10.1016/j.geoforum.2016.09.014
Jarvis, M. J. (1998). *Bermuda's architectural heritage: St. George's*. Hamilton: Bermuda National Trust.
Jarvis, M. J. (2002). Maritime masters and seafaring slaves in Bermuda, 1680–1783. *The William and Mary Quarterly*, *59*(3), 585–622. https://doi.org/10.2307/3491466
Jarvis, M. J. (2010). *In the eye of all trade: Bermuda, Bermudians, and the maritime Atlantic world, 1680–1783*. University of North Carolina Press.
Johnston, R., & Marwood, K. (2017). Action heritage: Research, communities, social justice. *International Journal of Heritage Studies*, *23*(9), 816–831. https://doi.org/10.1080/13527258.2017.1339111
Jones, R. (2004). *Bermuda: Five centuries*. Panatel VDS Ltd.
Jordan, L.-A. (2013). Managing built heritage for tourism in Trinidad and Tobago: Challenges and opportunities. *Journal of Heritage Tourism*, *8*(1), 49–62. https://doi.org/10.1080/1743873X.2013.765748
Jordan, L.-A., & Duval, D. T. (2009). Heritage management and tourism in the Caribbean. In Timothy, D. J. and Nyaupane, G. P. (Eds.), *Cultural heritage and tourism in the developing world: A regional perspective* (pp. 186–208). Routledge.
Jordan, L.-A., & Jolliffe, L. (2013). Heritage tourism in the Caribbean: Current themes and challenges. *Journal of Heritage Tourism*, *8*(1), 1–8. https://doi.org/10.1080/1743873X.2013.765735
Kelly, J. B. (1963). The Jamaican independence constitution of 1962. *Caribbean Studies*, *3*(1), 18–83.
Kerr, W. B. (1995). *Bermuda and the American revolution: 1760–1783*. Bermuda Maritime Museum Press.
Kerswill, M., & Mair, H. (2015). Big ships, small towns: Understanding cruise port development in Falmouth, Jamaica. *Tourism in Marine Environments*, *10*(3), 189–199. https://doi.org/10.3727/154427315X14181438892766
Kim, S. A., & Rodriguez, D. (2006). *Bermuda: Independence by any means; governance as an obsession*. Council on Hemispheric Affairs.
Knight, F. W., & Liss, P. K. (1991). *Atlantic port cities: Economy*, culture, and society in the Atlantic world 1650–1850. University of Tennessee Press.
Lawler, M., & Davenport, J. (2009). *Marcus Garvey: Black nationalist leader*. Infobase Publishing.
Levi, D. E. (1992). Fragmented nationalism: Jamaica since 1938. *History of European Ideas*, *15*(1–3), 413–417. https://doi.org/10.1016/0191-6599(92)90159-A
Lowry, R. (2011). Archaeology in Bermuda. *Post-medieval archaeology*, *45*(1), 7–17. https://doi.org/10.1179/174581311X12983864587855
Martha, W. J. H. (2019). Cruise tourism impacts on historic cities. In Honey, M. (Ed.), *Cruise tourism in the Caribbean: Selling sunshine* (pp. 52–72). Routledge.
McDowall, D. (1999). *Another world: Bermuda and the rise of modern tourism*. Macmillan Education.

McFarlane-Morris, S. (2019). "Come this close, but no closer!" Enclave tourism development and social change in Falmouth, Jamaica. *Journal of Tourism and Cultural Change*, 1–15. https://doi.org/10.1080/14766825.2019.1676253

McKee, D. L., & Chase, G. L. (2003). The economic impact of cruise tourism on Jamaica. *Journal of Tourism Studies*, *14*(2), 16–22.

Metzger, J. T. (1996). The theory and practice of equity planning: An annotated bibliography. *Journal of Planning Literature*, *11*(1), 112–126. https://doi.org/10.1177/088541229601100106

Miller, D. (2017). *Justice*. Mowforth, M., & Munt, I. (Eds.), Retrieved February 1, 2020, from https://plato.stanford.edu/archives/fall2017/entries/justice/.

Mullin, M. (2014). Laura Smalligan's Jonkonnu, a Jamaican slave dance: Contesting the African in African-American. *Slavery & Abolition*, *35*(1), 142–155. https://doi.org/10.1080/0144039X.2013.798172

Nelson, L. P. (2016). *Architecture and empire in Jamaica*. Yale University Press.

Nelson, L. P., & Chappell, E. A. (2014). *Falmouth, Jamaica: Architecture as History*. University of West Indies Press.

Palmer, C. A. (1994). Tourism and colonialism. The experience of the Bahamas. *Annals of Tourism Research*, *21*(4), 792–811. https://doi.org/10.1016/0160-7383(94)90084-1

Parrent, J. M. (2005). Preserving the rich cultural heritage of Falmouth, Jamaica: A case study. In 15th ICOMOS General Assembly and International Symposium. openarchive.icomos.org.

Pickett, D., & Brown, M. III (1998). Archaeology of the Tucker House kitchen revisited: Recent excavations of the kitchen. *Bermuda Journal of Archaeology and Maritime History*, *10*, 107–123.

Reardon, K. M., Ionescu-Heroiu, M., & Rumbach, A. J. (2008). Equity planning in post-hurricane Katrina New Orleans: Lessons from the ninth ward. *Cityscape*, *10*(3), 57–76.

Robertson, J. (2008). Late seventeenth-century Spanish town, Jamaica: Building an English city on Spanish foundations. *Early American Studies: An Interdisciplinary Journal*, *6*(2), 346–390.

Royle, S. (2010). Postcolonial culture on dependent islands. *Space and Culture*, *13*(2), 203–215. https://doi.org/10.1177/1206331209358230

Smalligan, L. M. (2011). An effigy for the enslaved: Jonkonnu in Jamaica and Belisario's sketches of character. *Slavery & Abolition*, *32*(4), 561–581. https://doi.org/10.1080/0144039X.2011.601620

Smith, L. (2010). Ethics or social justice?: Heritage and the politics of recognition. *Australian Aboriginal Studies*, *2*, 60–68.

UNESCO. (2000). *Historic town of St. George's and related fortifications*.

Waterton, E., & Smith, L. (2010). The recognition and misrecognition of community heritage. *International Journal of Heritage Studies*, *16*(1–2), 4–15. https://doi.org/10.1080/13527250903441671

White, D. L. (ed.) (1999). *Bermuda's architectural heritage: Sandys*. Bermuda National Trust.

White, D. L. (2015). *Bermuda's architectural heritage: Hamilton: Town and city*. Bermuda National Trust.

Wilkinson, H. C. (1950). *Bermuda in the old empire: A history of the island from the dissolution of the Somers Island company until the end of the American revolution*. Oxford University Press. https://doi.org/10.1086/ahr/56.1.97

Wilkinson, P. F. (2004). Caribbean tourism policy and planning. In Duval, D. T. (Ed.), *Tourism in the Caribbean: Trends, development, prospects* (pp. 81–98). Routledge.

Zapata, M. A., & Bates, L. K. (2015). Equity planning revisited. *Journal of Planning Education and Research*, *35*(3), 245–248. https://doi.org/10.1177/0739456X15589967

Representation of "mill girls" at a UNESCO World Heritage Site in Gunma, Japan

Naho Ueda Maruyama and Kyle Maurice Woosnam

ABSTRACT
Distribution and recognition injustices are often inextricably linked throughout history. Economic exploitation of a minority group by a dominant group may even transition into cultural domination of the former group by the latter. However, dual and historical analysis of these two forms of injustices is scarce regarding representation of women in the context of tourism. This study explores how "mill girls," young women who worked in the Tomioka Silk Mill, Japan, dating back to 1870–1940, are represented to tourists. A qualitative case study approach revealed that mill girls are represented in fantasized and sanitized forms. Also, only the short, most prosperous time of the mill is showcased at the expense of excluding the historic struggles of female workers. We argue based on our study that the touristic representations reflect persistent and changing forms of gender injustice that the mill girls have faced. While the mill girls were economically exploited as the mill was in operation, when the mill was turned into a cultural heritage, the form of economic injustice morphed into recognition injustice where the women's experiences find little expression. The study points to the importance of understanding the role of tourism in perpetuating historically engrained gender injustices.

Introduction

Women have long been objects of injustice, facing inequity in education, marriage, employment, and many other spheres of everyday life. This gender stratification can appear in both blatant and subtle forms, ranging from domestic violence to a "glass ceiling" when seeking managerial positions in the work place. Japan is one of the countries where women have been treated with widespread, subtle inequalities. Legal equality between men and women was spelled out in the new Constitution enforced in 1946, with numerous political reforms being passed, including suffrage and equal employment opportunities for women (Gender Equality Bureau Cabinet Office, n.d.). However, in reality, Japanese cultural and social norms perpetuate gender stratification (Barrett, 2004; King, 2009). Such an understanding raises many questions. In Japan, how are women, especially those who have made great contributions throughout history, represented to tourists? Are these individuals recognized as empowered women who have impacted society in general and as part of the workforce specifically, or is their femininity and submissiveness still championed? More importantly, are they represented in a way that demonstrate fair and socially just treatment to women in the context of tourism?

Understanding the historical context is especially important in studies of social representations of minority, disadvantaged or marginalized groups. A number of tourism studies (e.g., Barton & Leonard, 2010; Senju, 2006; Tegelberg, 2010) reveal that historically-produced economic divisions and colonization continue to influence present-day social representations of oppressed groups. Representations of the past in tourism marketing and interpretive materials are often selective, strategic, and aimed to promote favorable impressions of the place or even particularly ideologies ("worldmaking") to the audience (Hollinshead, 2007; Rivera, 2008). In the process of constructing positive impressions, historical conflicts and contestations can be altered through cultural reframing and covering (Buzinde & Santos, 2008; Goffman, 1963). Rivera (2008) argues that the process of directing attention away from the stigmatic past may become intensified particularly when an international audience is involved. However, there is little research on representations of Japanese women in tourism destinations that are focused on historical events, especially on how the past is being used to present the stories of women who have been subject to oppression or exploitation.

The goal of this study is to explore how "mill girls," who worked in the Tomioka Silk Mill (TSM), particularly between 1870 and 1940 (the time between the Meiji restoration and the beginning of WWII), are being currently represented to tourists. The Tomioka Silk Mill, designated as a World Heritage Site (WHS) by UNESCO, is one of the oldest mill factories in Japan. Young female workers known as "Mill Girls," who worked in cotton and silk mills of Japan, are considered to be the first female factory workers who made significant contributions to the growth of a modern industrial sector after the Meiji restoration that took place in the late 1860s (Hunter, 1993). At the same time, the female workers have also been known for their experiences of hard, exploitative working conditions. As Hunter (1993) argues, the burden of national growth and rapid industrialization were imposed on these women working in such an industrial position.

This study attempted to explore the representation of mill girls, specifically, whether and how the past exploitation and current representation interplay. The research presented below indicated that female workers are represented in a way that responds to tourists' expectations for innocent, submissive, and nostalgic women (Caton & Santos, 2007; Galbraith, 2009; Hinton, 2014). In addition, the arduous and exploitative work of the women are concealed throughout the process of showcasing the glorious history of technological development within the mill. Such touristic representations raise concern about how historical injustices tend to continue in subtle ways to influence the continuous and changing faces of gender injustice in Japan. It also points to the need for greater attention to how tourism might perpetuate historically situated gender injustices, and thus to greater research and theory building on justice and ethics in tourism. We offer a brief review of some approaches to justice in the literature review below that we felt were relevant to our study as it emerged. This is followed by the research study, findings and discussion. The paper concludes by reflecting on the study and its justice related aspects.

Literature review

Approaches to justice: Sensitizing concepts

While we did not explicitly draw upon on specific justice related theoretical approaches in our research study, our analysis was sensitized to the approaches we describe below, which facilitated identifying and discussing our findings in relation to representations of the mills girls in the touristic context described in our study below. Social justice has been approached in multiple ways, and we found the following especially helpful. Distributive justice, for instance, is concerned with the fair distribution of societal goods and benefits (i.e., wealth, social welfare, natural assets, etc.) and burdens (Rawls, 2009). Social justice has evolved significantly from a modernist focus on distributive justice to pluralistic approach to justice including *recognition justice* (see, for example, Fraser, 2009, 2010: Jamal, 2019; Young, 2011). Fraser (2009), for instance, argues for the importance of taking a cultural perspective on distributive justice, that includes recognition of the status of diverse social groups. Her approach is one of *perspective dualism*, integrating distribution justice and recognition justice, because injustices are typically multi-dimensional; various social groups suffer from both unjust distribution and misrecognition. For example, the division of labor can creates gender-specific forms of distributive injustice, e.g., paid labor that privileges men in higher-paid, professional occupations versus lower-paid occupations or invisible work of women as homemakers. Traits associated with masculinity are privileged while devaluing things associated with femininity. Fraser (2010) later supplemented her bivalent approach to justice with a *trivalent* approach that adds a political dimension of representation to distributive justice and recognition justice.

Fraser (2009) also discusses that the relationship among these need to be grasped historically. Her work points to the importance of empirical studies being able to trace how distributive and recognition justice are situated and develop over time. Like Fraser, Young (2011) is also concerned with recognition justice, and social patterns of representation, communication, and cultural interpretation. Among many important perspectives, she addresses exploitation and domination, stereotyping and objectification of underprivileged or lower status groups, including ethnic minorities, people with diverse sexual orientation (i.e., gays, lesbians, and transgendered), or women. Young (2013, p. 333) also addresses injustice as a form of "cultural imperialism" that involves "the universalization of a dominant group's experience and culture, and its establishment as the norm." Through various means including stereotyping, the cultural expression and interpretation of the dominant group receive wider dissemination and are considered "normal," while other marginalized or oppressed social groups come across as inferior. Cultural stereotyping can become so pervasive throughout society that it becomes unquestioned and unchallenged.

Yunseong (1999) argues that it is the power of hegemony that endows cultural imperialism with effectiveness. Hegemony is defined as "consent given by the great masses of the population… imposed on social life by the dominant fundamental group" (Gramsci & Hoare, 1971, p.12). The consensus is created and disseminated not by coercion, but through the influence of intellectuals and civic institutions. For instance, marginalized groups are often represented within a mainstream media in a way that favors the dominant group (Evans, 2002). Kurtz (1996) further maintains that representatives of marginalized groups generally behave in a way suitable to the cultural hegemony because cultural domination becomes so saturated within everyday life that individuals may subconsciously accept the consensus imposed on them without questioning.

Tourism research and justice

With the rise of research interests in sustainable tourism, the concept of justice has begun to receive substantial attention (Higgins-Desbiolles, 2008, 2010; Jamal & Camargo, 2014). The literature on tourism and justice has emphasized multiple dimensions, including economic distribution, environmental justice, and social and cultural justice (i.e., Bowitz & Ibenholt, 2009; Scheyvens & Momsen, 2008). For example, a number of studies (i.e., Sandrook, 2010; Yan & Wall, 2002) reveal that tourism produces unbalanced distribution of benefits between hosts and guests because of the high percentage of leakage, or even within a community due to the fact that some residents have better access to tourism resource than others. Similarly, environmental justice, that particularly concerns equitable distribution of environmental costs and benefits (i.e., exposure to a hazardous waste and dirty air, access to green space, etc.), has been a critical dimension of tourism research. Studies (Floyd & Johnson, 2002; Lee & Jamal, 2008; Porter & Tarrant, 2001; Schelhas, 2002) indicate that certain socioeconomic and racial groups suffer from unjust distribution of environmental costs and benefits through tourism development. (Fennell, 2017) terms this type of inequity as tourism development racism.

In research of both economic and environmental justice, the importance of fair procedure has also been emphasized (Camargo et al., 2007; Gezon, 2014; Karatepe, 2011). Studies similarly reveal the importance of stakeholders' participation in a decision-making process to reduce conflicts and enhance a sense of fairness (Dai et al., 2013; Huong et al., 2016).

Studies regarding local cultural and historical resources have also revealed multiple aspects of injustice. For example, studies (Santos & Yan, 2008; Shaw et al., 2004) reveal that tourism representation often emphasizes the differences of the local culture from the dominant culture and reinforce stereotypes, while ignoring how the locals perceive themselves and their own culture. The biased representation results in unfair distribution of benefits as well as a loss of ownership of their own culture. Kirtsoglou and Theodossopoulos (2004) also reveal the lack of opportunities for local indigenous residents to participate in the planning process.

Furthermore, Hollinshead (2007, p.168) argues that tourism representation is a process of "world making," that is an act of cultural selection or "the creative and often 'false' or 'faux' imaginative processes." Hollinshead and Suleman (2017) further describe that, through the world making process, narratives are negotiated, deconstructed, and reconstructed into a form that privileges a view of the dominant group, and then distributed through tourism promotion. Rivera (2008) similarly argues that tourism representation is often influenced by those who have power to institutionalize a particular narrative, such as a national government. Such institutions construct a master narrative by using several strategies (i.e., discrediting or covering) to respond to a stigmatic past and manage a positive reputation from the external audience.

A plethora of examples speak to tourism representation that highlights narratives of heritage sponsored by dominant institutions while concealing the alternative interpretation and experiences of subordinated and historically oppressed groups (Chang & Huang, 2005; Chronis, 2012; Goulding & Domic, 2009; Timothy & Boyd, 2006). Research by Buzinde and Santos (2008) demonstrate, for instance, representations of plantation heritage that work hegemonically towards unifying America into an imagined history that emphasizes economic prosperity and prowess of white gentry and purposefully makes invisible ongoing injustices related to its plantation heritage. Heritage representation here has much in common with dark tourism, which involves visits to sites related to death and disaster (Light, 2017). Like heritage tourism, dark tourism is often placed in a broader political project of national identity-making by constructing and promoting a collective memory (Alderman et al., 2016; Issac, 2018; Lischer, 2019). The commodification of "dark" heritage, such as through plantation tourism, reveals a common thread of commercial exploitation of a problematic history towards the end of profit.

Several studies have attempted to examine the interplay of distributive justice and recognition justice (e.g., Figueroa & Waitt, 2010; Whyte, 2016). Studies such as by McGregor (2009) and Senju (2006) demonstrate how social groups who were colonized by settlers or imperial powers continue to stay powerless even after the decolonization or recognition of the indigenous population, and their experiences are often not recognized under the norm set by the dominant group. For example, Figueroa and Waitt (2010) revealed how the historical relationship between Aboriginal and non-Aboriginal people have influenced touristic use of the local resource in Uluru, Australia, a sacred Aboriginal site. However, until a recent legal ban in 2019, tourists continued to climb the rock despite pleas by the Traditional Owners. McGregor (2009) argues that this may be because of "colonial expectation of the landscape" in which members of the dominant society frame landscape and indigenous people through a colonial framework. Similarly, Senju (2006) indicated that in Singapore, the fact that the Japanese military occupied the Raffles Hotel, a well-known colonial hotel in Singapore, during WWI is hidden from the Japanese tourists because such history may interfere with the experiences of tourists who might seek exotic experiences in Singapore. The representation of Angkor Watt (a designated WHS in Cambodia) by

Lonely Planet only focuses on exotic and auratic perspectives of the heritage, which articulates "colonial fantasies" while rendering local people invisible (Tegelberg, 2010).

Although the findings above indicate the importance of retrospective analysis that focuses on how the class divisions created for previous generations continue to shape representation of people in modern-day society, such a dual and retrospective approach has been lacking regarding representation of women. Several studies in tourism (Alessio & Jóhannsdóttir, 2011; Prichard, 2001) revealed that women are often represented as exotic or submissive "Others." Sirakaya and Sonmez (2000), for example, analyzed how women are represented in state tourism promotional materials based on Goffman's seminal framework. The study revealed that women tend to be portrayed shorter than men (relative size), lightly touching caressed objects or themselves (feminine touch), and cast in a supporting role than an executive role (rank order). In addition, women are portrayed as unable to master a situation, and inferior to men. Sirakaya and Sonmez (2000) warn that these gender stereotypes subtly embedded in tourism ads may perpetuate gender stereotypes and lead to negative consequences for destination development. Having these issues in mind, this study explores how women and their history are represented to current tourists.

Research study: Tomioka silk mill, Japan

Tomioka Silk Mill, or TSM, is located in Tomioka city within the Gunma prefecture. Gunma prefecture is located in the Kanto region, northwest of Tokyo. A large part of the prefecture is mountainous, but its central and southeast areas, including Tomioka city, have relatively easy access to Tokyo, owing to the highway and high-speed railway system connecting Gunma and Tokyo. The mill was established by the Meiji government in 1872 as the first government-owned silk-reeling factory (Tomioka City, n.2000d). Tomioka city was chosen as the location for Tomioka Silk Mill for numerous reasons. After terminating its isolation policy in 1854, the new government prioritized economic growth and industrial modernization to compete with western countries. Japan began to trade with Western countries where raw silk was in extensive demand. Due to this demand, mass-produced, low-quality products were often manufactured and traded. European countries who were in great need of raw silk, including Italy, England, and France, requested Japan to install the most advanced silk-reeling machines and hire Western directors to produce high quality raw silk, increase production, and train engineers. Several regions around Tokyo (the new capital after the Meiji revolution) were investigated as a site to build a modern factory. Among the regions investigated, Tomioka was chosen because sericulture in the area was flourishing and high-quality cocoons were produced in the area. Also, a large size of land and fresh water were available, and coal was produced in the nearby region.

Although Bukh (2007) argues that a large number of Korean and Chinese immigrants worked, often forcefully, at many plants and coal mines in Japan at the time, no record has indicated immigrant females from these countries worked at the TSM. Rather, mill girls at the TSM were from all over Japan, including Gunma prefecture, Shiga prefecture, and Nagano prefecture (Imai, 2014). Some of the workers were from prosperous families (i.e., Samurai families), and encouraged by their parents to hone their skills as mill workers and contribute to the national growth.

While the mill was run by the government for approximately 10 years, in 1893, the Meiji government sold the mill to the Mitsui Conglomerate, and the factory became privatized. In 1902, it was again transferred to the Hara Unlimited Partnership, and merged with the Katakura Silk Reeling and Spinning Company in 1939. Though the mill continued to thrive as a silk-reeling factory for all those years, because of the decline of the silk industry in Japan, it stopped its operation in 1987. Currently, the buildings are owned and maintained by Tomioka City.

In 2014, "Tomioka Silk Mill and Related Sites," that include TSM and three other sites that contributed to the silk production, were listed as World Heritage Sites by meeting two of the six criteria set by UNESCO (Tomioka City, n.2014d.). Firstly, TSM has played a significant role in enhancing the mutual exchanges between Japan and the world regarding mass production of high quality of raw silk. TSM introduced the Western silk-reeling technologies of mass production to Japan, improved the technologies within Japan, and redistributed the technologies to the worlds̀ silk industry. These facts were recognized to meet criteria (ii). In addition, the buildings of the silk mill were considered outstanding stages where the innovation of influencing silk reeling technology took place. This met the selection criteria (iv).

Study methods

This study employs a qualitative research design. Such a design is largely based on the assumption that knowledge is socially constructed rather than independent from the knowers, and that there are multiple realities that various social groups construct based on their own worldviews (Creswell & Creswell, 2017; Yilmaz, 2013). This study, therefore, aims to inquire about the images of mill women constructed by the dominant group and possible gaps in how such images may not reflect historical reality among the female workers. To increase the soundness of qualitative research, collecting multiple forms of data from different sources, or triangulating data, is an important strategy (Lincoln & Guba, 1986). Data for this study was collected from several mediums, including on-site investigation, analysis of historical documents and current tourism promotion materials regarding TSM, and a personal interview with an executive of the Tomioka Silk Mill Research Center.

First, extensive site visits were carried out by the first author to grasp the patterns of visual representation within and outside the TSM. The first author visited the TSM and the surrounding area eight times from 2016 to 2018. During these visits, pictures and notes were taken to record where and how mill girls are represented. In addition, the first author talked with local residents and shop owners to understand the locals' thoughts about the TSM and its representation. In addition to the TSM, an information center, souvenir shops, and restaurants were also visited to investigate how images of mill girls were used for consumer products. The data collected from the site investigation was then cross-referenced with historical records, reports, and current tourism promotional materials.

In addition, several historical records were used to investigate the working conditions of mill girls around 1870-1940. Two government reports published in 1933 and 1977 were used to explore the working hours, salaries, and sanitary conditions of the TSM and compare working conditions with other factories. Additionally, three documentary books were used to understand the working condition from the female workers' perspectives. *"Tomioka Diary"* is a memoir written by Ei Wada. This memoir reveals the working condition of the TSM under the government's management from the perspective of the female worker with a relatively privileged status. "*Nomugi* Pass" written by Shigemi Yamamoto published in 1968 and "Joko Aishi," written by Wakizo Hosoi published in 1925, on the contrary, revealed the experiences of female workers in the lower working class. These two books included interviews from the mill women who worked in different mills and cotton factories to expose exploitative working conditions under the Meiji government's policy to enrich the country.

Several researchers (Jenkins, 2003; Sirakaya & Sonmez, 2000) maintain that analyzing content of texts in tourism guidebooks and magazine articles allows us to systematically analyze how the nature of destination is promoted in a written form. Therefore, in this study, the textual and visual representation of mill girls of the TSM for current tourism was investigated. This included analysis of the official website of the TSM, five Japanese travel guidebooks, such as *Rurubu* and *Maple,* and 21 online travel guides both in Japanese and English, including Japan guide.com and Japan Monthly Web Magazine. An inductive content analysis was conducted to analyze textual representation. This analysis involved open coding and creating categories (Burnard, 1991; Elo &

Table 1.

[Category 1] Women's working conditions	
Workhours and Welfare	• Female workers were provided with a welfare package
	• Worked no more than about 8 hours a day and had Sundays off
Facilities	• The property had a dormitory and clinic
	• Evening classes were offered to learn sewing, reading, writing and abacus
[Category 2] Women's empowerment	
Social position	• Female worker take on a more active role in society
Having a well-respected career	• Female workers leant how to use the new, modern machines.
	• Female workers became silk-reeling supervisors in their hometowns

Kyngas, 2008). For the purpose of this study, the texts in the travel magazines and websites were carefully examined, and categories were generated. The lists of categories were then grouped under broader headings to reduce the number of categories (Table 1). This analysis was undertaken manually. To ensure the trustworthiness of the analysis, a table of categories and quotes is presented in the subsequent section. This demonstrates a link between the results and the data (Elo & Kyngas, 2008). Finally, a personal interview with a head of Tomioka Silk Mill Research Center was conducted by the first author in July of 2019 to investigate the history and current activities in place to preserve the TSM.

Addressing researchers' reflexivity is a critical task to ensure the trustworthiness of qualitative studies (Mauthner & Doucet, 2003). Particularly, researchers' positionality becomes salient when the research is concerned with those who are oppressed or socially marginalized (Hayfield & Huxley 2014; Pitman, 2002). More precisely, being an outsider allows the researcher to ask basic questions to explore topics and notice cultural patterns that insiders may overlook (Tang, 2007). The first author of this research, who conducted the data collection, was born and raised outside Tomioka city, and does not have any relative or friend who or whose family worked in a silk mill. This certainly provided her the advantages of being an outsider. At the same time, the first author, as a Japanese female, has experienced both obvious and subtle gender inequality both in work places and in personal lives in Japan. This allowed her to deeply understand the issues and provide culturally specific interpretations (Dwyer and Buckle, 2009).

Findings and discussion

Mill girls in Japan and TSM

Generally speaking, working conditions in silk mills were known for their harshness (Hunter, 1993; Tsurumi, 1991). Especially after the scale of production was enlarged and demand for labor was increased in the 1880s, a large number of women were employed in factories, which created a "new permanent working class" (Hunter, 1993, p.1). These workers were predominantly young daughters from impoverished families in rural areas, and bound by contract of one to three years with extremely low wages (Tsurumi, 1991). They were usually recruited through recruiting agents, or "slave dealers," who took high a percentage of commission. Government reports published in the early 1990s indicated that the female workers often worked longer than 12 hours per day, and the conditions inside the mill were extremely unhealthy and unsafe (Hunter, 1993). Because cocoons needed to be boiled for reeling, the air in the mill was constantly humid and dusty, and workers' hands were constantly exposed to boiled water. Because of these conditions, sickness, particularly tuberculosis, among the workers were prevalent. In the dormitory, the female workers were made to share bedding, and nutrition standards of the food was very low. Nonfiction books, titled "*Nomugi* Pass" published in 1968 and "*Joko Aishi*" published in 1929 revealed these histories to the public.

By contrast, the TSM has been described in its official promotional materials and in various travel magazines and websites as a "model recruiting factory" which offered privileged working conditions to female workers. In addition to achieving numerous technological innovations, the

Figure 1. Otomi-chan.

Japanese government attempted to train engineers and supervisors, and disseminate silk production skills nationwide. Therefore, the government recruited females from around the nation so that workers would learn the skills at the TSM and became silk reeling instructors once they returned to their homes.

Female workers in the TSM worked approximately eight hours per day, and were guaranteed to have one day off per week (Tomioka City, n.1993d.). Workers also had 10 days off for holiday seasons (i.e., summer, new year's). In addition, there was a clinic in the factory with a full-time doctor and nurse, and evening classes to learn writing and reading were also offered within the factory. The lifestyles of female workers in the TSM were recorded in a memoir titled "Tomioka Diary," written by Ei Wada (one of female workers who worked in the TSM from 1973 to 1974). Ei was in a leading position in the mill at the time, and became an influential instructor in her hometown after leaving the TSM (Yamazaki, 2016). The diary shows that Ei and other female workers were enthusiastic about learning new skills and took pride in their job.

Tourism representation of mill girls

Embracing the history as a model recruiting factory, a recurrent theme in tourism promotion of the TSM is the character "*Koujo*," meaning mill girls in Japanese. For example, visitors to TSM easily find a female character named "*Otomi-chan*" in the surrounding area of the mill. *Otomi-chan* is an imaginary character created by Tomioka City to represent the local area (Figure 1). The caricature is created using a stereotypical image of mill girls, wearing a kimono (a Japanese

Figure 2. An animated mill girl printed on souvenirs.

traditional clothing), hakama (a trouser type traditional clothing), and red tasuki (a cord used to tuck up the sleeves of a kimono). "*Otomi-chan*" is one of the "*gotouchi character*" mascots produced by local governments to represent their respective cities and towns. "*Gotouchi*" literally means "the place." The boom of using characters to promote local specialty began in the late 2000s in Japan (Tan, 2014). The characters are presented in a mascot-costume form as well as appear on numerous products (i.e., key chains, stuffed animals, food packages, etc.) especially near their hometowns (Sutera, 2016). *Otomi-chan* is printed on the various mediums in Tomioka city, including websites, posters, street signs, and packages of souvenirs.

Besides *Otomi-chan*, visitors may find several other animated mill girls printed on souvenirs and other products (Figure 2). These animated mill girls appear to be young, skinny girls with large eyes. They are wearing the similar outfit as *Otomi-chan*.

At the Takasaki station and Joshu-Tomioka station—the two train stations that visitors to the TSM typically use—, visitors may also find a female character printed on exhibition panels (Figure 3). This character is a young, skinny girl with pink cheeks and large eyes. Her name is "*Silk Tomioka*." She is a part of a character series titled "train girls-uniform collection," created by a Japanese toy factory. Similar to the *gotochi* character, this series is created to represent respective local areas. Although *Silk Tomioka* is not a main character in the official tourism promotion,

Figure 3. Silk Tomioka.

and not wearing the traditional outfit, she is given a role to support tourism in the Tomioka area (Tomica, n.d.). In addition, she is exhibited in the train station where many tourists form their first impression of what the town is like. Additionally, her image is printed on souvenirs sold at the train station.

However, once visitors step into the factory, the exhibits predominantly focus on very technical aspects of silk production and the facility, while exhibits regarding the life of mill girls is extremely limited. For example, visitors are provided an explanation of the construction methods used to build the factory buildings, and learn how advanced the methods were at the time and why the buildings have been maintained in the original shape. Visitors are also welcome to walk into the buildings where reeling machines were installed, and can learn the silk production process through tour guides and movies. The cocoon houses are also open for visitors as well as panels that explain the growing process. On the contrary, visitors are only allowed a view of the dormitory and teaching rooms used for evening classes from the outside.

"Moe-Okoshi" and TSM

Representing the TSM by using animated characters of mill girls may have the potential to transform the negative stereotypes of mill girls from selfless, powerless, and exploited individuals to more fashionable and somewhat empowered women who work in a privileged industry (Dewangga & Hendriyani, 2018). At the same time, however, using these characters may misrepresent the mill girls as "fantasy" or even "icon" without being historically accurate, particularly from the perspective of male tourists. More precisely, the representation of mill girls using female anime characters around the TSM is likely a manifestation of "*moe-okoshi*," Japan's emerging strategy of place branding and tourism promotion that adopts the concept of "*moe*." *Moe* is a slang Japanese word used to express a burning passion or imaginary love for attractive young girls (Galbraith, 2009; Morikawa, 2008). *Okoshi* means revitalization of regional towns that face depopulation and economic downturn.

Revitalization of regional towns by adopting anime characters has been triggered by the boom of Japanese contents tourism, or media-induced tourism, in which fans of media contents visit related sites (Seaton & Yamamura, 2015). Contents tourism was identified as a major economic force by the Japanese government in the early 2000s because of the international popularity of Japanese pop-culture. The national government put forth initiatives to promote regional tourism via popular culture, which resulted in creating a subculture that blended traditional and pop culture, attracting a younger generation. And, in the subculture, the objects of "*moe*" are the strong lure especially for male fans (Okamoto, 2015).

The objects of *moe* could be divided into several types. *Otomi-chan* is one of the *gotouchi* characters who generally have animal-like, round bodies and harmless personalities, and express innocence and childishness (Tan, 2014). On the other hand, other animated mill girls as well as "*Tomioka Silk*" can be examples of the "*bishoujo* character" type. *Bishoujo* literally means a pretty girl, although in the Japanese pop culture, it refers to a beautiful young female character generally in manga, anime, online games, and other media contents (Bloem, 2014). *Bishoujo* characters typically have large eyes, glossy skin, skinny body, and pure personality (Galbraith, 2009).

Although *gotouchi* characters (particularly those who are given the female roles) and *bishoujo* characters have different appearances, they have several elements in common that inspire a man's desire for "*moe*." Firstly, both types of characters are completely fictional and fantastical (Galbraith, 2009; Saito, 2014). The characters with animal-like appearance or unrealistically large eyes are deliberately separated from reality. In addition, the *moe* characters are generally in the forms of physically immature, young girls or animal-like beings who will not betray or judge viewers or who seem dependent on viewers for security. That is to say, *moe* characters allow male viewers the opportunity to escape from reality and enact their desire to feel more masculine in the virtual domain (Hinton, 2014).

Another element of *moe* is the strong focus on a character rather a narrative (Azuma, 2007; Hinton, 2014). The character that may have been originally embedded in a narrative is removed from the story and becomes an "icon." Galbraith (2009) describes this change as the shift from "narrative consumption" to a more fragmented, "database consumption." By isolating the characters from a complete story, viewers may reimagine the character as a mere object of *moe*. Lacking the actual exhibit in the mill may exacerbate such transformation of mill girls into fantasy or icons. Furthermore, not recognizing their historical contributions may reduce mill girls into simple images for consumption.

Lastly, these characters are given the personalities that are typically appealing for male viewers (Alessio & Jóhannsdóttir, 2011; Prichard, 2001; Sirakaya & Sonmez, 2000). For example, *Otomi-chan*'s personality is described as kind and reserved, and her goal is to cheer Tomioka city. The description may denote that she is submissive and assuming a supportive role. Similarly, *Silk Tomioka* is given a personality of "very knowledgeable about the TSM," but "still a novice in attending to tourists." This also connotes her inability to master the situation and appear as inferior to men.

Considering these elements of *moe*, several problems of using animated mill girls for tourism promotion may be identified. Firstly, visitors may perceive that mill girls are "fictional" beings rather than "real" workers in the past. Galbraith (2009) argues that individuals are reduced to characters and approached as fantasy once they become the objects of *moe*. If so, through *moe-okoshi*, mill girls may be reimagined as fictional characters who merely satisfy the viewers' desire for *moe*. By the same token, in turning the mill girls into *moe* characters, their historical contributions of advancing industrial development through arduous, exploitative work are disregarded, and instead, they become mere "icons" of innocent young girls from the past.

"Retro boom" and mill girls

The "downtown walk wearing mill girls' outfit" is an activity promoted to tourists in Tomioka city. This activity involves female tourists renting a mill girl outfit and walking throughout the downtown. The model walking course includes some old restaurants that mill girls used to visit. In these restaurants, the dishes that mill girls preferred (i.e., Japanese style curry and dessert) are highlighted as food loved by "Koujo."

Contrary to *moe-okoshi*, the downtown walk with the female workers' outfit is certainly targeting female tourists. The activity is influenced by the Japanese "retro room," or nostalgic boom, that began in 1970s and is still burgeoning (Brown et al., 2003). At the time, with rapid westernization and globalization, a collective sense of loss of tradition became a prevalent topic in popular Japanese culture. The boom is reflected in tourism marketing endeavors as rural areas or old towns become promoted to appeal to urbanites who are eager to "experience" the lost traditional lifestyle (Thelen, 2015).

However, the events, activities, and other touristic representations prepared for such tourists are often criticized as a sanitized, glorified, and thus reinvented past (Caton & Santos, 2007). The downtown walk in Tomioka may be an example of this. Mr. Saito argues that, while the popular image that mill girls wore a kimono, hakama, and red tasuki, there is no official record that indicates a dress code for mill girls to wear these outfits (personal communication, July 9, 2019). Rather, they might have worn their own clothing suitable for work in the mill, or only certain workers who were recognized for mastering higher skills were provided these taskis. In addition, their outfits most likely changed from a kimono to a more western-style of dress at some point. If so, the downtown walk with a kimono and hakama is a manifestation of a romanticized, "invented past" that responds to the desire of tourists to enjoy a retro style.

Issues with the narrative of TSM as a "model recruiting factory"

In addition to the visual representation of mill girls, text representation of female workers also indicated certain patterns (Table 1). Contents analysis of the official website, guidebooks, and on-line travel guides indicated two themes surrounding mill girls; namely, privileged working conditions in the TSM (workhours, welfare, and facilities) and women's empowerment (improved social position and having a well-respected career). For both themes, the narratives overtly stressed that the TSM was a model recruiting factory. This may also be a misrepresentation particularly from the perspective of female workers. Takase (2014) argues that such favorable working conditions in the TSM were available only for the first 10 years when the factory was owned by the government. After the mill became privatized in 1883, working conditions became exploitative because the goal was to generate profit rather than be a model factory. Coincidentally, the time of TSM's privatization matched the time when the demand for silk increased and working conditions worsened at many other silk mills (Hunter, 1993). The government report published in 1933 revealed that the sanitary conditions of the TSM after privatization became particularly poor, and thus sicknesses and diseases among the workers were more prevalent than in other factories. Although the clinic within the mill was still in operation after privatization, a nurse was only employed part-time. In addition, employees needed to work longer than eight hours per day, and were provided with only one or two days off per month. These working conditions were no longer privileged compared to other factories in the surrounding areas. Takase (2014) states a significant number of women committed suicide because of the harsh working conditions. On the contrary, Yusutani (2014) and Imai (2014) argue that, while the working hours grew longer after privatization, salaries were stable compared with other occupations for women at the time (i.e., agriculture or domestic work). Therefore, not all women working in the mill may have experienced extreme hardships and exploitation as some claim they had.

In any case, it might be problematic that only the first 10 years of the entire history of the TSM is narrated as the "official history." Such a narrative may have been created and stressed in the process of nominating the TSM as a WHS. According to Matsuura (2016), the nomination committee of the TSM in Tomioka city first attempted to nominate it by focusing on the fact that the TSM is the only Japanese factory built in the Meiji era that still maintains its original facilities. However, UNESCO began to reevaluate the WHS designation criteria that places values on scale, artistic expressions, uniqueness in architectural heritage (i.e., Palace of Versailles or Egyptian pyramids) to appreciate heritage representativeness of particular history or lost tradition that have influenced human civilization. As Matsuura (2016) states, how to compose history is very critical in gaining WHS status. Therefore, the committee may have needed to change its strategy and stress the history of innovative silk-reeling technology, international exchange of the innovation, and contributions Japan made to industrialization. Throughout the process of composing the history for UNESCO, exploitation of female workers might have been considered an "inconvenient history" that potentially would have hindered WHS nomination especially concerning the current western norm to support women's rights.

While many problems can be addressed, Mr. Saito said that the current representation of the TSM is one of many possible ways to represent, and certainly not a complete form (personal communication, July 9, 2019). According to Mr. Saito, the city government has been attempting to restore the building used as classrooms and open it to tourists in the near future. In addition, the research regarding female workers' outfit will be completed soon. He stated that because the TSM became open to tourists just a few years before its designation as a WHS, what and how to represent it to tourists are still under discussion and consideration.

Conclusion

The purpose of this study was to explore how the "mill girls" of Tomioka Silk Mill, a silk reeling factory that flourished around 1870-1940 of Japan and currently a designated WHS, are represented to tourists. The labor of these mill girls is considered to have made significant contributions to the growth of a modern industrial sector following the Meiji revolution in Japan, while exploitation of such workers was also documented (Hunter, 1993). This is the first study of its kind that retrospectively explored tourism representation of these female workers.

The analysis of visual representations that use mascots and animate characters around the TSM revealed that the current representation of mill girls seems intended to appeal to male viewers. The appearances of the characters (i.e., outfits, body structure, facial features) and their personalities are deliberately designed to portray submissiveness and immaturity, which may satisfy male viewers' desire to feel masculine and superior in the virtual domain (Galbraith, 2009; Sirakaya & Sonmez, 2000). These findings run parallel to previous studies (Alessio & Jóhannsdóttir, 2011; Prichard, 2001) that indicate tourism promotion often represents women as submissive "Others." As Sirakaya and Sonmez (2000) warn, these gender representations subtly embedded in tourism ads may enhance gender inequality and lead to negative consequences for the quality of destination development.

On the contrary, the downtown walk is certainly targeting female tourists eager to experience the life of mill girls. Ironically, those tourists themselves are members of a similarly marginalized group. This activity may be the case in which members of a marginalized group unconsciously play a role imposed on them by a dominant institution (Gramsci & Hoare, 1971; Kurtz, 1996). Female tourists at the TSM are actively consuming and perpetuating the romanticized version of mill girls without questioning the historical accuracy of the narratives and outfit of mill girls. This may be because the hegemonic idea of what mill girls in the TSM is strongly saturated and internalized by the female tourists.

The analysis of this study also indicated that the representation of mill girls in the TSM may be a case of cultural imperialism (Young, 2013) or "world making" (Hollinshead, 2007; Hollinshead & Suleman, 2017). More precisely, the narratives focus overtly on the bright side of the TSM, including women's privileged working conditions and empowerment. In addition, mill girls are represented as mascots and animated characters, which reduce them to a fictitious being or mere "icon" without narratives (Galbraith, 2009). Within the mill, the representation regarding mill women was very limited, and only technical aspects of silk production were the focus. Arguably, these representations universalize the experiences of a dominant society and established them as the norm, while concealing or even denying the history of the exploitation of women (Buzinde & Santos, 2008; Rivera, 2008; Young, 2013). As Rivera (2008) argues, the concealment of the contentious past might have been intensified in its attempt to be recognized as an internationally known heritage site. Although the contributions of the TSM currently narrated are not false, it may only be one side of the proverbial coin. Female workers suffered from harsh working conditions particularly after privatization of the mill, and their hard work has been taken for granted in light of their achievements (Takase, 2014). In the current tourism representation, however, the conflicting stories are reconstructed into a fantasized form (Hollinshead & Suleman, 2017). As Kurtz (1996) argues, the injustice in the reconstructed form of mill girls appear subtle, and the images are perpetuated through various mediums (i.e., posters and consumer products) without being questioned.

Finally, this study indicates that mill girls have suffered from both distribution and recognition injustice, and that this interplay between the two injustices emerges and comes into sharp focus by examining the historic context. The representation of mill women at the TSM reflects not only the past but also the current gender inequities that exist in Japan. Analysis indicated that female mill workers were exploited for the nationalistic goals of achieving economic growth and gaining an equal status with western countries during the Meiji era. When the mill stopped its operation as a factory and became viewed as "cultural heritage," the forms of economic and social injustice have morphed into "cultural imperialism" through which women's own experiences find little expression, and stereotypical imagery is imposed upon them. Representation of the TSM as a "model factory" appear to be favored in the name of celebrating past achievement and acquiring status as a globally-recognized heritage site. However, such representation can be problematic not just with respect to misrepresentation but also because they serve to exacerbate unresolved tension between social groups (see Buzinde & Santos, 2008; (Mellinger, 1994). Revealing the past exploitation of women might in fact be helpful to address the present gender inequality in Japan (Berlin, 2004). Furthermore, tourism research that traces the relationships between political interests and touristic representation such as of TSM can help to identify nationalistic agendas and work towards redressing the current social injustices throughout the country.

Implications and future research

This study of representations of mill girls and TSM offer useful insights for research, theory building, practice, and implications. The study reveals indicates the potential link between the history of institutionally embedded gender injustices and current gender representations in tourism. Examining both historical context and current representation can deepen our understanding of how and why certain social groups are represented to tourists in the way they are and how such representations are influenced by, and similarly influence, social recognition of the group (see Jamal, 2019). The study also presents some challenges from a practical perspective. Given that many of the female workers, particularly those who contributed to the development of the TSM in the pre-war era, have passed away and the majority of their descendants are dispersed all over the country, it makes it rather difficult to tell the full story of what life was like in the TSM from the perspectives of these women. In addition, because the mill ceased operation in 1987 and was not opened to the public until 2005, many local residents of Gunma only learned the history of the TSM in school and did not have an opportunity to develop interest in the cultural and social value of the TSM until mid-2000s (Matsuura, 2016). It makes it difficult to determine who "owns" the culture of mills girls and who has the right to decide how it is to be represented to society in general and tourists in particular. This relates to *representation* that Fraser (2009) discusses, with close connection to distributive and recognition justice, of course.

Collaborative, participatory approaches by which the mill girls of Tomioka can be fairly represented should therefore be considered and negotiated among the local stakeholders. Practitioners need to provide accurate historical facts to these stakeholders and invite open dialogue as to how mill girls should be represented, accurately reflecting them and their lives in history. These actions may heighten the locals' awareness and interest in women's representation. Sirakaya and Sonmez (2000) argue that printed image used for tourism promotion play an important role for destination choices. The subtle messages embedded in promotional materials may perpetuate gender stereotypes and subsequently influence the destination development. This might be particularly the case in Japan where cultural and social norms encourage the maintenance of gender stratification (Barrett, 2004; King, 2009). Therefore, tourism promoters also need to realize the importance of subtle, visual cues and design socially-responsible material to add a competitive advantage to the destination.

Though it was our intent to focus on the TMS and its specific immediate context, this study was only undertaken in one location. Additional studies may consider including other silk and cotton mills throughout Japan to gain a further sense of how mill girls are represented in promotional material. Exploring how the contribution of female workers are represented to tourists can deepen our understanding of social justice and tourism representation collectively considering additional cases. In addition, data for this study were collected from site-investigation, tourism promotion materials, and a personal interview with an executive of the Tomioka Silk Mill Research Center. If future studies are able to incorporate additional qualitative data from interviews with those who had family members employed in mills around 1870 and 1940, that would help us grasp a richer, fuller perspective of the female workers' experiences. Lastly, research is needed to explore tourists' perceptions and understanding of the contributions made by the female mill workers, as well as how they perceive modern-day representation of these women. This will enhance the mutual understanding between those who are represented and those who gaze upon those representations.

Timothy and Boyd (2006) maintain that heritage is selective; historical records are often incomplete, and society does not value all heritage equally: "society filters heritage through a value system that undoubtedly changes over time and space" (p. 2). Heritage representation is a result of negotiation and selection–which period of history to focus on, whose story to present, etc.–and it is highly political. The decisions of what to select and what not to select, who should tell the story and how, is often made by those in positions of authority and power at the time (Barton & Leonard, 2010; Senju, 2006; Tegelberg, 2010). Both tourism related scholars and practitioners need to realize that historical narratives have multiple faces and we need to be sensitive about whose stories are represented to tourists, and how such stories can continue to perpetuate historically engrained injustices (gender injustices in the case of mill girls and the TSM).

Disclosure statement

No potential conflict of interest was reported by the author(s).

Funding

This work was supported by JSPS KAKENHI Grant Number 15K01951.

References

Alessio, D., & Jóhannsdóttir, A. L. (2011). Geysers and 'girls': Gender, power and colonialism in Icelandic tourist imagery. *European Journal of Women's Studies, 18*(1), 35–50. doi:10.1177/1350506810386430

Azuma, H. (2007). *Kontentsu no shisō: manga, anime, raitonoberu* [The thought of content: Manga, Anime, Light Novels]. Seidosha.

Barrett, K. (2004). Women in the workplace: Sexual discrimination in Japan. *Human Rights Brief, 11*(2), 2.

Barton, A. W., & Leonard, S. J. (2010). Incorporating social justice in tourism planning: Racial reconciliation and sustainable community development in the Deep South. *Community Development, 41*(3), 298–322. doi:10.1080/15575330903444051

Berlin, I. (2004). American slavery in history and memory and the search for social justice. *Journal of American History, 90*(4), 1251–1268. doi:10.2307/3660347

Bloem, W. (2014). *Japanese fanspeak in the Anglophone manga and anime fan culture.* (Master's thesis, Leiden University, Leiden, Netherland). Retrieved from https://openaccess.leidenuniv.nl/bitstream/handle/1887/29529/Bloem_0912956_MAthesis.pdf?sequence=1.

Bowitz, E., & Ibenholt, K. (2009). Economic impacts of cultural heritage–Research and perspectives. *Journal of Cultural Heritage, 10*(1), 1–8. doi:10.1016/j.culher.2008.09.002

Brown, S., Kozinets, R. V., & Sherry, J. F. Jr, (2003). Teaching old brands new tricks: Retro branding and the revival of brand meaning. *Journal of Marketing, 67*(3), 19–33. doi:10.1509/jmkg.67.3.19.18657

Bukh, A. (2007). Japan's history textbooks debate: National identity in narratives of victimhood and victimization. *Asian Survey, 47*(5), 683–704. doi:10.1525/as.2007.47.5.683

Buzinde, C. N., & Santos, C. A. (2008). Representations of slavery. *Annals of Tourism Research, 35*(2), 469–488. doi:10.1016/j.annals.2008.01.003

Camargo, B., Lane, K., & Jamal, T. (2007). *Environmental justice and sustainable tourism: The missing cultural link* [Paper presentation]. Paper Presented at the the George Wright Forum.

Caton, K., & Santos, C. A. (2007). Heritage tourism on Route 66: Deconstructing nostalgia. *Journal of Travel Research, 45*(4), 371–386. doi:10.1177/0047287507299572

Chang, T. C., & Huang, S. (2005). Recreating place, replacing memory: Creative destruction at the Singapore River. *Asia Pacific Viewpoint, 46*(3), 267–280. doi:10.1111/j.1467-8373.2005.00285.x

Chronis, A. (2012). Between place and story: Gettysburg as tourism imaginary. *Annals of Tourism Research, 39*(4), 1797–1816. doi:10.1016/j.annals.2012.05.028

Creswell, J. W., & Creswell, J. D. (2017). *Research design: Qualitative, quantitative, and mixed methods approaches.* Sage.

Dai, Y.-D., Dai, Y.-Y., Chen, K.-Y., & Wu, H.-C. (2013). Transformational vs transactional leadership: which is better? A study on employees of international tourist hotels in Taipei City. *International Journal of Contemporary Hospitality Management, 25*(5), 760–778. doi:10.1108/IJCHM-Dec-2011-0223

Alderman, D. H., Butler, D. L., & Hanna, S. P. (2016). Memory, slavery, and plantation museums: The River Road Project. *Journal of Heritage Tourism, 11*(3), 209–218. doi:10.1080/1743873X.2015.1100629

Dewangga, R. G. L., & Hendriyani, H. (2018). *Ageism in Pop Culture: Comparison of Elderly People in Animated Superhero Made in America and Japan* [Paper presentation].Paper Presented at the Indonesia International Graduate Conference on Communication (IndoIGCC) Proceeding,.

Dwyer, S. C., & Buckle, J. L. (2009). The Space Between: On Being an Insider-Outsider in Qualitative Research. *International Journal of Qualitative Methods, 8*(1), 54–63.

Elo, S., & Kyngas, H. (2008). The qualitative content analysis process. *Journalof Advanced Nursing, 62*(1), 107–115

Evans, M. R. (2002). Hegemony and discourse: Negotiating cultural relationships through media production. *Journalism: Theory, Practice & Criticism, 3*(3), 309–329. doi:10.1177/146488490200300302

Fennell, D. A. (2017). Tourism ethics.Bristol, UK: Channel View Publications.

Figueroa, R. M., Waitt, G., & International Association for Envirnmental Philosophy. (2010). Climb: Restorative Justice, Environmental Heritage, and the Moral Terrains of Uluṟu-Kata Tjuṯa National Park. *Environmental Philosophy, 7*(2), 135–164.

Floyd, M. F., & Johnson, C. Y. (2002). Coming to terms with environmental justice in outdoor recreation: A conceptual discussion with research implications. *Leisure Sciences, 24*(1), 59–77. doi:10.1080/01490400252772836

Fraser, N. (1997). *Justice interruptus: Critical reflections on the" postsocialist" condition.* Routledge.

Fraser, N. (2009). Social justice in the age of identity politics. In G. Henderson & M. Waterstone (Eds.), *Geographic thought: A praxis perspective.* (pp. 72–91). Routledge.

Fraser, N. (2010). *Scales of justice: Reimagining political space in a globalizing world.* Columbia University Press.

Galbraith, P. W. (2009). Moe: Exploring virtual potential in post-millennial Japan. *Electronic Journal of Contemporary Japanese Studies,* from ttp://www.japanesestudies.org.uk/articles/2009/Galbraith.html.

Gender Equality Bureau Cabinet Office (n.d). What is a Gender-Equal Society? Retrieved October 2, 2019, from http://www.gender.go.jp/english contents/about _danjo/toward/society/index. html

Gezon, L. L. (2014). Who wins and who loses? Unpacking the "local people" concept in ecotourism: a longitudinal study of community equity in Ankarana, Madagascar. *Journal of Sustainable Tourism*, *22*(5), 821–838. doi:10.1080/09669582.2013.847942

Goffman, E. (1963). *Stigma*. Penguin.

Goulding, C., & Domic, D. (2009). Heritage, identity and ideological manipulation: The case of Croatia. *Annals of Tourism Research*, *36*(1), 85–102. doi:10.1016/j.annals.2008.10.004

Gramsci, A., & Hoare, Q. (1971). *Selections from the prison notebooks*. Lawrence and Wishart.

Hayfield, N., & Huxley, C. (2015). Insider and Outsider Perspectives: Reflections on Researcher Identities in Research with Lesbian and Bisexual Women. *Qualitative Research in Psychology*, *12*(2), 91–106.

Higgins-Desbiolles, F. (2008). Justice tourism and alternative globalisation. *Journal of Sustainable Tourism*, *16*(3), 345–364. doi:10.1080/09669580802154132

Higgins-Desbiolles, F. (2010). Justifying tourism: Justice through tourism. *Tourism and inequality: Problems and prospects*, 194.

Hinton, P. R. (2014). The Cultural Context and the Interpretation of Japanese'Lolita Complex'Style Anime. *Intercultural Communication Studies*, *23*(2)

Hollinshead, K. (2007). Worldmaking'and the transformation of place and culture: The enlargement of Meethan's analysis of tourism and global change. *The Critical Turn in Tourism Studies: Innovative Research Methodologies*, 165–193.

Hollinshead, K., & Suleman, R. (2017). Culture as seduction: Some liquid modern propositions on the palette of imagination of tourism. *Critical Tourism Studies Proceedings*, *2017*(1), 85.

Hunter, J. (1993). Japanese Women at Work, 1880-1920. *History Today*, *43*, 49–49.

Huong, L., Zheng, C., & Fujimoto, Y. (2016). Inclusion, organisational justice and employee well-being. *International Journal of Manpower*, *37*(6), 945–964. doi:10.1108/IJM-12-2015-0212

Imai, M. (2014). *Tomioka Seishijouto Kinusangyouisangun* [Tomioka Silk Mill and Related Industrial Heritage]. KK Best Sellers.

Jamal, T. (2019). *Justice and ethics in tourism*. Routledge.

Jamal, T., & Camargo, B. A. (2014). Sustainable tourism, justice and an ethic of care: Toward the just destination. *Journal of Sustainable Tourism*, *22*(1), 11–30. doi:10.1080/09669582.2013.786084

Jenkins, O. (2003). Photography and travel brochures: The circle of representation. *Tourism Geographies*, *5*(3), 305–328. doi:10.1080/14616680309715

Karatepe, O. M. (2011). Procedural justice, work engagement, and job outcomes: Evidence from Nigeria. *Journal of Hospitality Marketing & Management*, *20*(8), 855–878. doi:10.1080/19368623.2011.577688

King, R. B. (2009). Redefining motherhood: discrimination in legal parenthood in Japan. *Pac. Rim L. & Pol'y J*, *18*, 189.

Kirtsoglou, E., & Theodossopoulos, D. (2004). They are taking our culture away' tourism and culture commodification in the Garifuna Community of Roatan. *Critique of Anthropology*, *24*(2), 135–157. doi:10.1177/0308275X04042650

Kurtz, D. V. (1996). Hegemony and anthropology: Gramsci, exegeses, reinterpretations. *Critique of Anthropology*, *16*(2), 103–135. doi:10.1177/0308275X9601600202

Lee, S., & Jamal, T. (2008). Environmental justice and environmental equity in tourism: Missing links to sustainability. *Journal of Ecotourism*, *7*(1), 44–67. doi:10.2167/joe191.0

Light, D. (2017). Progress in dark tourism and thanatourism research: An uneasy relationship with heritage tourism. *Tourism Management*, *61*, 275–301. doi:10.1016/j.tourman.2017.01.011

Lincoln, Y. S., & Guba, E. G. (1986). But is it rigorous? Trustworthiness and authenticity in naturalistic evaluation. *New Directions for Program Evaluation*, *1986*(30), 73–84. doi:10.1002/ev.1427

Lischer, S. K. (2019). Narrating atrocity: Genocide memorials, dark tourism, and the politics of memory. *Review of International Studies*, *45*(5), 805–827. doi:10.1017/S0260210519000226

Matsuura, T. (2016). Tomioka Seishijou to Kinusangyouisangun [Tomioka Silk Mill and related sites: Path to the designation of WHS]. In T. C. U. o. E. R. S. R. Center (Ed.), *Tomoka Seishijou to Gunmano Youshigyou* [Tomioka silk MIll and silk reelking in Gunma]. (pp. 255–272). Nihon Keizai Hyouronsha Ltd.

Mauthner, N. S., & Doucet, A. (2003). Reflexive Accounts and Accounts of Reflexivity in Qualitative Data Analysis. *Sociology*, *37*(3), 413–431.

McGregor, D. (2009). Honouring our relations: An Anishnaabe perspective. *Speaking for Ourselves: Environmental Justice in Canada*, *27*, 27–41.

Mellinger, W. M. (1994). Toward a critical analysis of tourism representations. *Annals of Tourism Research*, *21*(4), 756–779.

Morikawa, K. (2008). *Shuto no tanjō: Moeru toshi Akihabara, zōhoban* [The birth of Hobby city:'Moe'city Akihabara, an expanded edition]. Gentōsha.

Okamoto, T. (2015). *Otaku tourism and the anime pilgrimage phenomenon in Japan* [Paper presentation]. Paper Presented at the Japan Forum,. doi:10.1080/09555803.2014.962565

Porter, R., & Tarrant, M. A. (2001). A case study of environmental justice and federal tourism sites in southern Appalachia: A GIS application. *Journal of Travel Research*, *40*(1), 27–40. doi:10.1177/004728750104000105

Rawls, J. (2009). *A theory of justice*. Harvard University Press.

Rivera, L. A. (2008). Managing "spoiled" national identity: War, tourism, and memory in Croatia. *American Sociological Review*, *73*(4), 613–634. doi:10.1177/000312240807300405

Saito, K. (2014). Magic, Shōjo, and metamorphosis: Magical girl anime and the challenges of changing gender identities in Japanese society. *The Journal of Asian Studies*, *73*(1), 143–164. doi:10.1017/S0021911813001708

Sandrook, C. G. (2010). Putting leakage in its place: The significance of retained tourism revenue in the local context in rural Uganda. *Journal of International Development*, *22*(1), 124–136. doi:10.1002/jid.1507

Santos, C. A., & Yan, G. (2008). Representational politics in Chinatown: The ethnic other. *Annals of Tourism Research*, *35*(4), 879–899. doi:10.1016/j.annals.2008.06.006

Schelhas, J. (2002). Race, ethnicity, and natural resources in the United States: A review. *Natural Resources Journal*, 42(4), 723–763.

Scheyvens, R., & Momsen, J. H. (2008). Tourism and poverty reduction: Issues for small island states. *Tourism Geographies*, *10*(1), 22–41. doi:10.1080/14616680701825115

Seaton, P., & Yamamura, T. (2015). *Japanese popular culture and contents tourism–introduction* [Paper presentation]. Paper Presented at the Japan Forum,. doi:10.1080/09555803.2014.962564

Senju, H. (2006). Kankoukara Yomitoku Post-Coloniality (Post-coloniality reflected in tourism). In K. Yasumura, H. Endou, & S. Teraoka (Eds.), *Kankou Shakai Bunkaron Kougi (Lectures for tourism, society, and culture)*. (pp. 61–70). Kunpuru.

Shaw, S., Bagwell, S., & Karmowska, J. (2004). Ethnoscapes as spectacle: reimaging multicultural districts as new destinations for leisure and tourism consumption. *Urban Studies*, *41*(10), 1983–2000. doi:10.1080/0042098042000256341

Sirakaya, E., & Sonmez, S. (2000). Gender images in state tourism brochures: An overlooked area in socially responsible tourism marketing. *Journal of Travel Research*, *38*(4), 353–362. doi:10.1177/004728750003800403

Sutera, J. R. (2016). Who is Hikonyan? The phenomenon of Japanese Yuru-Chara. *Sociology*, *6*(12), 775–782.

Takase, T. (2014). *Ikyonichitta Wakaiinochi* [Young lives died away from home]. Orion Sha.

Tan, C. S. (2014). Gotochikyara & Yurukyara)–the fusion of pop culture in place branding in japan. *UTCC International Journal of Business and Economics*, *6*(1), 73–89.

Tegelberg, M. (2010). Hidden sights: Tourism, representation and Lonely Planet Cambodia. *International Journal of Cultural Studies*, *13*(5), 491–509. doi:10.1177/1367877910372707

Thelen, T. (2015). Connecting Academia and Furusato-zukuri in Rural Japan: The" Festival Support Project" of Kanazawa University and its impacts on a Village Community in the Noto Peninsula. *Human and Socio-Environmental Studies*, *30*, 171–193.

Timothy, D. J., & Boyd, S. W. (2006). Heritage tourism in the 21st century: Valued traditions and new perspectives. *Journal of Heritage Tourism*, *1*(1), 1–16. doi:10.1080/17438730608668462

Tomioka City (n.d). Tomioka Silk Mill. Retrieved September 29, 2019, from http://www.tomioka-silk.jp/tomioka-silk-mill/

Tsurumi, E. P. (1991). Old wine in new bottles: Management practices in contemporary Japan. *Journal of Contemporary Asia*, *21*(3), 371–378. doi:10.1080/00472339180000251

Whyte, K. (2016). Indigenous experience, environmental justice and settler colonialism. *Environmental Justice and Settler Colonialism*, Retrieved from: https://ssrn.com/abstract=2770058

Yamazaki, M. (2016). Wada Ei to Tomioka Nikki: Higashi Ajia no Kindaikato Nihon [Wada Ei and Tomioka Diary: East Asia's modernization and Japan]. In T. C. U. o. E. R. S. R. Center (Ed.), *Tomoka Seishijou to Gunmano Youshigyou* [*Tomioka silk mill and silk reelking in Gunma*]. (pp. 69–86). Nihon Keizai Hyouronsha Ltd.

Yan, M., & Wall, G. (2002). Economic perspectives on tourism in China. *Tourism and Hospitality Research*, *3*(3), 257–275. doi:10.1177/146735840200300306

Yilmaz, K. (2013). Comparison of quantitative and qualitative research traditions: Epistemological, theoretical, and methodological differences. *European Journal of Education*, *48*(2), 311–325. doi:10.1111/ejed.12014

Young, I. M. (2011). *Justice and the Politics of Difference*. Princeton University Press.

Young, I. M. (2013). Five Faces of Oppression. In J. DeFilippis and S. Saegert (Eds.), *The Community Development Reader*. (pp. 328–337). Routledge.

Yunseong, K. (1999). Protestant Missions as Cultural Imperialism in Early Modern Korea: Hegemony and its Discontents. *Korea Journal*, *39*(4), 205–234.

Yusutani, R. (2014). *Sekaiisan Tomiokaseishijou* [Tomioka silk mill: World heritage]. Keisou Shobou.

Beyond accessibility: exploring the representation of people with disabilities in tourism promotional materials

Stefanie Benjamin, Ethan Bottone and Miranda Lee

ABSTRACT

Globally, over one billion people experience some form of disability. The number of people with disabilities (PWDs) continues to rise due to an ageing population, the spread of chronic diseases, and improvements in measuring disabilities. However, tourism promotional materials continue to perpetuate a homogenous gaze catering to non-disabled audiences. Thus, informed by critical disability theory, and an inclusive tourism approach, this study explores how PWDs are represented in tourism promotional materials, specifically tourism brochures, from the American Southeast. Through a content analysis of over 200 county-level brochures from nine southeastern states and interviews with state-level tourism marketing directors,three emergent themes were identified: ADA compliant is 'good enough'; 'Diversity' means including more people of color or 'ethnic' groups; and Pets are welcomed but how about PWDs? The findings offer insights for inclusive tourism and breaking down the physical and psychological barriers that hinder PWD participation in travel and tourism.

Introduction

Underrepresented populations in society, including people of color, LGTBQIA+, and people with disabilities (PWDs), are growing markets in hospitality and tourism and make a significant economic contribution to the industry. Focusing specifically on PWDs, globally one billion people, or fifteen percent of the world's population, experience some form of disability, and this number is increasing due to an ageing population, the spread of chronic diseases, and improvements in measuring disabilities (World Health Organization, 2019). Specifically, in the United States, the Open Doors Organization stated that in 2002, PWDs in the U.S. took 32 million trips and spent more than $13.6 billion on travel (Open Doors Organization, 2019). Furthermore, a study by McKinsey & Company stated that the baby-boomer generation commanded almost 60% of net U.S. wealth and 40% of spending (Court et al., 2007). In many categories, including travel, boomers will represent over 50% of consumption. Their impact on the travel and tourism sector is significant as over 40% of boomers will be retiring with some form of disability, raising the total value of this sector to over 25% of the market by 2020 (Open Doors Organization, 2019).

Most recently, a podcast from *Skift*, the travel resource center, shared that adults with disabilities in the U.S. spend $17.3 billion a year on leisure and business travel, and 26 million adults with disabilities took 73 million trips in 2016 (Samson, 2017). Yet, many marginalized groups, specifically PWDs, are not properly represented in tourism promotional materials that traditionally cater toward White, cisgender, heterosexual, and non-disabled men - perpetuating a White male touristic gaze (Alderman, 2013; Cloquet et al., 2018; Pritchard, 2001; Pritchard & Morgan, 2000a; Pritchard & Morgan, 2000b). The lack of inclusiveness disallows for a broader scope to address PWDs in tourism development and results in inequitable treatment of under-represented groups; they have little voice or ability to convey their needs and interests to disparate tourism service providers. PWDs have also been generally marginalized in tourism research, which has been slow overall to take up issues related to justice and fairness. The co-founders of the *Journal of Sustainable Tourism,* Bill Bramwell and Bernard Lane, noted that "while more researchers are beginning to look at the equality of tourism outcomes, there is still relatively little research on the wider issues of equity, fairness and social justice in tourism" (Bramwell & Lane, 2008, p. 2). More recently, scholars have begun attempting to explore and address disability issues within tourism studies (e.g., Cloquet et al., 2018; Darcy, 2002; Darcy & Taylor, 2009; Michopoulou et al., 2015; Nyanjom et al., 2018; Richards et al., 2010). However, disability is still predominantly a neglected subject within tourism inquiry (Aitchison, 2009) and an under-researched phenomenon (Darcy, 2002), despite the loyalty and growing economic power of this targeted group.

The recognition of PWDs in the tourism industry has been discussed in relation to *access* to tourism venues and services, but has yet to embrace an *inclusive* perspective where PWD travelers are recognized and actively involved in the ethical production, consumption, or sharing of tourism benefits (Cloquet et al., 2018; Scheyvens & Biddulph, 2018). Furthermore, promotion and advertising in marketing is an essential and crucial component of enabling and empowering PWDs' interests and motivations for planning and visiting touristic destinations. Though several recent studies have explored the inclusion of PWDs in tourism promotional materials and the accessibility of tourism information to PWDs (e.g., Buhalis & Michopoulou, 2011; Cloquet et al., 2018; Darcy, 2010; Nyanjom et al., 2018), these works focused on either European or Australian tourism markets. Thus, this paper shifts the focus to the United States to explore how people with visible physical disabilities are represented in tourism promotional materials.

Dr. Pauline Sheldon posited in her recent Travel Tourism Research Association Conference keynote address, "it's clear we need to shift our thinking about tourism ... we need a new relationship with capitalism ... [to] put the wellbeing of planet, people, and places-not just profits and growth-on the agenda" (personal communication, 2019). While recognizing the importance of economic arguments in justifying studies of inclusive tourism, as our own introduction does, we agree with Sheldon's assertion for a new relationship with capitalism. Through conceptions of tourism as a social force, empowered by the studies of humanities and their critical epistemologies (Higgins-Desbiolles, 2006), one can see how tourism has the power to change cultures and societies and, to quote Sheldon, the power to shift from a "me to we" economy. Consequently, our study aims to examine the extent to which the tourism industry has adopted an 'inclusive marketing approach,' specifically in the southeastern U.S., through a framework influenced by critical disability theory. Through an analysis of promotional literature from the American Southeast, we provide foundational insights to answer the following research questions: In what ways are people with physical disabilities included in tourism promotional materials in the southeastern United States? How can tourism promotional materials be more inclusive for PWDs? Our paper concludes that currently tourism marketing in the southeastern states of the U.S. cannot be regarded as *just* or *fair*, as it perpetuates an image in which PWDs are absent from the tourism and traveler populations.

The growing presence and influence of PWDs within tourism

Globally, it is estimated that over one billion people have a disability - defined as any condition of the body or mind (impairment) that makes it more difficult for the person with the condition to do certain activities and interact with the world around them (i.e. vision, movement, thinking, remembering, learning, communicating, hearing, mental health, social relationships) (The American Disability Association, 2019). Furthermore, within the European Union, people with access needs took 783 million trips in 2012, and this number is anticipated to grow to about 862 million per year by 2020 (GFK Belgium, 2014). Within the U.S., 58 million adults have a permanent physical disability, 70% of which are not visible, while 36 million adults have impaired mobility (Brettapproved.com, 2019). In 2008, the *UN Convention on the Rights of People with Disabilities* came into force, consisting of 165 Parties and 160 Signatories who recognized the right of PWDs to participate on an equal basis with non-disabled people within leisure, recreation, and tourism (Cloquet et al., 2018). However, it is contingent upon the priorities of governments and states to provide services and resources to PWDs, which, according to Cloquet et al. (2018), "comes with a risk of misrepresenting the higher complexity of developing inclusive tourism practices" (p. 223). While most aspects of American society have been, and still remain, organized around non-disabled citizenship, travel has been especially challenging for PWDs. Yet, with recent technological and legal innovations, more PWDs are traveling than ever before.

Challenges facing PWDs while traveling include untrained professional staff, inaccessible booking services and related websites, lack of inclusive and accessible airports, restaurants, hotel rooms, shops, public spaces, and unavailable information on accessible facilities, services, equipment rentals and tourist attractions (United Nations, 2019). Furthermore, according to Bill Forrester (2016) of TravAbility:

> Accessible tourism is no longer about building ramps and accessible bathrooms. It's about building products and services for a larger and rapidly growing market. This is no longer a niche but rather a segment that is approaching 25% of total tourism spending.

The significant growth of PWDs in tourism is beginning to be reflected in calls for tourist sites to move beyond being *accessible* or *compliant* toward *including* PWDs more directly in enjoying tourism's benefits. For instance, the United Nations recently begun working towards 'accessible tourism for all,' advocating for increasingly accessible cities and tourism provisions. Their push for accessible tourism calls on service providers and destination managers to 'go beyond the principle of universal design' (i.e., one-size fits all tourists) and ensure that "all persons, regardless of their physical or cognitive needs, are able to use and enjoy the available amenities in an equitable and sustainable manner" (United Nations, 2019, paragraph 11). However, as demonstrated further below, non-disabled-bodiness and its associated privileges remain deeply embedded in the institutional bedrock of touristic practices, environments, landscapes, and promotional materials, including travel brochures.

Critical disability theory and inclusive tourism

PWDs have been denied full inclusion in most nations around the world, a phenomenon that has been particularly documented in Western societies (Oliver, 1990; Oliver & Barnes, 2012). From this recognition of marginalization came the social model of disability that was first developed in the mid-1970s (UPIAS (Union of the Physically Impaired Against Segregation), 1976). This model maintains that disability is not a medical impairment, but a socially constructed barrier category used to other and exclude PWDs, leading to their systemic discrimination and oppression (Lang, 2001; Oliver, 1983). As such, "the social model gives precedence to the importance

[of] politics, empowerment, citizenship and choice" within the 'disabled' community (Lang, 2001, p. 4). Consequently, Devlin and Pothier (2006) developed an approach to critical disability theory, positing that "a system of deep structural economic, social, political, legal, and cultural inequality in which persons with disabilities experience unequal citizenship, a regime of dis-citizenship," exists within society today (p. 1). This idea of citizenship, which they define as the "capacity to participate fully in all the institutions of society - not just those that fit the conventional definitions of the political, but also the social and cultural," sits at the core of their critical approach (Devlin & Pothier, 2006, p. 1).

Due to its overarching nature, citizenship is a "practice that locates individuals in the larger community," creating a united and interconnected populace (Devlin & Pothier, 2006, p. 1-2). However, since many PWDs are "denied formal and/or substantive citizenship" as a result of their unequal access to everyday legal and social institutions, "they are assigned the status of 'dis-citizens,' a form of citizenship minus, a disabling citizenship" (Devlin & Pothier, 2006, p. 2). Supporting this view of 'dis-citizenship,' Kitchin (1998) asserted that "the dominant [non-disabled] group's cultural practices are promoted as the norm and the cultural practices of others [PWDs] are portrayed as deviant" (p. 346). Kitchin (1998) further argued that space is socially constructed and produced through social and spatial processes to ignore and discount PWDs in two main ways: by being organized to keep 'disabled people' "in their place" and through "social texts that convey to disabled people that they are 'out of place'" (p. 345). Space, and the society that utilizes it, then becomes actively situated against PWDs, further denying them access to full, inclusive citizenship.

Despite needed legal interventions like the 1990 Americans with Disabilities Act (ADA), the social inertia against the inclusion of PWDs remains in many aspects of daily life. It follows that to create a more just and equal tourism industry for PWDs requires not just accessibility, but also *meaningful* inclusion of PWDs in the landscape of travel and tourism. Tourism promotional materials, in particular, reproduce socially constructed inaccessible spaces by, respectively, a) presenting images that depict inaccessible spaces (i.e. no ramps) and b) not including images showing PWDs enjoying American Southeastern tourism destinations or touristic activities (e.g., dining at a restaurant). Devlin and Pothier (2006) above noted notion of *substantive citizenship* is helpful here as it includes the capacity to fully participate in all institutions of society. It offers a lens through which PWDs can be included in everyday life and the pursuit of travel.

An inclusive tourism framework for PWDs

One way to combat the exclusion of PWDs in tourism and promotional tourism literature is to advocate for inclusive tourism, which enables not merely access but also encourages the involvement of PWDs in travel and tourism. Inclusive tourism overcomes barriers to enable marginalized groups (PWDs in this instance) to participate meaningfully in tourism as producers or consumers (Scheyvens & Biddulph, 2018). It challenges stereotypes and calls for appropriate representations, as well as facilitating mutual understanding and respect. According to Scheyvens and Biddulph (2018), inclusive tourism builds upon two sets of crucial questions: (1) who is included (and excluded) and (2) on what terms are they included? (p. 593).

Incorporating inclusive language in such a framework, like Person-First Language, is essential because these words directly describe the inclusion of PWDs within the tourism industry, providing access to the same opportunities and benefits that tourism provides to non-disabled individuals. Outdated, inaccurate, or hurtful terms perpetuate negative stereotypes, feelings, or attitudes toward marginalized groups, like PWDs (Haller et al., 2006). Thus, the National Disability Authority (2019) argues for adopting Person-First Language when writing or speaking about PWDs in order to reflect the 'individuality, equality, or dignity' of PWDs. It is imperative that terms like 'the disabled', 'wheelchair-bound', 'cripple, spastic, victim', 'the handicapped', 'able-bodied' and 'normal' are no longer used. Instead, they suggest using language like 'people with disabilities', 'persons who use a wheelchair', 'wheelchair user', or 'non-disabled'. Person-First Language emphasizes the person, not the disability (The Arc, 2019) and fosters positive attitudes toward PWDs.

Informed by these frameworks of critical disability theory and inclusive tourism, we explore below not only how PWDs are situated in tourism promotional materials in the American Southeast, but how those in positions of authority, including marketing directors, address their ability to create inclusive tourism - understood as a space and environment where people of all abilities are accepted and welcomed as customers and guests (TravAbility.com, 2019). In unpacking how the American Southeast uses verbiage to talk about PWDs, we also explored the language used to describe accessible touristic spaces, places, and activities within our methodology.

A critical study of American Southeastern tourism brochures

Tourism promotional materials exist in diverse forms, including travel brochures, rack cards, websites, and online booking platforms, that influence a visitor's perception of a destination. These advertising and promotional materials are used by visitors for a variety of reasons, ranging from finding activities to learning about the culture of their destination (Morgan & Pritchard, 1998; Pritchard & Morgan, 2000a). As texts interspersed with images, they can also perform a social role by shaping the tourist imaginary–they provide their readers an opportunity to escape their own everyday life and imagine life as a tourist in an exotic destination (Burton & Klemm, 2011; Francesconi, 2011).

These promotional images are not just about displaying products and attractions, they also convey representations of social groups and societies. Some rely on racial representations and cultural norms to attract particular tourist markets to specific locations (Seiver & Matthews, 2016), and may reinforce stereotypes and exclude populations of difference (Buzinde et al., 2006). Extensive research shows that these representations favor non-disabled, White, cis-gendered, heterosexual people in the U.S. context (e.g., Alderman, 2013; Cloquet et al., 2018; Pritchard, 2001). Multiple studies have found that marginalized groups like PWDs are not well represented in travel brochures (Alderman, 2013; Burton & Klemm, 2011). Furthermore, analysis by Johan Edelheim (2007) shows the invisibility of PWDs within these materials, perpetuating a sense of an 'invisible minority.'

The American Southeast has captured the minds of many in the academy and the general public because of its unique cultural identity, rich and troubled history, and its diverse physical landscapes. The development of the region's tourism industry, resulting from economic shifts after the devastation of the American Civil War, has further entrenched the idea of the Southeast as an exotic location worthy of visiting (Stanonis, 2008). Since the mid-twentieth century, tourism has become a "multibillion-dollar-a-year industry, creating jobs, spawning new businesses, and generating much needed revenue," and developing into a "greater economic force than agriculture or manufacturing and ... one of the top three economic activities in every state of the former Confederacy" (Starnes, 2003, p. 1). Presently, according to a study conducted by a hotel booking company, four of the top ten most visited states by domestic tourists are located in the Southeast, and three more are found in the top 20, evidencing the continuing importance of tourism to the region (Polland, 2014). A governmental study also revealed the dominance of the American Southeast as a destination for international tourists, with the South Atlantic census region, centered along the southeast coast of the U.S., welcoming one-third of overseas visitors to the country (National Travel & Tourism Office, 2015). These visitors, both foreign and domestic, are influenced by representations found in popular culture and promotional tourism literature that depicts the South as steeped in sunshine, history, and hospitality, representations that produce ideas and landscapes that have permanently shaped the region's tourism industry (Cox, 2012; Stanonis, 2008).

Despite the importance of tourism to the economy and life of the American Southeast, research on how the region is represented in tourism promotional literature is sparse. Researchers have produced several edited volumes that provide brief glimpses into specific sectors of the South that recognize the region as complex and multifaceted (Cox, 2012; Stanonis, 2008). These anthologies reflect the importance of understanding the region's history, which has situated the American Southeast as unique and 'other' from the rest of the U.S. The continuing legacy of the Antebellum South and the Civil War, embodied in many tourism destinations across the region, plays perhaps the largest role in this othering process, also described as 'southering' by David Jansson (2017). Of course, much of this 'southering' is rooted in the exploitation and discrimination of marginalized populations, most notably Blacks/African-Americans.[1] Though overall a marginalized group in the U.S., comprising about 14% of the national population, Blacks/African-Americans disproportionately live in the American Southeast, accounting for 20% of the population in this region (U.S. Census Bureau, 2011). Given the similarities between these two marginalized populations in the American Southeast, we believe that PWDs are likely excluded and alienated within this region, which has historically ignored and/or exploited its most vulnerable communities. For instance, seven of the nine states sampled in our study contain more PWDs than the national average of 12.7% (Erickson et al., 2017) yet, are not proportionality represented mirroring the lack of representation of Blacks/African-Americans in tourism promotional material (Alderman & Modlin, 2013).

Methodology

According to Saldaña (2015), analyzing visual data is a "holistic, interpretive lens guided by intuitive inquiry and strategic questions" (p. 57). Thus, we conducted a qualitative content analysis (Saldaña, 2015) of tourism brochures from nine states located in the southeastern United States, including: Alabama, Florida, Georgia, Louisiana, Mississippi, North Carolina, South Carolina, Tennessee, and Virginia (See Figure 1).

Our research team analyzed 211 physical and/or online travel brochures, which promoted the tourism destinations of 228 counties in the above nine states, consisting of an estimated number of 6,462 pages, 13,600 images (including photographs, illustrations, and advertisements, with 9,427 of those images displaying people) published between January 2018–August 2018. While there were 228 total counties included in this study, twelve brochures (one from Mississippi, 11 from Virginia) featured multiple counties within the same publication, leading to the discrepancy between the number of brochures analyzed (211) and counties included (228) in this study. The second author reached out to the study counties via their official Destination Marketing Organization (DMO) websites to obtain the official county-level travel brochures which were mailed to his address in 2018. Each county produced their brochures differently - either through in-house publications or out-sourced companies. Consequently, when the first author contacted the state-level DMOs, she received many conflicting responses as to how and where the brochures were produced and content was decided. From our understanding, the state level provides their own promotional material compared to the county-level DMOs. Thus, each county has their own budget and resources that are not aligned with the state level.

Before we started the analysis, the first author held two coding workshops to help train our research team with the content analysis. For the purposes of this paper, our analysis focused on tourism promotional materials exploring people with *physical* disabilities. We understand that this limits the inclusion of people with disabilities that are not visible; however, as researchers, we could not decipher whether the person pictured had a disability that was not visible. Thus, only visible physical disabilities were counted, including people using wheelchairs, prostheses,

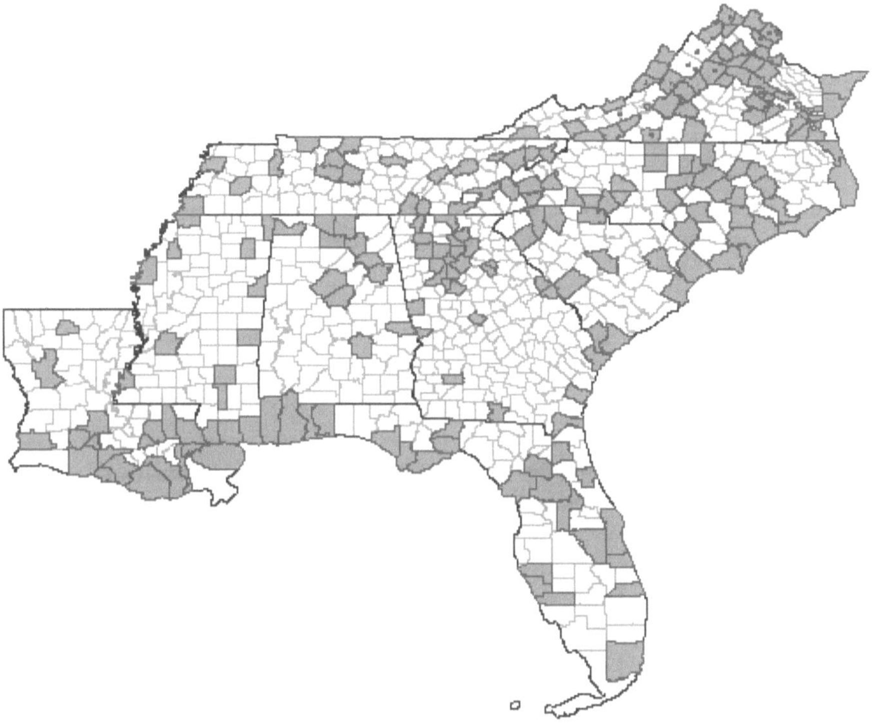

Figure 1. Map of counties analyzed.
This map illustrates the 228 counties (in blue) that were analyzed for this study.

Table 1. Example of code book for analysis of brochures.

County Name/Brochure Title	Number	Notes
Images		
Images of Physical Disabilities		
Images Depicting Accessible Spaces		
Text		
ADA Accessible or other mention of ADA		
Mentions of Accessible Ramps or parking		
Usage of Disabled vs. Handicapped		

and/or canes/walkers. Additionally, we coded for disability language and text to unpack how PWDs were represented or included in the touristic space.

During the second workshop, we created a code book where we individually counted the number of images with visible physical PWDs and analyzed text in a shared Google-drive Excel sheet with each tab correlating to the county examined. Within the code book, field notes and analytic memos were used to generate 'language-based data' that accompanied the visual data (Saldaña, 2015, p. 57). This process allowed for intercoder-reliability, generated dialogue, and confirmed the number of images consisting of people with visible physical disabilities in addition to any issues or concerns with the data (see Table 1). Lastly, as a research team, we relied on intensive group discussion, dialogical intersubjectivity, coder adjudication, and group consensuses as an agreement goal (Benjamin et al., 2017; Saldaña, 2015) via emails and face-to-face conversations to resolve any discrepancy with our coding.

Additionally, the first author invited a representative within each of the nine Southeastern state-level DMOs to participate in a semi-structured questionnaire in 2019. We were interested in understanding each state-level DMO's decision-making process and awareness regarding promotional collateral involving diversity, including efforts made toward representation of PWDs. The first author emailed and called each office several times (from January 2019 to May 2019), resulting in responses from South Carolina, Louisiana, and Tennessee. North Carolina and Florida declined to participate in the study, whereas the rest of the states never returned emails or phone calls. However, due to limited time from each representative, two state representatives asked to respond via email in May 2019 (Louisiana and Tennessee) and one over the phone (South Carolina). The questions asked during the interview and via email were as follows: Tell me about yourself (current role/position); Tell me about your promotional/advertising team; In what ways do you feel like your promotional collateral involves diversity or inclusion?; In what ways are you making any efforts with including PWDs in your tourism promotional material?; How do you feel like your state is accepting or willing to include more diverse advertisement campaigns around PWDs?; Is there anything else you would like to add about your tourism promotional efforts around the inclusion of PWDs?

We transcribed and coded the interviews with in-vivo and value coding, resulting in a thematic analysis (Saldaña, 2015). For our analysis, we blended together the results of the content analysis and semi-structured questionnaires with the DMOs, which yielded three major themes: American Disabilities Act (ADA) compliant is good enough; 'Diversity' means including more 'ethnic' groups; and Pets are welcomed but what about PWDs? Lastly, we would like to note that none of us, as authors, have a physical disability, but have hidden disabilities (i.e. anxiety disorder, depression disorder). Thus, as non-disabled scholars, we cannot truly understand what it is like to physically travel as someone with a physical disability. Furthermore, if one of us were to have a physical disability, we would be careful as to not be the 'token' representative of all PWDs. With this study, our hope is to be allies and advocates for PWDs in order to help highlight the absence of these voices within our scholastic literature and industry promotional collateral.

Findings

Partaking in tourism not only provides PWDs with the option to escape their usual roles as 'objects of care', but also improves their self-worth and increases confidence (Blichfeldt & Nicolaisen, 2011). However, historically, PWDs have not been widely embraced as a target market by the private sector (Hastings, 2009) and unfortunately, our findings confirm this statement. Out of the 211 brochures consisting of roughly 9,427 images of people (photographs, illustrations, and advertisements), there were twelve images depicting a person with a physical disability (see Table 2): One in Florida, one in Georgia, two in Mississippi, two in North Carolina, two in Tennessee, and four in Virginia.

Out of the twelve images depicting PWDs, seven portrayed elderly individuals who were either out of focus or not the main focal point of the photograph or image; in addition, all PWDs were White. These images suggest that the majority of PWDs are senior citizens/elderly, not active tourists capable of participating in touristic activities, and not diverse in terms of ethnic or racial identities. These findings are also supportive of critical disability theory's tenant that PWDs are not important enough to be included as part of the travel group or as a tourist (Devlin & Pothier, 2006).

In terms of language, our content analysis revealed that inclusion of and terminology about PWDs is still a contested issue. Of the 228 counties included in this study, brochures for only 114 (50%) mentioned disability in some manner (see Figure 2). This count includes terms directly relating to PWDs (e.g. "disabled," "handicapped") and language that promoted accessibility or inclusion (e.g. "accessible," "ADA-compliant," "wheelchair," "special needs"). Although there was mention of disability in half of the brochures analyzed, outdated words and verbiage were still the norm (see Figure 3). In those counties where PWDs were directly referred to (63, or 27.6%), only eight used the term "disabled." The majority of brochures that directly addressed PWDs (53, or 23.2%) used language that is considered harmful to the cause of disabled rights' (Haller et al.,

Table 2. List of PWDs in travel brochures.

County Name/ Brochure Title	Description of PWD	Notes
Escambia, FL	White, senior citizen (gender unknown) in a wheelchair	Blurred image at the right back
Clayton, GA	White, senior citizen female with cane	Part of a tourist group from a bus tour
Lee, MS	White, male (undisclosed age) in a wheelchair with his back turned to the camera	About to participate in a large running race in a large crowd at the center front
Lee, MS	White, senior citizen female with a cane	Pictured with a small group at a heritage site to the far right
Duplin, NC	White, senior citizen female with a walking cane	Blurred photo of a woman at the farmer's market to the left of the table
Wake, NC	White, senior citizen male in wheelchair	Being pushed in a wheelchair by a person at the Garner Veterans Memorial
Sumner, TN	White, female child in a wheelchair	One cartoon image for a park for children with disabilities
Sumner, TN	White, female child in a wheelchair	Photograph for a park for children with disabilities
Bristol, VA	White, male (age unknown) being pushed in a wheelchair in a large crowd	Blurred photo in a large crowd to the left
Bedford, VA	White, senior citizen female with a walking cane with her nurse	Advertisement for an assisted living facility
Prince William, VA	White, senior citizen male with a cane	Walking into a retail store in a crowd at the front left
Williamsburg, VA	Wheelchair at a venue	Unaccompanied wheelchair to the far back left at an event venue

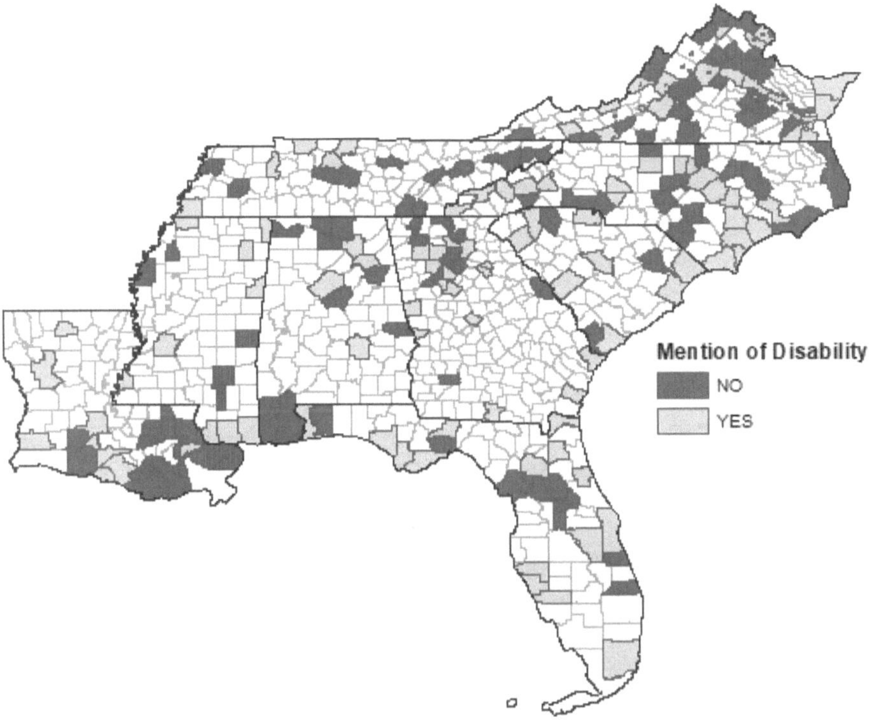

Figure 2. Mention of language of 'disability' in brochures.
This map illustrates the mention of disability within the text of the brochures.

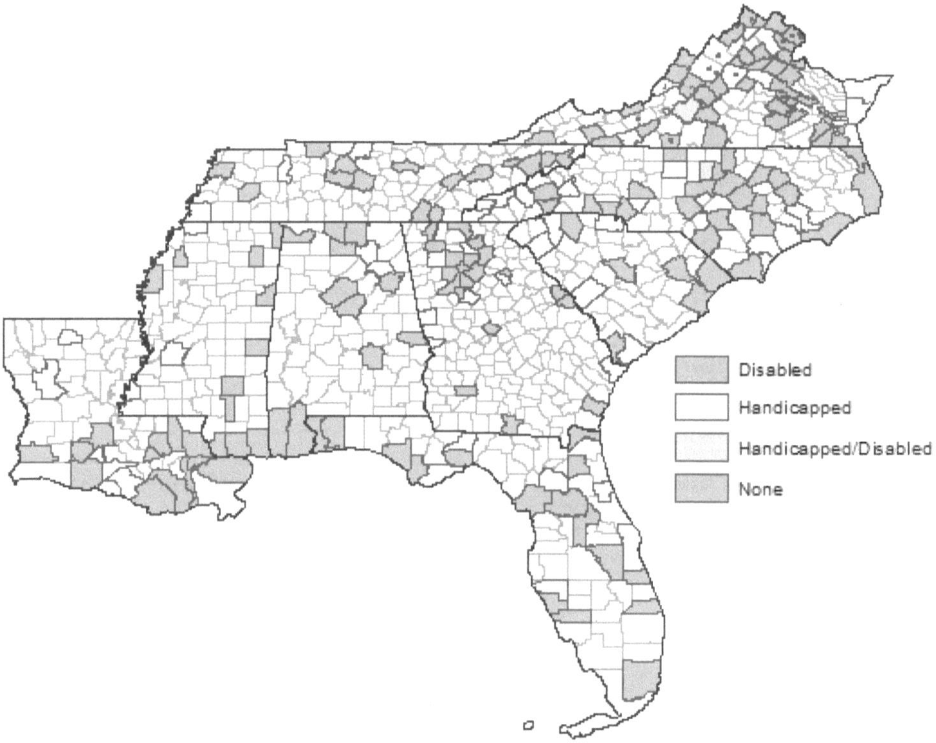

Figure 3. Mention of 'disabled' vs. 'handicap' language in tourism brochures.
This map illustrates how often the words 'disabled' vs. 'handicap' were used in the brochures.

2006), employing the words "handicap" or "handicapped" to describe PWDs. Two brochures, Orange County, NC and Bedford County, VA, used "disabled" and "handicapped" interchangeably throughout their publications; though the use of "disabled" by these DMOs is encouraging, the fact that the outdated term "handicapped" is found alongside more modern language gives the appearance that these tourist organizations are out of touch with the community. The overall trend towards the obsolete term "handicapped" found in our analysis demonstrates that the Southeastern tourism and hospitality industry remains far from inclusive towards PWDs.

Conversely, PWDs are not a solitary market, meaning they are surrounded by family members and friends who also understand the value in services, destinations, and attractions that accommodate all people in society (Yin et al., 2018). Within retail, in terms of economics and social justice, some businesses have recognized the market value of PWDs, including Walgreens, T.J. Maxx, Target, and SunTrust – all running advertisements showing people with physical disabilities engaging in everyday activities (Diament, 2017). Yet, the Southeastern tourism industry is slow to include or incorporate this market within their county-level brochures. What might the constraints, barriers, or hesitations causing this absence of PWDs be? After speaking with several representatives at the state-level DMO and analyzing 'disability' language within the brochures, it appears that being 'ADA' compliant is '*good enough*' for these institutions.

ADA compliant is 'good enough'

Our analysis revealed the noted usage of ADA-related terms, as well as direct mentions of the ADA within Southeastern tourism brochures. Of the 228 counties included in this study, 24 (10.5%) made some mention of the ADA. While this is a low number, these mentions were notable due to the stark and legal-sounding phrases that accompanied them. For instance, phrases such as "ADA compliant" and "ADA accessible" were found regularly in brochures, identifying to audiences that these locations meet the legal accommodations laid out by the federal government. Likewise, the brochure for Miami-Dade County, FL, referenced the ADA in a small section of fine print notifying PWDs to contact the Miami-Dade Office of ADA Coordination for information concerning travel to their area. Mentions of the ADA usually appeared in inventories of a county's attractions (i.e. trails and docks) or hotels, but were rarely expanded upon outside of the "compliant" and "accessible" phrases described above. Combined with the results demonstrating the extensive use of outdated language, our content analysis elucidates that American Southeastern tourism brochures do not promote an inclusive, or even accessible, image of their destinations.

The Department of Justice (DOJ) published the Americans with Disabilities Act Standards for Accessible Design in September 2010. These benchmarks indicated that all electronic and information technology must be accessible to PWDs. Adhering to this statement, the Tennessee DMO representative shared that the state is working toward achieving these standards:

> *TNvacation.com meets or exceeds federal guidelines for ADA accessibility compliance as well … As noted, TNvacation.com meets or exceeded federal guidelines for accessibility. We are also planning educational workshops for our partners to assist them in navigating accessibility laws and how to accommodate the differently abled … We are currently working on content that provides travel information for the differently abled such as including closed caption subtitles on videos and travel itineraries featuring hotels and attractions that meet ADA accessible requirements.* – Tennessee DMO Representative

Although it is essential and necessary to have ADA compliant websites and inventory, there is still an absence of PWDs within touristic promotional materials. Excluding PWDs perpetuates the stigma that this group does not travel, does not participate in dining or shopping, and that they are instead completely incapable or unable of being a tourist and belonging in a touristic setting. Even though imagery has the power to influence people's assumptions, opinions, and change mindsets (Cloquet et al., 2018), the argument surrounding ADA compliance as 'good enough' seems to be evident throughout this study:

> *[Our] consumer website meets regulations with PWDs … print or digital or television we stress diversity through ethnic diversity and geographic diversity across the state mountain areas to the coast … representing PWDs in advertising … we have not done that in the past and not doing it currently.* - South Carolina DMO Representative

Being ADA compliant is the basic – low hanging fruit – accessibility standard, helping to alleviate any legal concerns or issues. Furthermore, the argument around not 'selling' or promising a 'product' that cannot be sold can also be seen as an *argument* for not including PWDs in tourism promotional material. For instance, the Louisiana state DMO representative explained:

> *We feel we must first inventory which attractions can totally accommodate folks with a disability. You never want to promise something you cannot deliver! We need to make sure we have the product in-state … getting inventory from all 64 parishes in Louisiana …. BUDGET restraints, learning more about this segment (What are they looking for? How do they travel? Etc.). Right now our money must go to 'the middle of the bell curve.' Same reason we don't market for destination weddings.* – Louisiana DMO Representative

'Getting inventory' for accessible spaces, places, and attractions is essential for explaining and showcasing how a destination may be accessible, and the Louisiana DMO representative, through our ongoing email communications, expressed a passion and persistence toward working on gaining inventory for PWDs in their state. However, referring back to the crucial questions of the inclusive tourism framework (Scheyvens & Biddulph, 2018) – (1) who is included (excluded) and (2) on what terms – it is evident, through our study, that PWDs are excluded from the dissemination of inclusive touristic opportunities that welcome and attract this group to destinations. If PWDs are not being included currently, then who is?

'Diversity' means including more people of color or 'ethnic' groups

The term 'diversity' is perhaps one of the most popular buzzwords in both industry and academic settings today, resulting from the increased attention given to uplifting marginalized populations. However, despite its popularity, what 'diversity' actually entails is a contested subject. For PWDs, inclusion in diversity has been an often-neglected struggle, particularly in the academy, which informs decisions made by industry operators (Linton et al., 1995; Olkin, 2002; Taub & Fanflik, 2000). Our content analysis supported this finding, as PWDs were largely excluded in both visual and textual conditions from American Southeastern tourism marketing. Yet, increasing 'diversity,' in terms of diverse ethnicities and racial representation, appears to be a goal for many DMOs. When asked about the level of increasing diversity within their promotional material, disability was not mentioned:

> *It is something that we consider .... I can bring it up .... But it is not a subject that has been approached. We focus attention around the ethnic diversity across our advertising ... [disability] is not a subject that has come up with advertising other than the compliance online with websites.* – South Carolina DMO Representative

> *I feel strongly that our marketing collateral, especially in recent years, includes diverse offerings representing a wide variety of potential visitors of every background through the photography, video and articles we select to be featured on our website, all digital media platforms and vacation guide.* – Tennessee DMO Representative

Globally, travel and tourism has been overwhelmingly White-washed, leaving limited room to highlight the experiences of traditionally marginalized racial and ethnic groups. As a result, there is a near invisibility of Black/African-American and Latinx people in the travel sphere (Buzinde et al., 2006). Evidenced by promotional material research, the tourism industry has traditionally portrayed leisure vacations with all White faces (Burton & Klemm, 2011). However, Black/African-American travelers spend between $48 to $65 billion per year globally on travel (Dillette et al., 2019) whereas, the Latinx community contributes over $56 billion in leisure travel every year (Morse, 2019). Consequently, our conversations with the DMO representatives were parallel to recent articles and dialogue around including more people of color within the travel and tourism promotional spheres. For instance, Greater Miami's Convention and Visitor's Bureau has its own multicultural department that partners with 30 neighborhood businesses to help embed communities of color into their advertising (Greater Miami CVB, 2019). However, this definition of 'diversity' within travel and advertising is void of PWDs. For example, when a global hotel brand's head of diversity recently contacted Mandala Research to explore the possibilities of including certain 'diverse' groups in their promotion and advertising, they included African-American, Latinx, Muslim, and female travelers (Peltier, 2016), but there was no mention of PWDs.[2] Diversity, however, means more than just 'including people of color'; it also means including PWDs.

In 2050, it is projected that the majority of Americans will no longer be White, whereas Latinx and Asian populations are expected to nearly triple (Kotkin, 2010). Today in the U.S., 25% of children under age five are Latinx; by 2050, that percentage will be almost 40% (Kotkin, 2010). Despite these increasing racial and ethnic demographics, we found that when PWDs were included in images of Southeastern tourism, diversity was not represented. The twelve PWDs pictured in brochures were all White. However, the true intersectionality of a disability cuts through race, ethnicity, religion, sexual orientation, gender, socioeconomic status, etc. Thus, showcasing PWDs as multi-ethnic and from various age-groups is necessary in order to highlight that PWDs are not one race/ethnicity, one gender, or one age group. By doing so, this disrupts the predominantly White male, heteronormative, cis-gendered lens that informs and dominates the tourism promotional literature. Such an intersectional approach can foster a deeper, more socially inclusive representation of the changing demographics within the U.S.

This lack of inclusion reflects Harlan Hahn's (1987) seminal work around the role of advertising in disabilities studies. He argued that traditionally, advertising's emphasis on 'beauty and bodily perfection' has led to the total exclusion of PWDs (Hahn, 1987). Furthermore, he posited that non-disabled people feared becoming a person with a disability, which in turn, caused issues and hesitation to include PWDs within advertising. Consequently, PWDs 'inability' to ever fit within the context of beautiful bodies has caused their invisibility. However, the diversity argument does not compare to the amount of attention and acceptance of one special target audience that emerged from our analysis. This *more important* target market group has infiltrated not only the inventory of appendices of hospitality and tourism spaces and places, but visually dominated the promotional brochures – the special niche group of 'pets.'

Pets are welcomed but how about PWDs?

The 'welcoming of pets' while traveling is currently an upward trend within the tourism and hospitality industry. A recent article from *Skift* (Sheivachman, 2017) mentioned the increase of hotels *actively welcoming* pets through allowing boarding in rooms and providing small services including pet beds, feeding bowls, and dog treats. According to *Travelers Today*, pet owners can enjoy their vacations "without worrying about their pets" since so many destinations offer facilities such as pet-sitting and special pet themed events (Ara Lasco, 2017, paragraph 4). For example, being one of the first hotel chains to allow pets on their property, Kimpton Hotel Monaco in Colorado, announced that in August 2019 guests can book a private party with puppies and prosecco (Yasharoff, 2019). This package retails at $1,262 per night, and part of the proceeds go toward supporting a local no-kill puppy rescue facility.

Although we were not actively seeking this trend, throughout our analysis, the increase of inclusion and mention of pets (i.e. dogs) while traveling was outstanding compared to the lack of representation of PWDs. We noted whole sections devoted to 'pet friendly' touristic sites including hotels, restaurants, recreational sites, and tourism activities (ranging from pet friendly breweries to dog-friendly festivals). Our content analysis revealed that 109 (47.8%) counties included the mention of pets, such as "Pet Friendly" in hotel listings, but no mention of accessibility (even ADA) in said hotel listings. Furthermore, 36 (33%) of those counties did not mention disability in any form. However, while this study specifically focuses on American Southeastern tourism, we believe this disparity in inclusion is both a nationwide and global phenomenon.

Pets are now seen as 'part of the family' and leaving them behind is no longer an option when vacationing (Ara Lasco, 2017). DMOs, in addition to hospitality and tourism companies, are tapping into this target market. For instance, the Visit North Carolina DMO (2019) placed a call out for the 'first dog travel agent' in August 2019 in order to:

> Help tell the world about our state's woof-worthy destinations – as only a pupper can … No matter which trip you choose, Visit North Carolina believes vacations should be experienced with the pure, unbridled joy that our best friends bring to every moment.

In our analysis, we noticed a shift from 'family friendly' places and spaces to 'pet friendly' attractions, hotels, and touristic activities. This could be a consequence of declining fertility rates coupled with the increase of delaying marriage (Stone, 2017). This delay has influenced a boost in single people and child-free couples adopting pets, viewing them as 'family members' or 'furbabies', and self-identifying as 'pet-parents' (Stone, 2017). As such, pets seem to be replacing 'family friendly' promotional materials within brochures and contributing toward a niche market focused on advocating for animal rights, thus creating an inclusive environment for the *whole* family. However, there was no significant contribution toward creating an inclusive framework for the wellbeing and welcoming of PWDs. Therefore, pets were 'represented' and deemed as important or more valuable than PWDs. This finding aligns with the critical disability framework in that PWDs are denied substantive citizenship, or inclusion into society at large.

The exclusion of PWDs in both textual and visual aspects of tourism promotional literature reveals the "dis-citizenship" that many PWDs experience and demonstrates the banal nature that these forms of exclusion take on. The observation of animals as more 'important' than the mention or representation of PWDs is what emerged from our analysis. These brochures showed that pets are being humanized through their inclusion as part of the family, while PWDs are seen as less than human through their exclusion from the family unit. Currently, we argue that American society has shifted their attention and focus toward traveling with pets instead of traveling with a disability.

Discussion

Reflecting the economic bias in tourism, much of the past research in this field, especially before the mid-to-late-1990s and early 2000s, focused on increasing the bottom line of tourism-related businesses (Walle, 1997). An increasing tide of research being conducted through critical epistemologies seeks to "stimulate their audience to transform society and thereby to liberate themselves and others" (Bramwell & Lane, 2014, p. 2). In order to seek this 'just transformation' in tourism studies, tourism researchers have to recognize tourism as more than an industry – as a social force that has positive impacts (Higgins-Desbiolles, 2006).

Promotional tourism literature, such as we studied above, has the power to uplift people as both traveler and tourism provider, but it can also marginalize social and cultural groups. Our investigation into American Southeastern tourism promotion reveals this two-faced Janus of the industry where PWDs are remarkably absent from county-level brochures and the inner machinations of state-level DMOs. Through the use of outdated language such as "handicapped", the complete exclusion of PWDs from brochures, or the homogenous representation of PWDs as elderly and White, large swaths of the United States are portrayed as inaccessible and unwelcoming to this growing sector of the travel market. Moreover, it appears that animals, through pet-friendly hotel listings and abundant mentions of dog parks, are a more attractive and 'safer' market segment to target than potential travelers with disabilities. If destinations are able to source an inventory for pet-friendly places and spaces, why not source an inventory with accessible spaces and places?

Our study supports arguments for the social model of disability (Lang, 2001; Oliver, 1983) and Devlin and Pothier (2006) critical disability framework, through which PWDs are denied substantive citizenship, or inclusion into society at large. The exclusion of PWDs in both textual and visual aspects of tourism promotional literature reveals "dis-citizenship" and demonstrates the banal nature that these forms of exclusion take on. In the American Southeast, our interviews with personnel at state-level DMOs revealed one potential reason for the continuance of ableist tendencies in the region's promotional literature: differing definitions of diversity. Increasing ethnic and racial diversity was considered the main focus of most diversity campaigns by state DMOs, which conforms to traditional definitions of diversity (Olkin, 2002).

The rise in literature investigating the inclusion of African-Americans/Blacks in American Southeastern tourism over the last decade further reinforces the societal focus on raising awareness of racial/ethnic diversity in tourism (Alderman, 2013; Alderman et al., 2016; Dillette et al., 2019). Yet this focus, found in both the industrial and academic tourism communities, obfuscates other aspects of diversity, such as mental and physical disabilities (Linton et al., 1995; Olkin, 2002; Taub & Fanflik, 2000). Constructing diversity as only racially and ethnically based hides the similarities that PWDs share with other marginalized groups, including "prejudice, stigma, discrimination, and oppression" (Olkin, 2002, p. 134). Given the exclusion of PWDs demonstrated in this study, we argue that the American Southeastern tourism industry maintains an ableist hegemony through promotional tourism literature that only promotes a racially and ethnically based version of diversity, effectively erasing the disproportionately large community of PWDs that reside and travel within the region. As such, prejudice, discrimination, and oppression characterize the relationship that tourism operators in the American Southeast have with PWDs.

Furthermore, the noted humanizing of pets over PWDs as members of the family unit is another process by which the non-disabled hegemony is maintained within the American

Southeastern tourism industry. Through the inclusion of pet-friendly hotel listings and entire sections devoted to traveling with pets (primarily dogs) in lieu of accessibility, PWDs are relegated to the peripheries of tourists' imaginations as they flip through the pages of a selected brochure. Such marginalization is characteristic of the discrimination and oppression that the social model of disability (Lang, 2001; Oliver, 1983) and the critical disability framework (Devlin & Pothier, 2006) recognize as barriers to full societal inclusion for PWDs. By choosing to showcase accommodations for families with pets instead of families with disabilities, tourism spaces within the American Southeast are actively depicted as inaccessible and unwelcoming to a large swath of the region's population. Combined with their exclusion from diversity campaigns, travelers with disabilities are kept "in their place" as social others and portrayed as "out of place" in these tourism spaces (Kitchin, 1998, p. 345). Therefore, tourism promotional literature continues to uphold hegemonic views of disability despite their emancipatory and liberatory potential if informed by the tenets of inclusive tourism.

Thus, to combat the non-disabled hegemony, we argue for an *inclusive tourism* framework that includes involving PWDs in the ethical production and consumption of tourism (see Scheyvens & Biddulph, 2018). PWDs should be included in the decision-making processes and marketing of touristic destinations, activities, and spaces (e.g., serving on tourism advisory boards at the state, county, and local levels). Social justice activist Janaya Khan argued for a redefining of privilege, stating that "privilege isn't about what you've gone through, it's about what you haven't had to go through" (MSN. , 2019). Members of tourism marketing organizations must check their non-disabled privilege by including PWDs in their advertising and promotional campaigns. Having the voices of PWDs heard and valued at the decision making and production levels helps to create and foster a sustainable approach of inclusivity that leads the way towards full substantive citizenship and inclusive participation for PWDs in travel and tourism.

Conclusion

This paper contributes significantly to the critical study of PWDs in tourism promotional material. Through a detailed exploration of tourism brochures representing 228 counties in the American Southeast and interviews with personnel at state-level DMOs within the region, we demonstrated that PWDs are excluded from definitions of diversity used by tourism operators; that mentions of disability are couched in the stark institutional terms of the ADA; and that pets are humanized over family members with disabilities. Careful analysis is needed to disclose the structural and pervasive character of non-disabled hegemony, which, through economic and social constructions, influences almost every aspect of the tourism industry. Our findings support the social model of disability and the framework of "dis-citizenship" by displaying the social constructedness of disability and its varying inclusions/exclusions within the American Southeastern tourism industry. By narrowing and obfuscating the nature of disability, through both text and images, tourism promotional literature can create inaccessible and unwelcoming tourism spaces that place PWDs on the outside as potential visitors. Overall, our content analysis of American Southeastern tourism promotional materials revealed intense discrimination of PWDs in tourism promotional literature, evidencing the hegemonic relations that present tourism destinations as largely inaccessible to potential visitors with disabilities.

The study of access and inclusiveness for PWDs therefore advances the growing critical literature on tourism marketing. Our study showcased the dominance of non-disabled perspectives, as a large majority of brochures and travel guides lacked both visual and textual mentions of accessibility and PWDs. Further expansion and elaboration of critical disability theory alongside an inclusive tourism framework can help to bring attention to PWDs, an oft-forgotten community within the U.S. and the global population at large, and a growing market within the tourism industry (Aitchison, 2009; Cloquet et al., 2018; Darcy, 2002; Michopoulou et al., 2015; Nyanjom et al., 2018; Scheyvens & Biddulph, 2018).

Lastly, representation matters – and it is important to present a diverse and inclusive population of PWDs (including intersectionality of race/ethnicity, gender, sexual orientation, and age) in helping to shape tourism advertising and promotional media, ensuring that PWDs are welcomed, accepted, and included in touristic spaces and places. The inclusion of PWDs within the tourism industry aligns with notions of tourism as a positive social force (Higgins-Desbiolles, 2006). Through the incorporation of pictures displaying PWDs partaking in touristic activities and the use of text explaining accommodations in Person-First terms, tourists with disabilities can visualize themselves at destinations and plan more easily if they decide to travel, providing the opportunity to engage more fully with society and receive substantive citizenship.

Notes

1. Geographers and tourism scholars have noted troubling trends in African-American representation (or the lack thereof) in images and texts within brochures, websites, and on the grounds of Southeastern tourist destinations associated with slavery and the American Civil Rights Movement (Alderman et al., 2016). Other Southern tourism landscapes not directly related to slavery marginalize African-Americans largely through their absence in promotional materials and tourism development processes (Alderman & Modlin, 2013).
2. Mandala Research is a travel market research firm offering products and services - along with expert analysis - to Fortune 500 companies, non-profit organizations, and government agencies (Mandala Research, 2019).

Disclosure statement

No potential conflict of interest was reported by the author(s).

References

Aitchison, C. (2009). Exclusive discourses: Leisure studies and disability. *Leisure Studies, 28*(4), 375–386.

Alderman, D. (2013). Introduction to the special issue: African Americans and tourism. *Tourism Geographies, 15*(3), 375–379.

Alderman, D. H., Butler, D. L., & Hanna, S. P. (2016). Memory, slavery, and plantation museums: The River Road Project. *Journal of Heritage Tourism, 11*(3), 209–218.

Alderman, D. H., & Modlin, E. A. Jr. (2013). Southern hospitality and the politics of African American belonging: An analysis of North Carolina tourism brochure photographs. *Journal of Cultural Geography, 30*(1), 6–31.

American Disability Association. (2019). ADA Regulations. https://www.ada.gov/

Ara Lasco, C. (2017). Pet-friendly travel is now a trend this 2017. *Travelers Today*. https://www.travelerstoday.com/articles/37956/20170206/pet-friendly-travel-now-trend-2017.htm.

Belgium, G. F. K. (2014). *Economic impact and travel patterns of accessible tourism in Europe: Presentation of the key study findings*. https://www.accessibletourism.org/resources/toolip/doc/2014/06/09/guisette-and-li_eu-accessible-tourism_economic-demand-study.pdf.

Benjamin, S., Williams, J., & Maher, M. (2017). Focusing the lens to share the story: Using photographs and interviews to explore doctoral students' sense of well-being. *International Journal of Doctoral Studies*, *12*, 197–217.

Blichfeldt, B. S., & Nicolaisen, J. (2011). Disabled travel: Not easy, but doable. *Current Issues in Tourism*, *14*(1), 79–102.

Bramwell, B., & Lane, B. (2008). Priorities in sustainable tourism research. *Journal of Sustainable Tourism Research*, *16*(1), 1–4. doi:10.2167/09669580803489612

Bramwell, B., & Lane, B. (2014). The "critical turn" and its implications for sustainable tourism research. *Journal of Sustainable Tourism*, *22*(1), 1–8. doi:10.1080/09669582.2013.855223

Brettapproved.com. (2019). Sadly handicapped remain invisible to many. https://www.brettapproved.com/

Buhalis, D., & Michopoulou, E. (2011). Information-enabled tourism destination marketing: Addressing the accessibility market. *Current Issues in Tourism*, *14*(2), 145–168. doi:10.1080/13683501003653361

Burton, D., & Klemm, M. (2011). Whiteness, ethnic minorities, and advertising travel brochures. *The Service Industries Journal*, *31*(5), 679–693.

Buzinde, C. N., Santos, C. A., & Smith, S. L. (2006). Ethnic representations: Destination imagery. *Annals of Tourism Research*, *33*(3), 707–728.

Cloquet, I., Palomino, M., Shaw, G., Stephen, G., & Taylor, T. (2018). Disability, social inclusion and the marketing of tourist attractions. *Journal of Sustainable Tourism*, *26*(2), 221–237.

Court, D., Farrell, D., Forsyth, J. E. (2007). Serving aging baby boomers. *McKinsey Quarterly*. https://www.mckinsey.com/business-functions/strategy-and-corporate-finance/our-insights/serving-aging-baby-boomers

Cox, K. L. (2012). *Destination Dixie: Tourism & southern history*. University Press of Florida.

Darcy, S. (2002). Marginalized participation: Physical disability, high support needs and tourism. *Journal of Hospitality and Tourism Management*, *9*(1), 61–72.

Darcy, S. (2010). Inherent complexity: Disability, accessible tourism and accommodation information preferences. *Tourism Management*, *31*(6), 816–826. doi:10.1016/j.tourman.2009.08.010

Darcy, S., & Taylor, T. (2009). Disability citizenship: An Australian human rights analysis of the cultural industries. *Leisure Studies*, *28*(4), 419–442.

Devlin, R., & Pothier, D. (2006). Introduction: Toward a critical theory of dis-citizenship. In D. Pothier & R. Devlin (Eds.), *Critical disability theory: Essays in philosophy, politics, policy, and law* (pp. 1–22). UBC Press.

Diament, M. (2017, February 24). Walgreens features model with Down Syndrome. *Disability Scoop*. https://www.disabilityscoop.com/2017/02/24/walgreens-modeldown-syndrome/23367/

Dillette, A. K., Benjamin, S., & Carpenter, C. (2019). Tweeting the black travel experience: Social media counternarrative stories as innovative insight on #TravelingWhileBlack. *Journal of Travel Research*, *58*(8), 1357–1372.

Edelheim, J. (2007). Hidden messages: A polysemic reading of tourist brochures. *Journal of Vacation Marketing*, *13*(1), 5–17.

Erickson, W., Lee, C., von Schrader, S. (2017). *Disability statistics from the American Community Survey (ACS)*. www.disabilitystatistics.org

Forrester, B. (2016, July 3). Redefining accessible tourism as a profitable market. *TravAbility*. https://travability.travel/blog-node/redefining-accessible-tourism-profitable-market

Francesconi, S. (2011). Images and writing in tourist brochures. *Journal of Tourism and Cultural Change*, *9*(4), 341–356.

Greater Miami Convention & Visitors Bureau. (2019). Multicultural Tourism. https://www.miamiandbeaches.com/about-gmcvb/gmcvb-staff-directory/multicultural-tourism-development

Hahn, H. (1987). Advertising the acceptably employable image: Disability and capitalism. *Policy Studies Journal*, *15*(3), 551–570.

Haller, B., Dorries, B., & Rahn, J. (2006). Media labeling *versus* the US disability community identity: A study of shifting cultural language. *Disability & Society*, *21*(1), 61–75. doi:10.1080/09687590500375416

Hastings, R. R. (2009, September 29). Businesses target customers with disabilities. *SHRM*. https://www.shrm.org/resourcesandtools/hr-topics/behavioral-competencies/global-and-cultural-effectiveness/pages/businessestarget-customers.aspx

Higgins-Desbiolles, F. (2006). More than an "industry": The forgotten power of tourism as a social force. *Tourism Management*, *27*(6), 1192–1208. doi:10.1016/j.tourman.2005.05.020

Jansson, D. (2017). The work of southering: "Southern justice" and the moral landscape of uneven racism. *Southeastern Geographer*, *57*(2), 131–150.

Kitchin, R. (1998). Out of place', 'knowing one's place': Space, power and the exclusion of disabled people. *Disability & Society*, *13*(3), 343–356.

Kotkin, J. (2010). *The next hundred million: America in 2050*. Penguin.

Lang, R. (2001). *The development and critique of the social model of disability* (Unpublished Working Paper). Overseas Development Group.

Linton, S., Mello, S., & O'Neill, J. (1995). Disability studies: Expanding the parameters of diversity. *The Radical Teacher*, *47*, 4–10.

Mandala Research. (2019). A diversity report. http://mandalaresearch.com/

Michopoulou, E., Darcy, S., Ambrose, I., & Buhalis, D. (2015). Accessible tourism futures: The world we dream to live in and the opportunities we hope to have. *Journal of Tourism Futures*, *1*(3), 179–188. doi:10.1108/JTF-08-2015-0043

Morgan, N., & Pritchard, A. (1998). *Tourism promotion and power: Creating images, creating identities*. John Wiley & Sons.

Morse, P. (2019). The power of Hispanic consumers in the travel industry. *Forbes*. https://www.forbes.com/sites/forbesagencycouncil/2019/08/21/the-power-of-hispanic-consumers-in-the-travel-industry/#1389036c719f

MSN. (2019). #WeVoteNext Summit. https://www.msn.com/en-us/video/news/supernumberwevotenext-summit-activist-janaya-future-khan-on-redefining-privilege/vi-BBOL5Nd

National Disability Authority. (2019). Appropriate terms to use. http://nda.ie/Publications/Attitudes/Appropriate-Terms-to-Use-about-Disability/

National Travel and Tourism Office. (2015). *Overseas Visitation Estimates for U.S. States, Cities, and Census Regions: 2014*. https://travel.trade.gov/outreachpages/download_data_table/2014_States_and_Cities.pdf

Nyanjom, J., Boxall, K., & Slaven, J. (2018). Towards inclusive tourism? Stakeholder collaboration in the development of accessible tourism. *Tourism Geographies*, *20*(4), 675–697. doi:10.1080/14616688.2018.1477828

Oliver, M. (1983). *Social work with disabled people*. Macmillan.

Oliver, M. (1990). *The politics of disablement*. Macmillan.

Oliver, M., & Barnes, C. (2012). *The new politics of disablement*. Palgrave Macmillan.

Olkin, R. (2002). Could you hold the door for me? Including disability in diversity. *Cultural Diversity and Ethnic Minority Psychology*, *8*(2), 130–137. doi:10.1037/1099-9809.8.2.130

Open Doors Organization. (2019). *2015 market study*. https://opendoorsnfp.org/market-studies/2015-market-study/

Peltier, D. (2016). The business case for more diversity in tourism advertising. *Skift*. https://skift.com/2016/01/18/the-business-case-for-more-diversity-in-tourism-advertising/

Polland, J. (2014, October 30). A detailed look at how Americans travel within the US. *Business Insider*. https://www.businessinsider.com/the-most-popular-us-states-for-tourism-2014-10

Pritchard, A. (2001). Tourism and representation: A scale for measuring gendered portrayals. *Leisure Studies*, *20*(2), 79–94. doi:10.1080/02614360110068651

Pritchard, A., & Morgan, N. (2000a). Constructing tourism landscapes – gender, sexuality, and space. *Tourism Geographies*, *2*(2), 115–139.

Pritchard, A., & Morgan, N. (2000b). Privileging the male gaze: Gendered tourism landscapes. *Annals of Tourism Research*, *27*(4), 884–905. doi:10.1016/S0160-7383(99)00113-9

Richards, V., Pritchard, A., & Morgan, N. (2010). (Re)envionsing tourism and visual impairment. *Annals of Tourism Research*, *37*(4), 1097–1116.

Saldaña, J. (2015). *The coding manual for qualitative researchers*. Sage.

Samson, H. (Producer). Audio Podcast (2017, February 2). *Skift podcast: The evolution of accessible travel*. https://skift.com/2017/02/02/skift-podcast-the-evolution-of-accessible-travel/.

Scheyvens, R., & Biddulph, R. (2018). Inclusive tourism development. *Tourism Geographies*, *20*(4), 589–609.

Seiver, B., & Matthews, A. (2016). Beyond whiteness: A comparative analysis of representations of Aboriginality in tourism destination images in New South Wales, Australia. *Journal of Sustainable Tourism*, *24*(8–9), 1298–1314.

Sheivachman, A. (2017). Pet-friendly travel is growing yet still complicated. *SKIFT*. https://skift.com/2017/01/31/pet-friendly-travel-is-growing-yet-still-complicated/

Stanonis, A. J. (2008). *Dixie emporium: Tourism, foodways, and consumer culture in the American south*. The University of Georgia Press.

Starnes, R. D., (2003). Introduction. In R. D. Starnes, (Ed.), *Southern journeys: Tourism, history, & culture in the modern south* (pp. 1–14). University of Alabama Press.

Stone, L. (2017). Fewer Babies, more pets? Parenthood, marriage and pet ownership in America. *Institute for Family Studies*. https://ifstudies.org/blog/fewer-babies-more-pets-parenthood-marriage-and-pet-ownership-in-america

Taub, D. E., & Fanflik, P. L. (2000). The inclusion of disability in introductory sociology textbooks. *Teaching Sociology*, *28*(1), 12–23. doi:10.2307/1319418

The Arc. (2019). People first language. https://www.thearc.org/who-we-are/media-center/people-first-language

TravAbility. (2019). MyTravelNetwork. http://www.travability.net/

U.S. Census Bureau. (2011). *The Black Population: 2010*. https://www.census.gov/content/dam/Census/library/publications/2011/dec/c2010br-06.pdf

United Nations. (2019). Promoting accessible tourism for all. *Department of Economic and Social Affairs: Disability*. https://www.un.org/development/desa/disabilities/issues/promoting-accessible-tourism-for-all.html

UPIAS (Union of the Physically Impaired Against Segregation). (1976). *Fundamental Principles of Disability.* Union of the Physically Impaired Against Segregation.

VisitNC.com. (2019). *North Carolina travel tips by dogs for dogs.* https://www.visitnc.com/dog-travel-agent?fbclid=IwAR2VuD8bn8ndsNd2CRM7U6hKfcVI3ytF4U357Hnytc7KysAoS0-uwN5t8_A.

Walle, A. H. (1997). Quantitative versus qualitative tourism research. *Annals of Tourism Research, 24*(3), 524–536. doi: 10.1016/S0160-7383(96)00055-2

World Health Organization. (2019). *Disabilities.* https://www.who.int/topics/disabilities/en/.

Yasharoff, H. (2019). *USA Today.* https://www.usatoday.com/story/travel/news/2019/07/29/puppies-and-prosecco-denver-hotel-offers-cutest-party-ever/1857672001/

Yin, M., Shaewitz, D., Overton, C., & Smith, D. (2018). *A hidden market: The purchasing power of working-age adults with disabilities.* American Institutes for Research.

Tourism, animals and the scales of justice

David A. Fennell and Valerie Sheppard

ABSTRACT
Animals are used in tourism in many ways. Some uses are more benign such as ecotourism (e.g., viewing elephants in natural settings), while others induce significant levels of suffering, pain and distress with the denial of dignity and respect (e.g., elephants performing at a circus). What is missing is a comprehensive overview of how just these uses may or may not be. Justice, as a concept, is complex and controversial—even more so when applied to animals. In this paper, we propose a "scales of justice" framework for animals used in tourism (deep, intermediate, shallow, and no justice) based on two different normative ethics models. The first model is based on the conventional normative ethical domain that includes deontology and consequentialism (i.e., what we are told to do), while the second model is rooted in virtue ethics (i.e., what we ought to do as individuals). The intent is to "weigh" the just and unjust use of animals in the tourism industry based on these contrasting theoretical perspectives. Elephants are used as an example through the paper. We view this approach as a springboard for more intensive investigations on how animals are used in tourism in a variety of contexts.

Introduction

In early November of 2019, the story surfaced of a teenage elephant named Kanakota, who died of exhaustion after being forced, with legs shackled, to give numerous rides to tourists in the Sri Lankan heat (New Zealand Herald, 2019). The story goes on to describe how young elephants are often taken illegally from their mothers who are typically killed in the "kidnapping", and how their spirits are broken or "crushed" through torture in an effort to train elephants to obey human commands. The fact that elephants are so badly abused seems lost on tourists who pay for rides and then post selfies on Instagram and other forms of social media. The parallel fact that elephants are not meant to carry such weight, nor die at the age of 18, rather than the usual life expectancy of approximately 60 years, is also lost on most tourists. But what should our collective response be to these sorts of occurrences? Is this simply business as usual? Are these issues of improper welfare or improper management of animals? Or do we struggle to understand why abuse continues to happen, and what represents justice for such animals? (see Sheppard & Fennell, 2019). Ethical theory offers considerable promise in helping tourism agents better understand and evaluate justice for animals that are used so frequently in tourism.

Tourism is inherently a justice issue (Fennell, 2018; Jamal, 2019) because of the need to secure fairness, honour, and virtue (Hultsman, 1995), as well as to contribute to human wellbeing through peace, conflict resolution, and rights (Higgins-Desboilles & Blanchard, 2010), and the fair distribution of resources and benefits between and within societies (Jamal & Camargo, 2014; Mihalic & Fennell, 2015; Scheyvens, 2002; Smith & Duffy, 2003). Justice in tourism should also mean adherence to principles of sustainability (Issac & Hodge, 2011), especially given the emphasis placed on justice through the United Nations′ Sustainable Development Goals (SDG) for 2015 to 2030 (Fennell & Cooper, 2020). SDG number 16, *Peace, Justice and Strong Institutions*, articulates the need to promote "peaceful and inclusive societies for sustainable development, provide access to justice for all and build effective, accountable and inclusive institutions at all levels" (United Nations, 2019). However, justice interwoven into these major global sustainability goals is almost wholly human-centred. Such is the case with the UNWTO Global Code of Ethics for Tourism where consideration of animal interests is almost entirely absent (Fennell, 2014).

While there is an emerging base of literature in tourism studies that discusses animal ethics from several theoretical positions (Fennell, 2012), our purpose here is to build upon this base by introducing justice as an important tool to enhance our understanding of animal ethics issues in tourism. We do this in four main parts. The first part contains a discussion of five traditional normative theories/approaches that have universal appeal in informing the discussion on justice (contractarianism, ecocentrism, welfare, utilitarianism, and animal rights). The second part involves a discussion of virtue ethics as a departure point for Josef Pieper's (1955) forms of justice (commutative, distributive and legal). The third part involves a synthesis of the preceding two parts in the development of a "scales of justice" continuum for animals used in the tourism industry. The fourth section provides more of a connection to tourism industry practices through the aid of the scales of justice framework according to what it means t be a good tourist or tourism practitioner, and our duties and obligations regarding fairness to animals used in tourism. Elephants are used as an example throughout the paper.

Approaches to justice: conventional normative ethical theories

Justice is a basic principle of the fabric of human life, and deeply embedded in the nature of our social groups. Its importance to human goodness has long been recognised through the four cardinal virtues (prudence, courage, temperance, and justice) espoused by Plato, Aristotle, the Bible, and other theological doctrines, of which justice is said to be the most important. A longstanding core definition of justice comes from the *Institutes of Justinian*—the sixth century codification of Roman law—which states that justice is 'the constant and perpetual will to render each his due' (Pojman, 1997, p. 549). From this general perspective, Miller (2017) argues that there are important considerations inherent in this definition. First, justice is essential in determining how individual people are to be treated, i.e., 'to *each* his due'. Second, that just treatment is something *due* to each person, so that claims can be rightfully advanced against a person or institution granting justice. Third, justice is impartial and so the opposite of arbitrariness. Finally, justice requires action (or inaction) to alter the circumstances of those drawn into an issue.

A survey of the academic literature illustrates that there is a breadth of theoretical views on the meaning and application of justice. Some of these include classical liberalism based on justice and limited state intervention (Friedman, 1962); libertarianism based on entitlement (Nozick, 1974); utilitarianism in the protection of liberty (Mill, 2010); liberal egalitarianism from the perspective of justice as fairness (Rawls, 1971); individual choice and responsibility in the distribution of justice (Dworkin, 1977); communitarianism based on why people view justice and resource allocation from their own particular social context (MacIntyre, 2013); feminist approaches (Nussbaum, 1996; Young, 1985); and the narrative around human rights as a guiding force around justice (see Ignatieff, 2003). This variability has prompted scholars to observe that there

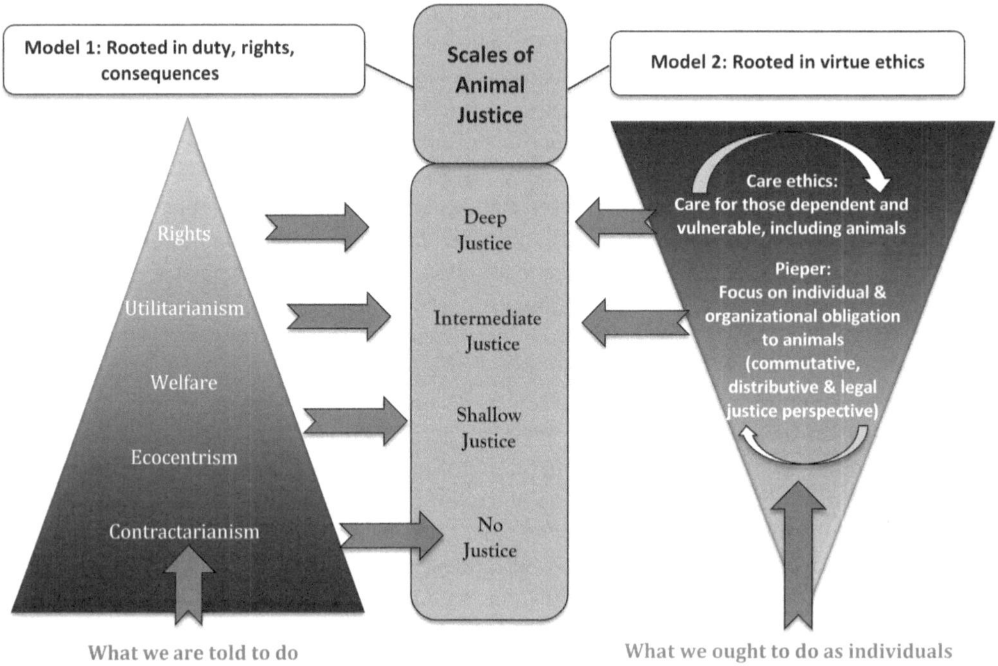

Figure 1. Contrasting theoretical frameworks for animals used in tourism.

is no single comprehensive theory of justice that can be relied upon to explain what justice requires us to do in all scenarios and circumstances (Miller, 2017; see also Regan (2004). An indication of how deep the justice discourse is may also be found in the sheer number of applications of justice in the literature and in practice. Examples of this depth include distributive justice, restorative justice, retributive justice, global justice, organizational justice, service recovery justice, procedural justice, justice as a virtue, recognition justice, and justice as the politics of difference (see Young, 2012).

In this section we briefly highlight a number of normative theories that have been the source of ethical guidance for hundreds of years, all of which have been adapted to address moral issues with animals. The focus with these theories lies in "What we are told to do" in deciding what is ethical or not. We order these according to what we judge as the level of justice that each offers to animals—from most basic, if at all, to the most advanced level. We provide only a brief overview of each within the confines of space, including the basic essence of these theories as well as some criticisms. Model 1 on the left hand side of Figure 1 reflects approaches based on duties, rights and consequences, while Model 2, which we discuss further below, is grounded in virtue ethics. As we argue below, the "scales of justice" framework provides guidance for the just and unjust uses of animals in tourism.

Contractarianism

Contractarianism follows from Hobbesian social contract theory, and assumes that because humans are largely self-interested, they will subscribe to a contract with others because: (1) it is the most rational way to maximise their chosen benefits; and, (2) we stand to benefit from entering into cooperative relationships (Narveson, 1988). Contractarianism has been used as an arena for justice, both human and animal studies, because it is premised on the notion that only rational individuals may enter this arena in deciding upon what principles of morality and justice are legitimate, i.e., what parties have agreed to, and therefore accept as fair (Mulgan, 2011). As such, there is an element of reciprocity built into contractarianism as agents both give and take according to the roles and responsibilities within cooperative relationships.

The most important contemporary doctrine on social contract ethics comes from Rawls (1971), who argued that the best way to govern society would be to make decisions that are fair for all people in society by adopting an "original position", where one could choose principles to govern society under ideal conditions (ideal theory). Fairness emerges from a thought experiment referred to as the "veil of ignorance," where distinguishing characteristics such as gender, race, social class, wealth, are unknown to the decision-maker. While Rawls argues that it is wrong to be cruel to animals and to eliminate whole species, we should have "duties of compassion and humanity in their case… [but]… they are outside the scope of the theory of justice, and it does not seem possible to extend the contract doctrine so as to include them in a natural way" (p. 512). There is no element of reciprocity, as noted above, and only moral (rational) beings are able to receive justice because only moral beings can act upon principles of justice (Garner, 2012). As such, animals from the Rawlsian perspective, would not be recipients of the equitable distribution of societal goods (Jamal, 2019).

Contractarianism, as a theory for advancing justice for animals, fails to recognize the inherent and autonomous nature of animals. Rather, it relies upon the goodwill of humans to do so (Garner, 2012); (see Abbey, 2007 and Rowlands, 1997, who contend that direct moral status can be extended to animals through contractarianism; however, these accounts push Rawls' work into ways that he himself might not necessarily agree with).

Ecocentrism

With the onset of environmental ethics some 50 years ago, questions emerged over which entities in the natural world should possess moral value, i.e., do we protected the interests of individuals only or whole ecosystems too (Varner, 1998). Curry's (2011) framework on environmental ethics advances three levels of environmental ethics based on different values and priorities. The first, light green or shallow anthropocentric ethics includes enlightened self-interest where the main goal is to preserve a healthy natural world for the purpose of commerce, aesthetics, and human health (Humphrey, 2000). The second, mid-green or intermediate ethics, ascribes moral value to not to ecosystems, but rather to individual animals within these ecosystems (see Varner's, 1998 work on biocentric individualism). The third, deep green or deep ecocentric ethics places moral value at the level of whole systems and ecosystems (e.g., Aldo Leopold's (1970) land ethic).

A driving force behind Leopold's land ethic is the stringent belief that we may use animals and resources in a manner that we see fit if, in our use, there is no harm to the "integrity, stability and beauty of the biotic community" (p. 262). We can fish, hunt, trap, dig, dredge, cut, clear, and so on, as long as the resources that we use remain in a state that does not compromise this integrity. For example, if deer populations rise in the absence of an apex predator like the wolf, and as a result vegetation suffers from over-grazing, culling deer would be morally praiseworthy in returning a proper balance to the ecosystem. As such, the interests of individual animals are not the locus of concern. If justice for animals is a matter of the right to life, protection of individual liberty, and prohibition of torture (Singer & Cavalieri, 1993), another theoretical perspective would need to be adopted. Indeed, ecocentrism has been criticised because anthropocentric values guide decision-making (Rowe, 1997) in subordination of the welfare of other beings for the integrity of the broader community (Regan, 2004). Both criticisms have compelled animal liberation theorists to argue that ecocentrism is overwhelmingly informed by a utilitarian doctrine based on the greatest good for the greatest number (Garner, 2004).

Welfare

Welfare is logically conceived as a constellation of approaches, and not one theory (although it does correspond to several other moral theories like utilitarianism, or the maximization of welfare across society). The argument around welfare, as a form of justice, is not new. Gearey (2013) explored welfare from an egalitarian account of distributive justice through ideas around communal and individual autonomy over centralized bureaucratic and market-based approaches.

Animal welfare is defined as "the state of an animal as regards its attempts to cope with its environment" (Hill & Broom, 2009, p. 532), and it entails the use of an animal as long as measures are taken to ensure that basic needs are taken care of in this use. The Five Freedoms of animal welfare, developed by the Farm Animal Welfare Council in 2009 (see https://www.humanecanada.ca/five_freedoms_of_animal_welfare), is a popular benchmark of proper welfare in the care and handling of animals. These freedoms include freedom: (1) from hunger and thirst; (2) from discomfort; (3) from pain, injury and disease; (4) to express normal behavior; and, (5) from fear and distress.

Welfare is widely embraced because it has improved the lives of countless animals. It is rejected because it opens the door to significant human exploitation. Those who embrace animal welfare contend that possession of an animal does not always work in opposition to its interests, as animals may have no objection to being possessed and cared for (Cochrane, 2009). Cochrane argues that forcing a domesticated animal into labour is nothing like doing the same for an autonomous human. The concern related to the possession of animals, Cochrane (2009) adds, ought to be more about better "*regulation* of the system of ownership rather than its *abolition*" (p. 22).

Utilitarianism

This ends-based theory is focused around weighing the balance of good and bad for those involved, in comparison to some other alternative (Frankena, 1963). Actions that lead to more happiness and less pain for the greatest number are argued to be best. Utilitarianism is positioned around notions of universality (equality on the basis of the interests of all), welfare (defining what is good based on interests), and consequentialism (evaluation of what is good based on the satisfaction of these interests in the end) (Matheny, 2006). Along these lines, justice is associated with utilitarianism in consideration of the best consequences for all involved. So, a just society is equated with an increase in the overall quality of life for citizens.

Singer (2009) is the animal ethics scholar most closely tied to the utilitarianism tradition. His main premise has a strong connection to justice through the concept of equal consideration of interests: that which is judged to be good or bad is a function of all those affected by the action or decision—including animals based on the fact that they, like humans, experience pain and suffering. On this account, an interest like suffering does not depend on whose interest it is. Human and nonhuman animals all have an interest in not suffering and the avoidance of pain. As such, the like suffering of one being (human) must be counted equally against the like suffering of another being (animal), and an animal that suffers more ought to be extended a higher level of concern than other beings that suffer less.

Where utilitarianism is criticized is in the belief that it is not the right type of equality according to Regan (1985). It does not adequately explain why the abolition of something like factory farming would increase net wellbeing. Furthermore, utilitarianism does not value equal inherent value—it is the measurement of pleasure and pain (the *interests*) of the individual that counts, not the person him/herself. In this sense, because utilitarianism is an aggregative theory, it does not give proper regard to individual rights. The focus is on the reduction of suffering in animals, not on securing animal rights, and not on the elimination of the use of animals. Bigger cages are seen to be the better option for Singer if they take away suffering. Singer's position is more akin to animal welfare in the eyes of rights-based scholars such as Francione and Charlton (2016), because animals can be used for the benefit of others, but "*only if* those benefits are greater than the loss inflicted" (Nozick, 1974, p. 37).

Rights

A right is "a moral or legal entitlement to have or obtain something or to act in a certain way" (Oxford Dictionaries, 2019). Entitlements mean that we have the wherewithal "to do, to demand, to enjoy, to be, to have done for us. Rights may also be defined as rights to act, to exist, to enjoy, to demand. We speak of rights as being "possessed, exercised, and enjoyed" (McCloskey, 1965, p. 117). Dworkin (1977) argued that rights are trumps over utilitarian reasoning that would place the interests of the collective over the interests of individuals, and institutions like government should treat individuals as equals on the basis of respect and dignity.

Regan (2004; 2003) argued that we have direct duties to animals because respect and harm ought to be integral components of our understanding of what it means to be just to animals. We should be egalitarian in our actions by extending the same respect equally, non-perfectionist where one individual is not better than others, we should not use others as means to our own ends, and we have a *prima facie* duty to protect victims of injustice from the actions of others (Regan, 2004). Regan extends these conditions to animals on the basis of consciousness, intentionality, and sentience, which render them subjects-of-a-life with inherent value.

Other rights-based theorists, such as Francione and Charlton (2016), argue that the most basic and important right that animals should possess is the right not to be treated as someone else's property (see also Francione, 1995). This will only happen if they matter morally, which means they are not viewed simply as things or resources for our use. Wyckoff (2014) contends that animals should be claimants of justice based on a radical abolitionism view. He takes this position by observing that the resource paradigm, defined as, "the cluster of beliefs, assumptions, and practices that take animals to be resources that humans may use in order to generate benefits for themselves and other humans" (p. 540), is manifestly unjust. It does not matter that we make the lives of animals better through welfare reforms. Animals are still slaves to humans.

Nozick's (1974) work on positive versus negative rights provides a different dimension to the discussion on justice for animals in tourism. Positive rights refer to the requirement of one to assist another in attaining their goals. We help a dog to better health, for example, by taking him or her to the veterinarian. By contrast, negative rights refer only to the fact that others must allow an agent to pursue their ends in a climate of noninterference. There is precedence for the implementation of positive rights in animal ethics based on work by Shooster (2017), who observes that positive rights need to be "expansive positive legal rights." In order not to be speciesist, Shooster argues, the same rights (e.g., provision of food, shelter, health) ought to be extended to animals that have many of the same needs as humans. On Shooster's account, governments must have a legal obligation to use all resources at hand to make the lives of animals (his focus was on wild animals) better. These rights should be extended even though they may be aspirational rather than achievable at the present time.

Where animal rights has be criticised is in the failure to recognise differences that exist between animals and humans (Garner, 2013). Furthermore, criticism comes from the illusion that each and every one of us ought to get something without regard to the costs that we would have to bear in attaining these things (O'Neill, 2000).

Virtue ethics for animals and justice

In this section virtue ethics is explored for the purpose of constructing an alternate model of animal justice (Model 2 in Figure 1). We enlist Josef Pieper's work on justice as a virtue in concert with care ethics, which has also been subsumed under the category of virtue ethics (Figure 2).

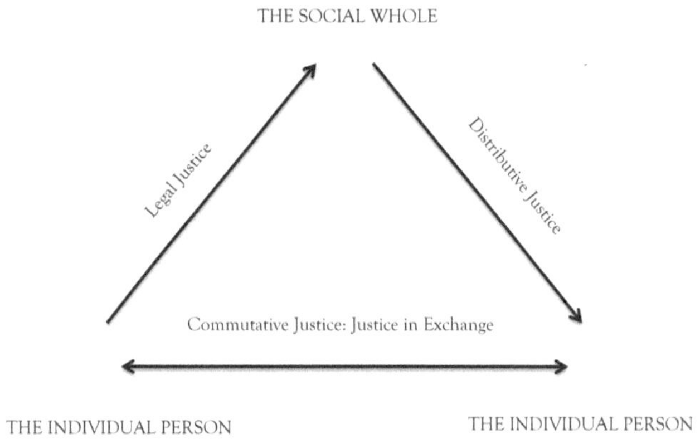

Figure 2. A schematic representation of the basic forms of justice (Adapted from Pieper, 1955).

Virtue ethics is focused not on what we are told to do, but rather on "What sort of person I should be." Virtues are positive traits of character such as altruism, compassion and loyalty, which, if practiced regularly in the proper shaping of our desires through reason and will, allow individuals to flourish (the Greek term eudaimonia). Negative virtues include stealing and lying (Garofalo & Geuras, 1999), or, in our case here, abuse and neglect of animals. The element of practice is important in virtue ethics because the qualities of a good person are not innate, but must be properly cultivated through habituation and training (Cooper, 1987). As such, lower order forms of happiness such as sex and popularity are typically short-lived and require others for their realization. Furthermore, although some pleasures are subjective and relative to certain cultures and individuals, eudaimonia ought to be pursued as an objective, universal good (Fennell & Malloy, 2007). There is thus an inherent ends-orientation consistent in virtue ethics in regards to the pursuit of the good life. The main criticism of virtue ethics is in trying to identify which character traits are virtuous and how they apply in different situations. Other normative ethical domains (deontology and teleology) are equipped with established sets of what is considered right and good without the necessity of deciphering characteristics that are indeed virtuous (Slote, 1992).

Pieper (1955) drew inspiration from the works of Aristotle and St. Thomas Aquinas on the four cardinal virtues (prudence, justice, fortitude, and temperance) as they relate to individuals, society and Christian religion. Although Pieper's treatise is highly theoretical, he makes an effort to explain these virtues in an applied sense, i.e., how do we realise the good in everyday practice because there are very few things in society that are not connected with justice, e.g., just wars, human rights, capital punishment, union strikes, and so on.

Pieper (1955) recognised that justice as virtue must be relegated to the scale of the individual within our families, organisations and state, because it is individuals who make decisions on behalf of these institutions. Furthermore, we build community through individual action. For example, the true character of a wrong or injustice, such as stealing some possession, is not the owner's loss of the possession, but rather "the implicit threat to the entire order of community life, affecting every member. Once this is recognized by everyone, justice can be said to prevail in that state." (Pieper, 1955, online version). Pieper argued that justice rules a community if three basic forms of justice are organised in the correct order (Figure 2, and right side of Figure 1). The first is commutative justice or justice in exchange (individuals and how they relate to each other), the second is justice as the fair distribution of resources (the relations of the social whole, or society, to individuals), and the third is legal or general justice (the relations of individuals to the social whole).[1]

Commutative justice

Commutative justice is nominally the form that we think of most because, at its foundation, it is premised on the assumption of equality between individuals entering into relationships. In the interests of equality each individual must take responsibility for accepting the burden of fairness within social exchanges. If we choose to cheat another by elevating the price of a product (taxi cab fare) or bargaining well below what is deemed proper for the sale of a souvenir, commutative justice has been violated. Restitution also falls within this realm and the expectation that if one cheats another, the perpetrator must make amends to the victim in efforts to right the ship.

Restitution is often considered through the restorative justice and service recovery justice literature. Restorative justice works towards repairing the harmful effects of actions between victim and offender, and efforts required in order to restore the relationship between both parties (Cossins, 2007; Gavrielides, 2007). It is also important as a mechanism to construct a better society for the present and the future (Marshall, 1999), as recognised by Pieper (1955). Restoring justice in our use of animals in tourism includes examples of the elimination of certain practices. Hughes (2001) argued for a shift away from viewing dolphins in captivity to viewing them only in the wild, which he reasoned was a structural shift in the nature of local policy.

In service recovery justice, consumers often evaluate an organisation's attempt to make things right, as being just or unjust. Those that are just strengthen the relationship between organisation and consumer (DeWitt et al., 2008). Furthermore, online rating organizations, such as TripAdvisor, have become an important vehicle for tourists who are dissatisfied with their service experience. Interestingly, while such rankings seem to be of concern to most tourism businesses, this appears not to be the case when it comes to animals. A case in point is the Namuang Safari Park in Thailand. One might assume the safari park would be encouraged by its TripAdvisor ranking (2.5 out of 5 stars overall) to take measures to remediate the consistently reported poor and inhumane treatment of animals over almost a decade (from 2010 onward). In other words, this park appears uninterested in service recovery, as long as tourists continue to patronize the business.

Distributive justice

If individuals were entirely responsible in their dealings with other agents we would not need other forms of justice. The fact of the matter is, as Piper reasoned, people are not entirely responsible, and so we therefore need other forms of justice to intercede. It is not the collective within our institutions that is not just or unjust, on Pieper's account, but rather the individuals holding power within these structures. Distributive justice focuses on who has more of something, e.g., money, food, health care, or natural resources, and the decisions around their allocation (Cohen-Charash & Spector, 2001, p. 279).

Most academic work on justice and tourism is grounded in distributive justice, as it signals the unequal allocation of tourism benefits among different stakeholders. Jamal and Camargo (2014) concept of the "Just Destination" places priority on an ethic of care, fairness, as well as equitable tourism policies that prevent marginalisation and peripherality of disadvantaged groups. Distributive justice has been also discussed in tourism in relation to greed over trophy hunting in Africa. Mkono (2019) writes that social media posts (by Africans) over foreign hunters to Africa demonstrates injustice not only because licences were being purchased by hunters outside the country (e.g., U.S.A.), but also that wealthy elites in Africa were realising most of the benefits. Mkono argues that the hunting industry is perpetuating unequal relations on territorial and cultural bases. What is subject to further consideration in Mkono's work is the belief that African local people "are considered the true owners and custodians of the wildlife resources" (p. 697). Somewhere lost in this issue is a more balanced debate around animals living free and not subject to contractarian discourses around ownership and servitude.

Legal or general justice

This third type of justice identified by Pieper (1955) refers to the relations of individuals to the social whole. Legal or general justice refers to the duties that we have to society in the creation of a better social whole. As such, when we as individuals focus too much on what is owed to us in a community, we fall short of understanding what our own duties are to society in creating a greater foundation of social justice. If all of us were to focus our energies on contributing to the common good, social justice would be much easier to attain. From this vantage point, Pieper (1955) contends that social justice should be more formally regarded as a matter of legal or general justice.

Some scholars argue that social justice advocates have not done a good enough job embracing animal rights as a core aspect of their activism (Jones, 2015; Shooster, 2015). Shooster makes this claim on the basis of the following three social justice values: (1) an opposition to arbitrary discrimination or justifying violence and inequality on the basis of sexism and racism, classism or speciesism; (2) the need to support the most severely marginalised or disadvantaged; and (3) respecting autonomy and individual rights. Regarding the third value, above, Jones (2015) argues that animals function as property or commodities that are freely and unconscientiously distributed amongst members of human society (see Mkono, 2019, above).

Legal justice is a problem in tourism practice because corporate justice does not often extend to areas where the rule of law and institutional oversight is weak (IPIS, 2014). Tourism companies operate around the world, in areas where rule of law and institutional oversight are weak and in areas where they are strong. Obviously, notions of justice are applied unequally to humans and animals. The fight to develop and strengthen the rule of law and institutional oversight for improving the welfare of animals is currently being fought by a number of justice-related organizations (see Author, in press).

While Pieper's views on justice are savoury, he maintains the long-standing Christian belief that it is only beings possessing a spiritual capacity (humans) that are owed justice:

> For stones, plants, and animals have also been created, yet we cannot say that they have their due in the strict sense of the word. For "being due" means something like belonging to or being the property of someone. A nonspiritual being, however, cannot properly have anything belonging to it; on the contrary it, itself, belongs to someone else, for instance, to man (Pieper, 1955, p. 47).

The Christian anthropocentric and instrumental tone used by Pieper cannot, therefore, be used exclusively to build a virtue ethics for animals. For this, we enlist care ethics and argue that merging the two—Pieper and care ethics—offers a new fortified approach to explaining how virtue ethics can be an agent in the pursuit of the good when it comes to animals used in tourism.

Care ethics

Gilligan's (1982) seminal work on moral decision-making represented a major paradigm change in diversifying the moral landscape. Gilligan argued that while men moralise according to normative, masculine concepts of rights, justice and law, women, by contrast, moralise according to responsibility and an ethic of care. For Held (2004), this means that empathy, responsibility and responsiveness may be more important as moral points of departure than abstract rules and principles. What is important from a relational standpoint are bonds of attachment rather than contracts between stakeholders (Robinson, 1997). While some contend that care ethics is independent of virtue ethics because the latter has a focus on the individual, and the former is concerned more with relationships (Held, 2004), there is consistency around care ethics as virtue because of the focus on addressing the needs of self and others. Care ethics as virtue surfaces as a motivational attitude towards empathy (Slote, 2007) or as a specific type of virtue ethics with care being the chief virtue (Halwani, 2003b). Care ethics benefits, Halwani (2003a) argues, by being affiliated with a more comprehensive moral framework (virtue ethics), with reason playing a regulative role in framing intimate relationships. Furthermore, there is precedence for linking care ethics with animal justice through a number of important contributions. Adams and Donovan (2007) argue that feminist care ethics can inform the relationship between people and animals on the basis of love and empathy. Kheel (2008) argues for an enlargement of the space for care to involve individual animals and larger populations, and this includes listening to other forms of life no matter how different they are from us (Donovan, 2006; Vance, 1995).

Synthesis: animal justice in tourism

Construction and description of the framework

Our purpose in this section is to construct a framework based on the theories and forms of justice discussed above in the creation of a scale that depicts the degrees of justice afforded to animals used in tourism. The scale contains four levels, ranging from the highest level of justice (deep justice), to intermediate justice, then to shallow justice, and ending at the lowest level (no justice). While we acknowledge that our organization of the scale is subjective and could be reorganized based upon the nature of the circumstance in play, we believe our efforts lay the foundation for future research and better practice in this area.

The first part of the framework as presented in Figure 1 positions the five conventional theories of ethics (Model 1) on the left-hand side, and virtue ethics (Model 2) on the right-hand side. Specifically, Model 1 theories of justice are positioned to represent the various levels of justice in the form of fairness and protection they provide animals, positioned within the grey triangle to indicate the degree of moral consideration associated with each theory. Indeed, this part of the framework represents a hypothetical indication of the numbers of animals used in tourism and the degree of justice that is extended to them. For example, contractarianism, would provide the least amount of fairness and protection (arrow indicates association with no justice), whereas rights would offer the greatest level of protection (arrow indicates the association with deep justice). Here it is suggested that the vast majority of animals are extended no justice while only very few animals are extended deep justice as outlined by the peak of the triangle.

Virtue ethics, on the right of Figure 1, is built on Pieper's (1955) three forms of justice, as discussed above. The intersection of care ethics opens the door to the possibility of an extension towards more advanced levels of justice (intermediate and deep) for animals as a marginalized group who are dependent and vulnerable. Model 2 indicates that this intersection between Pieper and care ethics would, in the best-case scenario, eliminate the presence of no justice and shallow justice if what we ought to do as individuals involves care for animals. Simultaneously it would increase the number of animals that would receive justice as more individuals (tourists and tourism providers) act in the interests of animals as indicated by the inverted triangle. Situated between Model 1 and Model 2 is the scales of animal justice representation, which depicts four levels of increasing justice (no justice, shallow justice, intermediate justice, deep justice). We provide more discussion around each of these four levels of justice below. We also illustrate the move from theory to practice in Table 1 and attempt to answer the following questions: *What does it mean to be a good tourist or good tourism broker" How should governance structures act (good governance, policy, etc.) when it comes to the proper treatment of animals used in tourism? What are our duties or obligations regarding fairness to animals?* We focus on the treatment of elephants in the tourism industry in an effort to attain consistency as we move through each level of justice.

Table 1. Connecting theory and practice in the scales of justice.

Scales of justice	Model 1: Conventional normative ethics	Model 2: Virtue ethics	Description of interactions between humans & animals
No justice	**Contractarianism** Animals fall outside the arena of moral consideration because they are not rational and they are unable to be reciprocal. Therefore, no duty to provide animals with an equitable distribution of societal goods. Harm to an animal is harm to the animal's owner.	Tourists are oblivious and or uncaring to condition of animals. Development policy is strictly economically and anthropocentrically focused. No ethic of care in the way service providers treat animals. CJ: There is no level of commutative justice for animals. No equality between humans and animals, and any harm to an animal means restitution for the animal's owner. DJ: Only distributive justice is applicable at this level, where all the benefits go to humans, at the expense of animals. LJ: There is no level of legal justice for animals, as they are considered to lack moral status. Laws are circumvented through corruption, bribery, and deceit.	Animals kept in conditions that are insufficient in terms of their physiological and psychological needs. Entertainment and commerce are the sole purposes behind their captivity. **Example:** Namuang Safari Park (Thailand). Elephants and other animals are forced to perform humiliating and degrading tricks for the entertainment of paying tourists (personal communication with a park visitor).
Shallow justice	**Group focus: Ecocentrism** Moral value placed at the level of whole ecosystems, communities and species. Animals can be used (e.g., hunted) in any way as long as the integrity, stability, and beauty of the natural world is maintained. The interests of individual animals are not the locus of concern. Animals are considered a resource that is available for the use of humans, and decisions about their wellbeing is considered from an anthropocentric perspective.	The extension of virtue and care are not the focus of the human-animal experience. Knowing that species are healthy and robust may be the only concern for tourists. Operators are aware of the need to appear to care for animals, but this is done to reassure tourists and does not reflect what transpires back stage. Destination/ tourism development policy is anthropocentric in nature. CJ: There is no equality in exchange between humans and animals, only care at broader levels. DJ: Limited distributive justice with animals receiving care that has benefits primarily for human beings based on profit for organizations. LJ: Care is extended through laws and regulations that focus on population numbers, e.g.,	**Examples:** Pinnawala Elephant Orphanage (Sri Lanka). While the elephants live in better conditions than other establishments in Sri Lanka, the "orphanage" has a breeding program and photo opportunities for the benefit of tourists. The prime focus is on the profitability of the organization. (see https://www.responsibletravel.com/holidays/elephant-conservation/travel-guide/why-we-dont-support-pinnawala-elephant-orphanage). Disagreements over trophy hunting in Africa based on greed. Local people who are owners of the wildlife should benefit over foreign hunters and operators.

(*continued*)

Table 1. Continued.

Scales of justice	Model 1: Conventional normative ethics	Model 2: Virtue ethics	Description of interactions between humans & animals
		catch and release, or bag limits.	
Shallow to intermediate justice	**Individual & group focus: Welfare** The focus is on measuring the state of an animal used for tourism purposes through the Five Freedoms. This may mean, for example, the use of bigger cages, but it still means that animals will be used for human instrumental use.	Virtue and care demonstrated through welfare standards, rather than a direct, personal ethic of care on how animals ought to be treated. CJ: Bonds of attachment and responsibility from individuals towards animals based on welfare guidelines. DJ: Habitat provided for animals that may replicate natural settings. The commercial focus is primary with more benefits for humans over animals. LJ: Abolition of the use of animals is not in question, but rather better regulation of the system (e.g., laws, codes, certification) of ownership and use in the reduction of suffering.	**Example:** Mason Elephant Park & Lodge (Indonesia). While the Park has attained Gold Certification from Asian Captive Elephants Standards, the elephants are forced to perform unnatural behaviour and tricks (see https://www.responsibletravel.com/holidays/elephant-conservation/travel-guide/elephant-sanctuaries-which-we-do-and-dont-support).
Intermediate justice	**Utilitarianism** The focus is on the elimination of suffering and pain in animals from the perspective of equal consideration of interests. The like suffering of one being must be counted against the like suffering of another. It is an aggregative theory that does not recognise individual rights. The weighing of costs and benefits for all involved (animals and humans)	Bonds of attachment are built through love, empathy and responsibility towards animals, but there are questions about the value of such because of the aggregative level of utilitarianism. Sometimes the ends do not justify the means. CJ: Individuals work together for animals to minimize their pain and suffering, but human interests may prevail over animal interests in assessment of costs and benefits. DJ: Distributive justice is served if tourists believe elephants benefit from the money they have paid to see them. LJ: Legal advocates work to provide oversight to ensure the organizations and individuals are meeting minimum standards of care and providing the necessary resources.	**Example**: Millennium Elephant Foundation (Sri Lanka). This family run organization seeks to raise awareness of the plight of elephants and improve welfare (i.e. campaigning against the use of the howdah—rider's throne). Although the organization still permits tourists to ride the elephants, they also encourage walking beside the elephants as an alternative (see https://www.responsibletravel.com/holidays/elephant-conservation/travel-guide/elephant-sanctuaries-which-we-do-and-dont-support)
Deep justice	**Animal rights** We have a duty to	Bonds of attachment with individual animals based	**Example:** Ecotourism: elephant viewing in the

(*continued*)

Table 1. Continued.

Scales of justice	Model 1: Conventional normative ethics	Model 2: Virtue ethics	Description of interactions between humans & animals
	respect and protect individual animals because they are subjects-of-a-life (inherent value, sentience, and intentionality, among other qualities). We should be egalitarian, non-perfectionist and not use animals as means to our instrumental ends. The most just way to treat animals, but also the most radical.	on strong relational connection to animals through respect, love, empathy and responsibility. CJ: Individuals work together to better the lives of animals (individual and populations), in their natural conditions. DJ: Resources are distributed to ensure that animal interests are protected as a priority even if such may radically disrupt current systems. LJ: Individuals work within and through the legal system to argue for animal justice on a case-by-case basis. Legal advocates work to include animals in the justice system, as sentient beings, entitled to protection through law, expanding the circle of morality.	wild. There are several ethical tourism operators in Africa and Asia. While there is minimal disruption and interference in the lives of the elephants and to their habitat, there is still a commercial element connected with tourists viewing elephants and other wild animals. However, the animals are free to disengage from their interactions with humans, at their instigation.

No justice for animals in tourism

Our first category, no justice, emphasizes situations where ownership, dominance and control of animals are primary features of human-animal relationships, and where respect, sympathy, autonomy, and dignity are universally denied. Examples include elephants that are trained to work (e.g., carrying tourists), and trained to perform (e.g., polo matches, painting pictures or performing in circuses). Tourists who pay to see these spectacles are oblivious and/or uncaring of the animals' plight. Government policy is focused on tourism development as an economic driver, also without regard for the interests of animals. As such, there can be no justice for elephants who must perform humiliating tricks for tourists that underscore submission and ridicule, based on the needs of a patriarchal society (Bone & Bone, 2016).

As Figure 1 demonstrates, there is no evidence that any stakeholders' behaviour is reflective of a virtue ethics perspective. Rather, we link this category with contractarianism, because this theory does not recognize the inherent value and autonomous nature of animals. Since animals are not rational and cannot understand their duties within cooperative relationships, they do not possess direct moral value. Such beliefs make it morally acceptable to manage them in conditions and by measures that are entirely insufficient in regard to their physiological and psychological needs.

Circuses and unregulated safari parks that position elephants as performers is a case in point as the antithesis of justice because of an overwhelming focus on economic grow (see also the Namuang Safari Park, above). In the case of such parks, commutative justice is reflected in the failure of human handlers to explain to tourists the methods employed to control and dominate animals for commerce and entertainment (e.g., negative reinforcement). There is no fairness in human exchange, therefore, tourists are not aware of how animals are prepared for their experiential consumption. From a distributive justice standpoint, benefits, apart from the satisfaction of basic physiological needs of animals, go entirely to human agents at a tremendous cost to animals. Legal justice is denied even if laws are present. In fact, measures may be taken to circumvent laws at every turn through corruption, bribery and deceit.

Shallow justice for animals in tourism

We view instances where animals are held in captivity, such as zoos, aquaria, and other commercial ventures, and where the focus is on providing entertainment for tourists as shallow justice. There is an instrumental element to this level: animals are captured and bred in captive environments. They are often traded and/or eliminated according to instrumental ends (Cohen & Fennell, 2016). To address growing public criticism against commercial operations displaying animals, many organizations seek out certification programs (see Box 1, below in reference to minimum standards). At this level, governments may involve themselves in the certification program, either encouraging business to pursue certification or holding them up as examples for other operators to follow; however, the certification programs often lack transparency and third-party checks. For example, as noted by Abraham (2019) in the context of the Thai elephant tourism situation:

1. The elephant trade in Thailand is controlled by people whose only interest is a desire to make as much money as possible;
2. There has been an increasing awareness among Western tourists about the abusive treatment of the elephants who work in the tourist trade;
3. This has led to unfavorable publicity which has cut into the revenues generated by elephants who service tourists;
4. In order to combat/refute the criticism/negative publicity a "certification" system/scheme has been developed by those owners who have the most to lose because they are the most abusive;
5. A for-profit company has been created who, for a fee that is paid to this company by the elephant camp owners, will "inspect" a trekking camp and then "certify" that it treats its elephants well.

We contend that ecocentric and a welfare approaches are both prevalent at this level, because stakeholder behaviour is not reflective of virtue ethics. Any care extended to animals is done to minimize rather than eliminate pain and suffering. For example, captive elephants are provided with the five freedoms, noted previously (see Box 1 where the emphasis is on wishing to ease the pain inflicted upon elephants, but not the willingness to end the practice). As Cohen (2009) contends, zoo and aquaria spaces are semi-contrived, where there is the appearance of wild, but where animals are tamed or habituated through contact with humans. Attention is paid to providing shelter that reflects their natural habitat, but elephants are on display mainly for entertainment purposes; although, there may be attempts made to provide educational, research and conservation components, entertainment and profits are still primary motivators (see for example ZooCheck New Zealand, n.d.). While many of these institutions claim a higher level of responsibility and care for elephants, they are essentially viewed as resources (the "resource paradigm" as explained by Wyckoff, 2014) that generate human benefits often at the expense of a good quality of life for the animal.

Commutative justice is constructed along assumptions of fairness between two parties, in this case the zoo as an organization and zoo visitors. Restitution, as part of commutative justice, was conceived as restorative justice in the discussion above. In the context of shallow justice this may simply mean that captive environments provide fairness (the provision of animals like elephants in a healthy state) for ocular consumption by visitors. It may also mean, however, the removal of elephants from zoos in cases where elephants are not with their conspecifics, as advocated by the Detroit Zoo and the Toronto Zoo because of climate, space and improper natural conditions (Good, 2014). If distributive justice means that we ought to be concerned with the fairness of outcomes, removing animals from zoos where these conditions, above, cannot be met seems just. The problem is that most zoos will not part with their elephants. Legal justice, or the duties that we have to society in the creation of a better social whole, suggests that we ought not to expect the community to owe, but rather what we owe to the community in making social progress. Even in cases where animal welfare laws have been enacted, such as the Cruelty Prevention and Welfare of Animal Act, B.E. 2557 in Thailand, these are often heavily criticised for being inadequate in protecting the welfare of animals, and individual operators do not follow them in good faith in creating positive social change (Dorloh, 2017). Good governance in its application to animals, therefore, is a function of competing interests between justice and commerce and entertainment.

Box 1

Responsible Travel: One of our suppliers, who wishes to remain anonymous

"I would suggest banning the use of chairs, which can be cruel for the animals. This would stop elephants standing around in the chairs all day long and support those camps which allow their animals more freedom. If every tourist insisted, they would ONLY pay for a ride without a chair, this could actually bring a real and positive change far more quickly than simply avoiding camps altogether."

https://www.responsibletravel.com/holidays/elephant-conservation/travel-guide/our-stance

. Box 1

Responsible Travel: One of our suppliers, who wishes to remain anonymous

"I would suggest banning the use of chairs, which can be cruel for the animals. This would stop elephants standing around in the chairs all day long and support those camps which allow their animals more freedom. If every tourist insisted, they would ONLY pay for a ride without a chair, this could actually bring a real and positive change far more quickly than simply avoiding camps altogether."

https://www.responsibletravel.com/holidays/elephant-conservation/travel-guide/our-stance

Intermediate justice for animals in tourism

Intermediate justice is connected to utilitarianism because while utilitarianism actively moves towards the elimination of suffering in animals, at least equal consideration of interests when it comes to suffering, there are questions that surround the profit motive and poor practice in captive or semi-captive environments like sanctuaries. For example, even though sanctuaries have larger spaces than zoos or other captive environments, practices like elephant riding still persist. And because utilitarianism is an aggregative theory it is not necessarily the interests of individuals that matter, but rather polices that are developed for entire populations of animals held. Furthermore, the greatest good for the greatest number must certainly mean that humans must be factored into the pool of recipients. As such, benefits are more likely realised for humans in terms of pleasure and entertainment (tourists) and operators (profits), then for elephants. Box 2 demonstrates the utilitarian component of the relationship between humans and elephants given the urge to preserve the practice, because it provides on-going benefits to mahouts, as well as to communities that benefit from the money generated from elephant tourism. The example also shows that there is a willingness to remove elephant suffering, which is an important component part of utilitarian reasoning for animal liberation, but under the conditions of human benefit.

In reference to Pieper's forms of justice, the cost-benefit calculation of utility (happiness) in utilitarianism, noted above, shows how the interests of humans take precedence over the interests of elephants (commutative justice). This scenario plays itself out over the heightened satisfaction that individual tourists experience by thinking that their money is going to happy elephants that are well taken care of through minimum standards (see PATA, 2019) as distributive justice. This is where utilitarianism takes on the appearance of welfare as noted by Francione and Charlton (2016), above, or even contractarianism in the absence of the willingness

Box 2

Bruce Haxton, from our supplier Starfish Ventures

"To say that 'elephant trekking is bad' doesn't pay consideration to the huge range of approaches. Whilst there are unfortunately those who don't treat their animals well, there are equally a huge number of mahouts who treat their elephants incredibly well and care for them as a member of their family.. the damage caused to these communities and their elephants could be immense..Rather than a ban we would suggest minimum guidelines. There needs to be a healthy dose of reality injected into the discussion in terms of what really would happen if elephant treks no longer take place."

https://www.responsibletravel.com/holidays/elephant-conservation/travel-guide/our-stance

of stakeholders to practice their individual duty in creating a better social world (legal or general justice).

It is also at this level where there is evidence of a virtue ethics perspective amongst stakeholders. Individuals and individuals within organizations acknowledge and accept their obligation to minimize the pain and suffering of animals. We see this in the Box 2 example where mahouts care for elephants as part of their immediate family. Whether this care is a function of empathy and responsibility is the subject of ongoing analysis in the face of the balance between "paternalistic protection and mutual exploitation in gauging responsible tourism (Sin & Minca, 2014). In the assessment of costs and benefits, from the utilitarian perspective, human interests triumph over animal interests.

. Box 2

Bruce Haxton, from our supplier Starfish Ventures

"To say that 'elephant trekking is bad' doesn't pay consideration to the huge range of approaches. Whilst there are unfortunately those who don't treat their animals well, there are equally a huge number of mahouts who treat their elephants incredibly well and care for them as a member of their family.. the damage caused to these communities and their elephants could be immense..Rather than a ban we would suggest minimum guidelines. There needs to be a healthy dose of reality injected into the discussion in terms of what really would happen if elephant treks no longer take place."

https://www.responsibletravel.com/holidays/elephant-conservation/travel-guide/our-stance

Deep justice for animals in tourism

Deep justice would provide the highest level of justice for animals. We view instances where animals are wild, free and not directly manipulated, controlled or owned by human interests as examples of deep justice (natural conditions that are completely unframed, according to Cohen, 2009) (see Box 3 below). Preservation (saving from use) is intentionally geared towards the well-being of individual animals, foremost, in addition to species and their ecosystems, so there is a direct attempt to make the lives of animals better. While traditional approaches to sustainability are largely anthropocentric in focus, deep justice would include protecting the interests of animals on the same level of human interests in efforts to be more holistic and responsible. Rehabilitation centres designed to treat animals and then release them back into the wild as the only option, may be a component part of deep justice.

Using the example of elephants, a rights perspective recognises that elephants are conscious, sentient and in the possession of inherent value. We should, therefore, be egalitarian in the sense that Individual elephants deserve the same respect as humans equally, there is an element of non-perfectionism such that one individual does not have higher moral value than others, we should not use elephants as a means to our own ends, we should not harm elephants, and we have a prima facia duty to protect elephants from injustice (Regan, 2004). Garner (2013) provides another contemporary position on rights for animals on the basis of moral pluralism (i.e., the state must protect individuals' right to pursue their own conceptions of the good), and because of an "enhanced sentience position." This means animals have a right not to suffer, they have a vibrant interest in continuing their lives, and it is only significant human interests that may be justified in ending an animal's life.

From a tourism perspective, a rights perspective would suggest that elephants have the ability to live their lives in freedom and dignity, within natural settings. Currently, there are few tourism-animal experiences that would fall within this category; however, ecotourism would be an exception. Fennell's (2013) first principle of ecotourism is perhaps the archetype of this form of justice, where elephants are able to disengage an interaction with humans on their own level in the absence of captivity and manipulation.

Box 3

Responsible Travel: Jo Cary-Elwes, campaigns manager at Elephant Family

"Energies urgently need to be channelled into protecting wild populations, reducing the demand for elephants in tourism and making sure that the elephants that are in captivity are well looked after. But, in an ideal world people would visit national parks and see these animals in the wild, which is where they belong."

https://www.responsibletravel.com/holidays/elephant-conservation/travel-guide/our-stance

As part of the commutative approach individuals and individual organisations would work towards enhancing the quality of lives of individual (primarily) and populations of elephants in their natural conditions (secondarily) through exchanges that place value on the interests of elephants. The distribution of resources must include efforts, in whatever measure, to ensure that not only human populations are taken care of (most would argue as a first priority), but that efforts are also employed to spread benefits across the species threshold that matters here. The work of governments, NGOs, and corporations moving in this direction is paramount in widening of the circle of morality to include all animals that are in possession of the traits noted above (following Regan, 2004). New legislation on animal rights in Turkey for crimes against living things close to humans is a step in the right direction (Erdem, 2019). This means opposing inequality based on speciesism, the support of marginalised populations, and respecting autonomy and individual rights (Shooster, 2015). As moral agents, lawyers and justice-focused organizations, based on Nozick's (1974) concept of positive rights, or the requirement of one to assist another in attaining their goals, would define justice and argue for animal justice (within and beyond tourism), on a case-by-case basis. As legal precedence is established (Shooster, 2017), progress would be made in ensuring that animals receive a deeper level of justice.

. Box 3

Responsible Travel: Jo Cary-Elwes, campaigns manager at Elephant Family

> "Energies urgently need to be channelled into protecting wild populations, reducing the demand for elephants in tourism and making sure that the elephants that are in captivity are well looked after. But, in an ideal world people would visit national parks and see these animals in the wild, which is where they belong."
> https://www.responsibletravel.com/holidays/elephant-conservation/travel-guide/our-stance

Virtue ethics, including care ethics, would guide human-animal interactions and decision-making. For example, animals that are held in sanctuaries or semi-natural conditions (Cohen, 2009) may receive deep justice, where the "keeping" of animals in large open spaces attempts to replicate natural conditions. In the best sanctuaries, elephants are able to live with their own kind, are cared for in a superior way by workers, volunteers and tourists, are allowed to practice normal behaviours that enhance wellbeing, and live out their lives completely in the absence of suffering. The Elephant Nature Park in Thailand is said to be one of the best examples of this type of model (Elephant Nature Park, 2019).

Deep justice also opens the door to a discussion on personhood in animals. What this means is that some higher-order animals such as great apes, dolphins, elephants and corvids (crows and ravens) have sufficient intelligence, social skills and sentience to be granted legal rights. The Great Ape Project of 1993 is a case in point, whereby chimpanzees, gorillas, bonobos, urangutans would be extended basic legal rights around the right to life, protection of individual liberty, and prohibition of torture (Singer & Cavalieri, 1993). The Balearic Islands was the first political entity to grant legal personhood status to apes in 2007 (Rose, 2007).

Conclusion

Justice is felt to be the most important of the four cardinal virtues and, as such, it is internal to each and every one of us. We understand the value of justice as individuals and within our societies because we confront it in so many different circumstances: fairness, honour, and virtue (Hultsman, 1995); human wellbeing (Higgins-Desboilles & Blanchard, 2010); the fair distribution of resources and benefits (Jamal & Camargo, 2014); and its adherence to principles of sustainability (Issac & Hodge, 2011). Justice, traditionally, has been relegated to the human world only because humans have almost exclusively been acknowledged as the only species possessing reason, intentionality, and reciprocity—humans alone have a conception of what is justly owed to them. And with justice is a virtue, it follows that individuals must play a central part in the bedrock of more ethical communities, as the sum of individual attitudes and behaviours should lead to an expanded view of social justice. Indeed, when individuals are unjust in their actions, they contribute to the erosion of social justice (Pieper, 1955). The literature illustrates, however, that justice is more than just a virtue in consideration of the broad and detailed manner in which it has been used from deontological and teleological perspectives. Indeed, in this regard, it has been a difficult task here to harness all of these varying perspectives in the presentation of a conceptual framework.

Given the firmly entrenched beliefs around the roles that animals are made to play in our varying social contexts, and in consideration of historical to contemporary realities, it is unsurprising that justice eludes animals in almost any situation. A considerable barrier in moving the justice agenda forward for animals is the belief that if justice is to be extended to animals, it must be non-comparative. For example, we extend "justice" to animals by providing them with proper and adequate levels of food, shelter, water, and freedom from pain and suffering. Under this model there is still asymmetry of power according to direct versus indirect moral treatment, which is very much tied to the welfare paradigm. In other words, people first over animals (Miller, 2017). Indeed, there is an embedded paternalism around our treatment of animals based on a view in which "just" is a function not only of current scientific knowledge on animal needs, but perhaps more importantly on human economic priorities.

The complete absence of justice in animal studies and tourism is a marker. It signifies clearly the role that tourism continues to play in our lives. Writing in the midst of the COVID-19 crises, we see the devastation that the virus has had on global tourism economies. Animals will no doubt suffer the consequences of this pandemic as service providers struggle to put food on their tables let alone feed and sustain the tools (animals) of their trade. Only time will tell how animals have fared at the hand of so many disruptive influences like COVID-19, environmental crises, and chaos in markets and the economy.

In the previous sections we attempted to conceptualize the theoretical and practical conditions around the just and unjust uses of animals in tourism, as the basis for arguing for more empirical investigations on this topic. Welfare, while not the most ethical or virtuous approach, is the current benchmark for measuring our treatment of animals; however, contractarianism and ecocentrism are likely the most prevalent practices within tourism. We argue that it is in the interests of the tourism industry in these changing times to move towards heightened levels of moral action when it comes to the use of animals in tourism. Our main conclusion, therefore, is that we need to place considerable effort into moving forward the benchmark of what is ethical in the use of animals in tourism, and what is not. This will necessitate a nod in the direction of virtue ethics (Model 2) along with utilitarianism and animal rights (Model 1) for a deeper level of justice for animals in tourism to better inform our actions.

In consideration of the foregoing discussion, it would seem that moral pluralism (see Garner, 2013) provides fertile ground for further investigation based on what is proposed here, Garner's (2013) work on enhanced sentiency, as well as work by Nussbaum's (2004; 2006a; 2006b). Nussbaum's capabilities approach is built around the notion that different species have unique capabilities that allow them to flourish (Aristotelian eudemonia). Our treatment of animals may be judged to be right or wrong, therefore, according to how such treatment catalyzes the ability to flourish or constrains it. Dignity and respect for individual animals is the primary locus of analysis, not species, with priority for species that have higher capacities than others (e.g., chimpanzees versus sponges). As such, we need to set proper benchmarks or species-specific norms while at the same time acknowledging our own limitations to fully comprehend animal lives because of our ignorance. Nussbaum also contends that every world nation should include in its constitution principles regarding nonhuman animals "as subjects of political justice and to" treat "them in accordance with their dignity" (2006b, p. 6). There is an obvious connection to virtue ethics in the capabilities approach in the manner in which Nussbaum follows Aristotle's lead around eudaimonia. Nussbaum is also indebted to deontological thinkers like Kant and Rawls on the basis of the intrinsic worth of beings and the fair distribution of goods.

This paper sought to demonstrate how different ethical theories intersect in the development of a framework for assessing the level of justice for animals used in tourism. We believe this is a valuable first step that may stimulate more animals-in-tourism research and instigate a deeper level of dialogue amongst tourism stakeholders, especially tourism operations that use animals. What prevents the realisation of this deeper state, however, is history and culture that lock societies into inevitabilities that are resistant to change. Money and the promise for a brighter future compel people to use animals in the most inhumane ways imaginable (Agoramoorthy, 2004). Even conservation, usually an agent for positive change, is problematic for animals like elephants because NGOs often have representatives of large corporations sitting on their boards. This deliberate approach translates into a more neoliberal focus on "corporate-led" conservation (Kopnina, 2016) built according to the commodification of nature.

As tourism scholarship and practice slowly matures, questions around what is responsible and sustainable will hopefully provide animals with a representative voice at the table. We believe that the tides are beginning to turn in the court of public opinion. This is evidenced by the existence of new organizations fighting for animal, ecosystem, and global justice. While we wait for the tides to turn, there is a need to more fully investigate what animal advocates, generally, and tourism organizations, specifically, are doing to advance animal justice. This seems like promising terrain for moving the tourism and animal ethics agenda forward. Indeed, the time is upon us to render each their due.

Note

1. There are differences between Pieper's commutative justice and communitarianism. While both move away from universal normative judgments, the latter equates virtue with community practices, i.e., the primary unit of analysis is the community over the individual (MacIntyre, 2007).

Acknowledgement

We wish to thank Dr. Tazim Jamal for her comments and encouragement of this paper, as well as the wisdom of three referees who helped to improved the paper considerably.

Disclosure statement

No potential conflict of interest was reported by the author(s).

ORCID

Valerie Sheppard http://orcid.org/0000-0003-0029-4339

References

Abbey, R. (2007). Rawlsian resources for animal ethics. *Ethics & the Environment, 12*(1), 1–22. doi:10.2979/ETE.2007.12.1.1
Abraham, N. (2019). Personal communication, November 19.
Adams, C., & Donovan, J. (2007). *The feminist care tradition in animal ethics*. Columbia University Press.
Agoramoorthy, G. (2004). Ethics and welfare in Southeast Asian zoos. *Journal of Applied Animal Welfare Science: JAAWS, 7*(3), 189–195. doi:10.1207/s15327604jaws0703_5
Author. (in press). Advocating for animal justice in tourism and beyond. *Tourism Recreation Research*
Bone, K., & Bone, J. (2016). The same dart trick: the exploitation of animals and women in Thailand tourism. In K. Markwell (Ed.), *Animals and tourism: Understanding diverse relationships* (pp. 60–74). Channel View Publications.
Cochrane, A. (2009). Ownership and justice for animals. *Utilitas, 21*(4), 424–442. doi:10.1017/S0953820809990203
Cohen, E. (2009). The wild and the humanized: animals in Thai tourism. *Anatolia, 20*(1), 100–118. doi:10.1080/13032917.2009.10518898
Cohen, E., & Fennell, D. A. (2016). The elimination of Marius the Giraffe: Humanitarian act or callous management decision. *Tourism Recreation Research, 41*(2), 168–176. doi:10.1080/02508281.2016.1147211
Cohen-Charash, Y., & Spector, P. E. (2001). The role of justice in organisations: a meta-analysis. *Organizational Behavior and Human Decision Processes, 86*(2), 278–321. doi:10.1006/obhd.2001.2958
Cooper, T. L. (1987). Hierarchy, virtue, and the practice of public administration: A perspective for normative ethics. *Public Administration Review, 47*(4), 320–328. doi:10.2307/975312
Cossins, A. (2007). Restorative justice and child sex offences. *British Journal of Criminology, 48*(3), 359–378.
Curry, P. (2011). *Ecological ethics: An introduction* (2nd ed.). Polity Press.
DeWitt, T., Nguyen, D. T., & Marshall, R. (2008). Exploring customer loyalty following service recovery: the mediating effects of trust and emotions. *Journal of Service Research, 10*(3), 269–281. doi:10.1177/1094670507310767

Donovan, J. (2006). Feminism and the treatment of animals: From care to dialogue. *Signs: Journal of Women in Culture and Society, 31*(2), 305–329. doi:10.1086/491750
Dorloh, S. (2017). The protection of animals in Thailand–an insight into animal protection legislation. *International Journal of Humanities and Cultural Studies, 4*(2), 58–63.
Dworkin, R. (1977). *Taking rights seriously*. Duckworth.
Elephant Nature Park. (2019). *Welcome to elephant nature park*. https://www.elephantnaturepark.org/
Erdem, U. (2019). *Turkeys' animal rights legislation underway*. https://www.hurriyetdailynews.com/turkeys-animal-rights-legislation-underway-148870
Fennell, D. A. (2012a). *Tourism and animal ethics*. Routledge.
Fennell, D. A. (2013). Contesting the zoo as a setting for ecotourism, and the design of a first principle. *Journal of Ecotourism, 12*(1), 1–14. doi:10.1080/14724049.2012.737796
Fennell, D. A. (2014). Exploring the boundaries of a new moral order for tourism's global code of ethics. *Journal of Sustainable Tourism, 22*(7), 983–996. doi:10.1080/09669582.2014.918137
Fennell, D. A. (2018). *Tourism ethics* (2nd ed.). Clevedon.
Fennell, D. A., & Cooper, C. (2020). *Sustainable tourism: Principles, contexts, practices*. Channel View Publications.
Fennell, D. A. & Malloy, D .C. (2007). Tourism codes of ethics: practice, theory, synthesis. Clevedon, UK: Channel View Publications.
Francione, G. L. (1995). *Animals, property and the law*. Temple University.
Francione, G. L., Charlton, A. (2016). The case against pets. *Aeon*. https://aeon.co/essays/why-keeping-a-pet-is-fundamentally-unethical
Frankena. (1963). *Ethics*. Prentice-Hall.
Friedman, M. (1962). *Capitalism and freedom*. University of Chicago Press.
Garofalo, C., & Geuras, D. (1999). *Ethics in the public service: The moral mind at work*. Georgetown University Press.
Garner, R. (2004). *Animals, politics and morality*. Manchester University Press.
Garner, R. (2012). Rawls, animals and justice: New literature, same response. *Res Publica, 18*(2), 159–172. doi:10.1007/s11158-011-9173-z
Garner, R. (2013). *A theory of justice for animals: Animal rights in a nonideal world*. Oxford University Press.
Gavrielides, T. (2007). *Restorative justice theory and practice: Addressing the discrepancy*. Criminal Justice Press.
Gearey, A. (2013). *Justice as welfare: Equity and solidarity*. Bloomsbury Academic.
Good, K. (2014). *One year later, former Toronto Zoo elephants are thriving at their sanctuary home*. https://www.onegreenplanet.org/animalsandnature/former-toronto-zoo-elephants-are-thriving-at-their-sanctuary-home/
Halwani, R. (2003a). Care ethics and virtue ethics. *Hypatia, 18*(3), 161–192. doi:10.1353/hyp.2003.0063
Halwani, R. (2003b). *Virtuous liaisons: Care, love, sex, and virtue ethics*. Open Court.
Held, V. (2004). Care and justice in the global context. Ratio Juris. *17*(2): 141–155.
Higgins-Desboilles, F., & Blanchard, L. (2010). Challenging peace through tourism: placing tourism in the context of human rights, justice and peace. In O. Moufakkir and I. Kelly (Eds.), *Tourism, progress and peace* (pp. 35–47). CABI.
Hill, S. P., & Broom, D. M. (2009). Measuring zoo animal welfare: Theory and practice. *Zoo Biology, 28*(6), 531–544. doi:10.1002/zoo.20276
Hughes, P. (2001). Animals, values and tourism: structural shifts in UK dolphin tourism provision. *Tourism Management, 22*(4), 321–329. doi:10.1016/S0261-5177(00)00070-4
Hultsman, J. (1995). Just tourism: An ethical framework. *Annals of Tourism Research, 22*(3), 553–567. doi:10.1016/0160-7383(95)00011-T
Humphrey, M. (2000). Nature' in deep ecology and social ecology: Contesting the core. *Journal of Political Ideologies, 5*(2), 247–268. doi:10.1080/713682940
Ignatieff, M. (2003). *Human rights as politics and idolatry*. Princeton University Press.
International Peace Information Service (IPIS). (2014). *The adverse human rights risks and impacts of European companies: Getting a glimpse of the picture*. http://corporatejustice.org/documents/ahrri_report_final-2.pdf
Issac, R. K., & Hodge, D. (2011). An exploratory study: Justice tourism in controversial areas: The case of Palestine. *Tourism Planning and Development, 8*(1), 101–108.
Jamal, T. (2019). *Justice and ethics in tourism*. Routledge.
Jamal, T., & Camargo, B. A. (2014). Sustainable tourism, justice and an ethic of care: Toward the just destination. *Journal of Sustainable Tourism, 22*(1), 11–30. doi:10.1080/09669582.2013.786084
Jones, R. C. (2015). Animal rights is a social justice issue. *Contemporary Justice Review, 18*(4), 467–482. doi:10.1080/10282580.2015.1093689
Kheel, M. (2008). Nature Ethics: An Ecofeminist Perspective. Lanham, MD: Rowman & Littlefield Publishers, Inc.
Kopnina, H. (2016). Wild animals and justice: the case of the dead elephant in the room. *Journal of International Wildlife Law & Policy, 19*(3), 219–235. doi:10.1080/13880292.2016.1204882
Leopold, A. (1970). *A sand county almanac with essays on conservation from round river*. Ballantine Books.
MacIntyre, A. (2007). *After virtue*. Notre Dame University Press.
MacIntyre, A. (2013). *Who's justice? Which rationality?* Duckworth Books.
Marshall, T. (1999). *Restorative justice: An overview*. Home Office, Information & Publications Group.

Matheny, G. (2006). Utilitarianism and animals. In Singer, P. (Ed.), *In defense of animals: The second wave* (pp. 13–25). Blackwell.

McCloskey, H. J. (1965). Rights. *The Philosophical Quarterly*, *15*(59), 115–127. doi:10.2307/2218211

Mihalic, T., & Fennell, D. A. (2015). In pursuit of a more just international tourism: The concept of trading tourism rights. *Journal of Sustainable Tourism*, *23*(2), 188–206. doi:doi:10.1080/09669582.2014.943763

Mill, J. S. (2010). *On liberty*. Penguin.

Miller, D. (2017). *Justice*. Stanford University of Philosophy. https://plato.stanford.edu/entries/justice/

Mkono, M. (2019). Neo-colonialism and greed: Africans' views on trophy hunting in social media. *Journal of Sustainable Tourism*, *27*(5), 689–704. doi:10.1080/09669582.2019.1604719

Mulgan, T. (2011). *Ethics for a broken world: Imagining philosophy after catastrophe*. McGill-Queen's University Press.

Narveson, J. (1988). *The libertarian idea*. Temple University Press.

New Zealand Herald. (2019). *Teenage elephant collapses and dies after carrying tourists*. https://www.nzherald.co.nz/travel/news/article.cfm?c_id=7&objectid=12282104&utm_source=facebook&utm_medium=social&utmc_campaign=nzh_fb&fbclid=IwAR0-C5sF-eDHC4TIGZuY3g5hj74lau4kJBm6zLmYudnLv4Bwxs_o23Schak

Nozick, R. (1974). *Anarchy, state and utopia*. Basil Blackwell.

Nussbaum, M. (1996). Patriotism and cosmopolitanism. In J. Cohen (Ed.), *For love of country: Debating the limits of patriotism* (pp. 3–17). Beacon Press.

Nussbaum, M. (2004). Beyond 'compassion and humanity': Justice for nonhuman animals. In C. R. Sunstein & M. Nussbaum (Eds.), *Animal rights: Current debates and new directions* (pp. 299–320). Oxford University Press.

Nussbaum, M. C. (2006a). *Frontiers of justice: Disability, nationality, species membership*. Harvard University Press.

Nussbaum, M. C. (2006b). The moral status of animals. *The Chronicle Review*, 52(22), B6. https://www.chronicle.com/article/The-Moral-Status-of-Animals/25792.

O'Neill, O. (2000). *Bounds of justice*. Cambridge University Press.

Oxford Dictionaries. (2019). *Right*. https://www.lexico.com/en/definition/right

Pacific Asia Travel Association (PATA). (2019). *Elephant camp animal welfare and sustainability standard assessment initiative*. https://sustain.pata.org/elephant-camp-animal-welfare-sustainability-standard-assessment-initiative/

Pieper, J. (1955). *The four cardinal virtues*. Pantheon Books.

Pojman, L. (1997). Equality and desert. *Philosophy*, *72*(282), 549–570. doi:10.1017/S0031819100062410

Rawls, J. (1971). *A theory of justice*. Oxford University Press.

Regan, T. (1985). The case for animal rights. In P. Singer (Ed.), *In defense of animals* (pp. 13–26). Basil Blackwell.

Regan, T. (2003). *Empty cages: Facing the challenge of animal rights*. Rowman and Littlefield.

Regan, T. (2004). *The case for animal rights*. University of California Press.

Rose, T. (2007). *Going ape over human rights*. http://www.cbc.ca/news/viewpoint/vp_rose/20070802.html

Rowe, J. S. (1997). From reductionism to holism in ecology and deep ecology. *The Ecologist*, *27*(4), 147–151.

Rowlands, M. (1997). Contractarianism and animal rights. *Journal of Applied Philosophy*, *14*(3), 235–247. doi:10.1111/1468-5930.00060

Scheyvens, R. (2002). *Tourism for development: Empowering communities*. Prentice Hall.

Sheppard, V., Fennell, D. A. (2019). Progress in tourism public sector policy: toward an ethic for non-human animals. *Tourism Management*, 73(8), 134–142. doi:doi:10.1016/j.tourman.2018.11.017

Shooster, J. (2015). Justice for all: Including animal rights in social justice activism. *The Harbinger*, *40*(39), 39–44.

Shooster, J. (2017). *Legal personhood and the positive rights of wild animals*. https://was-research.org/writing-by-others/legal-personhood-positive-rights-wild-animals/

Sin, H. L., & Minca, C. (2014). Touring responsibility: The trouble with 'going local' in community-based tourism in Thailand. *Geoforum*, *51*, 96–106. doi:10.1016/j.geoforum.2013.10.004

Singer, P. (2009). *Animal liberation (re-print)*. Harper Perennial.

Singer, P., & Cavalieri, P. (1993). *The great ape project: Equality beyond humanity*. Fourth Estate publishing.

Slote, M. (2007). *The ethics of care and empathy*. Routledge.

Slote, M. (1992). *From morality to virtue*. Oxford University Press.

Smith, M., & Duffy, R. (2003). *The ethics of tourism development*. Routledge.

United Nations. (2019). *UN sustainable development goals*. https://sustainabledevelopment.un.org/sdgs

Vance, L., (1995). Beyond just-so stories: Narrative, animals, and ethics. In Adams, C. J., & Donovan, J. (Eds.), *Animals and women: Feminist theoretical explorations* (pp. 163–191). Duke University Press.

Varner, G. (1998). *In nature's interests? Interests, animal rights, and environmental ethics*. Oxford.

Wyckoff, J. (2014). Toward justice for animals. *Journal of Social Philosophy*, *45*(4), 539–553. doi:doi:10.1111/josp.12077

Young, I. M. (1985). Impartiality and the civic public: Some implications of feminist critiques of moral and political theory. *PRAXIS International*, *5*(4), 381–401.

Young, I. M. (2012). *Justice and the politics of difference*. Princeton University Press.

ZooCheck New Zealand. (n.d.). *Zoos = conservation*. Retrieved January 30, 2012, from http://www.wildlife.org.nz/zoocheck/cons.htm

Intergenerational rights to a sustainable future: insights for climate justice and tourism

Dawn Jourdan and Jani Wertin

ABSTRACT

Citizens of the world are calling for climate action at the national and international levels. In those places slow to respond, climate litigation has emerged as a method that activists are using to force action. Of the 700 or so pending cases, the climate litigation presently garnering the most attention is Juliana v. United States where 21 youth have challenged the actions of the federal government for knowingly creating and exacerbating climate change. Others have taken to the streets to challenge government inaction on climate change. In both instances, courts are often left as de facto policy makers. This paper seeks to document the precedent-setting Juliana v. United States case and understand the implication of the actions of young climate advocates as they seek to defend their rights to a sustainable future. The paper also explores such youth and civic activism and inaction in relation to "sustainable" tourism. It calls for better consideration of the needs of children and young people in discussions of climate justice and tourism, particularly the future generations of children and youth who stand to bear the brunt of climate change impacts in the Global North and South.

Introduction

The response to climate change has varied across the globe, despite a steady stream of mounting scientific evidence and widespread consensus among climate scientists about anthropogenic global warming caused by human activity (see, for example: https://climate.nasa.gov/scientific-consensus/). In most regions of the world, policymakers have been slow to act. The basis for their reluctance to take action varies widely, from climate change denial to simply not knowing the best course of action. As the recent wildfires at the start of the New Year in 2020 demonstrate, severe losses to wildlife, ecosystems, natural and built habitats, are becoming "new normals" and the losses will continue to mount ecologically and socially, with severe impacts on tourism destinations as well. Blame is distributed widely; for instance, just a few days before the fires, Prime Minister Scott Morrison told a mining group that new laws were needed to crack down on climate activists and progressives who "want to tell you where to live, what job you can have, what you can say and what you can think" (Lane, 2020). Concerted global action so desperately needed to tackle this global climate "crisis" continues to lag.

In addition to ecological and environmental justice, climate justice arises here as a critical issue in relation to those who are most vulnerable and least able to take care of themselves as the severity of climate change impacts grow. This includes the poor and low-income, diverse populations (including communities of color, minority and oppressed ethnic groups), and other disadvantaged groups, including the elderly, people with disabilities, and young children, as well as groups facing discrimination or marginalization due to gender, religion, or other cultural orientations). Philip Alston, UN special rapporteur on extreme poverty and human rights, warned that lack of action on climate change will not only undermine basic rights to sustenance for hundreds of millions of people, it will also threaten democracy and the rule of law (Alston, 2019). Among the populations who stand to lose culture, community, and livelihoods as a result of such displacements, are another highly vulnerable group: young children and youth, not just the current generations but also future generations of children and youth. Barak Obama stated, "No challenge places a greater threat to our future generations than a change in climate" (State of the Union Address, 2014). The rights of children to participate in the decision-making that affects them was a conclusion drawn as a part of the Children's Rights and Habitat Declaration (UNICEF and Habitat, 1997). However, while youth participation is becoming a more recognized activity worldwide, the law generally does not provide real teeth for youth to pursue their interests through voting or legal challenges.

Given the complexity and gravity of the actions required, it is no wonder that policymakers have been slow to answer the call. While the current U.S. administration walked away from the Paris Accord, the general apprehension by lawmakers across the planet to take swift action has resulted in activism by a wide array of stakeholders. Climate litigation is becoming commonplace. In a new survey prepared by the Grantham Research Institute at the London School of Economics and Political Science, researchers chronicle 1,328 legal actions between 1990 and May 2019 across the globe (Setzer & Byrnes, 2019). Most notably, young persons have taken to the streets, staged walk-outs and even filed suit against their governments for their failure to respond to the climate crisis in a meaningful way. Their activism stems from a real fear that inaction will irreparably damage the planet, making it incapable of supporting the life and livelihood of current and future generations. And they recognize that widespread formalized actions by global policymakers and major greenhouse gas (GHG) emitting nations like the U.S., China, and India to ensure that the next generation will not perish is highly limited. It is the lack of concerted governmental action that the young people are protesting or litigating worldwide.

This paper seeks to chronicle these activities in order to establish a baseline for the year 2020. These activities are separated by type: litigation or protest. The Juliana v. United States case is documented in detail here, as it is a precedent-setting case of the rise of youth activism against the U.S. government, holding it responsible for violating their rights to a clean environment and a sustainable future. The paper aims to also offer a lens that appreciates the connection between climate change, youth activism, and climate justice in relation to tourism. "Climate justice" as used in this paper refers not only to the social inequalities related to climate change, but also to "principles of democratic accountability and participation, ecological sustainability and social justice and their combined ability to provide solutions to climate change" (Chatterton, Featherstone, & Routledge, 2013, p. 606, cited in Grant, 2019, p. 47). Addressing the needs of young children and youth (who cannot vote until they are of legal age) is a crucial issue for climate justice, not only with respect to the at-risk populations worldwide who stand to lose their culture, community, and livelihoods due to climate change, but also with respect to the rights and needs of future generations. Sustainable development, after all, is "development that meets the needs of the present without compromising the ability of future generations to meet their own needs (United Nations World Commission on Environment & Development [UNWCED], 1987, p. 8). However, the rights of current and future generations of youth in a world facing a "climate crisis" should surely have greater visibility here and in the context of "sustainable" tourism development. The United Nations World Tourism Organization's (UNWTO, 1994, p. 30) definition of sustainable tourism development draws on the Brundtland Commission's Report:

> *Sustainable tourism development meets the needs of present tourists and host regions while protecting and enhancing opportunity for the future. It is envisaged as leading to management of all resources in such a way that economic, social, and aesthetic needs can be fulfilled while maintaining cultural integrity, essential ecological processes, biological diversity, and life" support systems*

As the landmark case study Juliana v. United States and related discussion below illustrate, re-envisioning sustainable tourism in the context of climate justice, children, and youth requires close attention to actions and reactions occurring through social movements that are arising to protest and bring attention to the need for inclusive democracy that takes into consideration the voices of civic society and the needs of children and young people, particularly the future generations of children and youth who stand to bear the brunt of climate change impacts.

Civic disobedience and the youth protesters

Nobel Prize nominee in 2019 and Time Magazine's Person of the Year in 2019, Greta Thunberg has become the voice of climate change for young people across the globe. In August 2018, the 15-year-old Swedish youth began her solo peaceful protests before the Swedish parliament. Her weekly Friday Strike for Climate has become a global phenomenon and her skolstrejk för klimatet (school strike for climate) banner has been translated into a wide range of languages. As she expressed (Watts, 2019):

> "It's amazing," she says. "It's more than 71 countries and more than 700 places, and counting. It's increasing very much now, and that's very, very fun."

Young people across Europe and Australia have indeed followed suit. Youth now routinely protest on Fridays, calling for governments to take action to reduce carbon emissions. They call for governments around the world to take action to reduce the production of fossil fuels in hopes of slowing down or gradually reversing global warming. They ask for these changes in an effort to secure a future for themselves and future generations, as well their beliefs that the environment should be preserved for the sake of itself. For example, on Friday, March 15, 2019, more than a million youth protests took place in over 120 countries (see https://www.vox.com/energy-and-environment/2019/3/15/18267156/youth-climate-strike-march-15-photos). Other climate-related strikes have also sprouted up. A climate strike led by 14-year-old Uma Krieger in the UK has led to the Brighton and Hove City Council's decision to embrace the Green initiative which includes tenets such as: lowering the voting age to 16 and developing a national curriculum that stresses the importance of responding to the climate crises (Booker-Lewis, 2019). As Uma Krieger said: "If we want any chance of my generation or the next to have a planet that is habitable and remotely safe to live on then we have to act right now" (Booker-Lewis, 2019). Such small victories demonstrate that some decisionmakers understand and are willing to grapple with the complexities of responding to the climate crisis.

These victories have been occurring at the local level, however. Change at national levels is much more difficult to inspire, but recent IPPC reports stressing the urgency to reduce carbon emissions by 2030 have spurred additional activism, such as by Extinction Rebellion (XR) (https://rebellion.earth/). In these protests, younger and older people marched through streets, glued themselves to trains and buildings and staged "die ins" (Busby, 2019). In April 2019, XR protests in London (a major tourist destination!) blocked traffic and caused major disruptions, resulting in more than 1,000 arrests; XR protests around major landmarks have impacted both residents and visitors not only in London, but also destinations like New York, Sydney, Berlin, and Paris (Reuters, 2019). "Like the youth climate strikes, XR is an international movement representing a wide array of people who, in the face of political and social apathy surrounding climate change, have felt morally compelled to act" and "their actions have ultimately led the British government to formally declare a climate emergency, which was one of the group's key demands" (Miles, 2019). The issue of climate change is reaching the courts through the actions of youth protesters and activism by movements such as XR. However, in these cases, these courts are not being directly asked to rule on the issue of the culpability of governments in sanctioning activities that increase carbon emissions. They are simply arbitrating the disruptive actions of the protestors. By contrast, in court battles such as Juliana v. United States, youth protesters have taken to the courtroom, rather than the streets, to instigate changes in the law to litigate actions or inaction by governments that ultimately contribute to climate change. The next section details this case, drawing on legal documents and other secondary materials cited below. Juliana v. United States is important to document for it offers new precedent in giving youth of non-voting age a say in civic democracy on their rights to a sustainable future. Given that the U.S. is of course both an international and domestic tourism destination, and tourism is responsible for about 8% of global emissions (Lenzen et al., 2018), little concerted effort is evident by the tourism industry to undertake the challenge and influence the national or global policy level change that is needed (see, for instance, Becken, 2019), it can be argued that the youth have little recourse but to resort to civil disobedience (a constitutional right in the U.S.) and exercise their moral right to life, liberty, and the pursuit of happiness, a right that is currently threatened by the climate crisis. Small local-level actions are good but insufficient at this critical stage of the Anthropocene, where global policy and national policies must change in order to meet the IPCC recommendation of 50% reduction in global emissions by 2030, compatible with a temperature rise of no more than1.5 degreesCelsius above pre-industrial levels. As noted in the 2018 IPCC report, "lack of global cooperation, lack of governance of the required energy and land transformation, and increases in resource-intensive consumption are key impediments to achieving 1.5 °C pathways" (Rojelk et al, 2018: 95; see also https://www.ipcc.ch/sr15/).

Juliana v. United States

In the United States, legal cases and controversies challenging federal laws or the U.S. Constitution are typically brought in one of the federal district courts. For example, a plaintiff may petition a federal district court to hear their grievance if they can demonstrate that their rights have been negatively affected by an action of a federal agency. Juliana v. United States was originally filed in the Federal District Court of Oregon on August 12, 2015. The lawsuit was filed by a group of young people, ages 8-19, against the Executive Branch of the United States government (including the President and cabinet officials who have responsibility for policy that may have impacted global greenhouse emissions). In their claim, the Juliana plaintiffs contend that the United States of America continued its extensive utilization of fossil fuels, fully knowing the negative consequences the CO_2 released would have on the climate system, such as global warming and climate change (Hansen, 2015). The plaintiffs cite the White House Report of 1965, "Restoring the Quality of the Environment", to emphasize the importance of protecting the environment for the benefit of current and future generations (Hansen, 2015). In addition to this report, the plaintiffs refer to various environmental protection plans that both identify the imminent dangers associated with rising levels of CO_2 emissions in the atmosphere, as well as present strategies to be implemented that could reduce CO_2 emissions to a level that could stabilize the climate system and stop global warming (Hansen, 2015). These plans include the "Policy Options for Stabilizing Global Climate" submitted to Congress by the Environmental Protection Agency in 1990 and the "Changing By Degrees: Steps to Reduce Greenhouse Gases" submitted to Congress by the Congressional Office of Technology Assessment in 1991, though neither plan was implemented (Hansen, 2015). The plaintiffs argue that numerous studies have been published since 1965 (including the two aforementioned plans) that have repeatedly informed the defendants of the hazard associated with continued reliance on fossil fuels and that by permitting the use of fossil fuels and subsidizing the fossil fuel industry, the defendants have knowingly allowed concentrations of atmospheric CO_2 to rise to dangerous levels (Hansen, 2015). Approximately three months after the original lawsuit was filed (November 17, 2015), the defendants made a motion to dismiss the case from further review of the court. Among other challenges, the defendants argued that the plaintiffs lacked legal standing to file suit as minors and on behalf of future generations who have not yet been born.

What is special about the Juliana case, at least at this point in litigation, is that the case has not been dismissed for lack of standing. On April 8, 2016, Judge Coffin denied the government's motion to dismiss the case, stating that at the pleading stage of a case, "general factual allegations" are adequate so long as they are "concrete, particularized, and actual or imminent" (Coffin, 2016, p. 4). As the Juliana plaintiffs claimed injuries that are concrete and personal in nature, despite stemming from broader harms also claimed in this case and these injuries being shared by many, Judge Coffin ruled that these injuries could not be discounted and that gives the Juliana plaintiffs legal standing to sue on their own behalves and on behalf of future generations (Coffin, 2016). As Judge Coffin opined, the court must "accept the allegations of concrete particularized harm or imminent threat of such harm as true," making this more of a matter of whether the defendants' actions can be traced back to the injury caused and if the court can redress these injuries (Coffin, 2016). This motion was further denied by District Judge Anne Aiken (2016) who adopted Judge Coffin's findings.

After losing its preliminary motion to dismiss the case from consideration based on standing, the government next filed a summary judgement motion where it explained why none of the Juliana defendants' arguments merited further review by the District Court. The District Court both agreed and disagreed with the government's arguments, dismissing sitting President Trump as a defendant because he was not the acting U.S. President when the acts occurred (Aiken, 2018a). At the request of the Juliana plaintiffs, the District Court certified the issues in controversy for immediate appeal to the Ninth Circuit Court of Appeals, the appellate court responsible for hearing cases pertaining to federal issues in that region of the U.S. The Ninth Circuit rejected the federal government's argument for dismissal and sent the case back to the District Court for further review and proceedings. The defendants appealed the Ninth Circuit decision to the U.S. Supreme Court, the ultimate arbiter of legal disputes in the U.S. This body, siding with the Juliana plaintiffs, sent the decision back to the lower courts for a trial.

The Juliana case is still being reviewed by the U.S. court system. A trial has not yet occurred and it is stuck in a procedural morass. It is different from cases challenging the behavior of protestors like XR (described earlier), as the Juliana plaintiffs are challenging the actions of the executive branch of the U.S. government for their actions (or inactions as the case may be) on climate change and resource use. In both instances, the courts must look to the laws in place to determine the legal requirements which govern both behaviors. It is relatively straightforward to determine if a protestor has stepped over the proverbial legal line, i.e., if they have conducted protest in an unpermitted area, at a decibel level that is higher than the legal limit, or incited violence. Judges typically have wide discretion in the sanctions they apply to protestors. This means that punishments may vary and be influenced by the perceptions of individual judges. By contrast, judges in Juliana v. United States are being asked to review the actions of decision-makers (i.e., agencies of the federal government), and to consider both the language of the law and implementing regulations according to the spirit of the law. Some judges consider themselves strict textualists. This means that when reviewing a case before them, they will only consider the plain language and meaning of the documents or laws they are reviewing. If a term is undefined in the document, they rely on the standard dictionary definition to garner meaning. They do not consider the context in which a law was drafted or the spirit of the law, i.e., the intent of the lawmakers who drafted it.

The judges who are being asked to consider the legality of the action or inaction of decision-makers also have considerable discretion. They are, of course, bound by precedent from previous decisions made in their courts or others within their jurisdictions of review, but even here they have considerable discretion. These judges are being asked to make decisions about whether governments are doing enough to respond to climate change or, alternatively, if their inaction is contributing to the legal neglect of duties. In many cases, especially in the U.S. context, the judiciary is the only political body really engaging in decision-making related to climate change. As any student of basic civics might quickly point out, judges are not supposed to make the law. Rather, they should be interpreting it. However, judicial activism has sometimes served as a very important substitute for policymaking. Judges, through their rulings, have inspired those with executive and legislative functions to take necessary actions. Such was the case in the evolution of the civil rights movement in the U.S. This, too, was the case in the development of affordable housing policy following the infamous Mt. Laurel decision in New Jersey. The courts created a body to review immediate appeals of local decisions that negatively impacted affordable housing in New Jersey, until the State government stepped in to create an agency to take over this function (see http://fairsharehousing.org/mount-laurel-doctrine/). This may be where many countries across the world are headed in dealing with climate change. Such an approach, however, is piecemeal and may not lead to government action at a pace necessary for mitigating or adapting to the harmful effects of climate change.

Global action and global justice

Citizens around the world are continuing to protest while climate change continues to severely impact destinations and inhabitants and new groups of climate refugees are being created every day. In many cases, due to government inaction, they are being left to their own ingenuity and resources—to mitigate, adapt, or relocate. The rapidity of the changing climate is a particular threat to vulnerable populations, especially those with subsistence existences or economies built on the integrity of the natural environment. This issue of climate justice will continue to grow in significance for vulnerable groups both in the Global North and the Global South where the subtropics house large populations that stand to bear the brunt of extreme weather, droughts, sea level rise, ecosystem damage and species loss, along with other impacts of climate change. Many coastal destinations, including ones along the Arctic and in Small Island Developing States (SIDS), are already being severely impacted from sea level rise and extreme weather. Many are strongly dependent on tourism, and low-lying destination islands like Maldives, Marshall Islands, Vanuatu and Caribbean islands like the Bahamas are gravely at risk (see, for example, Moore, 2019). Climate refugees are on the move and many more will need to be moved from their preferred or territorial homes if carbon emissions do not comply with the Paris Accord's mandate to limit temperature rise to 1.5 degrees Celsius above pre-industrial levels (see American Geophysical Union, 2019; Rojelj et al., 2018; United Nations, 2020).

Issues of sea level rise and extreme weather challenges such as the SIDS are facing are exacerbated where governmental capacity is insufficient to deal with disaster events, let alone tackle the international community;the affected parties generally do not have time to protest nor the economic means to launch sustained actions to garner resources and redress from the Global North (Nalau, Handmer, & Dalesa, 2017). Among various actions to increase resiliency of all kinds, ecological and social as well cultural resilience that can help address climate injustice (Grant, 2019), it may mean calling on others to litigate or take actions to on their behalf. Better consideration of the needs of children and young people in discussions of climate justice and tourism is also needed, particularly the future generations of children and youth who stand to bear the brunt of climate change impacts in both the Global North and the Global South. Far too little has been studied in the context of sustainable tourism and climate action of this group whose democratic rights to influence policy are restricted until they are of voting age.

It is not that actions are not being taken, of course. Many destinations worldwide are reacting by reducing carbon emissions and striving to be sustainable, carbon neutral destinations (for example, see https://sustainabletravel.org/). The European Commission recently announced a European Green Deal, (https://ec.europa.eu/info/strategy/priorities-2019-2024/european-green-deal_en), popular destination countries like Ireland have initiated national tree replanting programs, and the UK has passed laws and initiated programs and strategies to limit carbon emission to net zero by 2050 (O'Reilly & O'Brien, 2008; https://www.gov.uk/government/news/uk-becomes-first-major-economy-to-pass-net-zero-emissions-law). Island nations like Kiribati (pop. 100,000) have preemptively signed treaties with other nations to ensure that their people will be welcomed when the time for migration ultimately arises (see the excellent documentary film, Anote's Ark: www.anotesark.com).

Actions are also being undertaken to invest in "green" transportation and curb air travel plus other environmentally unsustainable forms of transportation. A Swedish movement has arisen on flyskam or "flight shaming" those who rely on air travel for business or pleasure (see Timperley, 2019). Some governments, supported by organizations like Greenpeace, Friends of the Earth and the New Economics Foundation, are considering a tax levy on those who seek to travel by air more than once per year (Helm, 2015). Destinations that cater to the wealthy would be less impacted by such levies than lesser developed places that serve travelers with smaller budgets. If flight shaming or taxing becomes more commonplace, long-haul tourist destinations could be hard hit and so, too, would residents who rely on tourism for their livelihoods. Expanding public transit, mass rail, train, and boat transport are important considerations for 'sustainable transportation,' but significant infrastructure investment will be needed. As noted earlier, many developing destinations are already struggling to manage the impacts of climate change, so how will they secure significant additional resources, technologies and capacities to turn into "green" destinations?

The economic and equity issues raised above also extend to climate justice for vulnerable populations in tourism destinations experiencing the brunt of climate change (destinations in the sub-tropics, coastal, and polar regions being among these). International tourism growth and "overtourism" add a far greater burden on poorer populations, as the environmental costs and impacts of travel are generally externalized and inequitably distributed. Referring to tourism which is the flagship of the Bahamas, Moore (2019) notes the paradox of Anthropocene tourism: "anthropogenically vulnerable ecosystems and organisms are protected and promoted as valuable so that their tourism potential can be realized by attracting more tourists to the destination, sustaining the industry while perpetuating its anthropogenic effects" (p. 21). Carbon emissions by the wealthier Global North travelers not only disproportionately affect destinations in the Global South (see Lenzen et al., 2018), tourism's economic benefits tend to be inequitably distributed (Britton, 1982), and mobilities and the right to travel freely across border are not fairly apportioned either–consider the burden of proof laid on climate refugees versus the jet-setting, globetrotter tourist (Bianchi & Stephenson, 2014).

At-risk destinations like the SIDS are already suffering from climate-related injustices and finding it very difficult to obtain justice and garner necessary resources to build resilience for the future that awaits. Effective governance and regulatory bodies are lacking to address climate injustices in a global tourism system that is highly complex and fragmented in terms of stakeholders and control over tourism-decision making from the local to the regional, national, and international levels (Jamal, 2019). Organizations like the United Nations offer a range of guidance on tourism and climate change (see, for instance, Goal 13 in the Sustainable Development Goals: https://www.un.org/sustainabledevelopment/climate-action/), but have little regulatory force.

Effective institutional and governance structures to render global justice and a sustainable future for present and future generations are lacking overall. The gap is partially being filled by social movements addressing climate change and youth litigation efforts to hold their countries responsible for a sustainable future. They fill a vacuum of responsibility for climate justice in tourism and tourism destinations as well. Tourism is a global phenomenon that is interlinked with places of habitation and is implicated in terms of the GHGs that continue to warm the planet that the youth and children of present and futures generations will inherit. The burden being placed on them is far from equitable, but success in cases like Juliana v. United States will help advance governmental and corporate responsibility for climate justice and the challenging future that await current and future generations of children and young people in the destinations of the Anthropocene.

Conclusion

As we write this article, entering a new decade in the 21st century, our civilization stands on the brink of a climate crisis that will severely affect future generations and is already wreaking devastation in many areas worldwide. The extensive wildfires that heralded the New Year in 2020 is one of many "new normals" being set globally and were a late wake-up call for Australian citizens who are demanding action from their policymakers for mitigation and adaptation to build resiliency to future crises (Cave, 2020). The cost and burden on future generations will be immense. New paradigms are needed and new understandings of civic responsibility for planetary sustainability and resilience are needed. Immediate and aggressive action must be taken to curtail global warming, putting a halt to continuous market expansion and the "growth" paradigm entrenched with industrialization and globalization. Among other actions, citizens in the highest carbon-emitting countries (which are also global destinations, e.g., U.S., China, and India), must exercise their democratic rights and voice to secure climate actions from their governments. Meanwhile, unfair burden continues to be placed on those who stand to be most at risk to be impacted by climate change–vulnerable populations, including the low-income and poor (particularly in communities of color where environmental justice has long been an issue), and young children. Global youth movements, Friday Strikes for Future and court actions like Juliana v. United States are playing a vital role in advancing the cause for global climate action, filling a vacuum of government inactions on the Paris Accord, especially from dominant carbon-emitters, and too slow action on climate justice by an international tourism industry that contributes at least 8% of global GHG emissions (UN IPBES Report, 2019). Sustainable tourism development, if it is to truly address the needs of current and future generations in the Anthropocene, must take into better account climate justice and the rights of children and youth to a sustainable planet.

This article is written with the assumption that the current generation has a duty to future generations to act (e.g., Hubin, 1976). There are those that believe there is no duty either because we cannot predict the impact of climate change on future generations nor are we bound by law to protect their rights (Mank, 1996). And while the scope of environmental legal rights for children and future generations varies across political systems, this has not precluded policymakers from calling for climate action in the name of those very same stakeholders. During the Obama administration, the EPA proclaimed a moral obligation to "leave the planet healthy for future generations. How to accomplish this necessary action remains the question de jure. One thing is clear, failure to take actions to mitigate and adapt will have catastrophic effects for all global citizens from now into the distant future. It will affect our ability to enjoy those places to which we love to travel, the locations where the global food supply is grown, and even the very places where we domicile. Young people including youth not yet of voting age are rightfully challenging the inaction of policymakers and using the means available to them to nudge them into action. Other youth actions challenging their governments in court have arisen since Juliana v. United States. On October 9, 2019, the Alaska Supreme Court addressed Sinnok v. State of Alaska, a constitutional climate lawsuit brought by 16 young Alaska

Natives and Alaskans against the State of Alaska (https://www.ourchildrenstrust.org/alaska). Like the global climate strikes, youth climate lawsuits are arising globally, including in Canada, Colombia, Norway, Pakistan, and Uganda (West Coast Environmental Law, 2020). Whether they call for action through the courts or from the streets, these young future citizens may be crucial important agents of change necessary to change our climate future to a fairer and more sustainable world.

Disclosure statement

No potential conflict of interest was reported by the author(s).

References

Aiken, A. (2016). United States District Court for the District of Oregon, Juliana v. United States: Opinion and order, No. 6:15-cv-01517-TC, Document 83. Retrieved from http://blogs2.law.columbia.edu/climate-change-litigation/wp-content/uploads/sites/16/case-documents/2016/20161110_docket-615-cv-1517_opinion-and-order.pdf.

Aiken, A. (2018a). United States District Court for the District of Oregon, Juliana v. United States: Opinion and Order, No. 6:15-cv-01517-AA, Document 369. Retrieved from http://blogs2.law.columbia.edu/climate-change-litigation/wp-content/uploads/sites/16/case-documents/2018/20181015_docket-615-cv-1517_opinion-and-order-1.pdf

Aiken, A. (2018c). United States District Court for the District of Oregon, Juliana v. United States: Order, No. 6:15-cv-01517-AA, Document 444. Retrieved from http://blogs2.law.columbia.edu/climate-change-litigation/wp-content/uploads/sites/16/case-documents/2018/20181121_docket-615-cv-1517_order-1.pdf

Alston, P. (2019). "Climate change and poverty: Report of the special rapporteur on extreme poverty and human rights," Report presented to the UN Human Rights Council Forty-first session, 24 June–12 July, A/HRC/41/39. Retrieved from www.ohchr.org/EN/NewsEvents/Pages/DisplayNews.aspx?NewsID=24735&LangID=E

American Geophysical Union. (2019, April 23). New studies highlight challenge of meeting Paris Agreement climate goals. *ScienceDaily*. Retrieved from https://www.sciencedaily.com/releases/2019/04/190423114029.htm

Becken, S. (2019). Decarbonising tourism: Mission impossible? *Tourism Recreation Research*, *44* (4), 419–433. doi:10.1080/02508281.2019.1598042

Bianchi, R. B., & Stephenson, M. L. (2014). *Tourism and citizenship: Rights, freedoms and responsibilities in the global order*. New York, NY: Routledge.

Booker-Lewis, S. (2019). Brighton climate youth strikes win councillors' backing. *Brighton & Hove Independent*. Retrieved from https://www.brightonandhoveindependent.co.uk/news/brighton-climate-youth-strikes-win-councillors-backing-1-9015013

Britton, S. G. (1982). The political economy of tourism in the third world. *Annals of Tourism Research*, *9*(3), 331–358. doi:10.1016/0160-7383(82)90018-4

Busby, M. (2019, April 7). Extinction Rebellion activists stage die-in protests across globe. *The Guardian*. Retrieved from https://www.theguardian.com/environment/2019/apr/27/extinction-rebellion-activists-stage-die-in-protests-across-globe

Cave, D. (2020, January 2). As Australia burns, its leaders trade insults. *New York Times*, published Nov. 13, 2019, Retrieved from https://www.nytimes.com/2019/11/13/world/australia/fires-climate-change-pragmatism.html

Chatterton, P., Featherstone, D., & Routledge, P. (2013). Articulating climate justice in copenhagen: Antagonism, the commons, and solidarity. *Antipode*, *45* (3), 602–620. doi:10.1111/j.1467-8330.2012.01025.x

Coffin, T. M. (2016). United States District Court for the District of Oregon, Juliana v. United States: Order and findings & recommendation, No. 6:15-cv-01517-TC, Document 68. Retrieved from http://blogs2.law.columbia.edu/climate-change-litigation/wp-content/uploads/sites/16/case-documents/2016/20160408_docket-615-cv-1517_order.pdf

Grant. (2019). Climate justice and cultural sustainability: The case of Etëtung (Vanuatu Women's Water Music. *The Asia Pacific Journal of Anthropology*, *20*(1), 42–56. doi:10.1080/14442213.2018.1529194

Hansen, J. (2015). Exhibit A: Declaration of Dr. James E. Hansen in support of plaintiffs' complaint for declaratory and injunctive relief; In the matter Juliana v. United States, No. 6:15-cv-01517-TC, Document 1-1. Retrieved from http://blogs2.law.columbia.edu/climate-change-litigation/wp-content/uploads/sites/16/case-documents/2015/20150812_docket-615-cv-1517_complaint.pdf

Helm, T. (2015). Levy on frequent leisure flyers proposed to make airport expansion unnecessary. *The Guardian*. Retrieved from https://www.theguardian.com/world/2015/jun/20/frequent-flyer-tax-leisure-airfares

Hubin, D. C. (1976). Justice and future generations. *Philosophy and Public Affairs*, *6*(1), 70–83.

Jamal, T. (2019). Tourism ethics: A perspective article. *Tourism Review*. 10.1108/TR-05-2019-0184.

Juliana v. United States. (2015). Kelsey Cascadia Rose Juliana, Xiuhtezcatl Tonatiuh M. Et Al. v. United States, Barack Obama et al., No. 6:15-cv-01517-TC (D. Or. Aug. 12, 2015).

Lane, L. (2020). Australia's deadly, devastating wildfires: Affecting tourism, short and long term. *Forbes*. Retrieved from https://www.forbes.com/sites/lealane/2020/01/02/australias-deadly-devastating-wildfires-affecting-tourism-short-and-long-term/#2cc7f46c28bf

Lenzen, M., Sun, Y.-Y., Faturay, F., Ting, Y.-P., Geschke, A., & Malik, A. (2018). The carbon footprint of global tourism. *Nature Climate Change*, *8*(6), 522–528. doi:10.1038/s41558-018-0141-x

Mank, B. C. (1996). Protecting the environment for future generations: A proposal for a Republican "superagency. *NYU Environmental Law Journal*, *5*, 444–516.

Miles, M. (2019). Extinction Rebellion's First Big Act of Civil Disobedience in US Shuts Down Rush Hour Traffic in DC. *Earth Island Journal*. Retrieved from https://www.earthisland.org/journal/index.php/articles/entry/extinction-rebellions-first-big-act-of-civil-disobedience-in-us-shuts-down-rush-hour-traffic-in-dc?gclid=EAIaIQobChMIq6WgoJ_x5gIVSZyzCh1Qrg7XEAAYASAAEgJ-UvD_BwE.

Moore, A. (2019). *Destination anthropocene: Science and tourism in the Bahamas*. Oakland: CA. University of California Press.

Nalau, J., Handmer, J., & Dalesa, M. (2017). The role and capacity of government in a climate crisis: Cyclone pam in Vanuatu. In W. Leal Filho (Eds), *Climate change adaptation in pacific countries. Climate change management*. Cham: Springer.

O'Reilly, G., & O'Brien, P. (2008). Achieving a Greenhouse Gas Neutral Ireland," for the EPA CCRP Programme 2007-2013 (07-CCRP-3.1). Retrieved from https://www.epa.ie/pubs/reports/research/climate/GHG%20Neutral%20Ireland.pdf

Reuters. (2019, October 7). Extinction Rebellion holds protests around the world. *New York Times*. Retrieved from https://www.nytimes.com/video/world/europe/100000006755732/extinction-rebellion-climate-protests.html

Rogelj, J., Shindell, D., Jiang, K., Fifita, S., Forster, P., Ginzburg, V., … Vilariño, M. V. (2018). Mitigation pathways compatible with 1.5 °C in the context of sustainable development. In Masson-Delmotte, V., P. Zhai, H.-O. Pörtner, D. Roberts, J. Skea, P. R. Shukla, A. Pirani, W. Moufouma-Okia, C. Péan, R. Pidcock, S. Connors, J. B. R. Matthews, Y. Chen, X. Zhou, M. I. Gomis, E. Lonnoy, T. Maycock, M. Tignor, and T. Waterfield (Eds.), *Global Warming of 1.5 °C. An IPCC Special Report on the impacts of global warming of 1.5 °C above pre-industrial levels and related global greenhouse gas emission pathways, in the context of strengthening the global response to the threat of climate change, sustainable development, and efforts to eradicate poverty*. Retrieved from https://www.ipcc.ch/site/assets/uploads/sites/2/2019/02/SR15_Chapter2_Low_Res.pdf

Setzer, J., & Byrnes, R. (2019). *Global trends in climate change litigation: 2019 snapshot*. London: Grantham Research Institute on Climate Change and the Environment and Centre for Climate Change Economics and Policy. Retrieved from http://www.lse.ac.uk/GranthamInstitute/wp-content/uploads/2019/07/GRI_Global-trends-in-climate-change-litigation-2019-snapshot-2.pdf

State of the Union Address. (2014). Retrieved from https://obamawhitehouse.archives.gov/the-press-office/2014/01/28/president-barack-obamas-state-union-address

Thomas, Berzon, & Friedland. (2018). United States Court of Appeals for the Ninth Circuit, Juliana v. United States: Order, No. 18-80176, D.C. No. 6:15-cv-01517-AA. Retrieved from http://blogs2.law.columbia.edu/climate-change-litigation/wp-content/uploads/sites/16/case-documents/2018/20181226_docket-18-80176_order.pdf

Timperley, J. (2019, September 9). Why flight shame is making people swap plans for trains. *Future*. Retrieved from https://www.bbc.com/future/article/20190909-why-flight-shame-is-making-people-swap-planes-for-trains

UN IPBES. (2019). Nature's dangerous decline 'unprecedented'; Species extinction rates 'accelerating.' Global Assessment Report by the Intergovernmental Science-Policy Platform on Biodiversity and Ecosystem Services (IPBES). Retrieved from https://www.un.org/sustainabledevelopment/blog/2019/05/nature-decline-unprecedented-report/

UNICEF & Habitatat (UN Centre for Human Habitats). (1997). *Children's rights and habitat: Working towards child-friendly cities*. New York, NY: Author. Retrieved from https://digitallibrary.un.org/record/233327?ln=en

United Nations. (2020). The paris agreement. *UN Climate Change*, Retrieved from https://unfccc.int/process-and-meetings/the-paris-agreement/the-paris-agreement

UNWCED (United Nations World Commission on Environment and Development). (1987). Our common future. Brundtland Commission's Report. Oxford: Oxford University Press.

UNWTO (United Nations World Tourism Organization). (1994). Agenda 21 for Travel and Tourism: Towards Environmentally Sustainable Tourism. London, UK: WTO, WTTC and the Earth Council.

Watts, J. (2019). Greta Thunberg, school girl climate change warrior: 'Some people can let go of things. I can't,'. Retrieved from https://www.theguardian.com/world/2019/mar/11/greta-thunberg-schoolgirl-climate-change-warrior-some-people-can-let-things-go-i-cant

West Coast Environmental Law. (2020). Youth are leading the climate movement–in court and on the streets Environmental Law Blog. West Coast Environmental Law. Retrieved from https://www.wcel.org/blog/youth-are-leading-climate-movement-in-court-and-streets

Megaliths, material engagement, and the atmospherics of neo-lithic ethics: presage for the end(s) of tourism

Mick Smith, Siobhan Speiran and Peter Graham

ABSTRACT
Climate change spells the end of tourism, and tourism is just one, and one of the least important, things to be so very near its end. This situation has emerged because of the global dominance of forms of *material engagement* that completely misunderstand the distribution of agency in the world. Framing the world as just an external human 'resource' diminishes the many forms of agency that elude, transcend, effect, facilitate and subvert human intentions. Even archetypically *inanimate* objects, such as stones, are *actively* constitutive of worlds, thoughts, and lives, including human lives, in myriad transformative ways. This lithic agency is illustrated in terms of the *attraction* that draws some to visit stone circles, and the atmospheres created by, and felt in, such places. This rather different understanding of a 'visitor attraction' and its ontological and ethical implications exposes some of the inadequate presumptions of our dominant form of life and its destructive atmospheric consequences at local and global scales. The paper employs Material Engagement Theory (MET), originating in archaeology and anthropology, as a way to understand the differences between the reductive assumptions of the Anthropocene and the redemptive possibilities of what it terms a neo-lithic ethics.

"What broke from the cloud? Rain, Sun, the hawk A Stone walks under a cloud slowly"

George Mackay Brown The Forty-Fifth Stone from *Brodgar Poems*(Brown, 1992).

Introduction

This paper begins with the story of the destruction of a Neolithic standing stone, the Odin Stone at Stenness, Orkney, off the North coast of Scotland. It ends with some remarks about the future dissolution of tourism in what some have labelled a new geo-logical era, the Anthropocene.[1] It suggests the two issues are connected, but in unexpected ways, and via an argument that turns on very different ways of experiencing and understanding our place in the world as exemplified in our material engagements with stone (Figure 1).

We engage with stone in so many ways, (skimming, climbing, mining, building) yet these engagements are usually dominated by an ontology that considers stone as just inanimate raw

Figure 1. The Loch and Standing Stones of Stenness (Image from Barry, 1867, p. 210).

material at the service of human activities, especially economic activities. This modern understanding, as we shall see, underlay the historical destruction of the Odin Stone and, I argue, also leads inexorably to our current desperate situation. That is to say, the 'Anthropocene', both in terms of the appallingly destructive events now emerging under its umbrella, and the blinkered focus of its terminology, is the ultimate realization of an ontology that regards the whole earth as just a material/capital resource for (certain) humans. In Heidegger's (1993) terms, Earth is reduced to no more than 'standing reserve'.

Actually, though, stones are always much more than mere resources and certain rocks, stones and their settings are *active* participants in creating unique atmospheres, events, and experiences that become intimately woven into shared life-histories. Attending to this suggests the possibility of different ontologies, not of isolated valuing subjects consuming inanimate objects/resources but of 'ecologically' entangled people, stones, and places. Such ontologies (which make no pretense of providing a neutral description of a value-free 'external' reality) are associated with very different understandings of existence and very different concerns, including ethical concerns. I will refer to such approaches, which might take many diverse forms, as a 'neo-lithic ethics of place', that is as novel understandings of caring and considerate involvement that recognize contexts in which even stones might 'matter' in both the sense of their often unbidden materialization as constituents of the more-than-human world and their becoming things of concern.

This reference to *neo-lithic* ethics is not, then, a call to turn the clock back, nor an attempt to archaeologically recover how the Neolithic builders of these monuments thought about the stones and their world, though this is surely interesting. (It is also not as impossible as we might think if we take the phenomenology of *our* material engagement with the megaliths and their landscapes seriously (Tilley, 2004, DeMarrais, Gosden, & Renfrew, 2004)). Rather, *neo*-lithic, in this sense, refers to *contemporary* approaches that do not impose ontological limits on our worldly relations such that ethical concerns are deemed only to apply to humans or even just to those beings more usually defined as living/animate. A neo-lithic ethics, in contrast with almost all modern humanist ethical systems, is more openly environmental since it emphasizes differing response-abilities and responsibilities to all 'things' drawn into or manifesting themselves in, our existential horizons. This can, and often does, include paradigmatically inanimate things like mountains and megaliths.[2]

In this sense, forms of neo-lithic ethics do not fit readily with any straightforward application of a notion of justice, at least in its dominant Western humanist forms. Here 'environmental justice', for example, is usually framed precisely in terms of an exclusively human polity and 'nature' usually enters only as a resource to be justly or unjustly distributed. that is, as *means* to human ends. Nature is valuable largely "for the opportunities it offers to people" (Sen, 2009, p. 248). It actually seems difficult to speak of justice at all where, rocks, places or environments are concerned (although there are certainly attempts to include *sentient* non-human animals under this rubric). However, to say that the concept of justice is inadequate is not to presume that concerns like 'proportionality' of response to other beings are unimportant, or that we should not attend to the 'capacities' of more-than-just-human entities (quite the contrary). The difference lies in how neo-lithic approaches strive to reveal the experiential possibilities inherent in 'cultivating' different modes of attention to, and material engagement with, the world. They decline to adopt over-arching rationalistic or discursive principles for determining just deserts, but focus closely on ecological affordances offered, and the ethical values that arise, in particular encounters. We might say that rather than focusing on 'sharing out' (distributing) the world between 'us', neo-lithic ethics are concerned with different modes of sharing *in* the world, ways of being creatively co-engaged together. Perhaps this difference could be illustrated in terms of Rawls (1973) famous 'veil of ignorance' behind which abstract 'individuals' come to rational agreement about how to share out the real world from which they are experientially and situationally separated. A neo-lithic ethics turns this influential thought experiment on its head. Here the veil of ignorance would not be a description of an imaginary possibility, rather it might be a term for the reality we now inhabit, the operative fiction of a detached humanity produced through a form of material engagement that effectively blinds us to the very possibility of recognizing different ways of experiencing and existing in the world. This is the veil that needs to be parted – but more of this later.

This neo-lithic ethics can emerge and evolve through attending to entangled experiences and forms of engagement in specific places and situations that *expose* us to challenges concerning our self-understandings and our supposedly 'exceptional' human status (Ehrenfeld, 1978). I therefore focus on theorizing the phenomenology and ontology of visitor 'exposure' as it relates to stone circles and the many ways such standing stones might be said to exert or constitute an 'attraction'. This term, itself suggest a recognition of the agency of sites in drawing visitors to them. I write about the standing stones I am most familiar with, in Orkney, to illustrate how close attention to the kind of experiences/atmospheres discussed here might open passage from one dominant and destructive set of worldly relations (the Anthropocene) to other possibilities.

This point here is obviously not to encourage everyone to visit stone circles in order to attain ethical enlightenment! Rather this geontological (Povinelli, 2016)[3] turn, this attending to the agency of rocks, is exemplary of much broader ecological experiences and possibilities. It is, in any case, widely recognized that excessive visitor numbers at popular sites like Stonehenge make it exceedingly difficult to experience the presence and atmosphere of anything other than the carpark, interpretative center, and 'gift' shop. This too illustrates the ambiguous position of tourism, its instrumental promotion of the stone circle as tourist *resource* and quantifiable generator of revenue, but also the dependence of such promotion on allusions to the unique and unquantifiable ways that visitors *might* be personally exposed to potentially life-changing experiences, atmospheres, and events.

In Neolithic times, where climate, landscapes, and cultures were quite different these stone circles may have been amongst the first constructions to have attracted distant visitors (Card, 2018). Some of these same monuments may also survive into a near future where the only Anthropocene travelers are amongst the millions of refugees from climate change. In this paper, as in the world, there remains a certain ambiguity about whether this future is already here or might still be changed, about whether and how it might be *written in stone*.

The stone

> "The Odin stone, long the favourite trysting-place in the summer twilights of Orkney lovers, was demolished in 1814 by a sacrilegious farmer, who used its material to assist him in the erection of a cowhouse. This misguided man was a *Ferry-Louper* (the name formerly given to strangers from the south), and his wanton destruction of the consecrated stone stirred so strongly the resentment of the peasantry in the district that various unsuccessful attempts were made to burn his house and holding about his ears." (Gorrie quoted in Black & Thomas, 1903, p. 2)

The Odin Stone (see Figure 2) still stood when Sir Walter Scott visited Orkney although, he would be one of the very last tourists to see it. His diary for 16th August 1814, recording his travels and antiquarian interests, describes a "stone perforated with a round hole" (Lockhart, 1837, p. 198) which, he conjectures was used in "solemnly attesting the deity, [Odin] which the Scandinavians did by passing their head through a ring, *vide* – Eyrbiggia Saga" (198). Despite its name, though, the Odin Stone predated such Sagas by thousands of years, and local inhabitants had their own traditions connected to their own, more intimate, forms of material engagement with the stone. In Orkney folklore, a vow of love, made (with now forgotten words) as both parties clasped hands through the round hole that pierced the Odin stone was regarded as binding. The stone held other powers too. It was said that a small child passed through the hole would be "never shake with palsy in old age" (Black & Thomas, 1903, p. 2) and those who did suffer such ills might hope for healing if they thrust their limbs through the hole (Marwick, 2000, p. 59). Bread, cheese, rags, and small stones were often left as offerings.

Figure 2. View of Odin Stone drawn in 1805 by Lady Stafford (from Ritchie, & Marwick, 1976).

The Odin stone stood close by the Stones of Stenness, a Neolithic stone circle, dating to around 3000 B.C, that comprises one of the earliest parts of a huge complex of stone buildings and standing stones on Mainland Orkney (Card, 2018). The full extent of this unique complex is only now re-emerging with recent archaeological digs at the Ness of Brodgar. In more or less continuous use for more than a thousand years, this site "lay at the heart of an important ceremonial landscape" (anon, 2015, p. 30), a meeting place for Neolithic inhabitants from Orkney and probably from far beyond. Indeed, a good case can be made that these stone circles, erected on what is now considered the maritime margin of the European continent, were actually the earliest inspiration for many other immense Neolithic projects later undertaken across Britain and Northern Europe, projects that would eventually include Avebury and Stonehenge. Scott himself had declared that "Stonehenge excels these monuments, but I fancy they are otherwise unparalleled in Britain" (Lockhart, 1837, p. 199).

The farmer that destroyed the Odin stone, a Captain W. MacKay, seems to have had little concern about breaking apart links with the deep past, and no concern for the stone itself or the beliefs, experiences, and practices of those who visited to make pledges and offerings. His interests lay in 'agricultural improvement', in making the land he was renting from a private landowner, Mr. Riddoch, more productive and profitable.[4] He simultaneously felled two of the remaining upright stones of the Stenness circle, breaking one of them beyond any repair and severely damaging the other. Sherriff Peterkin's contemporary account describes how the stones were "torn from the spot on which they had stood for ages, and shivered to pieces" (Peterkin in Ritchie & Marwick, 1976, p. 5). However, as Ritchie & Marwick also note:

> "Peterkin decided not to take the legal matter further, when MacKay promised 'to desist from his operations'; Peterkin continued 'The occurrence is to be regretted as it may tend to hamper Capt. MacKay's comfort in Orkney and as it may lead the vulgar to confound every improving farmer with the idea of a desolator'" (Peterkin in Ritchie & Marwick, 1976, p. 5).

Malcolm Laing, a Justice of the Peace who objected to MacKay's activities held a rather different opinion.

> [I]n contrast to MacKay's feeling that the destruction of the stones was of concern only to himself and to the landowner; 'the offence', wrote Laing 'has not been given to Mr. Riddoch or to me, who interposed merely as Justices of Peace, but to the community at large, which has an interest in the preservation of public monuments and to men of letters and curiosity throughout the kingdom'. (Ritchie & Marwick, 1976, p. 5)

The positions result from and express very different forms of material engagement with the stones. Laing's analysis is certainly more broadly drawn, and yet the "community" he refers to is also hardly that of the incensed Orkney "peasantry" mentioned by Gorrie, but of an educated, "lettered", elite, like Scott. The moral outrage of the "vulgar" at the desecration of these standing stones, so intimate and longstanding an aspect of their lives, place, and traditions, is converted by Peterkin and by Laing into very different discourses circulating in their own class, namely, those of agricultural improvement and historical curiosity/heritage and their respective desires to protect continuing economic and/or epistemic 'progress'. There are, of course, ethical assumptions built into these positions, since 'improvement' and 'education' are such a morally-laden concepts. (There are also questions of private ownership/enterprise versus public goods.) However, polite rationalizations about a community of interests in the "preservation of public monuments" do not seem to come close to expressing the emotions unleashed by MacKay's act of vandalism, nor do they show any respect for previous forms of engagement with the stones. What is more, they do not seem to take any cognizance of the notion that, so far as the local residents were concerned *the "offence" was, primarily, done to the stones themselves.*

These same tensions persist in contemporary debates despite attempts to defuse tensions between public curiosity, heritage conservation, and economic 'progress' (or in neoliberal vernacular, 'growing the economy'). Orkney has ironically, but not unexpectedly, received a huge surge in visitors since the Ness of Brodgar (together with the Neolithic village at Skara Brae and the Maes Howe chambered tomb) was designated as the Heart of Neolithic Orkney World Heritage Site by UNESCO in 1999.[5] Tourist numbers increased to 124,400 in 2016 with 91,000 of these arriving on large cruise ships (an increase of "more than 160% (c. 60,000 passengers) over the period" 2011–2016 (Orkney Volume, 2017)). These new 'Ferry-Loupers' may not yet be as damaging to the stones as Captain MacKay's 'improvements', but there is certainly evidence of problems including soil compaction and footpath erosion at the closely associated Ring of Brodgar (Orkney Volume, 2017, p. 21). Vandalism too remains a problem with at least two instances of defacing the stones since 2015 (Guardian, 2019).

Within capitalism, of course, both the materiality of the sites and the experiences of residents and tourists tend to be subsumed under the rubric of economic growth. To this end visitor experiences are to be artificially enhanced by the provision of infrastructure 'improvements' (wider roads, more car parks, easier access, more hotels and restaurants, etc.) educational signage, visitor centers, and above all, more shopping opportunities. Heritage conservation, after all, often has to be justified as a *paying*, rather than an ethical, 'concern'. Yet, for all its faults, the term heritage also suggests important ways of relating current controversies to inter-generational concerns about the survival of material artifacts, cultural practices, or threatened ecologies. Here, what has existed, evolved, and been valued in the past should be conserved because of its *constitutive* role in creating the world we now inhabit and its potential future involvements.

These evaluative tensions became palpable in the public inquiry into a recent proposal to build a large windfarm within visual distance of the stones at the Ness of Brodgar. Here the issue was further complicated by arguments about the ethics and 'sustainability' of this particular form of development. This

> "conflict generated highly contentious arguments amongst all interested parties, and was steeped in moral rhetoric on both 'sides' from the outset about what the priorities of 'conservation' should be within a contemporary society that is concerned both with its history and heritage, but also the well-being of its future inhabitants. The various 'values' the site is claimed to embody in terms of its narrative worth to human history and origins, and its 'value' as a generator of tourism within the Orkney community (boosted by World Heritage status), were publicly set against arguments about the role and 'moral' responsibility that both the Orkney community and the Scottish nation should have in promoting and providing resources for renewable energy, how the economic future of the community and 'culture' could be ensured (and indeed, 'conserved'). (McClanahan, 2016)

The decision went against this windfarm development. However, McClanahan's comments reveal how both 'sides' feel the need to justify their 'moral' arguments economically as affecting the heritage site as a "generator of tourism" and ensuring the "economic future" of Orkney. As with Peterkin and Laing, certain anthropogenic modes of material engagement are prioritized and the ethical standing of the stones themselves seems to slip away even though they are the nominal focus of concerns that slide into a debate that typifies the tensions between the, somewhat ironically termed, Three Pillars of sustainability (WCED, 1987). Despite their solid materiality the stones, like the prevailing winds, risk being reduced to mere resources, whether economic, environmental, and social/cultural.

Something important is elided in the very structure of these debates and the lack of a language and an ethics that carries sufficient normative weight to express any respons(e)abilities to the stones themselves is deeply troubling. Interestingly, this same deficiency was explicitly addressed in an earlier (1994) public inquiry into a super-quarry proposal that would have completely destroyed another rock, Roineabhal, a mountain on the Scottish island of Harris. Here, one of the objectors, Alastair McIntosh, requested Chief Herney of the Mi'Kmaq First Nation of *Nova Scotia*, come to Scotland to speak to this missing element, that is, a mountain's 'animacy' and its ethical standing. The Mi'Kmaq, who were fighting a quarry application on their own Indigenous lands, expressly relate to certain stones in ways that recognize their entanglements with particular places and peoples, rather than reducing them to just inanimate resources. This alternative cosmology/ontology, voiced in Chief Herney's testimony, helped provide an "authentic alternative to a dominant capitalist discourse" (Milton, 2002, p. 145) and one that in alliance with resident objectors proved persuasive. This international sharing of lithic experiences and of different, but tragically entangled, histories of dispossession (settler colonialism and the Highland clearances) expressed a very different geontology (Povinelli, 2016) and created new ethical associations and alliances.

It clearly *matters*, that is, it makes a material *and* ethical difference, whether or not we attend to the kinds of experiences and ontologies that underlie objections to destructive acts and might counter the constant pressure to justify everything in terms of an economy that now threatens destruction on a planetary scale.[6] And if the idea of stones as being anything other than inanimate resources sounds ridiculous to modern ears then we need to remember the many ways we engage with stones, as rock climber, artist, alpine tourist and so on. We also need to recall that coming to regard mountains and rocks as inanimate raw material is itself a dubious accomplishment, one that required an immense ideological struggle, often undertaken precisely in order to facilitate the expansion of extractive industries like mining (Merchant, 1983). It required a radical change in world-views, including the replacement of organic (animate) metaphors with those appropriated from the same technical in(ter)ventions that were subsequently to be promoted, for example, the mechanics of clockwork, of hydraulics, steam engines, electricity, and now, of course, cybernetics and information systems. *Veins* of ore had to be dissociated from a framework, still widespread in Renaissance times, that took this metaphor 'too literally' regarding the earth as a living and fecund body. Early figures in this transition, like Georgius Agricola (1494–1555) in his *De Re Metallica*, (1950) had to contend with moral objections to mining based on such associations, together with, what today might count as ecological arguments about destruction by mining operations of woodlands and farm landscapes. His counter-arguments, in many ways, prefigure the assumptions of contemporary neoliberal economists:

> "The yearly profit of a lead mine in comparison with the fruitfulness of the best fields is three times or at least twice as great ... let them leave to miners the gloomy valleys and sterile mountains, that they may draw forth from these, gems and metals which can buy, not only the crops, but all things that are sold." (Agricola, 1950, p. 6)

Agricola, then, both prioritizes profit and assumes the economic substitutability of a world reduced to use and exchange values. "Metals", says Agricola, "are a creation of nature" but their value resides in how "they supply many varied and necessary needs of the human race." (18) Here in the pages of *De Re Metallica* the notion of a creative "provident and sagacious Nature" (12) is visibly slipping into the rhetoric of an entrepreneurial humanism, the anthropogenic agency of 'wealth-creation' that manufactures profits from otherwise "sterile mountains". Rocks are becoming no more than inanimate extractive resources, or even just 'overburden', useless waste material that needs to be removed to get to profitable minerals. Yet even here it seems that some of these activities depend on the residual and strange attractions of *precious* stones! A danger here, is that wherever heritage tourism too is reduced to a paying rather than an ethical concern, it becomes no more than a mode of squeezing profits from the residual matter left in the wake of this accelerating extractive commodification of every aspect of the planet. And yet, the visitors interest in experiencing heritage sites suggests that something does indeed continue to reside, to *live*, in these material remnants and still attract people to other possibilities of a world where much more than money matters.

Atmospheres and attractions

> "circles of stone exert a magnetic fascination to those who venture into their sphere of influence" (Richards, 2013, p. 3)

The growth of modern 'heritage' tourism, exemplified in Orkney's increasing visitor numbers, seems paradoxical in an age when the number of sites that can be accessed virtually has reached an unprecedented level. It seems that many of these visitors must feel that there is no substitute for the embodied experience of being as close as physically possible to, the "sphere of influence" of the sites that draw them. Explaining such 'attractions', though, is no simple task. No doubt there may be a sense of social 'distinction' (Bourdiou, 1986) or cultural kudos to be gained within certain social groups from visiting distant, or famous, locales. Perhaps it may just be a matter of fulfilling socially/psychologically generated expectations – ticking off another 'must see' destination. Yet many places, like the Naess of Brodgar, obviously do exude an 'atmosphere' that one has to be present to experience, to breathe in and feel, an atmosphere that is also, as the term suggests, almost impossible to pinpoint, quantify, or define, in ways that would seem anything but dismissively vague, spiritual, or poetic within predominant modern discourses.[7]

That said, we should also remember that this 'vagueness' is no barrier to 'atmosphere' being used as "a marketing tool for promoting tourist sites", (Bille, Bjerregaard, & Sørensen, 2015, p. 32) quite the contrary. And more, *or less*, successful attempts to create atmospheres are obviously the stock in trade of architects, restauranteurs, garden designers, organizers of sporting events, etc. (Edensor, 2015) Few of us are entirely immune to such 'staged productions'. No matter how enigmatic the sources of a particular atmosphere may seem, they can still affect us profoundly and intimately; making us tense, wary or relaxed, excited, engaged or alienated, communicating something of the exigencies and uniqueness of places and situations in ways that we never fully control or understand. They seem to make manifest an embodied awareness, sometimes intense, sometimes more diffuse, of what might be termed our 'ontological exposure' – a 'being placed' in and open(ed) to, the phenomenology of a place-specific intermingling of subject(s), object(s), and environ.

In other words, atmospheres appear

> "indeterminate [...] as regards their ontological status. We are not sure whether we should attribute them to the objects of environment from which they proceed or to the subjects who experience them." (Bille et al., 2015, p. 32)

Atmospheres and attractions (like ethical concerns) are never simply a matter of the subjective overwriting of a passive material setting, although they are made manifest through the *exposure* of the experiencing subject(s) to that place and therefore "contingent on subjective disposition, mood or state-of mind", a feature Sorenson (2015, p. 64) refers to as the "clause of subjectivity". What the visitor brings with them, namely the particular pre-formed biases that delimit that subjects readiness to be exposed to certain effects, clearly matters, but it never wholly determines what is felt. Atmospheres can often surprise, overwhelm, or confound expectations. The materiality of a place impinges on the visitor in ways that actively re-constitute their experiences, and sometimes make lasting impressions on the forms of their subjectivity.

> "The consequence of this [subjective] clause is that atmospheres cannot be passed down through representations, and that second-hand communication or inference of an atmosphere does not recreate the atmosphere, but only produces its mediation." (Sørensen, 2015, p. 64)

In short, and to again appropriate Heidegger: *You* have to *be there.*

What Richards, in the earlier quotation, terms a 'sphere of influence' seems very like an atmosphere and similarly difficult to define. It too combines felt *e*ffects and (emotional) *a*ffects, on bodies, moods, understandings, etc. in close, *shared*, relation with the material, yet often intangible, expressions of a place. In Richard's case this term is specifically related to the stone circles that are the focus of his own archaeological concerns. A sphere of influence also suggests a 'clause of subjectivity' but also, rather more explicitly, identifies a kind of 'agency' on the part of the stone circle and its material setting. Now, as Richards (2013, p. 27) points out, talk of agency may sometimes appear to be just an overblown way of recognizing "the effect an object has on an audience". However, if we attend to the generation of this 'influence' more closely, it does seem to suggest that even paradigmatically *inanimate* objects like stones must sometimes be recognized as active and intimate constituents of the world we inhabit and our subjective experience. They affect us materially, emotionally, and mentally, *all together* and we cannot just abstract our-*selves* from these affects. This might be regarded as a form of poeisis, that is, a bringing into being something that did not exist before.

Brodgar:

The skylark, ceasing song, lights upon the stone.

Walking the ring, hands held to lichen-crusted rock. Stones and sky circle slowly, time-turning backwards.

Midsummer dark. Awaiting dawn we perch against cold stone. Later, we will place bright flowers here.

The stones compose our ways of sharing with(in) the world and each other. Yet in modern Western metaphysics and 'common sense', stones are presumed lifeless and, as Heidegger claims, "worldless" (Heidegger, 1995, p. 185). Agency, from this dominant perspective, is almost entirely a human prerogative, something exhibited only in the decisions, interpretations, and actions of the stone circles' human builders and their visitors.[8] Any talk of non-human material agency (except occasionally in 'higher' mammals) is regarded as a category mistake based in a faux romanticism or mysticism, much like Wordsworth's 'lamentable' slips into lithic pantheism or that deep ecological 'gobbledygook' influenced by Aldo Leopold's famous call to "think like a mountain" (Leopold, 1949, p. 129). Mountains, the rational Western mind assumes, have no experiences of the world at all, still less do they think. They are just 'barren rock'.[9]

This reductive conclusion, though, seems entirely unwarranted. Stones do not think as (some) humans think, they may not walk as (some) humans walk,[10] but they do manifest their existence in forms that have innumerable effects; as poles to attract lightening, rock-tops where skylarks may land, footings where flowers might be placed, surfaces where lichens cling and grow, supports to sit against as sunrise exposes the world anew, places to celebrate love, concerns for, or loss of, others. They are things that orient worlds. They facilitate many different forms of material engagement by many different beings and forces creating lithic possibilities including the very kinds of thoughts and intentions the humanist seems to believe are generated solely from 'within' us, by purely human agency. But thoughts are never pure. They are not separate from things nor from forms of material engagement. So to simply label stones inanimate is to fail to recognize their innumerable roles and the affordances they offer to existence as the things they are. Rocks (quite literally in the case of the Earth) give birth to worlds.

Certainly the fact that many other cultures, like that of the Mi'Kmaq, *do* employ complex understandings of agency/animacy that blur the lines between ontological absolutes like living person/inanimate rock (see, for example, Hampton, 1999, Lewis-Williams, 2002, Richards, 2013. Povinelli, 2016) shows that the modern Western view is historically and culturally contingent. And as already indicated this metaphysical distinction is not even universally shared within Western cultural traditions. There is, for example, a longstanding philosophical tradition of *hylozoism* that goes back to pre-Socratic Greek philosophers such as Thales and emerges in various forms in Spinoza, Goethe, and even scientists like Humphrey Davy and Ernst Haeckel (who, perhaps *not* incidentally, coined the term 'ecology'). There is also now a very substantial theoretical and social philosophical literature that challenges any absolute division between human and material agency, including, for example, the actor-network theory of Latour (2017), 'Thing Theory' (see the collection by Brown, 2004), various revivals of Whitehead's 'process theory' (Stengers, 2011), the many forms of neo-materialism and neo-vitalist approaches influenced by theorists like Deleuze (Coole & Frost, 2010, Grosz, 2017). The re-emergence of such approaches *now* is not accidental. Indeed it might even be a consequence of social theorists/philosophers being *forced* by atmospheric changes beyond our control to re-cognize the lithic agency of that 'non-human' world modernity had, quite wrongly, believed it had superseded. In other words, the very idea of the *Anthropo*cene is bizarrely misplaced, since *rather than entering an* *anthropogenic epoch we are, more accurately, suffering the consequences of the modern suppression of the importance of non-human agency*, a blinkered modern ontology better characterized as the Great Derangement (Ghosh, 2016, p. 31).

Whatever their differences, most of these 'new' theoretical approaches undermine ideas of human exceptionality regarding agency. Rather, they emphasize the multiplicity of very different forms of agency exerted by non-human and by supposedly inanimate things caught up in constituting the world. They explicitly challenge a Cartesian and dualist ontology that sets the human mind over and against non-human inanimate matter. Interestingly, as with Richard's interests, some of these approaches have emerged directly from anthropological and archaeological attempts to understand prehistoric sites and the evolving relations between mind and matter – specifically stone. Colin Renfrew (who has also worked extensively on prehistoric Orkney (Renfrew, 1993)) is a key figure in this field. These approaches would include the kinds of phenomenological archaeology exemplified by Tilley (2004) and Ingold's (2000; 2006) phenomenologically inspired anthropology, together with what Watts (2013) refers to as *Relational Archaeologies* and Material Engagement Theory (Renfrew in DeMarrais et al., 2004, Malafouris, 2013).

The theoretical tendency here has been to move away from regarding archaeological sites/artifacts as inert material objects whose only meaning was that inscribed by past human activities; activities that therefore need to be interpreted and decoded to reveal those humans' original intentions. Rather, these *neo*-lithic approaches regard stone implements and monuments as things to be experienced in context and re-engaged with materially *and* thoughtfully as surviving *constituents* of a past world and particular places. In other words, such archaeology has moved away from a focus on the human mind as the sole maker and interpreter of meaning towards a recognition that all meaning (and thought) is created and held in the ways in which modes of materially engaged relations are, or are not, sustained and re-created over time. Thinking/meaning making is not something that only goes on 'inside' human heads. Minds are not immaterial processes that 'internally' represent external objects, rather thought *always* extends into the world. It is impossible to say where a thought stops and the material 'object' begins. We experience and think *with* things, they are, quite literally constitutive of our thinking, they enable, alter, and delimit, our patterns of thought.[11] Without certain things certain thoughts would be different, even impossible. Certain forms of material engagement with stone are actually likely to lie at the very beginning of human thinking – we probably came to think as we do through and with the manipulation of stone tools. Minds are shaped by, and are, in part composed of, 'things' (which is not to deny that they are also shaped by social/natural histories, interpretative frameworks, etc).

> "Material culture is not just the backdrop against which human cognition takes shape. Things mediate, actively shape, and constitute our ways of being in the world and of making sense of the world. Things also bring people together and provide channels for interaction. Things envelop our minds; they become us." (Malafouris, 2013, p. 44)

If our thoughts emerge through certain modes of engagement with aspects of the material world then it is obvious why it matters whether and how we experience the more-than-human world. This also explains why our 'being-there' matters, why we can only experience a stone circles *atmosphere* first-hand and subsequent upon entering its sphere of influence. This is precisely not to say that our subjective state is 'immaterial', rather our subjectivity *matters* quite literally, it exists *only* through being materially engaged. Of course, depending on what we bring to these encounters, and how open to being exposed to particular material influences we are, then we will be differently affected. A thing like the Odin Stone, places like Stenness and Brodgar, could become an intimate part of our being, thought, and concerns, at least as long as they remain there. We can be touched by them deeply, existentially. This existential aspect explains why the Orkney residents rioted on the stones destruction, it was an attack of their *and* the stones place in the world.

When schooled to approach stones as inert rock, as objects that have the makings of a prospective cow-shed and/or a quick financial return, we clearly engage with and think about them and the world in entirely different ways. In such cases the stones value is not a matter of their being there with us, it is not constitutive, nor ethical, but only of their being reduced to a means to another 'end'. From this, now globally dominant, perspective their value lies only in their being available to be set to work and transformed into something else by human agency. Now most of us slip between different modes of engagement with things and places in different circumstances but when the whole globe has become enframed by a mode of existence that dismisses things agency and has no interest in anything but their instrumental value, the inevitable result, later or sooner, is planetary disaster. Once again, this situation, mistakenly referred to by those who accept the dominant framing as the Anthropocene, obscures the key issue here, the imposition of a specific global mode of material engagement through capitalism and the claim that this is an incontrovertible given, that there is no alternative. But here we should recall that the stones at Brodgar also express a history, both natural and social, of the more than 4500 years before capitalism ever arrived on the scene. Almost all climate destruction has occurred in just the last 50 years, a mere 1% of the standing stones' existence. The stones are thus also material expressions of previous worlds and of the possibility of neo-lithic alternatives, albeit ones where capitalism and tourism have no place.

Presage to the end(s) of tourism

Stone circles, though their continuing existence and as experienced through certain modes of material engagement, might be regarded as providing a presage to the end of tourism. The atmospheres produced by tourism, in the sense of the desires and attractions it holds, the pleasures and the problems it creates are so varied, and yet it too is largely framed within a dominant mode of material engagement that has contributed to creating a global atmosphere that is becoming impossible to breathe. Tourism, and especially the burning of fossil fuels associated with it, contributes significantly to creating a regime of climate change (Hall, 2009) that is devastating cultures, species, and ecologies. Climate change will in turn, and in only a very few years, make current forms of tourism a thing of the past, a 'heritage' industry in a very different sense, and one that will have passed so many unwanted consequences on to future generations (however few of these there may, unfortunately, be).

To say that tourism is coming to an end in the Anthopocene might be understood in very different ways. For a few years yet rich people will continue to fly or float their way around the world as sea levels rise, storms increase in severity and unpredictability, hundreds of thousands of species become extinct, crops fail, drought and soil erosion and deforestation lead to mass starvation, mass migrations and economic and social collapse. Perhaps the human species will be one of those to survive, perhaps not. It may, *or may not*, be premature to say that modernity has utterly "failed to prevent unmanageable global warming and that global capitalist civilization as we know it is already over" (Scranton, 2015, p. 24). Ethically, it still seems impossible for most of those still living in this fading dream to ask, is it, or rather, was it worth it? Such a question will be immediately translated into the egocentric and hedonistic currency of neo-liberal individualism, treated as somehow equivalent to asking whether last night's party was worth todays hangover. But it is not the same question at all, for the consequences are not a passing headache and vaguely recalled embarrassment. For this is a party that burnt the entire neighborhood down to the ground and incinerated most of its occupants. It's an event that few may, quite literally, ever be able to live with.

Of course, this same tourism industry is economically 'adaptable' in the sense that it can even make a profit (in the very short term) from this appalling situation via 'last chance to see' excursions (Lamelin, Dawson, & Stewart, 2013) using the end of Nature in the Anthropocene as a capitalist fix (Fletcher, 2019). But this cannot continue and there will be no 'new normal' to be found in which to 'grow' a reformed sustainable tourist industry. There is no sense in which talk about 'resilience' and a little 'green' capitalism will save any of us. It is time to be realistic: Tourism is just one, and one of the least important, things to be so very near its end.

Realism requires that we recognize the many different senses that atmospheres are rapidly changing beyond recognition. Politically, of course, what seemed beyond imagining just a few years ago is now an everyday occurrence, Demagogues ratchet up toxic atmospheres of fear, intolerance, and hate, creating and then blaming ever more victims in order to protect the vested interests of bloated billionaires, laying waste to hope by deliberately burning and destroying the only things, like the Amazon rainforest, that offer any chance of a less than appalling future. It is not accidental that we have been drawn into this atmosphere of political crisis by those who, publically at least, still deny the material reality of the climate crisis. We seem stuck with a warped reflection of modernity's self-(mis)understandings, the last gasping breaths of *Homo economicus* still caught up in a transient consumptive materialism that can only think with things like automobiles and the internet, dollars and military hardware, flyaway holidays, fashions, fracking, and fossil fuels. It should seem obvious, but is still anathema to capitalism, to suggest that our dire and worsening situation shows that these are not good things to think with, that they involve forms of material engagement that ultimately lead to planetary disaster.

All this is an ethical matter, but there are many who think ethics is weak and powerless and many more who deny that ethics even has a place in business or in tourism – that it really doesn't matter how tourists engage with the world as long as they get individual subjective enjoyment from it. Butcher's argumentatively incoherent *Moralization of Tourism* (2003) for example, criticizes anyone who has any ethical concerns whatsoever about the effects of corporate capitalism in tourism, claiming that such criticism represents an elitist and effete 'moralization' of tourism by academics who really just want to stop all the fun, fun, fun. As I have pointed out elsewhere (Smith, 2009) this bizarre position is, perhaps, linked to Butcher's ongoing associations with a shape-shifting contrarian organization (founded by sociologist Frank Furedi) that helped pioneer the image based production of 'alternative facts'. This organization, although originally describing itself as Marxist, has, we might note, recently received very generous funding through an institute founded by fossil fuel billionaires the Koch brothers. Ethical denial it seems, is not unconnected with climate change denial, indeed this is another specialty of the same group.[12] Butcher's work either misunderstands and/or deliberately seeks to negate, the constituent role of ethics in all critique. Yet to provide an ethical critique of tourism for playing a significant part in rising CO2 levels is not to blame tourists for having fun. Rather it is to ask what ends tourism and tourists might have had in a differently constituted world.

As this paper has argued, it really does matter how visitors become engaged with the world and not least because certain modes of material engagement expose our understanding of 'subjectivity itself' to a reality test, drawing us away from hedonistic atomistic individualism towards concerned involvement in and for aspects of a much wider world. Some encounters can be transformative, we are changed in how we understand and experience other things, landscapes, atmospheres, not as ontological or economic externalities, but as creative co-constituents of what and how we think and value a world that exists with us, not for us. Such encounters, exemplified here in certain associations with stone circles, expose us phenomenologically and existentially to the more-than-just-human implications of unique atmospheres, and the animacy, agency, and intimacy of a world that is not, despite modernity's rhetoric, set over and against us but composes our very thoughts and experiences in myriad ways. Being drawn into the sphere of influence of a circle of stones is just one of many possible examples of the exposure of self to a world that calls forth a response and instigates ethical responsibilities. Not everyone is so affected, but those that are might find that they come to speak about such encounters and places in ways that far exceed any framework available in resource management.

Such neo-lithic ethical involvements will not, of course, occasion a bill of rights for rocks, this is not in any sense their intention. However, they may, and sometimes do, engender respect for other entities and places, and simultaneously reveal something of the enchantments offered by a world that has never been ours to own. They also signal something of the dangers inherent in more limited forms and understandings of material engagement. Like the weather, atmospheres, whether sensory, political or planetary, can change quickly and unexpectedly. This unpredictability is becoming more tangible with every storm, drought, fire, or flood. Tourism is, perhaps, an industry especially susceptible to such changes. Many now feel increasingly exposed to the changing dynamics of a planetary atmosphere that, for humans and so many other species, is potentially world destroying. Yet, tourism, like other consumer cultures of constant material distraction helps us bury our heads in business/busyness, holding concern and anxiety about the world's future at arms length.

The material impacts of climate change will not be denied. And so, still misrecognizing the real distribution of agency in the world, the danger is we flip from an inane and patently inadequate focus on the agency of individual (consumer) choice, to a feeling of utter powerlessness in the face of forces so far beyond individual or collective human control. Once again, we need to recognize that the affordances offered to us by the world depend upon the modes of material engagement in which we find ourselves involved. Some forms of material engagement (including some, but not many, aspects of tourism – the child's being drawn in wonder to breaking waves on beach sand, and to sea-shells and rock-pools) generate broader understandings of human involvements in a world that is certainly not 'inanimate' nor placed there at our service. The currently dominant form of material engagement militates against any such understandings, (Graham, forthcoming) but stones, waves, shells, rock-pools, skylarks, still *offer* experiential affordances that enact alternative 'moments of being', of living, valuing, and attending to the world. In this sense the world might be understood not as a resource but as an expression of ecological provid*a*nce (rather than providence in a puritanical religious sense) that is, as the source of material *provision* of possibilities for different modes of engagement. And if different forms of material engagement are so closely entwined with different affordances, lives, and values, this providence offers ways to reconsider our response-abilities and responsibilities, ways of "staying with the trouble" (Haraway, 2016) that our current patterns of material engagement have caused.

We will need forms, perhaps many diverse forms, of neo-lithic ethics for weathering the coming storm, here, together, on Earth, with other beings and things. They will not save the world in its entirety or current form, nor will they save tourism. However, if by some stroke of fortune, or unforeseen agency, our current planetary atmospherics play out differently than we have any right to expect then they might provide some answer to the question of whether, even after such an appalling disaster, 'we will have learnt anything' about where agency lies?

Notes

1. The term Anthropocene, which still awaits formal recognition as a new geological epoch, is commonly used to refer to that period where human agency (especially, but not limited to, anthropogenic climate change).has become the dominant force impacting the Earth's geology.
2. Even many modern ethical systems that explicitly regard themselves as environmental would also, on various grounds, e.g. their lacking a telos (Taylor, 1986) or their inability to 'flourish' (Attfield, 1991) deny rocks, mountains etc. per se, any ethical standing (see Smith, 2001 for a critique of such positions.)
3. Povinelli draws upon extensive anthropological work with Indigenous Australian cultures to develop a convincing critique of the tripartite Western ontological distinction between a) human *agent* b) *animate* beings and c) *inanimate* things. She argues that this Western construct has facilitated neo-liberal exploitation of the world. This division appears in many forms, for example, divisions between ethics / politics as a) a specifically human realm, b) biopolitics, as in Foucault, or in animal ethics) and c) an apolitical and amoral 'geo-logic'. It also occurs, of course, in Heidegger's (1995) famous, account of world-forming (human) *Dasein*, the 'world-poor' animal and the 'worldless' stone.
4. In this MacKay exemplified many others interested in agricultural improvement and / or wanting to turn a profit by demolishing monuments that had survived thousands of years. See, for example Burl's (1979) account of the massive destruction of stone circles, monuments and barrows around Avebury from the 17th century onwards.

5. "UNESCO World Heritage status is both a considerable benefit, and a difficulty. Such status arguably attracts a greater level of visitors to Orkney's World Heritage Sites, than would otherwise be the case." (Orkney Volume, 2017: 20)
6. Which, of course, is why Moore (2015) argues that we should refer to this epoch not as the Anthropocene, which suggests that humans per se are generating this global change, but the capitalocene.
7. Which is not to say that some do not try to account for certain atmospheres through scientific discourses. For example, Brennan (2004) provides a rather reductive and unconvincing argument about the production and communication of certain atmospheres based on pheromones.
8. This emphasis on the exceptional nature of human agency is even present in explicit forms of *materialism*, like Marxism, where "the living, shaping fire" of human labor is regarded as the 'active' motor, of the historical dialectic that transforms a "slumbering" nature (Marx in Smith, 2001).
9. Leopold, of course, did not actually consider mountains capable of engaging in economic calculus or political debate, rather he asked us to consider the anthropic limitation of our Western way of thinking of ourselves as *exceptional*, as being above and outside of the ways the more-than-just-human world works. His work exemplifies what Cohen (2015, p. 9) refers to as, "an ecological project of thinking beyond anthropocentricty", of "assuming a world irreducible to its human relations and not existing for any particular purpose" whilst stressing "alliance, continuity, and mutual participation over elemental solitariness and human exceptionalism".
10. Though there are many myths and stories of walking megalithic stones.
11. In this sense MET is also closely aligned with various philosophical notions of a materially extended and distributed mind, for example, the work of Andy Clark (1997).
12. For a critique of Butcher's 'arguments' see Smith (2009). As I pointed out in that article, Butcher explicitly associates himself with those circulating around an organization at one time known as the Revolutionary Communist Party that produced a magazine called Living Marxism (LM). "LM morphed, after losing a libel case concerning its accusation that ITN journalists had distorted the truth about film it shot of severely malnourished prisoners of war in Serbian concentration camps (Vulliamy, 2000) into several rather more innocuously named organizations, most notably the 'Institute of Ideas', launched in 2000 by previous editor of LM Claire Fox (aka Clare Foster) (lobbywatch, 2007)" and the 'spiked' web-site. Those contributing to these sites continue the LM tradition of contrarian politics, attacking environmentalism, supporting genetically modified foods, denying global warming and decrying anything that might constrain or oppose corporate capitalism. They were, for example, associated with this producing a notorious documentary for UK's Channel 4 called *The Great Global Warming Swindle*. Butcher's book explicitly acknowledges the Institute of Ideas as inspirational and he has produced many articles for the spiked website. https://www.spiked-online.com/author/jim-butcher/. George Monbiot (2018) recently revealed that spiked has been surreptitiously funded by the Koch Brothers. It is, perhaps, not entirely surprising then that an "article in 2016, when Spiked received $170,000 from the Charles Koch Foundation, attacked the Standing Rock protests against the Dakota Access Pipeline, in which the Koch brothers have a major interest." (Monbiot, 2018).

Disclosure statement

No potential conflict of interest was reported by the author(s).

References

Agricola, G. (1950). *De Re Metallica*. New York: Dover Publications.

Anon (2015). *The Ness of Brodgar*. Kirkwall: Ness of Bordgar Trust.
Attfield, R. (1991). *The ethics of environmental concern*. Athens: University of Georgia Press.
Barry, G. (1867). *The history of the Orkney Islands*. Pearce: Kirkwall.
Bille, M., Bjerregaard, P., & Sørensen, T. F. (2015). Staging atmospheres: Materiality, culture, and the texture of the in-between. *Emotion, Space and Society, 15*, 31–38. doi:10.1016/j.emospa.2014.11.002
Black, G. F., & Thomas, N. W. (1903). *Examples of printed folklore concerning the Orkney and Shetland Islands*. London: David Nutt.
Bourdieu, P. (1987). *Distinction: A social critique of the judgement of taste*. London: Routledge.
Brennan, T. (2004). *The transmission of affect*. Ithaca, NY: Cornell University Press.
Brown, B. (Ed.). (2004). *Things*. Chicago, IL: University of Chicago Press.
Brown, G. M. (1992). *Brodgar poems*. Oxford: The Perpetua Press.
Burl, A. (1979). *Prehistoric Avebury*. New Haven, CT: Yale University Press.
Butcher, J. (2003). *The moralization of tourism. Sun sand and saving the world*. London: Routledge.
Card, N. (2018). The Ness of Brodgar: Uncovering Orkney's Neolithic heart. *Current Archaeology*, No. 335, 20–28.
Clark, A. (1997). *Being there: Putting brain, body and world together again*. Cambridge MA: MIT Press.
Cohen, J. J. (2015). *Stone: An ecology of the inhuman*. Minneapolis: University of Minnesota Press.
Coole, D. & Frost, S. (2010). *New Materialisms*. Ontology, Agency and Politics. Durham: Duke University Press.
DeMarrais, E., Gosden, C., & Renfrew, C. (2004). *Rethinking materiality: The engagement of mind with the material world*. Cambridge: Macdonald Institute for Archaeological Research
Edensor, T. (2015). Producing atmospheres at the match: Fan cultures, commercialization and mood management in English football. *Emotion, Space and Society, 15*, 82–89. doi:10.1016/j.emospa.2013.12.010
Ehrenfeld, D. (1978). *The arrogance of humanism*. New York, NY: Oxford University Press.
Fletcher, R. (2019). Ecotourism after nature: Anthropocene tourism as a new capitalist fix. *Journal of Sustainable Tourism, 27*(4), 522–535. doi:10.1080/09669582.2018.1471084
Graham, P. (forthcoming) *Traces of (un-) sustainability: Towards a materially engaged ecology of mind*. New York, NY: Peter Lang.
Grosz, E. (2017). *The incorporeal. Ontology, ethics and the limits of materialisms*. New York, NY: Columbia University Press.
Guardian. (2019). World-renowned Ring of Brodgar stone circle vandalized in Orkney. Retrieved from https://www.theguardian.com/uk-news/2019/apr/11/world-renowned-ring-of-brodgar-stone-circle-vandalised-orkney
Hall, C. M. (2009). Degrowing tourism: Décroissance, sustainable consumption and steady-state tourism. *Anatolia, 20*(1), 46–61. doi:10.1080/13032917.2009.10518894
Hampton, O. W. "Bud" (1999). *Culture of stone: Sacred and profane uses of stone among the Dani*. College Station: Texas A & M University Press.
Haraway, D. (2016). *Staying with the trouble. Making kin in the Chthulucene*. Durham: Duke University Press.
Heidegger, M. (1993). *The question concerning technology in Basic writings*. San Francisco, CA: Harper.
Heidegger, M. (1995). *The fundamental concepts of metaphysics: World, finitude, solitude*. Bloomington: Indiana University Press.
Ingold, T. (2000). *The perception of the environment: Essays in livelihood, dwelling and skill*. London: Routledge.
Ingold, T. (2006). Rethinking the animate, re-animating thought. *Ethnos, 71* (1), 9–20. doi:10.1080/00141840600603111
Latour, B. (2017). *Facing GAIA*. Eight lectures on the new climatic regime. Oxford: Polity.
Lemelin, H., Dawson, J., & Stewart, E. J. (Eds.). (2013). *Last chance tourism: Adapting tourism opportunities in a changing world*. London: Routledge.
Leopold, A. (1949). *A Sand County almanac: And sketches here and there*. Oxford: Oxford University Press.
Lewis-Williams, D. (2002). *The mind in the cave*. London: Thames and Hudson.
Lobbywatch. (2007). Online document Retrieved from http://www.lobbywatch.org
Lockhart, J. G. (1837). *Memoirs of the life of Sir Walter Scott Bart* (Vol. 3). Edinburgh: Robert Cadell.
Malafouris, L. (2013). *How things shape the mind: A theory of material engagement*. Cambridge MA: MIT Press.
Marwick, A. (2000). *The folklore of Orkney and Shetland*. Edinburgh: Birlinn.
McClanahan, A. (2016). The ethics of landscape: Discourses of cultural and environmental sustainability in the Heart of Neolithic Orkney World Heritage Site. In M. Robinson, M. Gravari-Barbas, & L. Bourdeau (Eds.), *World heritage, tourism and identity: Inscription and co-production*. Burlington, VT: Routledge.
Merchant, C. (1983). *The death of nature: Women, ecology, and the scientific revolution*. New York, NY: Harper & Row.
Milton, K. (2002). *Loving nature: Towards an ecology of emotion*. London: Routledge.
Monbiot, G. (2018, November 7). You want it darker? *The Guardian*. Retrieved from https://www.monbiot.com/2018/12/10/you-want-it-darker/
Moore, J. W. (2015). *Anthropocene or capitalocene: Nature, history, and the crisis of capitalism*. Oakland CA: Kairos.
Orkney Volume. (2017). *Towards a theory of material engagement as here*. Retrieved from https://whc.unesco.org/en/list/514/multiple=1&unique_number=2125

Povinelli, E. A. (2016). *Geontologies: A requiem to late liberalism*. Durham, NC: Duke University Press.
Rawls, J. (1973). *A theory of justice*. Oxford: Oxford University Press.
Renfrew, C. (2004). Towards a theory of material engagement. In E. DeMarrais, C. Gosden, & C. Renfrew (Eds.), *Rethinking materiality: The engagement of mind with the material world*. Cambridge: McDonald Institute Monographs.
Renfrew, C. (Ed.). (1993). *The prehistory of Orkney BC4000–1000*. Edinburgh; Edinburgh University Press.
Richards, C. (2013). Interpreting stone circles. In C. Richards (Ed.), *Building the great stone circles of the North*. Oxford: Windgather Press.
Ritchie, J., & Marwick, E. W. (1976). The Stones of Stenness, Orkney. *Proceedings of the Society of Antiquaries of Scotland*, *107*, 1–60.
Sen, A. (2009). *The idea of justice*. Cambridge, MA: Harvard University Press.
Smith, M. (2001). *An ethics of place: Radical ecology, postmodernity and social theory*. Albany: State University of New York Press.
Smith, M. (2009). Development and its discontents: Ego-tripping without ethics or idea(l)s. In J. Tribe ed. *Philosophical Issues in Tourism*. Bristol: Channel View Publications.
Sorenson, T. F. (2015). More than a feeling: towards an archaeology of atmosphere. *Emotion, Space and Society*. *15*, 64–73.
Stengers, I. (2011). *Thinking with Whitehead: A free, wild creation of concepts*. Cambridge, MA: Harvard University Press.
Taylor, P. W. (1986). *Respect for nature: A theory of environmental ethics*. Princeton, NJ: University of Princeton Press.
Tilley, C. (2004). *The materiality of stone: Explorations in landscape phenomenology*. Oxford: Berg.
Vulliamy (Ed.). (2000, March 15). Poison in the well of history. *The Guardian*. Retrieved from http://www.guardian.co.uk
Watts, C. (2013). *Relational archaeologies: Humans animals things*. London: Routledge.
World Commission on Environment and Development. (1987). *Our common future*. Oxford: Oxford University Press.

Tourism, inclusive growth and decent work: a political economy critique

Raoul V. Bianchi and Frans de Man

ABSTRACT

This paper interrogates the ideas of 'sustained' and 'inclusive' growth that are intrinsic to one of three UN Sustainable Development Goals (SDG 8 - Decent Work and Growth) adopted by the UN World Tourism Organisation's (UNWTO) 2030 sustainable tourism agenda. It provides a Marxian-inspired political economy critique of the UNWTO's embrace of SDG8 and highlights the blind spot within the UNWTO's inclusive growth-led SDG agenda with respect to questions of equity and social justice. The paper contends that the UNWTO's SDG-led agenda is contradicted by the logics of growth, competitiveness and profit-making that drive the continued expansion and development of tourism. Rather than addressing the structural injustices that entrench inequalities and reproduce exploitative labour practices, the notion of sustained and inclusive growth reinforces the primacy of capital and market notions of justice and continues to perpetuate a growth driven tourism development model. The paper contributes to a critical theorization of sustainable tourism and offers an informed critique of the current political agenda for sustainable tourism and its potential outcomes.

Introduction

Despite the innumerable policies and strategies devoted to advancing sustainable development agreed since the 1992 Rio Earth Summit a systemic paradigm shift towards sustainable tourism development remains hindered by the existence of "defence and delaying mechanisms in the economic and political system" (Müller, 1994, p. 134). The launch of the UNWTO's sustainable tourism development agenda in 2015, framed by the United Nations Agenda 2030 and associated 17 Sustainable Development Goals (SDGs) seemed to "indicate the need to rethink the current economic growth ideology in the context of social and environmental needs in development" (Saarinen, 2018, p.33). UN Agenda 2030 allegedly signals a major step towards building a global, integrated multi-stakeholder framework for ending extreme poverty, fighting inequality and injustice, and combatting climate change. Meanwhile however, growth-led, extractive, profit-driven economics continues to exacerbate an inter-connected global environmental, climate and development crisis (Klein, 2014).

The UNWTO and its partners contend that the managed growth of tourism can help engineer a transition to a "green economy" and contribute to "decent work creation, poverty alleviation, improved efficiency in resource use and reduced environmental degradation" (Stroebel, 2015,

p. 2226). This paper presents an analysis and critique of the UNWTO's embrace and application of the SDGs with an explicit focus on SDG 8 (Decent Work and Growth) namely, the UNWTO's desire to harness tourism in order to "promote inclusive and sustainable economic growth" and "full and productive employment and decent work for all" (UNWTO , 2016, p. 33). Specifically, it addresses the neglect of work and labour in the tourism sustainability debate (see Baum et al., 2016), with due regard to the lack of attention to the critique of political economy in the UNWTO's SDG agenda (Hall, 2019, p. 7).

This paper argues that the UNWTO's failure to address, much less challenge, the systemic processes of accumulation and exploitation that shape and constitute the competitive dynamics and industrial structures of tourism capitalism (Bianchi, 2018), undermines its ability to put forward a genuinely fair and progressive sustainable tourism agenda centred on concerns of environmental and social justice. It contends that earlier critiques of capitalist tourism development in the Global South, in which principles of economic justice and solidarity between nations were foregrounded (e.g., Shivji, 1973; WTO, 1980), have largely been superseded by market-friendly notions of 'sustained' and 'inclusive' growth that offer little by way of a challenge to the accelerated monetisation of nature and dehumanisation of labour. The injustices of enclosure and dispossession are furthermore often obscured by normative global sustainability discourses in which conservation rather than the just and equitable use and control of resources is foregrounded (Banarjee, 2003).

The argument put forward in the paper is premised upon the historical materialist methodology drawn from Marxist political economy (e.g., Marx, 1973, pp. 100–108; Marx, 1977). A materialist approach interprets capitalism as a historically specific, dynamic and contradictory mode of production driven by the pursuit of profit and the accumulation of capital which further intensifies the competitive dynamics of capitalist growth (Marx, 1974, p. 555). Here, *capital* does not refer to simply the existence of commerce and markets, rather, it is a system with a distinctive set of logics that can only be understood "in the context of a particular social relation between appropriator and producer" (Wood, 2016, p. 24). The historical-geographical reproduction of capitalist social relations is played out through the continuous separation of humans from their *means of production* in tandem with the *social reproduction* of workers via the (unpaid) household labour of women (usually) and the wider relations of solidarity and mutual exchange that sustain the social fabric of communities (De Angelis, 2007, p. 37). A third component of capitalist expansion enabling the continuous production of a labour force is constituted via the *enclosure* and appropriation of natural assets and dispossessing once self-sustaining inhabitants of their independent means of obtaining a livelihood. The 16th and 17th century English acts of enclosure which drove peasants off their land and extinguished customary use rights as well as the processes of settler colonialism witnessed in the US, both of which gave rise to capitalist property relations, represent similar logics of appropriation and dispossession (see XXX in this issue), which have played out in various ways in colonial and imperial histories worldwide (see Wolf, 1982).

The resultant antagonism between capital and labour arises out of the exploitative nature of capitalist production and more specifically, the treatment of human labour as a factor of production (Selwyn, 2015, pp. 516). Systemic forces of injustice arise out of the commodification of human labour power and the enclosure of natural wealth and resources necessary to sustain human livelihoods, but which are also integral to the expansion of capital and endless growth upon which it depends (Hickel, 2019a). Challenges to the 'normal' orderly workings of the market and the attendant need to ensure an adequate return to capital are characteristically viewed as 'disruptive' and contrary to the "formal justice of the market" (Streeck, 2014, pp. 60–61).

Section one identifies the logics of growth, process of capital accumulation and role of labour in the tourism political economy as the basis for the analysis and critique of sustainable tourism development as construed by the UNWTO. The paper then interrogates the UNWTO's SDG agenda in the context of the historical-political struggle of the UNWTO to define, develop and lead a global sustainable tourism agenda since 1992. This section draws on an analysis of key

UNWTO reports[1] to interrogate the ideological and linguistic framing that shapes the UNWTO's engagement with SDG 8 together with insights drawn from the personal involvement of the second author of this paper in many of the events recounted here. As director of the Retour Foundation he was a core member of Tourism European Network (TEN) and the Transforming Tourism Initiative both of which have lobbied the UNWTO for greater civil society representation in relation to discussion and elaboration of the *Global Code of Ethics for Tourism* (WTO, 1999b), the *Framework Convention on Tourism Ethics* (UNWTO, 2017b) and the UNWTO's SDG agenda. Between 1999 and 2002 he was the elected Northern co-chair of the tourism caucus of the CSD-NGO steering committee, the official civil society partner of the United Nations Commission on Sustainable Development (UNCSD) from its 7th meeting in 1999 (UNCSD-7), when tourism was introduced into the UN sustainable development agenda, until the 2002 World Ecotourism Summit. It was at this point that the UNWTO brought the officially sanctioned NGO involvement to an end as a result of growing disagreement over the content and direction of UNWTO policy. Following this NGO participation became increasingly ad hoc and sanctioned only at the UNWTO's discretion. The final section contributes to a political economy critique of the principles of 'sustained' and 'inclusive' growth that are integral to the UNWTO's global sustainable tourism development agenda.

The political economy of sustainable tourism

Scholarly attention in sustainable tourism has moved away from a conservationist resource management focus towards diverse and rigorous theoretical inspection of concerns related to *inter alia*, destination governance (Bramwell, 2011; Dredge & Jamal, 2013) sustainable livelihoods (Tao & Wall, 2009), poverty reduction (Neto, 2003; Scheyvens, 2007), empowerment and social justice (Jamal & Camargo, 2014; Coria & Calfucura, 2012; Scheyvens, 1999), environmental ethics (Holden, 2003), the tourism commons (Briassoulis, 2002), and increasingly, degrowth (Hall, 2009; 2010; Higgins-Desbiolles, Carnicelli, Krolikowski, Wijesinghe, & Boluk, 2019). And yet studies of tourism's complex relationship to sustainability and the forces of economic development are typically reliant upon static and descriptive conceptions of political economy in which tourism capitalism is construed as the aggregate of commercial tourism activities or "set of assets" (Sharpley, 2015, p.448), rather than one driven by the competitive struggle between capitals, the exploitation of labour and the energy-intensive use and commodification of nature.[2] Contradictory capital/labour relationships are neither static nor universal (Burnham, 1994). Rather they are shaped by the daily struggle of labouring classes to defend their standards of living (including, resistance to enclosure of the resources and eco-systems upon which such livelihoods depend) and the right of capital to organize the labour process in accordance with the imperatives of profit-making (Selwyn, 2015). The precise social coordinates of struggle are nevertheless shaped by geographic variances of capitalist development and local social structures together with the collective capacity of labouring classes to organise and resist exploitation (Chibber, 2013).

The growing consensus surrounding the imperative of sustainability in tourism that permeate policy and corporate pronouncements conceals unresolved contradictions between sustaining *tourism* and sustainable *development* (Higgins-Desbiolles, 2018; McCool, Moisey, & Nickerson, 2001). At the heart of this protracted debate lies a tension between "weak" and "strong" interpretations of sustainability (Hunter, 1997, pp.853–854). The former is "industrially-oriented" and tends towards an adaptive stance, permitting a managed and regulated expansion of tourism, while the latter advocates stricter "limits to growth" and constraints on the depletion of natural capital (see Saarinen, 2006, pp.1126–1129). This apparent dichotomy nevertheless avoids a more fundamental interrogation of the meaning of *sustainable development* in the light of the forces making for the relentless growth of tourism and the conditions under which value is created and surpluses appropriated.

The political ecology of tourism employs a similar conceptual armoury to examine the struggle between different actors and institutions over access to and use of environmental resources for tourism (Bramwell, 2015, p. 211). Such conflicts are evident in Cole's (2012) work on tourism and water in Bali as well in recent analyses of the intersections between the restructuring of tourism capitalism, territorial planning and socio-environmental struggles in the Balearic Islands (Blázquez, Artigues Bonet, & Yrigoy Cadena, 2015; Hof & Blázquez-Salom, 2015). Governance perspectives examine the struggle between different state actors to define and advance the adoption of policies and planning mechanisms for sustainable tourism (Bramwell, 2011). Advocates of tourism degrowth have argued have that to redesign and align tourism economies with sustainability requires a more profound challenge to the current model of growth-led tourism that goes beyond simply managing and/or balancing the growth of tourism with resource management and conservation (Hall, 2009; Higgins-Desbiolles et al., 2019). However, with notable exceptions (see Büscher & Fletcher, 2017; Duffy, 2015; Fletcher, 2011), the analysis and critique of the complex and contradictory relationship between tourism, capitalist development and sustainability remains divorced from more searching and systematic analyses of its political economy much less the class dynamics of capital-labour relations in the struggle to define and implement sustainable tourism.

Tourism development and growth

The period between 1990 and 2008 witnessed 260 per cent growth in international tourist arrivals and the consolidation of tourism as a major category of international trade, with the fastest growth in arrivals taking place between 2003 and 2007 (UNCTAD, 2010, p. 2). Largely unperturbed by the 2008 financial crisis, by 2018 growth in international tourism had reached 6 per cent per annum, exceeding the average rate of global economic growth (3.7 per cent) (UNWTO, 2019). During the same year international tourism accounted for 10.4 per cent of global output with an estimated total economic value of US$8.3 trillion (WTTC, 2018). Meanwhile, since 1990 the number of people living on less than US$5 a day increased by more than 370 million while global GDP per capita simultaneously grew by 65 per cent (Hickel, 2017, p.56). Despite the emergence of increasingly diverse measures of human well-being and measures to account for externalities since the OECD first mandated the pursuit of GDP growth as the priority of government in the 1960s, it remains the principal strategy of governments and international development agencies for delivering prosperity (Raworth, 2018, p.38). Indeed, the SDGs themselves prescribe growth rates of 7 per cent per year as the principal means of eliminating extreme poverty in 'developing' countries.

Bramwell and Lane (1993, p.2) argue that sustainable tourism is not "anti-growth" but rather is linked to the need, identified in the 1987 Brundtland Report (WCED, 1987), to bring about "fairness" and "convergence" between rich and poor nations in the global trading system. Notwithstanding the variants of capitalist development worldwide, since its adoption as a major plank of economic development by newly-independent 'Third World' states commencing in the 1950s tourism has been construed as an ideologically-neutral, technical enterprise whose success or failure is predominantly judged in terms of annual increases in arrivals and tourist receipts and their contribution to GDP-led economic growth (UNCTAD, 1973, 2010).

During the 1960s and early 1970s when environmentalists began to highlight tensions between economic growth and the environment, tourism seemed to offer 'developing' countries a viable alternative to becoming suppliers of raw materials and agricultural commodities to the industrialized West, with the potential to help transform them into modern industrialized economies (Peters, 1969). Since then much has been made of the contribution of tourism-led growth to economic development (WTTC, 2018), including the claim that market-driven tourism has functioned as an engine of wealth redistribution from rich to poor countries (Keller & Bieger 2011, p.1).

While the view persists amongst international development agencies that tourism is uniquely placed to contribute to economic development and reduce poverty in the Global South (UN News, 2011), revenue leakages have continued to plague tourism development UNCTAD (2010, p.9). The UNWTO and other international institutions have recognized such shortcomings but remain committed to the logics of export-led tourism albeit one whose adverse consequences merely need to be better managed (Rifai, 2017). However, the very *need* to trade (in tourism) and the terms under which such trade is undertaken is "often a sign that an economy is unsustainable" (Douthwaite, 1999, p.171). The contribution made by tourism to improvements in national income, living standards and poverty reduction[3] often lack deeper interrogation of the social and political dynamics of tourism development. The imperative to earn a wage and the very fact that tourism may provide jobs for people who don't have many alternatives (see Steiner, 2006: 170), is indicative of the disempowerment of labouring classes and often coexists with exploitative labour regimes within destinations. Although many low-income states did experience a boost to national income and employment the development of tourism destinations often entailed the disruption of native economies and consequent emergence of socio-economic inequalities (Britton, 1980). Tourism investment and economic growth in recent years continues to accelerate land dispossessions, loss of self-reliance and the intensification of ecological destruction (Holden, 2013, pp.108–111), with little discernable decrease in inequalities (Alam & Paramati, 2016).

The acceleration of global trade and consolidation of neoliberal market economics in the 1990s saw tourism emerge as a major avenue of capitalist development and pillar of economic growth strategies for advanced and emerging economies alike (Britton, 1991). In 1991 GNP was superseded by Gross Domestic Product (GDP) as the principal benchmark against which the economic strength and wealth of nations would be measured (Fioramonti, 2013, pp. 41). This seemingly innocuous change attributes income generated by foreign-owned firms to the countries in which they were generated and not to the country in which the firm is headquartered. This gives the misleading impression that national incomes are growing and the conditions of development improving despite continuing economic privations and revenue leakage in the Global South (see Hickel, 2017). Not only does GDP shape the understanding of value, given the weighting of tourism in the export sectors of many low-income economies in which foreign-based transnational corporations and off-shore investors play a significant role, it potentially inflates the economic benefits provided by tourism (see FDI Intelligence, 2018). The question of value, how it is produced and by whom it is appropriated in tourism economies, is therefore a critical one for the evaluation of any sustainable tourism development agenda.

Tourism capitalism and the creation of value

Tourism development catalyses new forms of value and reshuffles the internal distribution of capital, land use and organisation of labour as they become structured around the demands of tourism capital accumulation. The industrial organisation and concentration of tourism industries emerges from the interaction between the degree and scope of integration of domestic economies into global markets, firm strategies and the character of state intervention in a given destination (Clancy, 2001, pp. 26-27). Considerable corporate concentration exists in the various tourism sub-sectors and increasingly amongst online digital platforms (Gössling & Hall, 2019, p.12). Destinations however often comprise a multitude of different firms of varying scale operating across a spectrum of "capitalist and non-capitalist and formal and informal sectors" (Gibson, 2009, p.529). Nevertheless, local firms in the Global South are often subordinate to lead tourism firms based in the advanced capitalist countries, who are able to monopolise revenues through their governance of global commodity chains (Christian, 2016; Mosedale, 2006).

Given the difficulty of exerting "property rights over tourism experiences" (Williams, 2004, p.62), competition between diverse constellations of tourism enterprise over access to and use of resources fuels the relentless privatization and commodification of lands, ecosystems and labour in continuous cycles of "destructive creation" (Büscher & Fletcher, 2017). In the absence of robust mechanisms for reconciling competing uses, tourism may precipitate the enclosure and exploitation of "common pool resources" (Briassoulis, 2002), to make way for the construction of resort infrastructures and privatized tourism activities. The very destruction of customary livelihoods and monetization of these habitats often results in the expansion of wage-labour upon which such growth in fact relies (Mittal & Fraser, 2018). Tourism economies are also often vulnerable to unproductive 'rent seeking' and speculative real estate investment as means of sustaining profitability and capital accumulation, such as fuelled the frenzy of real estate-led tourism growth along the Spanish littoral from the 1990s until the 2008 financial crash (Murray Mas et al., 2017; Hof & Blázquez-Salom, 2015).[4]

Financialization has fuelled growth in mergers and acquisitions in the global tourism and hotel sectors which is increasingly augmented by the entrance of private equity and 'real estate investment trusts' (REITs) in the purchase of major tour operations, hotel and resort assets (ILO, 2010, pp. 29–32; Yrigoy, 2016). The resultant complexity of corporate ownership and financial flows in tourism has been further magnified by the rise of corporate on-line travel agencies (OTAs) and online rentalplatforms, leading to the disruption of existing patterns of corporate dominance in key tourism sub-sectors whilst exacerbating new patterns of online market concentration (Gössling & Hall, 2019).[5] Despite their ability to unlock value in idle assets, the rise of online rental platforms creates additional challenges for achieving 'decent work' in many areas of tourism and hospitality (Dredge & Gyimóthy, 2015). Evidence nevertheless does exist for the emergence of alternative business models and distinctive modes of value creation in tourism (Cave & Dredge, 2018). However, collaborative and non-monetary forms of tourism value creation thus far present little threat to the dominance of corporate and financialized systems of capital accumulation that drive tourism growth.

Tourism and labour: the hidden dimension of sustainability

Given the prevalence of low pay, exploitation and weak labour protections across the tourism and hospitality industries (Cañada, 2014; ILO, 2010, 2017; Teberga De Paula, 2018), it is unsurprising to find that 'decent work' (SDG 8) is integral to the UNWTO's SDG agenda (UNWTO, 2017). Studies of sustainable tourism are rarely however accompanied by an examination of the intersecting axes of exploitation and discrimination arising out of the class character of tourism capitalist development and attendant patterns of resource use and appropriation. Despite a substantial literature on employment practices and working conditions consideration of the systemic nature of labour exploitation in tourism capital accumulation has been overshadowed by orthodox managerial and economic perspectives (Ladkin, 2011). Critical engagement with SDG 8 from the perspective of dignity in tourism employment also construes the tourism workforce as a factor of production, disregarding the class relations of struggle that underpins exploitative labour regimes (Winchenbach, Hanna, & Miller, 2019).

Tourism industries are marked by "diverse labour market tensions, varied subsistence strategies, and complex dynamics of power relations" (Madsen Camacho, 1996, p.33). These are shaped as much by a nation's overall political economy (Williams 2004, p.66), as they are the "hybrid and contradictory economic formations" of tourism capitalism (Gibson, 2009, p.530). The myriad small-scale entrepreneurs that underpin tourism economies are also often exposed to precarity and chronic insecurity resulting from market volatility, seasonality and unfavourable regulatory regimes (see Jamal, 2019, pp.43-45). That said, it is those at the lower ends of the labour market with little more than their labour or "free floating" capital to sell (Crick, 1995, p.163), who are most exposed to the insecurities and injustices of tourism work.

The political economy of global tourism and associated regimes of accumulation is produced and reproduced at different spatial scales partly through the ability to draw on a vast pool of globalised and socio-ethnically-differentiated labour amongst whom ethnic minorities, women and migrant workers are disproportionately represented (Chin, 2008; Chok, 2009; ILO , 2012). Hence, the divisions of labour within large, transnational firms (particularly in global cruise companies, large-scale resorts and hotel chains) are often structured according to ethnicity, nationality and gender (Chin, 2008; Hampton, 2010; Wood, 2000). Moreover, despite the growing popularity and high yields associated with specialist 'niche' tourisms systemic labour precarity and poverty is widespread in these contexts (see Hampton, Jeyacheya, & Lee, 2018). Ample evidence also exists pointing to a correlation between the construction of luxury tourism installations and exploitative working conditions, including in all-inclusive resorts in the Seychelles (Lee, Hampton, & Jeyacheya, 2015), luxury hotels in Pakistan (Sheikh, 2010) and integrated mega-resorts in South-East Asia (Chok, 2009; Hampton, 2010).

Collective bargaining in the tourism and hospitality sectors is undermined by weak trade union representation as well as the fragmented and geographically dispersed structure of tourism producers. Notwithstanding strong labour unions in key support services (e.g., air traffic control and airlines), the fragmentation of tourism divisions of labour and perishability of tourist services themselves constrains the ability of labour to slow down the speed of production or disrupt supply-chains. The complexity of globalised corporate ownership structures has enabled the out-sourcing and sub-contracting of work to employment agencies undermining collective bargaining and eroding employee protections (Grossman & Greenfield, 2006; Sheikh, 2010).

While low-skilled poorly paid tourism-hospitality sector jobs may be preferable to the arduous nature of work in other export sectors or traditional agrarian and/or fishing occupations, this does not negate its frequently exploitative nature. As attested by the prevalence of precarious conditions amongst tourism workers in the Global South 'decent work' in tourism can be hard to find (Lee et al., 2015, p.198). Neither does the fact that foreign-owned tourism corporations may pay higher wages than many small-scale locally owned firms or that wages may rise in line with growth and profits negate the need to interrogate the political-economic conditions under which surpluses are produced and distributed.

The UNWTO and sustainable tourism

The Brundtland Report (WCED, 1987) is commonly regarded as the launch pad for global political action to achieve sustainable development.[6] Although it fell short of mapping out an agenda for radical political reform it explicitly recognised the relationship between the "short-sighted" pursuit of prosperity, overuse of environmental resources and, poverty (WCED, 1987, p.27). However, the emphasis on continued growth as means of relieving poverty meant that tensions remain between economic growth and the equitable distribution of "life-sustaining resources and opportunities" (Goldman, 2011. p.2).

Agenda 21, a template for practical action on sustainable development agreed at the Rio Earth Summit in 1992, sought to address questions of equity through encouragement of civil society involvement in the design and implementation of a global sustainable development agenda. Somewhat paradoxically, trade liberalization across all economic sectors was endorsed as means of stimulating sustainable economic growth (UN, 1992). Meanwhile, proposals to set up a centre to monitor corporate malpractice and references to over-consumption in the industrialized countries were deleted from the draft convention as a result of lobbying by leading industrial nations (Hilary, 2013, pp.61–63).

During the period from 1992 until the 2012 Rio + 20 Earth Summit in Johannesburg the United Nations Commission on Sustainable Development (UNCSD) coordinated a series of annual multi-stakeholder meetings with representatives of civil society and other 'major groups' (labour unions, local governments, local communities and industry bodies etc.) in an attempt to forge a consensus around key issues and challenges for sustainable development and identify ways for governments to address these. In the lead up to the UNCSD-7 meeting in 1999 NGOs involved in campaigning on tourism-related issues were invited to communicate a common position on sustainable tourism that respected the divergent stances of Northern and Southern NGOs (CSD, 1999a).

Although the WTO provided some input into the UNCSD (see WTO, 1999a), at this stage it was unable to take a leading role at the UNCSD-7 meeting due to being neither an industry body nor an NGO. While differences emerged over the Industry-specific Agenda 21, supported by the World Travel and Tourism Council (WTTC)[7] and the International Hotel and Restaurant Association (IHRA) (see CSD, 1999b), versus Local Agenda 21 (a joint initiative of UNEP and ICLEI (Local Governments for Sustainability)[8] and supported by the NGO Tourism Caucus and Local Authorities, the WTO refrained from taking a firm position.

The multi-stakeholder dialogue on tourism at the UNCSD-7 meeting marked the zenith of NGO and civil society participation following the Rio summit (Dodds, 2019). It also marked a move away from the WTO's earlier commitment to a social vision of tourism development set out in the Manila Declaration (WTO, 1980) and Acapulco Document (WTO, 1982), towards an unequivocal commitment to neoliberal market-led principles (see Higgins-Desbiolles, 2006). Tensions between the NGO Tourism Caucus and the WTO over the neoliberal direction of its sustainability agenda at the World Ecotourism Summit in 2002, resulted in a permanent rift, marking the end of formal NGO participation in the high-level agenda setting framework. This signaled a major turning point for the WTO's engagement with an emerging global agenda for sustainable tourism. Hitherto, the WTO had been primarily concerned with the promotion of tourism as an instrument of economic development. In its report to UNCSD-7 the WTO (1990a, p.5) states that "market forces alone do not guarantee that tourism will be sustainable". This is then somewhat contradicted in the *Global Code of Ethics for Tourism* in which tourism is directly addressed as a "factor of sustainable development" (Article 3) but within the framework of a "market economy, private enterprise and free trade" (WTO, 1999b). The UNWTO's reinforced its commitment to trade liberalization in tourism services following World Trade talks in Doha in 2001 and Cancún in 2003, by which time it had become a specialized agency of the UN (Ferguson, 2007).

The UNWTO and the sustainable development goals: SDG 8

The United Nations 2030 Agenda for Sustainable Development is the basis of UNWTO's attempt to establish a common framework of action centred on principles of sustainable tourism. The SDGs represent a compromise negotiated and agreed between governments, the implementation of which falls to major industry associations and multi-stakeholder groups representing different sectors, in which corporate influence is significant (see Gleckman, 2016). Of the five over-arching sustainable development goals set out in the preamble of Agenda 2030[9] economic growth is not in fact addressed as an explicit goal (UN, 2015). However, the proclaimed necessity of growth is clearly stated in SDG 8 with the resolve to: "create conditions for sustainable, inclusive and sustained economic growth, shared prosperity and decent work for all" (https://sustainabledevelopment.un.org/sdg8). Here growth is reframed as an explicit *goal* of sustainable development, not just a means to achieving it, although no specific rationale is given for doing so other than the taken-for-granted assumption that growth and productivity improvements are essential for driving progress towards implementing the SDGs.

While the SDGs are construed as "integrated and indivisible" (UN, 2015), the UNWTO has chosen to focus specifically on SDG 8 with the aim to "Promote sustained, inclusive and sustainable economic growth, full and productive employment and decent work for all" along with SDGs 12 and 14 (UNWTO, 2017a, p.99). Tourism is explicitly addressed in Target 8.9: "to devise and implement policies to promote sustainable tourism that creates jobs and promotes local culture and products" (UNWTO, 2017a, p.99). This target is linked to two indicators:

- 8.9.1: Tourism direct GDP as a proportion of total GDP and in growth rate
- 8.9.2: Proportion of jobs in sustainable tourism industries out of total tourism jobs.

When charged by the UN to formulate measures for Indicators 8.9.1 and 8.9.2 by 2020, the UNWTO proposed to replace them with a single indicator (combing growth and decent work) entitled "Progress towards sustainable tourism" along with three additional sub-measures "that provide a good (conceptually precise and feasible) indication of the three dimensions of sustainable tourism (economic, social and environmental)" (UNWTO, 2018a). Additionally, the UNWTO proposed that the *economic* dimension of sustainability be measured by tourism's contribution to GDP, using the UNWTO's Tourism Satellite Accounting methodology. By way of justification the UNWTO claims that "While Target 8.9 has many parts, within the context of Goal 8 it is understood that its main focus is "promoting sustainable tourism" (UNWTO, 2018a, p.6). In addition to misrepresenting the parent goal, the relationship between growth and sustainable tourism remains unclear if not outright contradictory. The proposed measures themselves consist of little more than a restatement of growth-led economic orthodoxies through which tourism's contribution to economic development has been conventionally interpreted.

The prioritization of "sustained economic growth" demonstrates the degree to which the SDGs are framed by business-focused market principles in which the "priorities" of the corporate sector are often foregrounded (UNWTO , 2016, p.24). Constant reference is made to supporting and incentivizing the private sector and promoting a "pro-competitive and effective policy framework" in order to stimulate further growth (UNWTO, 2017a, p.19). Both the wider Agenda 2030 and the UNWTO's specific interpretation express and consolidate the institutionalization of a "normative neoliberalism" through which markets and competition have become the defining principle of a single, global political-economic system (Davies, 2016, pp.127-129).

Agenda 2030 demonstrates the continued hold of this logic through its explicit support (see SDG17.10) for "a universal, rules-based, open, non-discriminatory and equitable multilateral trading system under the World Trade Organization" (https://sustainabledevelopment.un.org/sdg17). The UNWTO (2017a, p.12) makes frequent reference to the need to "remove barriers to trade" with little evidence given to demonstrate how this enhances sustainable tourism development other than by virtue of tourism's potential to enable developing countries to "take part in the global economy" (UNWTO, 2017a, p.24). The *Framework Convention on Tourism Ethics* similarly encourages tourism multinationals to "promote local and sustainable consumption and production" and to avoid "excessive repatriation of their profits" in return for the "freedom to invest and trade" (UNWTO, 2017b, Art. 12.5, p.88). Nowhere is there any mention of intractable global inequalities of wealth and income (Hickel, 2017, pp.37-43), or the link between tourism growth, uneven distribution of wealth and excessive levels of consumption by a privileged minority of travellers.

Sustainable tourism and decent work

The idea of 'decent work' emerged in parallel to the broader debates on sustainability during the 1990s. In 1999 the International Labour Office defined decent work as "jobs of acceptable quality" and "respect for the fundamental rights of work" (ILO, 1999, pp.4-7). While such rights are seen as essential to enable workers to "claim a fair share of the wealth they have helped to generate" the ILO's definition remains framed by the contribution decent work can make to growth alongside the need to accommodate "business and employers concerns" (ILO, 1999, p.10). Decent work was subsequently incorporated into the MDGs and is now integral to UN Agenda 2030 as part of SDG 8, thanks in part to the persistent lobbying of the ILO itself.

The ILO's (2017, p.11) recent guidelines on decent work identify key challenges to be addressed in the areas of tourism and hospitality employment including, the high incidence of informal working arrangements, insecurity, poor working conditions, low wages, long working hours, high turnover rates, limited social protections and incidences of discrimination, exploitation and sexual harassment. It goes on to propose ways in which sustainable tourism policies could contribute to a decent work agenda related to Targets 8.9, 12.b and 14.7 (ILO, 2017, pp.15–16). However, despite acknowledging the need to strengthen labour protections, promoting social dialogue and the rights to collective bargaining the ILO's conception of decent work remains aligned with "enhancing tourism enterprise performance" and policies that promote increased "efficiency and productivity" (ILO, 2017, p.36, p.43). There is little engagement with the underlying causes of 'indecent' work much less an attempt to understand how increased productivity may in fact result from intensified forms of capitalist exploitation and integration of weaker, low income economies into global markets.

The UNWTO has, in its various declarations (WTO, 1980, 1999a, 2017), stressed that due attention be given to the rights of workers and the equitable distribution of wealth in tourism. In 2008 the UNWTO and ILO jointly agreed to support "dignified work" and advance the application of international labour standards but failed to identify an explicit programme of action to achieve this (Gascón, 2019). Despite the participation of trades unions in UNCSD-7 the systemic forces of labour exploitation in tourism remain unaddressed (CSD, 1999c). Much of the UNWTO's focus has been on developing statistical formulae for calculating tourism employment with little regard to develop more robust criteria for the evaluation of decent work. Not only does Indicator 8.9.2 reduce the interpretation of decent work merely to the "number of jobs in tourism industries as a proportion of total jobs and growth rate of jobs, by sex", the UNWTO dilutes the meaning of sustainable tourism even further by suggesting that the *social* dimension of tourism sustainability can be comprehensively assessed through measures of tourism employment (UNWTO, 2018a, p.5)

To reduce the evaluation of decent work to a simple calibration of the numbers employed in the tourism industries disregards the interplay between the organization of work in tourism economies and the class dynamics of capitalist labour relations (see Selwyn, 2014). The limited horizons of the UNWTO's interpretation of dignified and decent work is reflected in Target 8.7 which deals with the elimination of forced labour, modern slavery and human trafficking and securing the prohibition and elimination of the worst forms of child labour (UNWTO, 2017a, p.99). As serious as these issues are, this construes decent work predominantly as a moral issue disregarding how such human rights abuses and exploitative labour regimes may be linked to the hollowing out of social protections and dismantling of workplace rights as part of the 'normal' workings of tourism capitalism.

The UNWTO nevertheless recognizes that, "tourism's ability to bolster decent work is a complex issue" and acknowledges the challenges presented by the fragmented character of tourism labour regimes and conditions of work (UNWTO, 2018b, pp. 55-56). However, the recent Framework Convention suggests that the protection of job security and social protection for workers can only be guaranteed "so far as possible" (UNWTO, 2017, p.87). This exempts the tourism industry from respecting universally acknowledged workers' rights by confining these within "the specific constraints linked in particular to the seasonality of their activity, the global dimension of their industries and the flexibility often required of them by the nature of their work" (UNWTO, 2017b: Art 12.1, p. 87). As if to further demonstrate the attenuated definition of decent work and its subordination to growth the current UNWTO Secretary-General failed to include decent work amongst the most pressing challenges facing the tourism industry (Fingar, 2018), while not one of the 21 recommendations in the Chengdu Declaration on Tourism and the SDGs address decent work (UNWTO, 2017c).

Discussion and critique

The UNWTO's SDG agenda prioritizes the sustained growth of a 'green' tourism economy that will contribute to "economically viable and robust growth, decent work creation, poverty alleviation, improved efficiency in resource use and reduced environmental degradation" (Stroebel, 2015, p.2226). As if to reinforce the limited prospects for a genuine rethinking of the tourism political economy the notion of green growth has stimulated a "green gold rush" of foreign investment in "responsible" and "sustainable" tourism assets (Mitchell, 2017). Meanwhile global spending on 'ecotourism' – still largely associated with sustainability (UNWTO, 2017a, p.26) – outstrips aggregate investment in the industry (UNEP & UNWTO, 2012, p. viii).

For the UNWTO "growth is not the enemy" (Rifai, 2017). Rather, the adverse consequences of tourism growth can be addressed through the sustainable management of growth and concomitant embrace of "more inclusive" and "sustainable" business models (UNWTO, 2017a, p. 12). As indicated by Target 8.4 it is expected that the negative externalities of tourism growth can be addressed through "decoupling" growth from ecologically unsustainable resource use (UNWTO, 2017a, p. 28). While there has been some evidence of the relative decoupling of growth from resource degradation between 1980 and 2002, the material footprint of growth has since accelerated (Hickel, 2019b, p.3). Moreover, claims that many rich nations have achieved lower carbon emissions despite higher GDP, ignores the degree to which emissions have been out-sourced through global supply chains or that aviation and shipping emissions are not included in these calculations.

The UNWTO (2017a, p.27) makes explicit mention of the potential for "innovation and new technologies" to stimulate new business models and "efficiency gain". There is however no mention of the associated costs in terms of unsustainable resource use, residential displacement and gentrification pressures brought about by the rise of digital rental platforms and the attendant concentration of corporate power in the so-called 'sharing economy'. As welcome as they are, proposals to account for the hidden costs or externalities produced by tourism (see Epler Wood, Milstein, & Ahamed-Broadhurst, 2019), construe nature as an "ecosystem service" (Sala, 2011), reducing the natural world and all its complexity to merely another form of capital that is substitutable for another (Fioramonti, 2013, p.89).

The model of "inclusive green growth" espoused by the UNWTO (2017a, p.7) is rooted in the World Bank's paradigm of "inclusive growth" which calls for accelerated growth to reduce poverty (World Bank, 2009, p.1). The UNWTO (2017a, p.7) is also at pains to draw connections between the SDGs and "sustainable business operations that can spur competition and increase profit". There are numerous references to the "priorities" of the private sector, including those of the Spanish tourism sector (UNWTO , 2016, p.24). But there is little indication of the diverse capitals that have shaped the structure of Spanish tourism, much less acknowledgement of the damage to the environment and social fabric wrought by the speculative investments of real estate and construction capital that drives the growth of tourism (Murray Mas, Cadena, & Blázquez-Salom, 2017).

In its continued support for trade liberalization the UNWTO's SDG-led agenda envisages the further integration of small-scale producers in developing countries into global markets as a means of enhancing the competitiveness and export-led tourism growth (UNWTO, 2017a, p.24). Inclusiveness is envisaged in terms of facilitating access of producers to global markets and increasing employment opportunities for the poor without challenging existing hierarchies of wealth and power. There is no acknowledgement of how trade liberalization and increased foreign investment, the logic of which pervades inclusive growth discourses (Saad-Filho, 2010), may accentuate the flow of benefits to large private capital at the expense of such small-scale producers and workers (Schilcher, 2007).

Governed by liberal market pragmatism the UNWTO disregards the tenuous link between tourism growth and reduction of poverty and/or inequalities (see Hampton, Jeyacheya, & Long, 2018), as well as the threat to the livelihoods and resources of peasant and labouring classes entailed by the drive for the sustained growth of tourism (Devine, 2017). The idea that competitiveness is intrinsic to sustainable tourism ignores the fact that open markets and liberal trade regimes enables mobile transnational capital to seek absolute advantage, benefitting global investors and corporations at the expense of domestic enterprise and indebted states in the Global South. The superior bargaining power of the former enables them to weaken the regulatory environment or acquire lucrative assets at a favourable price. This is particularly so where debt restructuring has impoverished national governments as in the case of the sale of state-owned Hotel Montelimar to the Barceló group in Nicaragua (Buades, 2009, pp. 69–72).

The emphasis on inclusiveness also implies a democratic and participatory approach to economic development. However, the UNWTO merely urges tourism companies to "encourage multi-stakeholder initiatives" (UNWTO , 2016, p. 39). Despite calls for the active participation of civil society "partners" alongside other stakeholders in formulating SDG-framed policy for tourism, it is telling that the UNWTO sees the role of civil society as one of largely supporting and encouraging the private sector to incorporate sustainability in their business models (UNWTO, 2017a, p.19). The prioritization of corporate interests is further reinforced in the *Framework Convention on Tourism Ethics* which encourages the development of partnerships between enterprises of generating and receiving countries and corporate support "the equitable distribution of the benefits of its growth" rather than address the complex relationship between tourism growth and poverty itself (UNWTO, 2017b: Art 12.6, p. 9). There is no discussion of the considerable constraints to achieving such an equitable distribution imposed by a coercive rules-based international trade regime nor critical scrutiny of how unequal economic relations are produced and reproduced throughout the world trade system and global corporate commodity chains.

The concept of inclusive growth also betrays limited scope of ambitions regarding the role of labour and decent work within a sustainable (tourism) economy. The horizons of decent work are limited to being "inclusive of the large part of the country's labor force" and "productive employment rather than income redistribution" (World Bank, 2009, p.4). This ignores the fact that low wages are significantly determined by the weak bargaining power of workers and the concomitant ability of businesses to impose low wages and harsh working conditions on them (Selwyn, 2014). The conceptual invisibility of contested class relations in sustainable tourism are reflected in the praise given for the corporate social responsibility policies of the NH Hotel Group (UNWTO , 2016, p.41), which has been heavily criticized for outsourcing its cleaning operations and making redundancies while simultaneously increasing profits by €76 m between 2014 and 2016 (Burgen, 2017).

Growth serves both as the principal mechanism through which capitalism's continually expands as well as an ideological tool foreclosing demands for redistributive justice by aligning the interests of labouring classes with capital. Indeed, challenges to restrictive environmental measures by developers have often been justified in the name of job creation and economic development. On occasion this has led to clashes between environmentalists and workers as occurred in the Hilton construction project in Malta during the 1990s (Boissevain & Theuma, 1998). However, to interpret such conflicts as simply a trade-off between environmental conservation and economic development ignores how the schism between workers and environmentalists is produced out of unequal processes of capitalist development in which workers' ability to make meaningful choices over their livelihoods are constrained. Furthermore, it is often peasant farmers, fishers, indigenous peoples and workers who bear the brunt of the impact of environmental degradation resulting from urban development and tourism.

Conclusion

This paper has argued that the UNWTO's SDG agenda, specifically the notions of 'sustained' and 'inclusive' growth intrinsic to SDG 8, is marked by unresolved tensions between sustainable development and sustainable tourism. The analysis and critique presented here contends that the principles of 'inclusive growth' and 'decent work' espoused by the UNWTO do little to address the contradictions and tensions inherent in the logics of growth and processes of capital accumulation that drive the growth and expansion of tourism. In its belief that sustainability can be reconciled with a growth-led 'inclusive' market capitalism, the UNWTO's proposed revisions to SDG 8 and its broader sustainability agenda remain blind to the injustices that are intrinsic to the systemic processes of exploitation characteristic of tourism capitalism. The horizons of inclusiveness envisaged in their agenda is reliant upon the market to deliver wealth redistribution through sustainable growth. This leaves unaddressed the question of widening and deepening democratic and cooperative forms of ownership, the control and use of productive assets, including in the workplace itself, and the potential for fomenting diverse arrangements of tourism production, thus naturalizing capitalist property relations and inequalities in the distribution of resources. Moreover, there is little to indicate that the UNWTO's SDG agenda has begun to grapple with the increasingly predatory modes of profit extraction and class struggles that characterise emergent variants of authoritarian statist capitalism and its role in driving tourism growth in emerging economies.

The paper has also highlighted the UNWTO's role in marginalising discordant voices from civil society in the process of shaping and subsequently leading the global sustainable tourism agenda that culminated in its current SDG-led agenda. Despite the involvement of civil society in the UNCSD-7 the UNWTO has consistently failed to create an institutional framework that would advance the democratic participation of disempowered groups in ways that not only foster dialogue but would also enable such groups to exert influence on the policies and strategic direction of the institution. By promoting the idea that the interests of corporations can be balanced with workers and local communities without addressing the forces sustaining existing corporate ownership structures and the class dynamics of capital accumulation that drives endless growth, the UNWTO's SDG agenda promises to do little more than reinforce existing hierarchies of political-economic power. In this regard, the authors have highlighted the limited ambitions of decent work as construed by the UNWTO's SDG-agenda. To date, the UNWTO has yet to devise more convincing and robust indicators for such a critical element of sustainable tourism. In so doing the UNWTO remains conceptually blind to the exploitative relations of capitalist development that produce and sustain indecent work in tourism.

The ability of workers to negotiate decent working conditions is furthermore seen as something to be delivered by a benevolent state and/or humane employers. Workers are regarded as a factor of production and/or one amongst many interest groups whose demands are to be set alongside those of states, corporations and entrepreneurs whose superior power and ability to shape the organization of production is rarely addressed. This shifts the focus away from the exploitation of labour and expropriation of the commons in the pursuit of profit and the constellations of class power that drive the growth of tourism. With regard to possible challenges to this agenda, tourism *degrowth* advocates (Higgins-Desbiolles et al., 2019; Hall, 2009), rightly emphasise the need to reconcile degrowth strategies with fairness, equity and justice to avoid the trap of elitist environmentalism that marginalises the poor and labouring classes. Accordingly, tourism degrowth strategies need to be coupled with a class-relational conception of justice and radical redistribution of power and wealth (not merely a critique of neoliberalism) that involves a more robust interrogation of the industrial organisation and processes of capital accumulation that drive the growth of tourism.

Language and verbal imagery too (note the compelling visual logos for the SDGs) exercise a profound influence on human cognitive understanding, particularly in the realm of abstract economic ideas (Raworth, 2018). Through its prolific use of such adjectives as 'sustained', 'inclusive' and 'decent' to qualify its fully-fledged drive for continual growth the UNWTO's SDG-led agenda is not only misguided but dangerous. Discourses of inclusive growth and decent work shape the distinctive grammar and communicative logics through which markets and growth are normalized in the UNWTO's SDG agenda. This of course is no accident and can perhaps also be interpreted as a political attempt by the UNWTO to steer the dialogue on sustainability away from the questions of environmental and social justice in tourism that would necessitate a radical reorganization of the political and economic structures that drive and sustain the growth-led dynamics of tourism capital accumulation. Further to the critique presented here of the market orthodoxy that frames institutional discourses of decent work and its application to the UNWTO's sustainable tourism agenda, there is scope for closer inspection of the diverse cultural understandings of decent work that may be excluded from such discourses. However, recognition of the diverse social contexts and cultural differences that shape tourism labour markets must not come at the expense of affirming the universal struggle of tourism workers and destination communities to defend their well-being and live free from exploitation.

Notes

1. The analysis draws on recent UNWTO reports on the SDGs while also identifying important discursive and ideological continuities with other key reports approved by the General Assembly of the UNWTO such as the UNWTO *Global Code of Ethics* (WTO, 1999b) and its successor the *Framework Convention on Tourism Ethics* (UNWTO 2017b). It is important to note however that these documents do not always reflect the official position of the UNWTO.
2. Recent calculations suggest that international tourism accounts for 8 per cent of global GHG emissions (Lenzen et al., 2018).
3. SDG1 uses the standard international benchmark for 'extreme poverty' of US$1.90 (updated from US$1.25 per day in 2015). Even if the logics of growth-driven market capitalism are accepted Woodward (2015) contends that to eradicate global poverty (at the previous poverty line of US$1.25 per day) would take at least 100 years at a rate of 2.35 per cent per capita growth.
4. Simon Kuznets, the economist who devised GDP metrics in the 1930s, later to sought to exclude the gains from speculative financial transactions from national income (Fioramonti, 2014: 57-8).
5. There are few major tourism, hospitality and resort corporations amongst the world's top 100 corporations. An exception is Booking Holdings, the world's largest on-line travel retailer whose market value (US$93bn) increased by a remarkable 3,233 percent between 2009 and 2015 (PwC, 2018).
6. US industrial interests rejected any reference to 'sustainability' at the 1972 UN Conference on the Human Environment held in Stockholm (Fioramonti, 2013, p.84).
7. The WTTC is a global coalition of corporate executives formed in London in 1990 representing 170 of the world's largest tourism and hospitality companies.
8. ICLEI is a global network of 1,500+ cities, towns and regions committed to building a sustainable future.
9. The five major goals are: to end poverty and hunger everywhere; to combat inequalities within and among countries; to build peaceful, just and inclusive societies; to protect human rights and promote gender equality and the empowerment of women and girls; to ensure the lasting protection of the planet and its natural resources.

Disclosure statement

No potential conflict of interest was reported by the authors.

References

Alam, M. S., & Paramati, S. R. (2016). The impact of tourism on income inequality in developing economies: Does Kuznets curve hypothesis exist? *Annals of Tourism Research, 61* (November), 111–126. doi:10.1016/j.annals.2016.09.008

Banarjee, B. (2003). Who sustains whose development? Sustainable development and the reinvention of nature. *Organization Studies, 24*(1), 143–180.

Baum, T., Cheung, C., Kong, H., Kralj, A., Mooney, S., Nguyễn Thị Thanh, H., … Siow, M. (2016). Sustainability and the tourism and hospitality workforce: A thematic analysis. *Sustainability, 8*(8), 809. doi:10.3390/su8080809

Bianchi, R. V. (2018). The political economy of tourism development: A critical review. *Annals of Tourism Research, 70* (May), 88–102. doi:10.1016/j.annals.2017.08.005

Blázquez Salom, M., Artigues Bonet, A. A., & Yrigoy Cadena, I. (2015). Crisis y planificación territorial turística en las Islas Baleares. *Investigaciones Turísticas, 9*, 24–49.

Boissevain, J., & Theuma, N. (1998). Contested space: Planners, tourists and environmentalists in Malta. In S. Abram & Waldren, J. (Eds.), *Anthropological Perspectives on Local Development* (pp. 96–119). London: Routledge.

Bramwell, B. (2011). Governance, the state and sustainable tourism: A political economy approach. *Journal of Sustainable Tourism, 19*(4-5), 459–477. doi:10.1080/09669582.2011.576765

Bramwell, B. (2015). Theoretical activity in sustainable tourism research. *Annals of Tourism Research, 54* (September), 204–218. doi:10.1016/j.annals.2015.07.005

Bramwell, B., & Lane, B. (1993). Sustainable tourism: An evolving global approach. *Journal of Sustainable Tourism, 1*(1), 1–5. doi:10.1080/09669589309450696

Briassoulis, H. (2002). Sustainable tourism and the commons. *Annals of Tourism Research, 29*(4), 1065–1085. doi:10.1016/S0160-7383(02)00021-X

Britton, S. G. (1980). The evolution of a colonial-space economy: The case of Fiji. *Journal of Historical Geography, 6*(3), 251–274. doi:10.1016/0305-7488(80)90081-X

Britton, S.G. (1991). Tourism, capital, and place: Towards a critical geography of tourism. *Environment and Planning D: Society and Space, 9*(4), 451–478. doi:10.1068/d090451

Buades, J. (2009). *Do not disturb Barceló*. Barcelona: Icaria.

Burgen, S. (2017). Chambermaids' revolt aims to shame Spain's rogue employers. *The Guardian*, 15 July, Retrieved 3 September 2019, from https://www.theguardian.com/travel/2017/jul/15/spanish-hotel-cleaners-fightback-exploitation.

Burnham, P. (1994). Open Marxism and vulgar political economy. *Review of International Political Economy, 1*(2), 221–231. doi:10.1080/09692299408434277

Büscher, B., & Fletcher, R. (2017). Destructive creation: Capital accumulation and the structural violence of tourism. *Journal of Sustainable Tourism, 25* (5), 651–667. doi:10.1080/09669582.2016.1159214

Cañada, E. (2014). Los que limpian los hoteles. *El País*, 30 July, Retrieved 20 July 2018, from https://elpais.com/elpais/2014/07/30/alterconsumismo/1406706574_140670.html.

Cave, J., & Dredge, D. (2018). Reworking tourism: Diverse economies and in a changing world. *Tourism Planning & Development, 15*(5), 473–477. doi:10.1080/21568316.2018.1510659

Chibber, V. (2013). *Postcolonial theory and the specter of capital*. London and New York: Verso.

Chin, C. (2008). *Cruising in the global economy: Profits, pleasure, and work at sea*. Farnham: Ashgate.

Christian, M. (2016). Tourism global production networks and uneven social upgrading in Kenya and Uganda. *Tourism Geographies, 18*(1), 38–58. doi:10.1080/14616688.2015.1116596

Chok, S. (2009). Mega development projects and labour supply chains: Whose responsibility is it? *CSR Asia Weekly, 5*(1), 4–5.

Clancy, M. (2001). *Exporting paradise: Tourism and development in Mexico*. Oxford: Elsevier.

Cole, S. (2012). A political ecology of water equity and tourism: A case study from Bali. *Annals of Tourism Research, 39*(2), 1221–1241. doi:10.1016/j.annals.2012.01.003

CSD. (1999a). *Tourism and sustainable development: A non-governmental organizational perspective*. Background Paper No. 4. UNCSD NGO Steering Committee, Seventh Session of the Commission on Sustainable Development, New York, 19-30 April, United Nations Department of Economic and Social Affairs. Retrieved 14 December 2018, from https://sustainabledevelopment.un.org/content/documents/398wttc.pdf.

CSD. (1999b). *Tourism and sustainable development: The global importance of tourism*. Background Paper No. 1, World Travel and Tourism Organization and International Hotel and Restaurant Association, Seventh Session of the Commission on Sustainable Development, New York, 19-30 April, United Nations Department of Economic and Social Affairs. Retrieved from https://sustainabledevelopment.un.org/content/documents/398wttc.pdf.

CSD. (1999c). *Tourism and sustainable development: Workers and trade unions in the web of tourism*. Background Paper No. 2, International Confederation of Free Trade Unions and Trade Union Advisory Committee to the OECD, Seventh Session of the Commission on Sustainable Development, New York, 19-30 April, United Nations Department of Economic and Social Affairs. Retrieved 14 December 2019, from https://sustainabledevelopment.un.org/content/documents/398wttc.pdf.

Crick, M. (1995). *Resplendent sites, discordant voices: Sri Lankans and international tourism*. Chur: Harwood Academic Publishers.

Coria, J., & Calfucura, E. (2012). Ecotourism and the development of indigenous communities: The good, the bad, and the ugly. *Ecological Economics*, *73*, 47–55. doi:10.1016/j.ecolecon.2011.10.024

Davies, W. (2016). The new neoliberalism. *New Left Review*, *121*, 121–134.

De Angelis, M. (2007). *The beginning of history: Value struggles and global capital*. London and Ann Arbor, MI: Pluto Press.

Devine, J. A. (2017). Colonizing space and commodifying place: Tourism's violent geographies. *Journal of Sustainable Tourism*, *25*(5), 634–650. doi:10.1080/09669582.2016.1226849

Dodds, F. (2019). *Stakeholder democracy: Represented democracy in a time of fear*. London and New York: Routledge.

Douthwaite, R. (1999). Is it possible to build a sustainable world? In R. Munck & D. O'Hearn (Eds.), *Critical Development Theory: Contributions to a New Paradigm* (pp. 157–177). London: Zed Books.

Dredge, D., & Gyimóthy, S. (2015). The collaborative economy and tourism: Critical perspectives, questionable claims and silenced voices. *Tourism Recreation Research*, *40*(3), 286–302. doi:10.1080/02508281.2015.1086076

Dredge, D., & Jamal, T. (2013). Mobilities on the Gold Coast, Australia: Implications for destination governance and sustainable tourism. *Journal of Sustainable Tourism*, *21*(4), 557–579. doi:10.1080/09669582.2013.776064

Duffy, R. (2015). Nature-based tourism and neoliberalism: Concealing contradictions. *Tourism Geographies*, *17* (4), 529–543. doi:10.1080/14616688.2015.1053972

Epler Wood, M., Milstein, M., & Ahamed-Broadhurst, K. (2019). *Destinations at risk: The invisible burden of tourism*. Bristol: The Travel Foundation.

FDI Intelligence. (2018). *Tourism investment 2018*. London: Financial Times.

Ferguson, L. (2007). The United Nations World Tourism Organisation. *New Political Economy*, *12*(4), 557–568. doi:10.1080/13563460701661587

Fingar, L. (2018). UNWTO secretary general puts employment first. *FDI Intelligence*, 12 April. Retrieved 18 July 2018, from https://www.fdiintelligence.com/Sectors/Hotels-Tourism/UNWTO-secretary-general-puts-employment-first.

Fioramonti, L. (2013). *Gross domestic problem: The politics behind the world's most powerful number*. London: Zed Books.

Fletcher, R. (2011). Sustaining tourism, sustaining capitalism? The tourism industry's role in global capitalist expansion. *Tourism Geographies*, *13*(3), 443–461. doi:10.1080/14616688.2011.570372

Gascón, J. (2019). Tourism as a right: "frivolous claim" against degrowth? *Journal of Sustainable Tourism*, *27*(12), 1825–1838. doi:10.1080/09669582.2019.1666858

Gibson, C. (2009). Geographies of tourism: Critical research on capitalism and local livelihoods. *Progress in Human Geography*, *33*(4), 527–534. doi:10.1177/0309132508099797

Gleckman, H. (2016). Multi-stakeholder governance: A corporate push for a new global governance. *Open Democracy*, 19 January. Retrieved 10 November 2019, from https://www.opendemocracy.net/en/multi-stakeholder-governance-corporate-push-for-new-global-governance/.

Goldman, B. A. (2011). Equity and the 1992 Rio Earth Summit. *Fordham Environmental Law Review*, *4*(1), 1–8.

Gössling, S., & Hall, C.M. (2019). Sharing versus collaborative economy: How to align ICT development and the SDGs in tourism?. *Journal of Sustainable Tourism*, *27*(1), 74–96. doi:10.1080/09669582.2018.1560455

Grossman, P., & Greenfield, G. (2006). Financialization: New Routes to Profit, New Challenges for Trade Unions. *Labour Education, The Quarterly Review of the ILO Bureau for Workers' Activities*, 1/2006, No. 142. Retrieved 20 June 2016, from http://www.iufdocuments.org/www/documents/Financialization-e.pdf.

Hall, C. M. (2009). Degrowing tourism: Décroissance, sustainable consumption and steady-state tourism. *Anatolia*, *20*(1), 46–61. doi:10.1080/13032917.2009.10518894

Hall, C. M. (2019). Constructing sustainable tourism development: The 2030 agenda and the managerial ecology of sustainable tourism. *Journal of Sustainable Tourism*, *27*(7), 1044–1060. doi:10.1080/09669582.2018.1560456

Hampton, M. P. (2010). Enclaves and ethnic ties: The local impacts of Singaporean cross-border tourism in Malaysia and Indonesia. *Singapore Journal of Tropical Geography*, *31*(2), 239–253. doi:10.1111/j.1467-9493.2010.00393.x

Hampton, M. P., Jeyacheya, J., & Lee, D. (2018). The political economy of dive tourism: Precarity at the periphery in Malaysia. *Tourism Geographies*, *20*(1), 107–126. doi:10.1080/14616688.2017.1357141

Hampton, M. P., Jeyacheya, J., & Long, P. H. (2018). Can tourism promote inclusive growth? Supply chains, ownership and employment in Ha Long Bay, Vietnam. *The Journal of Development Studies*, *54*(2), 359–376. doi:10.1080/00220388.2017.1296572

Hickel, J. (2017). *The divide: A brief guide to global inequality and its solutions*. London: William Heinemann.

Hickel, J. (2019a). Degrowth: A theory of radical abundance. *Real World Economics Review, 87*, 54–68. Retrieved 17 January 2020, from http://www.paecon.net/PAEReview/issue87/Hickel87.pdf.
Hickel, J. (2019b). The contradiction of the sustainable development goals: Growth versus ecology on a finite planet. *Sustainable Development*, 1–12. doi:10.1002/sd.1947
Higgins-Desbiolles, F. (2006). More than an industry: Tourism as a social force. *Tourism Management, 27*(6), 1192–1208. doi:10.1016/j.tourman.2005.05.020
Higgins-Desbiolles, F. (2018). Sustaining tourism: Sustaining tourism or something else? *Tourism Management Perspectives, 25*, 157–160. doi:10.1016/j.tmp.2017.11.017
Higgins-Desbiolles, F., Carnicelli, S., Krolikowski, C., Wijesinghe, G., & Boluk, K. (2019). Degrowing tourism: Rethinking tourism. *Journal of Sustainable Tourism, 27*(12), 1926–1944. doi:10.1080/09669582.2019.1601732
Hilary, J. (2013). *The poverty of capitalism: The economics of meltdown and the struggle for what comes next*. London: Pluto Press.
Hof, A., & Blázquez-Salom, M. (2015). Changing tourism patterns, capital accumulation, and urban water consumption in Mallorca, Spain: A sustainability fix? *Journal of Sustainable Tourism, 23*(5), 770–796. doi:10.1080/09669582.2014.991397
Holden, A. (2003). In need of new environmental ethics for tourism? *Annals of Tourism Research, 30*(1), 94–108. doi:10.1016/S0160-7383(02)00030-0
Holden, A. (2013). *Tourism, poverty and development*. London: Routledge.
Hunter, C. (1997). Sustainable tourism as an adaptive paradigm. *Annals of Tourism Research, 24*(4), 850–867. doi:10.1016/S0160-7383(97)00036-4
ILO. (1999). Decent work. Report of the Director-General. Report of the 87th session. Geneva: International Labour Office. Retrieved 11 November 2019, from https://www.ilo.org/public/english/standards/relm/ilc/ilc87/reports.htm.
ILO. (2010). *Developments and challenges in the hospitality and tourism sector*. Geneva: International Labour Office.
ILO (2012). *Migrant workers in the international hotel industry*. International Migration Paper No. 112. Geneva: International Labour Office.
ILO. (2017). *ILO Guidelines on decent work and socially responsible tourism*. Geneva: International Labour Office.
Jamal, T. (2019). *Justice and ethics in tourism*. London and New York: Routledge.
Jamal, T., & Camargo, B. (2014). Sustainable tourism, justice and an ethic of care: Toward the Just Destination. *Journal of Sustainable Tourism, 22*(1), 11–30. doi:10.1080/09669582.2013.786084
Keller, P. & Bieger, T. (Eds.). (2011). Introduction. In P. Keller, & Bieger, T. (Eds.), *Tourism development after the crises global imbalances – Poverty alleviation*. (pp. 1–25). Berlin: Erich-Schmidt Verlag.
Klein, N. (2014). *This changes everything: Capitalism vs the climate*. London: Penguin.
Ladkin, A. (2011). Exploring tourism labor. *Annals of Tourism Research, 38* (3), 1135–1155. doi:10.1016/j.annals.2011.03.010
Lee, D., Hampton, M., & Jeyacheya, J. (2015). The political economy of precarious work in the tourism industry in small islands states. *Review of International Political Economy, 22*(1), 194–223. doi:10.1080/09692290.2014.887590
Lenzen, M., Sun, Y.-Y., Faturay, F., Ting, Y.-P., Geschke, A., & Malik, A. (2018). The carbon footprint of global tourism. *Nature Climate Change, 8*(6), 522–528. doi:10.1038/s41558-018-0141-x
McCool, S. F., Moisey, N., & Nickerson, N. P. (2001). What should tourism sustain? The disconnect with industry perceptions of useful indicators. *Journal of Travel Research, 40*(2), 124–131. doi:10.1177/004728750104000202
Madsen Camacho, M. E. (1996). Dissenting workers and social control: A case study of the hotel industry in Huatulco, Oaxaca. *Human Organization, 55*(1), 33–40. doi:10.17730/humo.55.1.76086h6l775457g1
Marx, K. (1973). *Grundrisse*. Harmondsworth: Penguin.
Marx, K. (1974). *Capital: A critical analysis of capitalist production*. Vol. 1. London: Lawrence and Wishart.
Marx, K. (1977). The German Ideology. In D. McLellan (Ed.), *Karl Marx: Selected writings* (pp. 159–191). Oxford: Oxford University Press.
Mitchell, J. (2017). Tourism's green gold rush. FDI Intelligence, 14 December. Retrieved 14 August 2019, from https://www.fdiintelligence.com/Sectors/Hotels-Tourism/Tourism-s-green-gold-rush.
Mittal, A., & Fraser, E. (2018). *Losing the Serengeti: The Maasai land that was to run forever*. Oakland, CA: The Oakland Institute.
Mosedale, J. (2006). Tourism commodity chains: Market entry and their effects on St. Lucia. *Current Issues in Tourism, 9* (4-5), 436–458. doi:10.2167/cit271.0
Murray Mas, I., Yrigoy, I. Y., & Blázquez-Salom, M. (2017). The role of crises in the production, destruction and restructuring of tourist spaces. The case of the Balearic Islands. *Revista Investigaciones Turísticas, 13*, 1–29.
Müller, H. (1994). The thorny path to sustainable tourism development. *Journal of Sustainable Tourism, 2*(3), 131–136.
Neto, F. (2003). *A New approach to sustainable tourism development: Moving beyond environmental protection*. DESA Discussion Paper 29. New York: United Nations.
Peters, M. (1969). *International tourism: The economics and development of the international tourist trade*. London: Hutchinson.

PwC. (2018). *Global top 100 companies by market capitalisation*. London: Price Waterhouse Cooper. IPO publications.
Raworth, K. (2018). *Doughnut economics: Seven ways to think like a 21st century economist*. London: Random House.
Rifai, T. (2017). Tourism: Growth is not the enemy; it's how we manage it that counts. *UNWTO Press Release*. August 15. Retrieved 14 December, from http://media.unwto.org/press-release/2017-08-15/tourism-growth-not-enemy-it-s-how-we-manage-it-counts.
Saad-Filho, A. (2010). *Growth, poverty and inequality: From Washington Consensus to inclusive growth*. UN/DESA Working Paper 100. November. New York: United Nations Department of Economic and Social Affairs.
Saarinen, J. (2006). Traditions of sustainability in tourism studies. *Annals of Tourism Research*, *33*(4), 1121–1140. doi: 10.1016/j.annals.2006.06.007
Saarinen, J. (2018). Beyond growth thinking: The need to revisit sustainable development in tourism. *Tourism Geographies*, *20*(2), 337–340. doi:10.1080/14616688.2018.1434817
Sala, E. (2011). Ocean ecosystem services can increase — But only if we take less. *National Geographic*, 5 April. Retrieved from https://blog.nationalgeographic.org/2011/04/05/ocean-ecosystem-services-can-increase-but-only-if-we-take-less/.
Scheyvens, R. (1999). Ecotourism and the empowerment of communities. *Tourism Management*, *20*(2), 245–249. doi: 10.1016/S0261-5177(98)00069-7
Scheyvens, R. (2007). Exploring the tourism-poverty nexus. *Current Issues in Tourism*, *10*(2), &3), 231–254. doi:10.2167/cit318.0
Schilcher, D. (2007). Growth versus equity: The continuum of pro-poor tourism and neoliberal governance. *Current Issues in Tourism*, *10*(2-3), 166–193. doi:10.2167/cit304.0
Selwyn, B. (2014). How to achieve decent work? Global Labour Column, No. 161, January.
Selwyn, B. (2015). Twenty-first century International political economy: A class-relational perspective. *European Journal of International Relations*, *21*(3), 513–537. doi:10.1177/1354066114556659
Sharpley, R. (2015). Tourism: A vehicle for development? In R. Sharpley & D. J. Telfer, (Eds.) *Tourism and Development: Concepts and Issues* (pp. 3–30). Clevedon: Channel View.
Sheikh, A. (2010). Outsourcing of employers' responsibility in the false name of flexibility. *Employee Relations*, *32*(5), 495–508. doi:10.1108/01425451011061658
Shivji, I. (1973). *Tourism and socialist development*. Dar es Salaam: Tanzania Publishing House.
Steiner, C. (2006). Tourism, poverty reduction and the political economy: Egyptian perspectives on tourism's economic benefits in a semi-rentier state. *Tourism and Hospitality Planning & Development*, *3*(3), 161–177. doi:10.1080/14790530601132286
Streeck, W. (2014). *Buying time: The delayed crisis of democratic capitalism*. London and New York: Verso.
Stroebel, M. (2015). Tourism and the green economy: Inspiring or averting change?. *Third World Quarterly*, *36*(12), 2225–2243. doi:10.1080/01436597.2015.1071658
Tao, T. C. H., & Wall, G. (2009). Tourism as a sustainable livelihood strategy. *Tourism Management*, *30*(1), 90–98. doi: 10.1016/j.tourman.2008.03.009
Teberga De Paula, A. (2018). Un panorama sobre la precarización del trabajo turístico en Brasil. *Alba Sud*, 23 June. Retrieved 23 July 2018, from http://www.albasud.org/noticia/es/1039/un-panorama-sobre-la-precarizaci-n-del-trabajo-tur-stico-en-brasil.
UN. (1992). *Agenda 21*. United Nations Conference on Environment & Development. Rio de Janerio, Brazil, 3-14 June 1992. Retrieved 23 October 2019, from https://sustainabledevelopment.un.org/content/documents/Agenda21.pdf.
UN. (2000). *United Nations millennium declaration*. Resolution adopted by the UN General Assembly, 55/2. Retrieved 21 September 2018, from http://www.un.org/millennium/declaration/ares552e.htm.
UN. (2015). *Transforming our world: The 2030 agenda for sustainable development*. New York: Resolution adopted by the General Assembly on 25 September 2015. Retrieved 21 September 2018, from https://sustainabledevelopment.un.org/post2015/transformingourworld.
UN News. (2011). Tourism an important force to reduce poverty and foster global solidarity – UN. Retrieved 10 September 2018, from https://news.un.org/en/story/2011/09/389282-tourism-important-force-reduce-poverty-and-foster-global-solidarity-un.
UNCTAD. (1973). *Elements of tourism policy for developing countries*. New York: United Nations Publications TD/B/C.3/89/Rev.1.
UNCTAD (2010). *The contribution of tourism to trade and development*. 3-7 May 2010. Geneva: United Nations Conference on Trade and Development.
UNEP & UNWTO. (2012). *Tourism in the green economy – Background report*. Madrid. UNWTO.
UNWTO (2016). *The tourism sector and the sustainable development goals*. Madrid: UNWTO.
UNWTO. (2017a). *Tourism and the sustainable development goals – Journey to 2030*. Madrid: UN World Tourism Organisation.
UNWTO. (2017b). Framework convention on tourism ethics. Madrid: UN World Tourism Organisation. Retrieved 20 September, from http://cf.cdn.unwto.org/sites/all/files/docpdf/ares707xxiiconventionontourismethics.pdf.

UNWTO. (2017c). Chengdu Declaration on tourism and the sustainable development goals. General Assembly, 22nd session, Chengdu, China, 11-16 September. Retrieved 11 November, from http://cf.cdn.unwto.org/sites/all/files/pdf/13.09.17_unwto_ga22-hl_chengdu_declaration_final_clean_.pdf.

UNWTO. (2018a). *Contribution from the countries in the Committee on Tourism Statistics and Tourism Satellite Account of the UN WTO (UNWTO)*. Seventh Meeting of the Inter-Agency and Expert Group on the Sustainable Development Goal Indicators. 10-12 April, Vienna: Austria.

UNWTO. (2018b). *Tourism for development – volume I: Key areas for action*. Madrid: United Nations World Tourism Organisation.

UNWTO. (2019). *International tourist arrivals reach 1.4 billion two years ahead of forecasts*. PR19003, 21 January. Madrid: United Nations World Tourism Organisation Retrieved 13 August, from https://www2.unwto.org/press-release/2019-01-21/international-tourist-arrivals-reach-14-billion-two-years-ahead-forecasts.

WCED. (1987). *Our common future. World Commission on Environment and Development*. Oxford: Oxford University Press.

Winchenbach, A., Hanna, P., & Miller, G. (2019). Rethinking decent work: The value of dignity in tourism employment. *Journal of Sustainable Tourism*, *27*(7), 1026–1043. doi:10.1080/09669582.2019.1566346

Wolf, E. (1982). *Europe and the people without history*. Berkeley, CA: University of California Press. doi:10.1086/ahr/89.1.89

Wood, R. E. (2000). Research, Caribbean cruise tourism: Globalization at sea. *27*(2), 345–369.

Wood, E. M. (2016). *Democracy against capitalism: Renewing historical materialism*. London and New York: Verso.

Woodward, D. (2015). *Incrementum ad absurdum:* Global growth, inequality and poverty eradication in a carbon-constrained world. *World Economic Review*, *4*, 43–62.

World Bank. (2009). *What is inclusive growth?* World Bank Note 10. Retrieved 10 September 2018, from http://siteresources.worldbank.org/INTDEBTDEPT/Resources/468980-1218567884549/WhatIsInclusiveGrowth20081230.pdf.

WTO. (1980). *Manila Declaration on World Tourism*. Madrid: World Tourism Organisation. Retrieved 14 November, from https://www.e-unwto.org/doi/pdf/10.18111/unwtodeclarations.1980.01.01.

WTO. (1982). *Acapulco Document*. Madrid: World Tourism Organisation. Retrieved 14 November, from https://www.e-unwto.org/doi/abs/10.18111/unwtodeclarations.1982.02.01.

WTO. (1999a). *Tourism and sustainable development*. Seventh Session of the Commission on Sustainable Development, New York, 19–30 April 1999, United Nations Department of Economic and Social Affairs. Retrieved 20 September 2019, from https://www.un.org/ga/search/view_doc.asp?symbol=E/CN.17/1999/5/Add.1&Lang=E.

WTO. (1999b). *Global code of ethics for tourism*. Madrid: Word Tourism Organisation. Retrieved 14 September 2018, from http://www.unwto.org/ethics/index.php.

WTTC. (2017). *Coping with success: Managing overcrowding in tourism destinations*. London: McKinsey & Company and WTTC.

WTTC. (2018). *Travel and tourism: Economic impact 2018*. London: World Travel and Tourism Council.

Yrigoy, I. (2016). Financialization of hotel corporations in Spain. *Tourism Geographies*, *18*(4), 399–421. doi:10.1080/14616688.2016.1198829

Humanism, dignity and indigenous justice: the Mayan Train megaproject, Mexico

Blanca A. Camargo and Mario Vázquez-Maguirre

ABSTRACT

Indigenous groups continue to experience injustices in relation to tourism development, management, and marketing despite calls for equity, justice, and fairness in sustainable tourism. Economic interests continue to dominate and, consequently, the social, environmental, and cultural wellbeing of minority and Indigenous groups tend to take a secondary position in tourism development plans. Progress has been made in developing frameworks to examine justice for Indigenous groups, but they do not take into account the concepts of dignity, which we argue is a core principle in a humanistic approach that seeks fairer outcomes for Indigenous and minorities groups. Drawing upon humanism and humanistic management theory, we examine the Mayan Train megaproject in southern Mexico, propose guiding principles and identify responsibilities for key actors that prioritize the restoration, protection, and promotion of the dignity of groups and prevent potential injustices resulting from tourism projects. We also suggest transformative actions that revalue the cultural identity and status of Indigenous groups, increase their capabilities and self-esteem, and promote their autonomy and wellbeing.

Introduction

Indigenous groups continue to experience injustices in relation to tourism development, management, and marketing despite calls for equity, justice, and fairness in sustainable tourism policy and practice. These injustices include, among others, the use of their culture as part of a destination tourism product without permission or fair compensation (Robinson, 1999; Wanda George, 2010), tourism developments in their land and sacred sites (Finnerty, 2015; Peoples Dispatch, 2019), the omission of their narrative in tourism interpretations (Harvey Lemelin et al., 2013), their marginalization in the distribution of resources for tourism development (Camargo, 2019a), and their exclusion from participation in tourism governance (Jamal & Camargo, 2018). Such injustices can have devastating effects on their economic opportunities, culture, and relationships with the natural environment, affecting their well-being and dignity.

Slowly, issues related to justice and equity are being addressed in sustainable tourism research. The first theoretical discussions of justice began with an environmental focus (Floyd & Johnson, 2002; Lee & Jamal, 2008) and evolved to address cultural injustices affecting low-income, minority, and Indigenous groups (Jamal et al., 2010). Missing from these discussions of justice is the concept of dignity. Tourism can and has perpetuated or exacerbated postcolonial dynamics that affect Indigenous people's sense of self-worth and self-respect, but as Winchenbach et al. (2019) point out, dignity has received very little attention in tourism studies. In this paper, we argue that justice for Indigenous groups in relation to tourism needs to incorporate not only the distributive, procedural, and recognition dimensions in justice frameworks for tourism, but also a humanist dimension that protects and elevates their dignity and promotes their flourishing.

We illustrate the need for a humanist approach through a case example of the Mayan Train, the tourism megaproject proposed by the federal government of Mexico that will connect five states in southern Mexico. This area hosts Mayan archaeological sites, protected natural areas, and Indigenous communities that will suddenly and inevitably be exposed to mass tourism. The government promises to deliver development opportunities for historically marginalized groups, but many locals fear that the Mayan Train will mostly benefit private investors and the government by exploiting Mayan cultural and natural resources for the purposes of tourism (Tourliere, 2019a), adversely affecting the dignity, culture, and lifestyle of these Indigenous communities (Tello, 2019). We draw upon humanism and humanistic-management theory (Melé, 2016; Pirson et al.,, 2017) to propose a new approach to help address the particular injustices that surround Indigenous groups, especially in postcolonial settings where they have been subjected to historic and systemic racism and marginalization, such as in the case of Mexico. We pay particular attention to the government and private industry responsibility in recognizing the human dignity of the Mayan residents in communities around the Mayan Train stations.

The paper begins with a discussion of approaches to justice for Indigenous groups in relation to tourism (for an overview of principles and conceptions of justice that apply to tourism see Jamal, 2019), followed by discussion of humanism and its key concepts of dignity and why it is important for explorations of justice for Indigenous people. A short overview of humanistic management is then presented. A case example of the Mayan Train will subsequently illustrate how tourism megaprojects can violate people's dignity and establish the context for our proposal of an alternative approach for Indigenous justice. In the last section, we suggest a set of guiding principles as well as highlight government and industry responsibilities to restore, promote, and protect the dignity and well-being of Indigenous groups through tourism.

Literature review

Approaches to indigenous justice in tourism studies

General references to justice in tourism studies emerged in the 1990s. Hultsman (1995) proposed a *just tourism* paradigm for tourism providers and future tourism professionals to evaluate the ethicality of their behaviour. Hultsman argued that all tourism services should be conducted in a principled manner, with a basic ethicality that could help the profession advance and be considered legitimate. Just and equitable forms of tourism were also advocated by Higgins-Desbiolles (2008, 2010) and Scheyvens (2002) to counterbalance the severe impacts of mass tourism, overcome inequalities caused by the global trade system, and promote social change and political activism in oppressed regions (see also Isaac & Hodge, 2011). Although these works brought attention to issues of justice for oppressed groups, greater theory-building is needed to guide future research and identify important principles to guide practice.

Only a few theoretically grounded justice studies examine injustices in tourism settings. Inspired by the environmental justice movement (see Bullard, 1993, 2000), Lee and Jamal (2008) proposed an environmental justice framework for sustainable tourism to address injustices related to human-environmental relationships, particularly affecting disadvantaged, minority, and Indigenous groups in a destination. Their framework incorporated dimensions to examine the distribution of environmental impacts of tourism development among different groups (distributive justice), residents' participation in environmental decision-making processes (procedural justice), and whether inequities in the distribution of environmental impacts were influenced by geography, class, race, or ethnic background (environmental discrimination and racism). Empirical research on environmental justice in tourism explored, among other things, inequities in residents' access to tourism and campsites in Southern Appalachia, United States (Porter & Tarrant, 2001), access to freshwater among tourists and residents in Bali and Labuan Bajo, Indonesia (Cole, 2012, 2017), and the distribution of wildlife tourism impacts in pastoral communities in Namibia (Schnegg & Kiaka, 2018).

Cultural justice is another area of study that addresses injustices in relation to Indigenous groups' tangible and intangible cultural heritage (see Johnston 2006 for a detailed discussion of Indigenous peoples and tourism). Robinson (1999) drew early attention to cultural consent, cultural compensation, allocation of cultural rights, and the right of the individual to determine their actions as key issues in achieving cultural equity in tourism. Jamal et al. (2006) discussed equity with respect to the use, protection, and valuation of cultural resources for ecotourism and call for the "fair consideration of the changes being brought to the community's cultural fabric" (p. 164). Camargo et al. (2007) defined cultural justice as "the active involvement of low-income and minority groups in decision-making related to their (eco) cultural goods and their (human-environmental) relationships" (p. 75) and suggested examining issues of cultural justice in natural area destinations from the perspectives of distributive justice, procedural justice, cultural discrimination (a disproportionate number of tourism impacts borne by cultural minority groups), and cultural racism (the exclusion or prevention of minority and Indigenous groups from conducting their traditional practices by tourism initiatives, laws, policies, etc.). Empirical work on cultural justice has revealed the unequal allocation of government resources for tourism development and marketing for Indigenous Mayan groups in Mexico as well as their marginalization from tourism decision-making (Camargo, 2011, 2019a), and the uncompensated government appropriation of images of Afro-Colombian women for tourism purposes (Camargo, 2019b).

Rights-based notions of distributive justice and procedural justice at a societal level, such as those advocated by Rawls (1971), dominated early contemporary theoretical explorations of justice in the context of tourism. Based on feminist political theory and social justice studies, subsequent justice related tourism research has addressed issues pertaining to minority and Indigenous groups. After Fraser (2000) and Figueroa (2006), Whyte (2010) suggested including a recognition justice dimension in environmental justice frameworks for Indigenous tourism to give fair consideration to the social circumstances and cultural terms of Indigenous groups within institutional processes and transactions; the opposite, recognition injustice, will take place when the social circumstances and cultural terms of the majority are, without justification, represented and privileged within social institutions and transactions, including tourism. Recognition justice seeks the direct participation of Indigenous groups in decisions that protect their environmental identity and heritage (Figueroa, 2006).

Figueroa and Waitt (2010) further expand environmental justice for Indigenous tourism by adding a restorative justice dimension. Restorative justice focuses on repairing injustices and harms through amends and transformative remedies that bring together victims and perpetrators to find resolution and reintegration within their communities (Restorative Justice, 2019). It is based on four principles (Van Ness, 2000): encounter – the meeting between the victim, the offender, and community members who have been affected by crime or injustices; amends – the tangible steps taken by the wrongdoer to make amends for the harm inflicted; reintegration – restoring the affected parties to a position where they once again become whole, contributing members within their communities; and inclusion – the meaningful participation of victims, offenders, and the community in the justice process.

Another important aspect in the tourism concept is reconciliation and efforts to enact restorative justice for Indigenous groups (Figueroa & Waitt, 2010; Higgins-Desbiolles, 2003). For instance, the tourist centre at Camp Coorong in South Australia has been the medium of bringing together Indigenous and non-Indigenous Australians to foster learning and discussion of past and contemporary injustices suffered by Aboriginal groups and disseminate knowledge about the Ngarrindjeri culture. Tourism has also served as a vehicle of reconciliation, hence contributing to restorative justice, for historically divided groups in non-Indigenous settings in the United States (Barton & Leonard, 2010; Skipper, 2016), Rwanda (Alluri, 2009), and Colombia (Alsema, 2018).

The above approaches to justice emanate from the Eurocentric West (i.e., the Global North) and much work is needed to explore other (e.g. Asian, Islamic, African, Indigenous) perspectives of justice. It is also important to explore other fields of enquiry that perhaps can inform theories of justice. In the next section we explore how humanism, through its key principle of dignity, can help better approach justice for Indigenous groups in the context of tourism.

Humanism

Although humanism is often associated with European Renaissance thinking that mostly valued the intellectual and artistic contributions of the humanities (Caton, 2016), its underlying beliefs can be traced back to ancient India, China, Greece, and Rome, where great thinkers developed the ideas that constitute humanist thinking (von Kimakowitz et al., 2009). Buddhism, Confucianism, ancient Greek, and Medieval Islamic thought emphasized central humanist principles such as tolerance, reason, individual free thought, or the need to work for the common good (Caton, 2016). From the 19th century onwards, the term humanism was used by educators, intellectuals, and scholars of classical Latin and Greek (Campana, 1946); the notion of humanism in this period, which is also how it is commonly understood today, emphasizes knowledge and concern for things related to the human person (Melé, 2016). In the 20th century, humanism encouraged progressive social causes that challenged the status quo (e.g. labour rights, women's suffrage, civil rights for racial and ethnic minorities); its underlying belief focuses on the capacity of humans to improve their own lives (Caton, 2016).

Humanism has a diversity of approaches and understandings, e.g., Marxist humanism, Existentialist humanism, or Christian humanism. Some approaches emphasize transcendence and human dependence on God, while others maintain that it is opposed to, or even may be a substitute for religion (Melé, 2016). Erich Fromm (1961) defines humanism as a system centered on the human being, on human integrity, development, dignity, and liberty: each person bears within herself humanity as a whole and is herself the bearer of her own end. Therefore what it is to be human has to be safeguarded and respected, since it expresses the fundamental human condition of our dignity (Dion, 2017).

In tourism studies, Caton (2016) emphasizes the unique character of humans as moral creatures that seek to make improvements in the world, and have the responsibility to use compassion-based moral reasoning for the greater good. The author calls for alternatives to epistemic literalism[1], which focuses on using methods and tools that produce verifiable knowledge, but evades questions of foundations. Humanism offers a very different standpoint to speak from for it is based on an axiological claim that our actions are tied to moral obligations to promote individual flourishing and decrease the suffering of humans and other beings, thus, making progress as a society.

The diversity of approaches to humanism raises a challenge in adopting a single definition. However, we can say that our approach in this paper is not of the kind of humanism that arose in Europe during the Renaissance and gave priority to the human individual and rational thinking, rejected the church and supernatural belief, and involved a distinction between a human and non-human world.[2] Rather, we embrace the larger principles and values that guide the study of the humanities and humanistic enquiries, among them, the search for dignity, happiness, and well-being.

Dignity

At the core of human rights and humanism discourses is the notion of dignity. References to human dignity[3] are found in the Universal Declaration of Human Rights: "All human beings are born free and equal in dignity and human rights" (UN, 1948, Article 1), the United Nations Declaration of Rights of Indigenous People: "The rights recognized herein constitute the minimum standards for the survival, dignity and well-being of the Indigenous peoples of the world" (UN, 2007, Article 43), the International Labour Organization Constitution: "All human beings irrespective of race, creed or sex, have the right to pursue both their material wellbeing and their spiritual development in conditions of freedom and dignity, of economic security and equal opportunity (International Labour Organization, 1944), and several other declarations and public discourses. But what constitutes dignity is ambiguous, theoretically complex (Debes, 2009; Kateb, 2011), and even contested (Mattson & Clark, 2011), so providing an explicit definition may be elusive or conducive to many interpretations. However, important abstractions can be made from the literature on the notion of dignity.

First, dignity is related to the inalienable worth of the individual, a worth that cannot be earned, bequeathed, or compared to other values (Debes, 2009); it cannot be assigned or subjected to the goodwill, intentions, or attitudes of other moral subjects. Immanuel Kant, one of the most prominent scholars on dignity, defined it as the intrinsic, unconditional, incomparable worth or value of every person that warrants respect (Stanford Encyclopedia of Philosophy [SEP], 2018).

Second, dignity is an object of recognition and respect (Debes, 2009, Hodson, 2001; Nussbaum, 2011; Schachter, 1983; Taylor, 1991). For Schachter (1983), dignity requires a recognition that people are entitled to their own feelings, beliefs, and attitudes, and have a capacity to make choices. Kateb (2011) argues that dignity is an existential value imputed to the identity of the person: it means recognizing the proper identity of an individual, and what that individual is in relation to others. A person's dignity is at stake when we fail to treat her as a unique individual who is irreplaceable and not exchangeable for another. For Kant, to respect a person's dignity is to respect her worth by treating her as an end in herself, that is, acknowledging her reasoning, her desires, her interests, and her affections, not as a mere means to further our own interests. It also requires respecting her capacity to be autonomous, make rational choices in her life, and set her own ends. Autonomy for Kant refers to the capacity to be self-legislating, self-governing, and to freely determine the meaning of one's own life (Dillon, 2018). When we restrict the capacity to be autonomous, we violate a person's dignity.

Schachter (1983) argues that dignity has a psychological significance, and for this reason, we must avoid imposing beliefs and attitudes on people and treating them in a demeaning and humiliating matter. Affronts to dignity include, among others, statements that demean and humiliate groups, derision of beliefs that people hold in reverence, denial of the capacity of a person to assert claims to basic rights, attacks on personal and communal beliefs and ways of life, attacks on the groups and communities with which individuals are affiliated, dissemination of negative stereotypes, denial of education or employment opportunities based on membership in groups or their beliefs, or restriction on equal participation in political process because of beliefs, status, or origin (p. 852).

Nussbaum (2011) discusses dignity in relation to creating capabilities that allow people to actively thrive. A capability is the combination of a person's abilities and the opportunities available "to do and to be" (p. 18). She too espouses the principles of each person being treated as an end in herself, not as a means to an end[4] and proposes ten capabilities that can help human beings to lead a life of dignity and worth: 1. life, 2. bodily health, 3. bodily integrity, 4. senses, imagination, and thought, 5. emotions, 6. practical reason, 7. affiliation, 8. other species, 9. play, and 10. control over one's environment. These capabilities are conceived as substantial freedoms that need to be developed, respected, and protected from being abused and subjected to the arbitrary will of others.

The recognition and respect of the individual value of each person, in other words, her dignity, is the base of a democratic society (Taylor, 1991), and it is strongly linked to justice. Distributive justice requires recognition of the equal value of all people in the allocation of basic resources and societal goods; procedural justice calls for meaningful participation in decision-making processes and respect of people's will, consent, and their capacity to assert justice claims; in other words, equality and autonomy, the core of human dignity (Solum, 2004). For Nussbaum, delivering the ten capabilities to all citizens is a necessary condition for social justice: governments should be accountable for the presence (or lack) of the ten capabilities if the society is to be considered just. Developing capabilities requires government expenditure and allocation of economic and social rights for all citizens.

Research on humanistic management also takes on board the notion of dignity and offers important insights into how to promote a more humane and just approach to tourism for Indigenous groups., We therefore introduce humanistic management and its potential contribution to the justice turn in tourism studies, before proceed to apply these insights to the Mayan train case.

Humanistic management

Humanistic management seeks a change in the current economic paradigm that privileges business profits and financial indicators and goes beyond these metrics to focus on upholding the unconditional human dignity of every individual (Spitzeck, 2011). The rationale for humanistic management is that economistic management approaches have left a void in the way human dignity is protected, human flourishing promoted, and ethical development encouraged, and have not adequately resolved issues related to social justice and social welfare (Sen, 2001; Pirson et al., 2017). Sen (2009) argues that the idea of human beings motivated only by self-interest (economic paradigm) does not incorporate people's notions of fairness and the desire to eliminate injustices that could be remedied by taking a different approach to social and economic issues. Humanistic management starts from the idea that human beings are both key to, and the ultimate focus of, any organization (Pirson et al., 2017), and these entities must recognize, respect, and value their individual qualities (Bédard, 2003). Humanist management discourses coincide in that the ultimate goal of an organization is to foster unconditional human dignity (Spitzeck, 2011), which is a necessary condition for well-being (Pirson et al., 2017).

Under circumstances of injustice or unfair access to basic resources, dignity can provide a principle of morality (Waldron, 2013): all decision-making should privilege the human dignity of individuals above any other concern. In this sense, humanistic management may provide a framework that guides managerial practices towards fairer outcomes for humanity. Melé (2016) suggests seven propositions that are implied in humanism, and consequently, in humanistic management, that should be considered as a starting point of the discipline: (1) *wholeness*: the human being is complex and embodies myriad motivations, emotions, personality, character, capacities, and spirituality, among other qualities; (2) *comprehensive knowledge*: human beings should be understood in a holistic and integral way in their different dimensions (spiritual, rational, emotional); (3) *dignity*: dignity is an intrinsic, absolute, and non-transferable element of the human being that should be defended; (4) *development:* organizations should provide the necessary conditions and resources for people to grow and flourish in an integral and holistic way; (5) *common good:* humanism looks for the harmonious balance between individual freedoms and common welfare; (6) *stewardship–sustainability:* humanism recognizes the interconnection of the human being with the environment and all living things and promotes a sense of stewardship; and (7) *transcendence:* the human being is a spiritual and transcendent entity who seeks meaning and is respectful of others' religions and beliefs.

Pirson and Steckler (2018) propose a model of four independent innate drives that, when balanced, can guide managerial practices and ensure basic human dignity and increase well-being. These drives (to acquire, to defend, to bond, and to comprehend) depend on moral rules that guide responsible management. These rules mandate seeking fair and honest exchanges, helping to enhance rather than steal or destroy another's property, sharing useful information, respecting the other's beliefs even when disagreeing, and helping to protect rather than harm or abandon the other. In addition, Pirson et al. (2017) argue that different economic and social dynamics (e.g. injustice, exclusion, marginalization, instrumentalization, lack of recognition) have affected the dignity of particular groups. However, humanistic management can contribute to remedying the injustices done to these individuals through dignity restoration, dignity protection, and dignity promotion, as organizations adapt different humanistic practices towards these ends.

Departing from these approaches, and using the case example of the Mayan Train, in the next section we explore how issues of dignity, justice, and equity played out for Indigenous groups through tourism; the case highlights the need for a different approach to guide tourism policy, development, and management, which we discuss in the last section of this paper.

Case example: the Mayan Train project

This exploratory case examines the Mayan Train project in southern Mexico as a case example to illustrate the need for a different approach to tourism, one based on humanistic principles that prioritize dignity protection of Indigenous groups and prevent potential injustices derived from this type of project. We follow a case study example strategy based on secondary data from multiple sources. Case studies are central to most research in the humanities and social sciences, including tourism research (Beeton, 2005). They enable the study of phenomena within their actual circumstances (Eisenhardt, 1989), and is the method of choice when the phenomenon being studied is not readily distinguishable from its context (Yin, 2003).

The selection of the case example was based on four elements: (1) its visibility, since it is perhaps the most important tourism project proposed by the Mexican government since the first integral development in Cancún in the 1970s; (2) its relevance, as this case is situated in a context where Indigenous communities have been subjected to colonial and postcolonial discrimination, marginalization, and injustices that have built structural barriers preventing their self-determination and development; (3) its potential impact, because the project will affect the different Mayan communities, their ecocultural resources, and the largest tropical forest in America after the Amazon; and (4), as noted by Beeton (2005), cases need not be so specific as to preclude their being applied to other situations. This case describes the common phenomenon of generating tourism megaprojects to bring mass tourism and development to a certain region, ultimately affecting the way of life and values of local communities.

Background: Indigenous groups and development in Mexico

Mexico is the eighth country in the world with the largest number of Indigenous people (República, 2017). It also records the largest inequalities in household income among member-countries of the Organisation for Economic Co-operation and Development (OCDE), with individuals in the top 20 per cent earning 10 times more than those in the lowest 20 per cent (OCDE , 2017). Most Indigenous people are part of the lowest quintile; 72 per cent of the 11 million Indigenous people living in Mexico are poor (Molina, 2018). Poverty is 3.3 times higher among Indigenous peoples, compared with the rest of the population (United Nations, 2009). Since colonial times, Indigenous groups in Mexico have suffered from discrimination, prejudice, and belittlement which has not only resulted in displacement and loss of rights, but also had severe consequences for their identity, culture, and language (Castellanos-Guerrero, 2000, 2003). Although Mexico is now an intercultural and multicultural country, none of these elements has contributed to changing existing racial hierarchies and privileges, or reducing the effects of racism and racial injustice on Indigenous people and their communities (Saldivar, 2018). Often Indigenous and cultural groups are portrayed as noble and hard-working, but uneducated and unprepared, a stereotype that has led to their marginalization in political life and exclusion from society's decision-making processes, including those concerning tourism. Their names, resources, landscapes, and images, are commonly used and exoticized in tourism marketing and promotional materials with no fair compensation being paid (See Camargo, 2011).

The efforts destined to create favourable conditions for the development of Indigenous communities in Mexico had often ended up reproducing colonial and postcolonial patterns of exploitation; for example, in tourism projects in Cobá (Quintana Roo) and Tajín (Veracruz), places rich in archaeological sites and Indigenous culture, local Indigenous people compete with large hotels and restaurants in highly unfavourable conditions; local traditions have also been turned into shows for mass tourism (Tello, 2018). Wind energy megaprojects in southern Mexico have also resulted in some form of exploitation, resembling an unjust extractive model, with no consideration for local Indigenous cultures or organizations (Zárate-Toledo et al., 2019). Also, in August 2019, eleven Indigenous communities presented a legal appeal brought on grounds of violation of rights and liberties for the construction of the Mexico City International Airport in their lands, which constitutes a megaproject that also aims to propel mass tourism. The communities argue that they have not been consulted and they do not know of any formal study regarding the social and environmental impacts of this project, especially how it is going to affect their way of life and ecosystems (Lastiri, 2019). Recently, the government has been accused of faking consultations with Indigenous communities regarding the 'Corredor Transístmico' megaproject, which aims to be an alternative to the Panama Canal. The invitations to roundtable discussions, which lasted less than a month, were issued by government bodies that did not have legal status to do so, and they only reached 11 out of 28 Indigenous communities in Oaxaca and Veracruz (Meza, 2019). These antecedents further reinforce the idea of injustice and discrimination towards Indigenous communities' rights and self-determination, and also bring into question how Indigenous communities will participate and be affected by the Mayan Train project which we explore below.

The emergence of the Mayan Train project

The Mayan Train may be the newly elected left-wing Mexican president's most important infrastructure project. Scheduled for completion in 2023, it originally consisted of a route spanning 1,525 kilometres with 15 stations[5] (figure 1) distributed through five of the country's southern states that form part of the Mayan jungle: Chiapas, Tabasco, Campeche, Quintana Roo, and Yucatán (Gobierno de México, 2018). Its construction will require an investment of between US$6,000 million and US$8,000 million, 90 per cent of which will be financed by private investment under the public-private partnership scheme. Tourism companies are expected to develop facilities and services in the areas around each train station which will mean that Indigenous communities will have to grant access to *ejidos* (areas of communal land) and collaborate with the developers (Tourliere, 2019b). The government expects the train to attract three million people in its first years of operation, which, it is hoped, will redistribute tourism – currently concentrated primarily in Cancún and the Riviera Maya – throughout the region (Varillas, 2019).

The Mayan Train will cover one of the most culturally diverse areas yet historically marginalized regions of Mexico. Tourism is the most important economic activity in the region; in Quintana Roo, for example, almost one out of every four companies and 44.7 per cent of the

Figure 1. The Mayan Train route. Source: Reuters (https://theconversation.com/mexico-quiere-construir-un-tren-en-el-corazon-de-la-region-maya-deberia-de-hacerlo-121861).

workers are engaged in tourism (SECTUR. , 2016). Rural communities in the Yucatán Peninsula are mostly Indigenous and have a fluctuating income that is based on different activities throughout the year, which are mainly agricultural, although also involve the raising of livestock (pigs and poultry), fishing, elaboration of handcrafts (natural fibre products, basketry, manufacture of ceramic goods, manufacture of cotton garments, weaving of hammocks), and seasonal migration of people to be employed in tourist sites along the Riviera Maya (Comisión Nacional para el Desarrollo de los Pueblos Indígenas, 2018; Programa Selva Maya, 2019).

The Mayan ethnic group largely inhabits the jungle region that will be affected by the Mayan Train project. There are 1.5 million Mayans, of whom almost 10 per cent are illiterate (Comisión Nacional para el Desarrollo de los Pueblos Indígenas, 2018). The Mayans have polytheistic beliefs that have also been mixed with elements of Christianity; their universe is composed of three elements or regions: the sky, the earth, and the underworld, which is inhabited by different, generally anthropomorphic, supernatural beings, who are identified as the guardians of nature (Comisión Nacional para el Desarrollo de los Pueblos Indígenas, 2018). Other large Indigenous groups that inhabit the region are the Ch'ol (328,867 inhabitants), the Chontal (57,792), the Tseltal (689,797), the Tsotsil (614,105), and the Zoque (104,321) (Comisión Nacional para el Desarrollo de los Pueblos Indígenas, 2018). The cultural heritage of this region, combined with tropical rainforests and jungle ecosystems, form attractive destinations for mass tourism.

In relation to the natural environment, the train will pass through the Selva Maya (Mayan Forest) and several protected areas, including the UN-designated Calakmul Biosphere Reserve, the second-largest expanse of tropical forest in the Americas, after the Amazon. Covering 42,300 km^2, the Selva Maya is the largest tropical rainforest in North and Central America and the second largest in the continent as a whole. It hosts an extraordinary biological diversity with twenty different ecosystems and a large number of endemic species (Comisión Nacional para el Conocimiento y Uso de la Biodiversidad, 2017). The Calakmul Biosphere Reserve, which spans an area of 7,232 km^2, recognized as a Mixed World Heritage Site by UNESCO, has one of the largest jaguar populations in Central America (Programa Selva Maya, 2019).

Politically, Article 2 of the Political Constitution of Mexico guarantees Indigenous people the right to self-determination and the autonomy to decide their internal forms of coexistence and social, economic, political, and cultural organization. In addition, Article 6 of the National Institute of Indigenous Peoples' Law (Ley del Instituto Nacional de los Pueblos Indígenas, 2018) guarantees their right to consultation and free and informed consent whenever the federal government promotes legal reforms and administrative acts likely to affect them. Internationally, Article 32 of the United Nations Declaration on the Rights of Indigenous Peoples (2007) and Convention 169 of the International Labour Organization (1989) also support Indigenous people's right to self-determination and consultation, and require that their free and informed consent be obtained before approval is given to any project that may have an impact upon them.

Controversies and conflict over the mayan train

The Mayan Train has met with strong opposition from environmental and civil society groups, activists, researchers, and the private sector (Tello, 2019). They are concerned that legislation put into place to protect development projects in Indigenous communities will be disregarded, and that the government's overly optimistic view of the project stems from a lack of environmental and sociocultural impact assessments.

Current legislation indicates that the future of the Mayan Train is likely to depend on the Indigenous people giving their consent before a project of this scale can be built on their territory (Morgan, 2019). However, there is no indication that the government undertook meaningful consultation or sought informed consent with the Indigenous groups that inhabit the five states affected by the Mayan Train prior to launching the project and presenting it to private investors; their participation was limited to attending informative sessions about the potential benefits of the project or workshops to finesse their handicraft skills (e.g. TrenmayaMX, 2019d). While the government has invited national and international investors to finance the project (TrenmayaMX, 2019b), the College of Notaries to protect government decisions regarding the Mayan Train (TrenmayaMX, 2019a), and international organizations such as the United Nations to oversee the project (TrenmayaMX, 2019c), local communities have been given a passive role in the planning process. Iván Franco, a researcher from INAH (National Institute of Anthropology and History), believes there is a huge risk that this megaproject will end up, as have previous projects, producing low-skilled jobs for Indigenous people: cleaning bathrooms, washing floors, painting facilities, or serving food and coffee in national and international chains, which is altogether different from an authentic megaproject that promotes adequate alternatives, based on principles of sustainability and citizen participation (Tello, 2019).

Although the federal government promotes the Mayan Train as an inclusive and sustainable tourism development, environmentalists are concerned about the impact of a train going through ecological reserves such as Sian Ka'an and Lake Bacalar (Fuentes, 2018), the destruction of jungle, and the loss of ecosystems and cultural elements that make local communities unique (Reportur, 2019; Tello, 2019). According to official projections from FONATUR (National Fund for Tourism Promotion), Calakmul Biosphere Reserve receives 40 thousand tourists a year, but it is expected that, of the almost 17 million tourists who visit Cancún, three million will also visit Calakmul when the project is completed (Tello, 2019). In this sense, the government claims the environmental impact will be minimal, as 90 per cent of the infrastructure the project needs already exists and it will also ensure the development of animal corridors and reforestation areas (Gobierno de México, 2018). Furthermore, the train will encourage mainstream agricultural activity, since it offers cheaper transportation of agricultural products to urban areas, creating additional pressure on the forest and jungle (Fuentes, 2018). The Mayan Train can also stimulate activities that may negatively affect existing ecosystems such as wind farms, agribusiness, pig farms, and "green" tourism, threatening Indigenous groups' traditional productive activities such as beekeeping (Muñoz, 2019). Ultimately, any scheme to move tourism from the Riviera Maya to

Table 1. Guiding principles for humanistic tourism.

Principle	Objective
Self-determination and self-control	The right of Indigenous and ethnic groups to freely pursue their economic, social, and cultural development (United Nations, 2007).
Decent and meaningful tourism work	Providing decent jobs that allow people to escape poverty and live a dignified life.
Prioritization of entrepreneurial opportunities for Indigenous people	Ensuring the availability of the resources and support needed to generate enterprises that naturally create social impact and promote local value.
Recognition	Transforming how Indigenous groups are valued and represented in tourism; transformative actions to revalue cultural identity and status.
Oneness	Unity of the human being with nature. A worldview that relates to achieving environmental, economic, and sociocultural sustainability.

Source: Camargo et al. (2020).

the jungle must take into account the capacity of the jungle to support mass tourism, something that is still unknown despite the project having already been approved (Molina, 2019).

Insights gained from the Mayan Train suggest that Indigenous communities along the route are facing increasing affronts to their individual and collective dignity. As these groups have partially lost the capacity to assert claims to basic rights (such as self-determination, or informed consent), to exert control over their environment, and to participate in conditions of equality in political processes (such as consultations), the need for a different approach that protects and privileges dignity becomes urgent. Besides, potential future outcomes of this megaproject, such as the use of Indigenous people for menial work, and the alteration of Indigenous communities' social and economic dynamics, also call for a new scheme that treats people as an end, and that is centred on each individual's intrinsic value.

Dignity: a new focus for indigenous justice in tourism

A humanistic approach to tourism can complement existing justice frameworks that seek to achieve fairer outcomes for Indigenous and minority groups who are most vulnerable to the impacts of tourism. Environmental and cultural justice frameworks seek the fair distribution of economic, environmental, and cultural benefits and costs of tourism among all groups in a destination, and the recognition of cultural, minority, and Indigenous groups as equal partners in tourism decision-making; humanism puts people – in this case, Indigenous groups – as well as their dignity, their well-being, and their flourishing at the forefront of any tourism decision-making. Caton (2016) called for a humanist paradigm for tourism studies and challenged scholars to outline principles by which to promote happiness and human dignity through tourism. Taking on this challenge, Camargo et al., (2019) draw upon Pirson and Steckler's work (2018) to propose a set of guiding principles for humanistic tourism that can ensure basic human dignity and increase the well-being of Indigenous groups: self-determination and self-control, decent and meaningful tourism work, prioritization of entrepreneurial opportunities for Indigenous people, recognition, and oneness (Table 1).

The above principles can contribute to a sense of justice for Indigenous groups by restoring, protecting, and promoting their dignity. Pirson et al. (2017) argues that one of the main goals of humanistic management is to rethink principles and practices of the current economic paradigm and formulate new elements that aim towards improving the dignity threshold. To this end, and inspired by these humanistic principles, we suggest government and industry responsibilities that aim to restore, protect and promote dignity and justice in tourism (Table 2). These responsibilities target specific dignity thresholds (Pirson et al., 2017) that need to be identified to

Table 2. Humanistic principles and government and industry responsibilities that aims to restore, protect and promote dignity and justice in tourism.

Dignity threshold	Principles involved	Government responsibilities	Industry responsibilities
Restoration	Decent and meaningful tourism work.	Develop schemes to integrate previously excluded Indigenous groups into the labour market. Enforce private industry compliance with labour law.	Promote formal employment of Indigenous groups. Respect labour laws. Provide professional development programmes to Indigenous groups
	Prioritization of entrepreneurial opportunities for Indigenous people.	Promote the creation of community-based tourism enterprises, cooperatives, and Indigenous social enterprises Prioritize community-based enterprises in destination-marketing initiatives Help to build an entrepreneurial ecosystem with institutions that foment local micro-business development and fair competition.	Include Indigenous businesses in the value chain of corporations.
Protection	Self-determination and self-control.	Enforce agreements, laws, and norms that safeguard human and Indigenous rights. Respect Indigenous groups' wish regarding whether or not to participate in tourism development. Maximize the participation of Indigenous people in tourism governance bodies Support for Indigenous supracommunity bodies that protect the autonomy of Indigenous groups and counterbalance the economic interests of corporations aiming to exploit mass tourism in Indigenous communities.	Respect Indigenous groups' choice to accept or reject tourism projects. Seek permission to incorporate Indigenous cultural elements in tourism offerings. Promote respectful, symmetrical, and transformational relationships with communities and local businesses. Consult with local communities to establish tourism quotas, entrance fees, environmental regulations, and restricted perimeters of Indigenous areas for tourists.
	Oneness.	Conduct serious participatory studies that focus on the living conditions of local people and the potential impacts of tourism megaprojects. Evaluate potential threats that may alter economic and cultural patterns and the communities' relationship with nature.	Develop environmental awareness programmes for tourists through activities that protect and regenerate the ecosystems. Maintain and improve the ecological base in Indigenous lands

(*continued*)

Table 2. Continued.

Dignity threshold	Principles involved	Government responsibilities	Industry responsibilities
Promotion	Recognition.	Develop cultural projects that reconnect local people with their cultural heritage and ecosystems.	Acknowledge Indigenous groups as key partners in tourism planning, development, and management.
		Help advance an agenda of development according to local world views.	Promote cultural products in the tourism market (e.g. alternative medicine, music, celebrations)
		Promote cultural heritage and Indigenous knowledge internationally as a source of national pride.	Recognize Indigenous communities as legitimate stakeholders.

generate the transformative actions needed to revalue the cultural identity and status of Indigenous groups and increase their capabilities and self-esteem.

Dignity restoration

Dignity restoration refers to remedying violations to people's unconditional worth, freedom, autonomy, and identity. The tourism industry violates (Indigenous) people's dignity, as Kant argues, when it treats them as a means to an end, which often materializes in exploitative menial work. Megaprojects such as the Mayan Train can contribute to dignity restoration by promoting formal employment that provides benefits and adequate working conditions, and offer entry-level tourism employees scholarships and ongoing professional development courses, on-the-job training, and management shadowing programmes to assist them in obtaining supervisory and managerial positions in the tourism industry. Also, the industry should prioritize building a strong local value chain through inclusive business strategies that strengthen the creation and development of Indigenous enterprises. In this form, Indigenous people might have better opportunities to overcome the structural barriers that prevent them from obtaining decent jobs.

Another mechanism to restore dignity is through the creation of community-based and social enterprises. There is evidence that Indigenous social enterprises, particularly in Latin America, apply principles of humanistic management: they seek sustainability and integrate previously excluded minorities into the labour market, respect employees' dignity, and promote the autonomy and self-governance of the community (see Vázquez-Maguirre, 2018, 2019). These ventures usually prioritize holistic health and well-being (Vázquez-Maguirre et al., 2018), social innovation mechanisms to reach their purpose (Vázquez-Maguirre & Portales, 2018), generate the development of entrepreneurial ecosystems in the community (Vázquez-Maguirre, 2018), and have a strong link to social justice agendas[6] (Waterloo Institute for Social Innovation and Resilience, n.d.). This last element is fundamental to restoring dignity violations among these groups. The government has a fundamental role to promote the creation of social enterprises by providing funding and prioritizing these entities in destination-marketing initiatives. The government also needs to prioritize infrastructure development, capacity building, subsidies, and marketing of community-based enterprises and social enterprises in such a way that they can fairly compete with private businesses. This is especially important in Cancún and Riviera Maya, where large tourism companies may be willing to expand their operation throughout the Mayan Train route with the risk of implementing traditional managerial practices that tend to produce menial work and exploitative working conditions.

Protection of dignity

Dignity protection has its conceptual basis in human rights and refers to safeguarding individuals' intrinsic worth, freedom, autonomy, and identity. Kant emphasizes that dignity refers to the capacity to be self-legislating, self-governing, and to freely determine the meaning of one's own life (SEP, 2016). In this sense, the government would need to enforce the self-determination of Indigenous communities in Mexico that is guaranteed by the Political Constitution of Mexico, National Institute of Indigenous Peoples' Law, United Nations Declaration on the Rights of Indigenous Peoples, and Convention 169 of the International Labour Organization. Also, when projects such as the Mayan Train are discussed, the government must promote serious participatory studies that focus on the living conditions of local people and evaluate potential threats that may alter economic and cultural patterns and the communities' relationship with nature.

The government is also responsible for protecting the main governing body that safeguards the self-determination and self-control at a community level: The Comisariado de Bienes Comunales (Commissariat of Common Goods, CCG). This body is democratically elected through public general assemblies. The CCG manages and distributes land ownership among families and regulates the use of land for productive purposes (Comisión Nacional para el Desarrollo de los Pueblos Indígenas, 2018). In this case, the CCG in each community must approve a change of the use of land to allow projects such as the Mayan Train to be built in their territory. In this sense, the government should respect Indigenous communities' wishes regarding whether or not to participate in tourism developments.

Although the CCG promotes the communities' self-determination and self-control, this body lacks the scale and jurisdiction to counterbalance the economic interests of corporations aiming to exploit mass tourism in Indigenous communities. For this end, the government must support the creation of Indigenous supracommunity bodies that work along with major tourism corporations towards establishing respectful, symmetrical, and transformational relationships with tourists (avoiding situations of servility). Tourism quotas, fixed prices, environmental restrictions, delimitation of areas and sites that communities wish to open to tourists, and the cultural aspects these groups are willing to share with the visitors are other factors that need to be considered and supported at government level.

The tourism industry should respect the Indigenous group's choice to accept or reject tourism projects and seek permission to incorporate Indigenous cultural elements in tourism offerings. Dignity protection includes allowing Indigenous people to reaffirm control of their ecosystem by increasing their capacity to protect and regenerate local flora, fauna, and water sources. In this way, Indigenous communities affected by the Mayan Train will be in a better position to protect themselves from the vulnerabilities that may compromise their right to self-determination and self-control.

Promotion of dignity

The third threshold is dignity promotion: it is geared towards well-being creation and human flourishing (Pirson, 2017). Dignity promotion includes the recognition of Indigenous communities by revaluing their cultural heritage at a local, national, and international level. This cultural heritage may have been previously affected in these communities by economic or social stress, including lack of funding for cultural or religious celebrations, inadequate public infrastructure and schools, and a paucity of mechanisms to preserve elements such as language, traditional clothing, dishes, economic activities, and forms of organization. The promotion of Indigenous heritage can help generate deeper meaning in these places, revalue the sense of belonging to an Indigenous group, and allow local people to perform meaningful and valuable activities that lead to overall human development and happiness. To this end, the government is responsible for fomenting cultural projects locally within Indigenous communities, and for promoting the cultural heritage and knowledge of these groups to an international audience. The government should also help advance an agenda of development according to the worldview of every Indigenous group.

The tourism industry should recognize Indigenous communities as legitimate stakeholders acknowledging them as equals in tourism planning, development, and management. In addition, the industry has the responsibility for revaluing cultural products and services (crafts, foods, music, celebrations, alternative medicine, and other cultural expressions) in the tourism market. Major tourism players are also important institutions that should foment the involvement of Indigenous groups in interpreting their heritage and advancing an agenda of development according to local world views. This agenda should include developing the necessary capabilities that will allow Indigenous people to obtain physical, mental, and social well-being.

Conclusion

These are challenging times of growing inequality, xenophobia and racism, climate change, loss of biodiversity, and global pandemics like COVID-19 that has further exacerbated inequity and discrimination (Ahmed et al., 2020). Indigenous groups are more vulnerable than ever to economic, environmental, and cultural injustices that can damage their dignity and well-being. Tourism megaprojects like the Mayan Train tend to replicate historical, structurally embedded racism that treats Indigenous people, their culture, and their land as a means to an end, thereby invalidating their dignity. But tourism can be planned and managed differently. In this paper, we propose a theoretical approach for justice frameworks in tourism that is grounded in humanism in order to situate human dignity as an integral principle in tourism development. We argue that dignity must be part of a justice framework for Indigenous tourism because it will protect their inalienable worth, their freedom, and their capacity to be autonomous. We also propose five guiding principles for a dignity-based, humanistic approach to tourism: self-determination and self-control, respectable, meaningful work, prioritization of entrepreneurial opportunities for Indigenous people, recognition, and oneness. These principles aim to restore, protect and promote human dignity and can help redress past injustices, improve Indigenous groups economic and ecocultural well-being, restore self-worth and stand in their society, and promote their wellbeing.

Much is needed to conceive tourism in a humanist way. Economic interests dominate and, consequently, the social, environmental, and cultural wellbeing of minority and Indigenous groups take a secondary position in tourism development plans. Profits should not come at the expense of Indigenous groups' dignity and wellbeing; in a humanist approach to tourism the government and the private industry have key responsibilities in restoring, protecting, and promoting the dignity of Indigenous groups. One of the biggest challenges is to come to terms with the idea that Indigenous people might not even wish to participate in tourism. Justice starts with respecting Indigenous people's dignity, in other words, their freedom and autonomy, and recognizing their social and cultural circumstances in all political processes (Whyte, 2010). Dignity must reject all utilitarian justifications (Debes, 2009), and the potential benefits of tourism projects cannot override the consent and autonomy of Indigenous groups. Dignity cannot be traded off against other values, especially not economic ones.

In the case of willing participation, the government and private industry need to make significant efforts to improve their conditions in the labor market, which is usually exploitative, and integrate their enterprises into the tourism system, all of which are steps towards promoting dignity at work (See Bolton, 2007; Hodson, 2001; Winchenbach et al., 2019) and justice in accessing tourism benefits. At the same time, Indigenous groups need to actively participate and have a permanent representation in governing boards and councils of public and private organizations. Thus, their needs, norms, and visions of development must be taken into consideration in decision-making regarding tourism product development, marketing, management, and treatment of their ecocultural resources, accomplishing the goal of recognition justice as forwarded by Whyte (2010). Two future avenues for research include exploring dignity from the lived experience of Indigenous groups in tourism destinations, and the role and responsibilities of tourists in dignity restoration, protection, and promotion in these communities.

Notes

1. Tourism research has been mostly approached from a correspondence-based theory of truth, which implies there is an objective reality "out there" that can be captured and verified. Thus, it has focused on building methods that better capture the reality (Caton, 2016).
2. Humanism has been criticized for its over-emphasis on the (male) human as the centre of life, and the measure of all things to the detriment of non-human entities. These critiques led to a posthumanist movement that reacts to anthropocentrism and human primacy and seeks mediation and reconciliation of existence from a non-hierarchical standpoint with other beings (See Ferrando, 2019).
3. Most academic works focus on human dignity, but dignity has also been applied to posthuman beings (Bostrom, 2005), non-human beings (Gruen, 2014), and the living environment (Huber, 1991).
4. Nussbaum recognizes the dignity of animals and other non-human beings, but unlike Kant, she sees our obligation to protect as not deriving from duties to ourselves.
5. In August 2019 the Mexican government added three more stations to facilitate access to the vast number of tourists visiting Cancun and increase its attractiveness via a duty-free station on the border with Belize and one in Tabasco where tourists could easily access more archaeological sites and natural wonders (Yucatan Times, 2019).
6. Governmental and non-governmental organizations (NGOs) have supported ecotourism projects in the Selva Maya region as a strategy to achieve nature conservation and poverty alleviation (Daltabuit et al., 2007). There are successful cases of indigenous social enterprises formed by dozens of communities (see Salvatierra, 2016) and community-based tourism (see Beavers, 1995; Daltabuit et al., 2007; Ramayo, 2017) as a result of these efforts. However, there are also communities where limited economic resources, lack of knowledge, and the poor negotiation skills of local leaders have prevented the emergence of community-based ecotourism ventures (Mendoza-Ramos & Prideaux, 2014). The Mayan Train project should include strong collaboration with the NGOs that have promoted the formation of community-based tourism in the region; these organizations should continue helping local indigenous communities to develop the necessary skills to start these ventures.

Disclosure statement

No potential conflict of interest was reported by the author(s).

References

Ahmed, F., Ahmed, N., Pissaride, C., Stiglitz, J. (2020). Why Inequality could spread CoVID-19. *The Lancet Public Health*. Retrieved 13 April from https://doi.org/10.1016/S2468-2667(20)30085-2

Alluri, R. (2009). *The role of tourism in post-conflict peacebuilding in Rwanda*. Working Paper 2/2009. Swiss Peace Foundation.

Alsema, A. (2018). Tourism helps reconciliation in Colombia formers conflict zones. Colombia Reports. https://colombiareports.com/tourism-helps-reconciliation-colombias-former-conflict-zones/

Barton, A., & Leonard, S. (2010). Incorporating social justice in tourism planning: Racial reconciliation and sustainable community development in the Deep South. *Community Development*, *41*(3), 298–322. https://doi.org/10.1080/15575330903444051

Beavers, J. (1995). *Community-based tourism in the Maya Forest: Six case studies from communities in México, Guatemala and Belize*. The nature conservancy USAID/MAYAFOR Project.

Bédard, R. (2003). Los Fundamentos del Pensamiento y las Prácticas Administrativas: El rombo y las cuatro dimensiones filosóficas. *AD-minister*, *3*, 68–88.

Beeton, S. (2005). The case study in tourism research: a multi-method case study approach. In B.W. Ritchie, P. Burns and C. Palmer (Eds.), *Tourism research methods: integrating theory with practice* (pp. 37–48). CABI Publishing.

Bolton, S. ((2007). ed.). *Dimensions of dignity at work*. Routledge.

Bostrom, N. (2005). defense of posthuman dignity. *Bioethics*, *19*(3), 202–214. https://doi.org/10.1111/j.1467-8519.2005.00437.x

Bullard, R. (2000). *Dumping in Dixie: Race, class, and environmental quality* (3rd ed.). Westview Press.

Bullard, R. (ed.). (1993). *Confronting environmental racism: Voices from the grassroots*. South End Press.

Camargo, B. (2011). *Justice and fairness in tourism: A grounded theory study of cultural justice in Quintana Roo, Mexico* [Doctoral dissertation]. Texas A&M University.

Camargo, B. (2019a). Mayan cultural tourism in Quintana Roo, Mexico. In T. Jamal (ed.), *Justice and ethics in tourism* (pp. 51–59). Routledge.

Camargo, B. (2019b). Palenqueras in Colombia: A case of cultural injustice in tourism. In T. Jamal (ed.), *Justice and ethics in tourism* (pp. 72–77). Routledge.

Camargo, B., Lane, K., & Jamal, T. (2007). Environmental justice and sustainable tourism: The missing cultural link. *The George Wright Forum*, *24*(3), 70–80.

Camargo, B. A., Vázquez-Maguirre, M., & Benito, A. (2020, accepted). Humanistic tourism: New directions for justice for Indigenous groups in tourism destinations. In E. Giudici, Della Lucia, M. & Pettinao, D. (Eds), *Humanistic Tourism: The path of The Future or Utopia?*. Routledge.

Campana, A. (1946). The origin of the word 'Humanist. *Journal of the Warburg and Courtauld Institutes*, *9*, 60–73. https://doi.org/10.2307/750309

Castellanos Guerrero, A. (2000). Antropología y racismo en México. *Desacatos*, *4*(2000), 53–79.

Castellanos Guerrero, A. (2003). Imágenes racistas en ciudades del sureste. In A. Castellanos Guerrero (ed), *Imágenes del racismo en México* (pp. 35–142). Plaza y Valdés.

Caton, K. (2016). A Humanist Paradigm for Tourism Studies?. In A. M. Munar and T. Jamal (Eds.), *Tourism Research Paradigms: Critical and Emergent Knowledges* (pp. 35–56). Emerald Group. https://doi.org/10.1108/S1571-504320150000022009

Cole, S. (2012). A political ecology of water equity and tourism: A case study from Bali. *Annals of Tourism Research*, *39*(2), 1221–1241. https://doi.org/10.1016/j.annals.2012.01.003

Cole, S. (2017). Water worries: An intersectional feminist political ecology of tourism and water in Labuan Bajo. *Annals of Tourism Research*, *67*, 14–24. https://doi.org/10.1016/j.annals.2017.07.018

Comisión Nacional para el Conocimiento y Uso de la Biodiversidad. (2017). Selva Maya: Monitoreo de Biodiversidad y Cambio Climático. Gobierno de México. https://www.gob.mx/conabio/prensa/selva-maya-101932

Comisión Nacional para el Desarrollo de los Pueblos Indígenas. (2018). Atlas de los Pueblos Indígenas de México. http://atlas.cdi.gob.mx/

Daltabuit, M., Hernández, A., Barbosa, S., & Valdez, A. (2007). Reflexiones en torno al ecoturismo en Yucatán. *Península*, *2*(1), 103–124.

Debes, R. (2009). Dignity's gauntlet. *Philosophical Perspectives*, *23*(1), 45–78. https://doi.org/10.1111/j.1520-8583.2009.00161.x

Dillon, R. S. (2018). Respect. *The Stanford Encyclopedia of Philosophy*. Retrieved 13 April 2020 from https://plato.stanford.edu/archives/spr2018/entries/respect/

Dion, M. (2017). Corporate Citizenship, Social Responsibility, and Sustainability Reports as "Would-be" Narratives. *Humanistic Management Journal*, *2*(1), 83–102. doi: https://doi.10.1007/s41463-017-0022-x

Eisenhardt, K. (1989). Building theories from case study research. *Academy of Management Review*, *14*(4), 532–550. https://doi.org/10.5465/amr.1989.4308385

Ferrando, F. (2019). *Philosophical Posthumanism*. Bloomsbury Academic.

Figueroa, R. (2006). Evaluating environmental justice claims. In J. Bauer (Ed), *Forging environmentalism: Justice, livelihoods and contested environments* (pp. 360–376). M. E. Sharpe.

Figueroa, R., & Waitt, G. (2010). Climb: Restorative justice, environmental heritage, and the moral terrains of Uluṟu-Kata Tjuṯa National Park. *Environmental Philosophy*, *7*(2), 135–164.

Finnerty, M. (2015, 13th March). Compromise complicated in debate over faith, water, land. Retrieved 14 October 2019 from https://www.azcentral.com/story/travel/2015/03/13/navajo-nation-files-human-rights-protest-snowbowl-snow-making/70214892/

Floyd, M., & Johnson, C. (2002). Coming to terms with environmental justice in outdoor recreation: A conceptual discussion with research implications. *Leisure Sciences*, *24*(1), 59–77. https://doi.org/10.1080/01490400252772836

Fraser, N. (2000). Rethinking recognition. *New Left Review*, *3*, 107–118.

Fromm, E. (1961). *Marx's concept of man*. Frederick Ungar.

Fuentes, Y. (2018). Tren Maya: así es el ambicioso proyecto que propone AMLO y tiene un costo de miles de millones de dólares para México. BBC Mundo. Retrieved 12 May 2019 from: https://www.bbc.com/mundo/noticias-america-latina-45254080

Gobierno de México. (2018). *¿Qué es el Tren Maya?* Retrieved 15 July 2019 from https://www.gob.mx/cms/uploads/attachment/file/455316/TM_Folleto_Espa_ol.pdf

Gruen, L. (2014). Dignity, captivity, and an ethics of sight. In L. Gruen (ed), *The ethics of captivity* (pp. 231–247). Oxford University Press.

Harvey Lemelin, R., Powys Whyte, K., Johansen, K., Higgins Desbiolles, F., Wilson, C., & Hemming, S. (2013). Conflicts, battlefields, indigenous peoples and tourism: addressing dissonant heritage in warfare tourism in Australia and North America in the twenty-first century. *International Journal of Culture, Tourism and Hospitality Research, 7*(3), 257–271. https://doi.org/10.1108/IJCTHR-05-2012-0038

Higgins-Desbiolles, F. (2003). Reconciliation tourism: Healing divided societies. *Tourism Recreation Research, 28*(3), 35–44. https://doi.org/10.1080/02508281.2003.11081415

Higgins-Desbiolles, F. (2008). Justice tourism and alternative globalisation. *Journal of Sustainable Tourism, 16*(3), 345–364. https://doi.org/10.1080/09669580802154132

Higgins-Desbiolles, F. (2010). Justifying tourism: Justice through tourism. In S. Cole & N. Morgan (Eds.), *Tourism and inequality: Problems and prospects*, (pp. 194–211). CABI.

Hodson, R. (2001). *Dignity at Work*. Cambridge University Press.

Huber, W. (1991). Rights of Nature or Dignity of Nature? *The Annual of the Society of Christian Ethics, 11*, 43–60.

Hultsman, J. (1995). Just tourism: An ethical framework. *Annals of Tourism Research, 22*(3), 553–567. https://doi.org/10.1016/0160-7383(95)00011-T

International Labour Organization. (1944). ILO Declaration of Philadelphia. Retrieved 17 March 2020 from https://www.ilo.org/legacy/english/inwork/cb-policy-guide/declarationofPhiladelphia1944.pdf

International Labour Organization. (1989). C169 – Indigenous and Tribal Peoples Convention. Retrieved 23 June 2019 from https://www.ilo.org/dyn/normlex/en/f?p=NORMLEXPUB:12100:0::NO::P12100_ILO_CODE:C169

Isaac, R., & Hodge, D. (2011). An exploratory study: Justice Tourism in controversial areas. The case of Palestine. *Tourism Planning & Development, 8*(1), 101–108. https://doi.org/10.1080/21568316.2011.554048

Jamal, T. (2019). *Justice and ethics in tourism*. Routledge.

Jamal, T., & Camargo, B. (2018). Tourism governance and policy: Whither justice? *Tourism Management Perspectives, 25*, 205–208. https://doi.org/10.1016/j.tmp.2017.11.009

Jamal, T., Borges, M., & Stronza, A. (2006). The institutionalisation of ecotourism: Certification, cultural equity and praxis. *Journal of Ecotourism, 5*(3), 145–175. https://doi.org/10.2167/joe120.0

Jamal, T., Camargo, B., Sandlin, J., & Segrado, R. (2010). Tourism and cultural sustainability: Towards an eco-cultural justice for place and people. *Tourism Recreation Research, 35*(3), 269–279. https://doi.org/10.1080/02508281.2010.11081643

Johnston, A. M. (2014). *Is the sacred for sale: Tourism and Indigenous peoples*. Earthscan.

Kateb, G. (2011). *Human dignity*. Harvard University Press.

Lastiri, D. (2019). Ahora, pueblos indígenas contra Santa Lucía. El Universal. Retrieved 30 August 2019 from https://www.eluniversal.com.mx/nacion/ahora-pueblos-indigenas-contra-santa-lucia

Lee, S., & Jamal, T. (2008). Environmental justice and environmental equity in tourism: Missing links to sustainability. *Journal of Ecotourism, 7*(1), 44–67. https://doi.org/10.2167/joe191.0

Ley del Instituto Nacional de los Pueblos Indígenas. (2018). Comisión Nacional de los Derechos Humanos. http://www.cndh.org.mx/sites/all/doc/Normatividad/Ley_INPI.pdf

Mattson, D. J., & Clark, S. G. (2011). Human dignity in concept and practice. *Policy Sciences, 44*(4), 303–319. https://doi.org/10.1007/s11077-010-9124-0

Maya, P. S. (2019). Selva Maya. http://selvamaya.info/es/

Melé, D. (2016). Understanding Humanistic Management. *Humanistic Management Journal, 1*(1), 33–55. https://doi.org/10.1007/s41463-016-0011-5

Mendoza-Ramos, A., & Prideaux, B. (2014). Indigenous ecotourism in the Mayan rainforest of Palenque: empowerment issues in sustainable development. *Journal of Sustainable Tourism, 22*(3), 461–479. https://doi.org/10.1080/09669582.2013.828730

Meza, M. A. (2019). Indigenismo simulado. Retrieved 12 May 2019 from https://elpais.com/internacional/2019/05/09/mexico/1557364435_294505.html

Molina, H. 2019. (Tren Maya no tendrá consecuencias ambientales. Retrieved 12 May 2019). https://www.eleconomista.com.mx/estados/Tren-Maya-no-tendra-consecuencias-ambientales-20190514-0013.html

Molina, H. (2018). En pobreza, 71% de los indígenas en México. Retrieved 12 August 2019 from https://www.eleconomista.com.mx/politica/En-pobreza-71-de-los-indigenas-en-Mexico-20180809-0145.html

Morgan, M. (2019). Mexico to 'speed up' Maya train route. Retrieved 12 May, 2019 from https://www.bbc.com/news/blogs-news-from-elsewhere-47454359

Muñoz, G. (2019). Tren Maya: "me va a costar matar la vida y la paz de mi territorio y las costumbres de mi pueblo". Aristegui Noticias. Retrieved 20 September 2019 from https://aristeguinoticias.com/2009/mexico/tren-maya-me-va-a-costar-matar-la-vida-y-la-paz-de-mi-territorio-y-las-costumbres-de-mi-pueblo/

Nussbaum, M. (2011). *Creating Capabilities- The Human Development Approach*. Belknap.

OCDE. (2017). How's life in Mexico? Retrieved 28 May, 2019 from https://www.oecd.org/statistics/Better-Life-Initiative-country-note-Mexico.pdf

Peoples Dispatch. (2019). Protests intensify over proposed telescope on Native Hawaiian sacred site. https://peoplesdispatch.org/2019/07/24/protests-intensify-over-proposed-telescope-on-native-hawaiian-sacred-site/

Pirson, M. (2017). *Humanistic management: Protecting dignity and promoting well-being*. Cambridge University Press.

Pirson, M., Martin, K., & Parmar, B. (2017). A humanistic perspective for management theory: Protecting dignity and promoting well-being. *Journal of Business Ethics, 145*(1), 1–19. https://doi.org/10.1007/s10551-017-3755-4

Pirson, M., Steckler, E. (2018). A humanistic ontology for responsible management learning. https://doi.org/10.2139/ssrn.3162851

Porter, R., & Tarrant, M. A. (2001). A case study of environmental justice and federal tourism sites in southern Appalachia: A GIS application. *Journal of Travel Research, 40*(1), 27–40. https://doi.org/10.1177/004728750104000105

Ramayo, M. M. (2017). *How they do it: Community Based Tourism; Case study of the community participation of Kikil, Yucatan, Mexico* [Master dissertation]. https://edepot.wur.nl/422175

Rawls, J. (1971). *A theory of justice*. The Belknap Press of Harvard University.

Reportur. (2019). Tren Maya: los indígenas critican que AMLO destruirá selvas. https://www.reportur.com/hoteles/2019/04/29/tren-maya-los-indigenas-critican-amlo-destruira-selvas/

República, S. D L. (2017). Población indígena de México con mayor probabilidad de caer en pobreza, destaca el IBD. Retrieved 23 August 2019 from http://comunicacion.senado.gob.mx/index.php/informacion/boletines/38021

Restorative Justice. (2019). Lesson 1: What is restorative justice. http://restorativejustice.org/restorative-justice/about-restorative-justice/tutorial-intro-to-restorative-justice/lesson-1-what-is-restorative-justice/encounter/

Robinson, M. (1999). Collaboration and cultural consent: Refocusing sustainable tourism. *Journal of Sustainable Tourism, 7*(3-4), 379–397. https://doi.org/10.1080/09669589908667345

Saldivar, E. (2018). Uses and abuses of culture: mestizaje in the era of multiculturalism. *Cultural Studies, 3*(32), 438–459. doi: https://doi.10.1080/09502386.2017.1420092

Salvatierra, H. (2016). *Chicza Rainforest: la cooperativa que vende chicles orgánicos a Europa.* Forbes México. https://www.forbes.com.mx/chicza-rainforest-la-cooperativa-que-vende-chicles-organicos-a-europa/

Schachter, O. (1983). Human dignity as a normative concept. *The American Journal of International Law, 77*(4), 848–854. https://doi.org/10.2307/2202536

Scheyvens, R. (2002). *Tourism for Development: Empowering Communities*. Pearson.

Schnegg, M., & Kiaka, R. D. (2018). Subsidized elephants: Community-based resource governance and environmental (in) justice in Namibia. *Geoforum, 93*, 105–115. https://doi.org/10.1016/j.geoforum.2018.05.010

SECTUR. (2016). *Estadística Turística Derivada de los Censos Económicos 2014.* https://www.datatur.sectur.gob.mx/Comite%20de%20Estadisticas%20del%20Sector%20Turismo/CENSOS_ECONOMICOS_2014.pdf.

Sen, A. (2001). *Development as freedom*. Oxford University Press.

Sen, A. (2009). *The idea of justice*. Harvard University Press.

Skipper, J. (2016). Community development through reconciliation tourism: The behind the Big House Program in Holly Springs, Mississippi. *Community Development, 47*(4), 514–529. https://doi.org/10.1080/15575330.2016.1146783

Solum, L. B. (2004). *Procedural justice*. University of San Diego Public Law and Legal Theory Research Paper Series. 2. Retrieved 19 March 2020 from http://digital.sandiego.edu/lwps_public/art2

Spitzeck, H. (2011). An integrated model of humanistic management. *Journal of Business Ethics, 99*(1), 51–62. doi: . https://doi.org/10.1007/s10551-011-0748-6

Taylor, C. (1991). *The Ethics of Authenticity*. Harvard University Press.

Tello, J. A. (2018). El Tren Maya y su impacto en las comunidades. Proceso. Retrieved 12 May 2019 from https://www.proceso.com.mx/561332/el-tren-maya-y-su-impacto-en-las-comunidades

Tello, J. A. (2019). A quién beneficiará el Tren Maya. Retrieved 12 August, 2019 from https://www.proceso.com.mx/597159/a-quien-beneficiara-el-tren-maya

Tourliere, M. (2019a). Grandes empresarios, prestos a beneficiarse con el Tren Maya. Retrieved 20 January 2020 from https://www.proceso.com.mx/611874/grandes-empresarios-prestos-a-beneficiarse-con-el-tren-maya

Tourliere, M. (2019b). Tren Maya provocará un "choque cultural" en algunas comunidades, afirma el titular del Fonatur. Retrieved 12 May 2019 from https://www.proceso.com.mx/584103/tren-maya-provocara-un-choque-cultural-en-algunas-comunidades-afirma-el-titular-del-fonatur

TrenmayaMX (2019b, October 12). @FonaturMX presentó el proyecto #TrenMaya a la población de rieles del paraíso, impartido por el enlace territorial @CarlosBarceloR, quien dio a conocer los beneficios y avances del mismo. Durante el evento se contó con la participación de comerciantes y empresarios de la zona. [Tweet]. https://twitter.com/TrenMayaMX/status/1183177293100912640

TrenmayaMX (2019c, October 8). El Director de @FonaturMx Rogelio Jimenez Pons se reúne con la Secretaria General Adjunta de las @ONU_e y Directora Ejecutiva de @ONUHabitatMex @MaimunahSharif para dialogar sobre el #TrenMaya. [Tweet]. https://twitter.com/TrenMayaMX/status/1181592669736226817

TrenmayaMX (2019d, October 20). Les invitamos a la presentación del proyecto del #TrenMaya en el centro comunitario @CENEPRED de seguridad pública, en Villahermosa Tabasco, a partir de las 10:00 hrs. [Tweet]. https://twitter.com/TrenMayaMX/status/1185951322534748162

TrenmayaMX. (2019a, October 16). @FonaturMx y el Colegio de @Notarios22 firmaron un convenio de cooperación para unir esfuerzos y dar seguridad jurídica en asuntos relacionados con el #TrenMaya y con los Centros

Integralmente Planeados a cargo de #Fonatur. [Tweet]. https://twitter.com/TrenMayaMX/status/1184635669827805184

United Nations (1948). Universal Declaration of Human Rights. Retrieved 2 February 2020 from http://www.verklaringwarenatuur.org/Downloads_files/Universal%20Declaration%20of%20Human%20Rights.pdf

United Nations. (2007). United Nations declaration on the rights of Indigenous People. Retrieved 24 July 2019 from https://www.un.org/development/desa/indigenouspeoples/wp-content/uploads/sites/19/2018/11/UNDRIP_E_web.pdf

United Nations. (2009). State of the world's Indigenous Peoples. United Nations Department of Economic and Social Affairs. Retrieved 2 September 2019 from https://www.un.org/esa/socdev/unpfii/documents/SOWIP/en/SOWIP_web.pdf

Van Ness, D. (2000). *The shape of things to come: A framework for thinking about a restorative justice system [Paper presentation]*. Paper Presented at the Fourth International Conference on Restorative Justice for Juveniles, Tubingen, Germany.

Varillas, A. (2019). Everything you need to know about the Maya Train project. Retrieved 18 May 2019 from https://www.eluniversal.com.mx/english/everything-you-need-know-about-the-mayan-train-project

Vázquez-Maguirre, M. (2018). Sustainable ecosystems through Indigenous social enterprises. In J. Leitão, H. Alves, N. Krueger, & J. Park (Eds.), *Entrepreneurial, innovative and sustainable ecosystems. Applying quality of life research (Best practices)* (pp. 173–189). Springer.

Vázquez-Maguirre, M. (2019). El desarrollo sostenible a través de empresas sociales en comunidades indígenas de América Latina. *Estudios Sociales. Revista de Alimentación Contemporánea y Desarrollo Regional*, *29*(53), 2–22. doi: http://dx.doi.org/10.24836/es.v29i53.617

Vázquez-Maguirre, M., & Portales, L. (2018). Profits and purpose: Organizational tensions in indigenous social enterprises. *Intangible Capital*, *14*(4), 604–618. https://doi.org/10.3926/ic.1208

Vázquez-Maguirre, M., Portales, L., & Velásquez, I. (2018). Indigenous social enterprises as drivers of sustainable development: Insights from Mexico and Peru. *Critical Sociology*, *44*(2), 323– 340. https://doi.org/10.1177/0896920516688757

von Kimakowitz, E., Pirson, M., Spitzeck, H., Amann, W., & Khan, S. (2009). Humanism in business – Concluding observations. In H. Spitzeck, M. Pirson, W. Amann, S. Khan, & E. von Kimakowitz, (Eds.), *Humanism in business* (pp. 413–427). Cambridge University Press.

Waldron, J. (2013). The paradoxes of dignity – About Michael Rosen, dignity: Its history and meaning (Harvard University Press, 2012.). *European Journal of Sociology*, *54*(3), 554–561. https://doi.org/10.1017/S0003975613000404

Wanda George, E. (2010). Intangible cultural heritage, ownership, copyrights, and tourism. *International Journal of Culture, Tourism and Hospitality Research*, *4*(4), 376–388. https://doi.org/10.1108/17506181011081541

Waterloo Institute for Social Innovation and Resilience. (n.d.). Indigenous innovation. University of Waterloo. Retrieved 29 August 2019 from https://uwaterloo.ca/waterloo-institute-for-social-innovation-and-resilience/research/indigenous-innovation

Whyte, K. P. (2010). An environmental justice framework for indigenous tourism. *Environmental Philosophy*, *7*(2), 75–92.

Winchenbach, A., Hanna, P., & Miller, G. (2019). Rethinking decent work: the value of dignity in tourism employment. *Journal of Sustainable Tourism*, *27*(7), 1026–1043. https://doi.org/10.1080/09669582.2019.1566346

Yin, R. K. (2003). *Applications of case study research*. SAGE.

Yucatan Times, (2019, August 23). Three more stations added to the Maya Train Route (FONATUR). *Yucatan Times*. Retrieved 23 August 2019 from https://www.theyucatantimes.com/2019/08/three-more-stations-added-to-the-maya-train-route-fonatur/

Zárate-Toledo, E., Patiño, R., & Fraga, J. (2019). Justice, social exclusion and indigenous opposition: A case study of wind energy development on the Isthmus of Tehuantepec. *Energy Research and Social Science*, *54*, 1–11. https://doi.org/10.1016/j.erss.2019.03.004

Indigenous food sovereignty and tourism: the Chakra Route in the Amazon region of Ecuador

Verónica Santafe-Troncoso and Philip A. Loring

ABSTRACT
This research applies the concept of food sovereignty as a framework to explore the impacts of tourism on Indigenous food systems in the Chakra Chocolate and Tourism Route (referred to as the "Chakra Route" in the paper), a tourist destination in the Amazon region of Ecuador that aims to improve the livelihoods of Kichwa people. Using a qualitative and collaborative research approach, we examine how Kichwa and non-Kichwa people in this destination area understand food sovereignty, particularly concerning tourism development. Findings show that *chakra* gardens, a traditional agroforestry method, offer a symbolic and practical embodiment of food sovereignty for local people. Participants expressed a variety of values and concerns regarding tourism and *chakra,* including on destination branding; the role Indigenous women and their traditional knowledge play in tourism; the food choices promoted to tourists; self-determination and the level of participation of Indigenous people in governance of the route. Overall, our research contributes to a pluralistic notion of justice in Indigenous tourism and illustrates how the study of food sovereignty in this Amazonia destination can serve as a holistic and collaborative frame for exploring the multidimensional impacts of tourism on communal well-being, food security, and biodiversity and cultural conservation.

Introduction

Tourism is often discussed as a desirable development strategy for Indigenous communities, but is also seen to have a generally poor track record when it comes to its impacts on local communities and ecosystems (e.g. Devine, 2017; Gascón, 2015; Johnston, 2006). This study offers an alternative perspective on tourism and development in Indigenous contexts, specifically through a focus on tourism's relationship with food sovereignty. We contest the common assumption that increasing local income through tourism is, on its own, sufficient for improving food security and other food-related outcomes for Indigenous peoples (Ambelu et al., 2018; Ashley et al., 2001; Richardson, 2010). We use the concept of food sovereignty to draw attention to the complexities of development and ways that tourism can undermine or improve local livelihoods and ecosystems. Our study focuses on a tourist route in the Amazon region of Ecuador called the

Chakra Route, where we undertook extensive qualitative research in close collaboration with Indigenous Kichwa people who have deep traditional relationships with Amazonia. Ethnographic research was done by the first author over two field seasons with people associated with the Chakra Route, which was co-developed with participants to elicit local values and concerns regarding tourism's positive and negative impacts on local food systems and food sovereignty.

Food plays a critical role in tourism (Henderson, 2009; Quan & Wang, 2004), and tourism is a growing industry in Indigenous territories worldwide (Weaver, 2010; Whitford & Ruhanen, 2016; World Tourism Organization -UNWTO, 2019). However, research exploring the nexus between Indigenous food, tourism, and food sovereignty is rare. Of the studies that have been done in this area, many have focused on the commodity value of Indigenous foods, but overlooked essential cultural, political, and environmental elements of food and food security (Du Rand et al., 2003; Rogerson, 2012), such as impacts on the availability of traditional food and the continuance of traditional food practices (Loring & Gerlach, 2009).

We focus below on two related research questions: 1) How do local participants in the Chakra Route understand food sovereignty (this concept is addressed in Kichwa terms in our interviews and workshops) and 2) how does the Chakra Route as a tourism destination affects the food sovereignty of local Kichwa people? The paper is structured as follows. First, we review key concepts in the literature on Indigenous tourism as they relate to food systems and sovereignty, and then describe the collaborative process by which we developed the methodology and implemented this research. This is followed by the findings and discussion of how this research expands our understanding of Indigenous food sovereignty and the usefulness of food sovereignty as a "boundary concept" for collaborating with Indigenous people on tourism as a sustainable and just development strategy (cf. Brand & Jax, 2007). Directions for future research are included in the concluding section.

Indigenous food systems

There are approximately 370 million Indigenous people residing in over ninety countries worldwide. Historically, these Indigenous peoples have maintained close biocultural connections with their traditional territories, stewarding biodiversity through subsistence and other environmental management activities (e.g. Balée, 2013; Berkes, 2008). According to the World Bank (2019), Indigenous people are presently the stewards of 80% of the world's remaining biodiversity. However, over the last few centuries, Indigenous people and their lands have been (and continue to be) threatened by violent colonization, land dispossession, and the advance of extractive industries in their territories (Chellaney, 2019; Ulloa, 2014). From the Arctic to the Amazon, similar patterns have been witnessed by Indigenous people living under the oppression of colonialism, including the criminalization of traditional food system practices (Padoch & Pinedo-Vasquez, 2010; Trosper, 2002), a nutrition transition from traditional to Western diets (Popkin & Gordon-Larsen, 2004; Leatherman & Goodman, 2005), and loss of languages and cultural expertise (Maffi, 2001; Regan, 2010).

Despite these tragic impacts, traditional food systems continue to be essential to the overall health, well-being, and cultural continuity of Indigenous People worldwide, and especially embody their biophysical, spiritual, and cultural connections to their land (Loring & Gerlach, 2009; Toledo & Barrera-Bassols, 2008). Following the approach used by Kuhnlein and Receveur (1996), we define Indigenous food systems as encompassing all food and food practices valued by Indigenous people. This includes not only material elements such as land, water, animals, or seeds, but also the knowledge and practices that Indigenous people have ancestrally developed about their food (Toledo & Barrera-Bassols, 2008).

Tourism and Indigenous food systems

To date, studies done at the intersection of Indigenous food and tourism have primarily focused on the economic potential that Indigenous food has in the tourism market, essentially as a commodified object of cultural experience for tourists (Du Rand et al., 2003; Rogerson, 2012). Traditional foods and food practices contribute to the health and well-being of many Indigenous peoples around the world and often in ways that extend beyond their nutritional and commodity values (Kuhnlein et al., 2009; Loring & Gerlach, 2009). It is therefore essential that research in this area expands its focus to include the nuanced nutritional, cultural, and spiritual values of Indigenous food and foodways, as well as the complex social circumstances that make many Indigenous peoples food insecure (Loring, 2017). Tourism development adds further complexity and challenges to food security on Indigenous lands.

A few key revelatory studies exist that highlight the problematic social and ecological complexities at the nexus of Indigenous food systems and tourism (e.g. Leatherman et al., 2010; Rosenberg, 2018; Torres, 2003). For instance, a long-term study in the Yucatan Peninsula in Mexico shows that although the tourism industry has provided jobs and incomes for Mayan people, it has also negatively impacted their traditional food systems and nutritional health, by shifting diets from traditional to imported foods and driving increased malnutrition, diabetes, and obesity among Mayan people (Leatherman et al., 2010; Leatherman & Goodman, 2005). Likewise, Rosenberg (2018) reports cases of water and land grabs associated with tourism development in Bali, Indonesia, which has negatively impacted people's food sovereignty. While the details vary from place to place, a clear pattern is evident: Tourism has a mixed track record at best when it comes to supporting Indigenous livelihoods and food systems.

Increasingly, scholars working with Indigenous communities to understand the impacts of tourism are also raising concerns regarding social and environmental justice (Higgins-Desbiolles, 2010; Jamal, 2019; Jamal & Camargo, 2014; Whyte, 2010), and self-determination (Johnston, 2006; Nielsen & Wilson, 2012; Williams & Gonzalez, 2017). Some critical scholars argue that tourism development cannot escape the broader historical patterns of colonization and dispossession of Indigenous peoples from their traditional lands and results in contributing further to such entrenched injustices (Devine, 2017; Johnston, 2006; Whyte, 2010; Williams & Gonzalez, 2017). By contrast, others argue that Indigenous foods can be central to a new community-driven, rights-centered paradigm for a "just" tourism that sheds colonial trappings and empowers Indigenous people to reclaim autonomy and self-determination, and restore their relationships with their lands, kin, spirituality, and others (Grey & Newman, 2018; Grey & Patel, 2015; Lee, 2018). Our work seeks to contribute directly to this ongoing debate, specifically with respect to whether food sovereignty as a framework or guidepost for tourism development can help Indigenous peoples realize such resurgence and paradigm shift toward justice and well-being.

Food sovereignty for just tourism

Several authors argue that to make tourism both just and sustainable for Indigenous host communities, it is necessary to complement economic goals with a pluralistic and participatory approach to identifying goals, visions, and benchmarks for development (Whitford & Ruhanen, 2016; Whyte, 2010). Jamal (2019) argues that a justice approach in tourism is pluralistic but also sensitive to the particular situation – it is integrated (tourism as part of a broader socio-ecological and political system), contextual (situated in place, time, and identities), and relational (recognizing the relationships among material and symbolic factors). A fair and just approach to tourism also attends to conservation of ecological, social, and cultural goods for sustainability, and to the well-being of its human and non-human inhabitants.

Following this, we argue that the concept of food sovereignty, which is often included in the political and justice agendas of Indigenous people worldwide (Grey & Patel, 2015; Whyte, 2018), offers a poignant framework for pursuing "just sustainability" in the development of tourism in Indigenous contexts (Agyeman, 2008).

A landmark report from the Nyéléni (2007) Forum for Food Sovereignty defines food sovereignty as "people's right to healthy and culturally appropriate food produced through ecologically sound and sustainable methods, and their right to define their own food and agriculture systems" (p. 9). We use this definition as a starting point, recognizing that the concept is evolving as people engaged in struggles around the world grapple with what it means to them (Edelman et al., 2014). In other words, food sovereignty takes on important place-based meanings that reflect the histories, politics, ecologies, and cultures of a region as well as the struggles in which people are presently mired (Schiavoni, 2017). With these nuances in mind, the Nyéléni report sets out a framework for food sovereignty known as the Six Pillars Framework. This framework describes food sovereignty as involving six core features: 1) a focus on food for people; 2) valuing food providers; 3) localizing food systems; 4) placing control locally; 5) building knowledge and skills; and 6) working with nature (Nyéléni, 2007). Far from being a rigid checklist, these pillars are intended as guidelines that signify how food sovereignty is not only about food production and farmers' struggles, but also about justice and connections among all the varied actors and elements in a food system. The principles presented in the Nyéléni report's Six Pillars Framework became a crucial reference during the data analysis of our research below.

Many proponents of food sovereignty also consider it to be a vehicle for pursuing self-determination and participatory governance, and for building international solidarity (Edelman et al., 2014; Grey & Patel, 2015). It is in the latter sense that food sovereignty was first promoted and popularized by La Via Campesina, a transnational social movement of peasants, small-scale farmers, Indigenous people, conscious consumers, and others affected by industrial agrobusiness and free trade. La Via Campesina advocated for a focus on food sovereignty during the 1996 World Food Summit in Rome as an alternative to neoliberal strategies for addressing food insecurity and poverty.

Many Indigenous communities similarly see food sovereignty not just as a matter of food and diet, but as a holistic vision for autonomy and self-determination: a pathway to reverse, on their own terms, the varied impacts of colonization and heal their connections to themselves, the land, and others (Coté, 2016; Whyte, 2018). In other words, many communities are wielding the language of food sovereignty as a conceptual reclaiming of cultural authority and governance around all aspects of Indigenous livelihoods, from spirituality to natural resource management, commerce, and even international trade (Coq-Huelva et al., 2018; Grey & Patel, 2015; Whyte, 2018). Despite varied interpretations and definitions, food sovereignty has gained also adherents among diverse social movements, as well as scholars, policymakers, and multinational governing agencies from both the Global South and the Global North. Though much of the scholarship around food sovereignty has been in the field of agrarian and rural studies, growing interest from scholars in other fields in enriching this subject ranges from studies related to gender (Patel, 2012) and health equity (Weiler et al., 2015). With respect to tourism, the issue of food sovereignty has received limited attention in both research and application. Gascón and Cañada (2012) argue that food sovereignty should guide the future development of tourism in the rural areas of the Global South, where past tourism development has largely undermined the ability of rural peoples to access to land, water, and fair wages (Devine, 2017; Gascón, 2015; Hall, 2007). Our research study below aims to contribute toward filling this gap.

Case study: the Chakra Route in the Amazonia of Ecuador

This research used an in-depth qualitative, case study approach to explore the relationship between tourism and food sovereignty in Indigenous contexts. Yin (2009) recommends the case study methodology when the goal of the research is to understand real-life phenomena and the contextual conditions that make this phenomenon unique, meaningful, or otherwise important. We selected as our case a tourist route in the upper basin of the Amazonia, in the Ecuadorian province of Napo, as a single, "revelatory" case for analysis – meaning that it offers a window into otherwise unexplored phenomena (Yin, 2009). As we discuss below, this route is unique in bringing together indigenous food systems and tourism and giving the Indigenous Kichwa people a platform to revitalize traditional practices and to contest dominant narratives about food security, poverty, and economic development. The high level of interest demonstrated by the local Kichwa representatives of the route when we discussed the research idea was a determining factor in developing this research, and close collaboration on developing and implementing the study was immediately established. Community leaders argued that the participatory dynamic of this research could create positive changes as the project evolved and the inclusion of policymakers in the study could influence local policies for tourism development. The methodology reflects sensitivity to decolonizing research such as advocated by Linda Tuhiwai Smith in *Decolonizing Methodologies* (Smith, 2012), though we developed our own particular way to undertake this with the Kichwa participants as described further below.

Setting and background

The Chakra Route was officially launched in 2017 under the name of the "Cacao Route." Its antecedents go back to 2010 when the Cacao Roundtable, a group of local actors engaged in cacao farming, created a project proposal for this route. The primary goals of this route were to diversify the local economy and extend the participation of Kichwa communities in the value chains of cacao farming and tourism. Although this route initially focused only on Kichwa entrepreneurs engaged in cacao farming or tourism, it has expanded its focus over the years, by including non-Kichwa local entrepreneurs and enlarging its focus from cacao cultivation to the entire Indigenous system of *chakra* gardening. In 2018, members of the route changed the name of the route from "The Cacao Route" to "Chakra: Chocolate & Tourism" (hereafter, the "Chakra Route") to reflect this expanded focus. Chakra is a traditional agroforestry system that Kichwa people have used for centuries, primarily for subsistence but more recently for cultivating high-quality cacao to support local livelihoods and families (Coq-Huelva et al., 2018). It refers to a form of local agroecological farming that uses traditional knowledge passed down over generations to grow crops and use its plants in a way that maintains biodiversity and forest health (Coq-Huelva et al., 2017; Irvine, 1989; Perreault, 2005). Kichwa women farmers, also known as Chakra Mamas, are bearers of this ancient knowledge and lead the sustainable agroecological practices in the *chakra* gardens (Castellón, 2015; Coq-Huelva et al., 2017).

The Chakra Route is governed under a collaborative system facilitated by the *Fondo Ecuatoriano de Cooperacion para el Desarrollo* (FECD). Members of this collaborative group include several Kichwa organizations, government agencies, NGOs, educational centers, and non-Kichwa entrepreneurs. Four Kichwa cacao farming cooperatives and ten Kichwa community-based tourism associations (CBTs) are the primary stakeholders in this route (See Table 1). Altogether the membership comprises nearly 1,500 members (Fondo Ecuatoriano de Cooperación para el Desarrollo [FECD], 2017).

Table 1. Kichwa organizations that are members of the Chakra Route.

Member Organization	Primary activities
Wiñak	Cacao and Guayusa (*Ilex guayusa*) farming and exporting Chocolate production "Bean to bar" tourism
Kallari	Cacao farming and exporting Chocolate production "Bean to bar" tourism
Tsatsayaku	Cacao farming Chocolate production "Bean to bar" tourism
Amanecer Campesino	Cacao farming "Bean to bar" tourism
CBT Sinchi Warmi	*Chakra* tours, lodging, traditional cuisine, Guayusa tea ceremonies, chocolate spa, artisanal chocolate making
CBT Shiripuno	*Chakra* tours, artisanal chocolate making, traditional dances
Kamak Maki	*Chakra* tours, handcrafts, environmental education
CBT Shandia	*Chakra* tours, lodging, traditional cuisine, biking, artisanal chocolate making
CBT Cotundo	Archaeological tourism
CBT Cavernas Templo de Ceremonia	*Chakra* tours, hiking
CBT Santa Rita	*Chakra* tours, artisanal chocolate making, traditional dances and music
CBT AMUPAKIN	Traditional medicine, *chakra* tours, lodging, traditional cuisine, environmental education
CBT Tamia Yura	Cave tours, *chakra* tours, environmental education
CBT Sacha Waysa	Birdwatching, *chakra* tours, handcrafts, artisanal chocolate making.

Source: FECD 2017. CBT = Community-based Tourism.

Cacao cooperatives include tourism as a complementary activity to their cacao business by providing tours of their farms and engaging visitors in a "bean to bar" experience where they learn the process of making chocolate. The CBTs are compromised of Kichwa members and their families, who work together to control the development of tourism in their territories and governed their entrepreneurship in a communitarian ethos. They are led primarily by the Kichwa women, who are further innovating their services by providing tours in the *chakra* gardens guided by "Chakra Mamas" and incorporating wellness and food experiences exhibiting their cultural values (see Figure 1). Non-Kichwa entrepreneurs also participate in the route in a variety of ways; restaurateurs, for example, are integrating products from the *chakra* gardens into their menus and tour guide operations are complementing their own offerings with field trips to CBT *chakras*.

Figure 1. Chakra Mamas describing and interpreting their chakra gardens to visitors at Amupakin in Archidona, Ecuador. Photo Credit: Author 1 (Summer 2018).

One critical aspect of the Chakra Route's heritage is that it overlaps with the ancestral territory of the Kichwa Napo Runa people, a cultural group described as the embodiment of resistance, resilience, and entrepreneurship (Coca Pérez, 2016; Coq-Huelva et al., 2017; Perreault, 2005). *Chakra,* the traditional method of agroforestry among the Napo Runa, has long been a key component of local food security and household economic diversity. In the last decade, however, high demand for organic cacao and other crops has led some of the local people to reduce or abandon *chakra* practices (Coq-Huelva et al., 2018). Another aspect is that the Kichwa were pioneers in developing communitarian models of business in the chocolate and tourism sectors in Ecuador (Coca Pérez, 2016; Coq-Huelva et al., 2018). However, Houck et al. (2013) offered evidence that increased market integration is negatively affecting the health of Kichwa people by leading them to become overly dependent on lower-quality market foods.

Another important feature of this case is that at a national level Ecuador recognized food sovereignty as a key principle in their 2008 constitution. Likewise, at the provincial level, the government of Napo recently introduced legislation that promotes the *chakra* garden as a food sovereignty strategy.

Lastly, this tourism route is part of the Amazon, which is a hugely important biome from a global perspective for ecosystem services such as carbon sequestration, capacity for slowing climate change, and biodiversity (Foley et al., 2007; Malhi et al., 2008). The Amazon rainforest is the home of numerous Indigenous groups, all of which adapted their ways of life to this ecosystem over centuries (Balée, 2013; Coq-Huelva et al., 2017). However, Amazonia, particularly the Ecuadorian Amazon Region, have been subject to intense ecological and political pressures from extractive industries and urbanization (Rudel et al., 2002). Indigenous people worldwide share experiences of globalization and industrialization (Adams & Hutton, 2007); thus, our case study has relevance for Indigenous tourism and food sovereignty initiatives around the world.

Methods

In the Summer of 2018 and the Spring of 2019, Author 1, who is originally from Ecuador, conducted 28 semi-structured interviews and two workshops with locals associated with the Chakra Route. In total, Author 1 spent 24 weeks in the field, 16 weeks in the first year, and eight weeks in the second. Our approach to the research and analysis is primarily interpretivist in philosophy (Moon & Blackman, 2014) as we are seeking to learn from how a specific set of people understand and value food sovereignty, especially with respect to the challenges and opportunities that they see in their participation in the Chakra Route. As described below, we collaborated with Kichwa leaders to define specifics regarding participant recruitment and our research activities.

Spanish was the primary language of communication between Author 1 and the participants. In total, 52 participants, including members of community-based tourism, policymakers, entrepreneurs, health experts, NGOs' officers, and academics, collaborated in this research. Thirty-three participants were females and the rest males; 37 self-identify as Kichwa (here the term "Kichwa" includes people who identify as Kichwas, Amazonian Kichwas, or Napo Runas) and 15 as non-Kichwa. Among non-Kichwa participants, 12 self-identify as mestizos, and three as foreigners. The term mestizo describes a mixed ethnicity in Latin American society of both Spanish and Indigenous descent.

Purposive recruitment of participants was done with the aid of a local research coordinator and then asking each interviewee for additional suggestions for whom to contact. Our goal was to foreground cultural diversity and pursue a meaningfully collaborative research approach; as such, Author 1 engaged Kichwa community leaders in conversations to identify key participants; the best locales to hold workshops and interviews; the most appropriate activities to be incorporated as part of the workshops; and the best ways to translate the results to the community.

Table 2. Main research focus and guiding questions.

Research Focus	Guiding questions for interview guides and workshops
1. Food sovereignty from a Kichwa perspective	• What does food sovereignty mean to you? • Why are *chakras* central to the tourist route? • What is "healthy" food? • What are the hazards of development, including tourism, on traditional food?
1. Impacts of tourism on the food sovereignty of Kichwa people along the Chakra Route	• Are local people producing and consuming traditional foods? • What is the role of women in the Chakra Route? • Are traditional foods being marketed outside of tourism? • What is the role of traditional knowledge in the Chakra Route? • How does the Chakra Route affect or contribute to intercultural relationships and community building? • What are the environmental concerns related to the impacts of the Chakra Route?

Most of the encounters between Author 1 and participants were held in community spaces along the route.

Following the recommendations of Kichwa leaders, the 15 non-Kichwa participants were invited to participate because they play key roles in connecting Kichwa food systems and tourism along the route. Kichwa leaders also helped in planning the data collection meetings to allow Kichwa participants multiple culturally appropriate ways to express their ideas. For instance, traditional food was shared during the meetings for this research, which recognized the importance of ceremony in research with Indigenous people (Wilson, 2008).

The data collection process was done in two stages. The first stage involved 20 interviews and a workshop session with ten participants to collect information related to local people's values regarding food sovereignty and the concerns that participants have about their traditional food system. For the second stage, there were eight additional interviews and a second workshop with 14 participants was held to explore how tourism influences food sovereignty in the area. Table 2 shows the main research focus (see research questions in the Introduction section) and the kinds of guiding questions used during conversations with participants.

Data analysis

Recordings and other materials (e.g. drawings) collected during the two field trips were transcribed and then analyzed using a thematic analysis and the NVIVO program (version 12). For the first research topic, Author 1 coded transcripts inductively; the Six Pillars Framework noted earlier provided a helpful guide to understand and situate emergent themes as they arose. For the second research topic, Author 1 used inductive data analysis too. Inductive thematic analysis followed the procedure suggested by Braun and Clarke (2006), whereby codes are first generated from the text and then organized into emergent themes. Emergent themes were discussed with participants (i.e. member checking, see below). Author 2 also performed code checking. Both authors discussed the codes and coding scheme as it developed.

Trustworthiness

Thematic analysis in interpretivist research does not simply rely on the frequency of occurrence of themes in transcripts, notes, and other "texts" as an indicator of conceptual importance or centrality, because these texts are necessarily disembodied from the intersubjective research engagement between the researcher and participants that produced them (Agar, 2013; Clifford, 1983). Instead, we identified and validated key themes through a set of activities for achieving trustworthy analysis in qualitative research, including member checking, journaling, and triangulation of research methods (Amankwaa, 2016; Hays & Singh, 2012). Member checking involves receiving feedback from participants at multiple stages in the research from data collection to analysis. Participants had the opportunity to check the transcripts of the interviews before our initial coding, and Author 1 also met with Kichwa leaders and other key participants to share preliminary results, obtain feedback on the emerging themes, and discuss saturation (whether sufficient interviews had been completed). These ongoing interactions increase the trustworthiness of the results by ensuring that our findings are fully informed by the understandings and interpretations of the participants.

Journaling was a supportive research method in this study (Amankwaa, 2016). As a woman descendant of Kichwa people of Cotopaxi and a long-time participant in grassroots organizations, Author 1 has extensive experience with the challenges faced by people in these communities. Author 1 kept a research journal to acknowledge and account for her standpoint and empathy in this regard (Lee, 2018; Smith, 2012; Wilson, 2008). This journal provided the basis for dialogue and code checking with Author 2 on decisions about data collection, analysis, identification of emergent themes in the data, and the preparation of this manuscript.

Finally, triangulation describes the use of multiple forms of research data to improve overall understanding, not for corroboration or validation of the data per se, but to ensure that our account is "rich, robust, comprehensive and well-developed" (Amankwaa, 2016, p. 122). Here, Author 1 triangulated among interview transcripts, the outcomes of workshops, feedback received from member checking, and the material captured in her journal entries.

Results

Our findings follow two primary narratives: The first explores how Kichwa people understand food sovereignty, the concerns that people presently have regarding food sovereignty, and how these meanings and concerns related to the Six Pillars Framework described earlier. Going back and forth between the themes that emerged from our coding and the Six Pillars Framework helped us to more fully appreciate and investigate the chakra gardens as an embodiment of food sovereignty for the Kichwa people. The second narrative explores how the inclusion of *chakra* gardens and other Kichwa food traditions in the Chakra Route influences the food sovereignty of Kichwa people. This influence includes symbolic and social benefits as well as emerging concerns that people raised about how tourism has developed in the route thus far. As a matter of style, we offer direct quotes as exemplars of key emergent themes and, on occasion, note parenthetically the demographic details of participants who offered a particularly informative perspective that we opt to summarize.

Kichwa perspectives on food sovereignty

We started this project with an assumption that many participants would be familiar with the concept of food sovereignty because food sovereignty features centrally in the Ecuadorian Constitution of 2008 and also features heavily in the platforms of Indigenous and peasants' organizations in Latin America (Grey & Patel, 2015). However, in the first phase of this research, we quickly learned from Kichwa leaders that, while food sovereignty is a term Indigenous leaders and policymakers use, it is not commonly discussed by Kichwa people in general. In its place, we found that participants often spoke of their *chakra* garden or "*chakra* system" to capture and communicate their diverse values and concerns regarding Indigenous food and food systems. As we show below, participant narratives regarding the *chakra* system corroborate and inform the Six Pillars Framework well.

Figure 2. Kichwa women from Tzawata (Aso. Tsatsayaku) describing a poster that they created during a workshop to explain what chakra means for them. Values listed are Kichwa women identity, variety of products, culture, unity and empowerment, sharing, community development, and biodiversity conservation. Photo Credit: Author 1 (Summer 2018).

When discussing *chakra,* Kichwa participants discussed material and non-material values and benefits: Material includes traditional food, income, and income diversity, while non-material includes identity, empowerment, and culture. In addition to these social and economic values, people also raised ecological values, such as biodiversity conservation (Figure 2). Among these diverse values, however, the most heavily emphasized by participants was the strong connection between Kichwa women and their *chakra* gardens. As noted earlier, Kichwa women who manage *chakra* are known as Chakra Mamas, a symbolic title which, according to Uzendoski (2005), signifies these women hold traditional wisdom received from their ancestors on how to ask *Nunkui,* the spirit of the *chakras*, to bless and provide food in their *chakras*. Many people spoke at length about the importance of Chakra Mamas. One participant described this important spiritual and ancestral connection between women, *chakra*, and Kichwa culture as follows,

> *Chakra* is more than crops. It is the place where Kichwa women learn from the land and restore their energy … We do not see our *chakras* as a task, instead our *chakra* is the place where we restore and heal ourselves. (Policymaker01, Female, Kichwa, August 09, 2018)

This quote illustrates the strong multifaceted connection between Kichwa women and their *chakras* that we observed, and many people discussed. It also exemplifies a sentiment expressed by many interviewees: Among Kichwa people, *chakra* gardens represent more than just a piece of land useful for growing food. Indeed, the *chakra* gardens emerged as a powerful symbolic embodiment of food sovereignty; they are a source of a diversity of foods, and a platform for agency, building community, teaching traditional knowledge, expressing cultural identity, empowering women, stewarding the environment, and maintaining spiritual wellness. Interestingly, many of these values for *chakra* came up in discussions with participants about threats to or worries for their *chakra* and traditional culture (see Table 3), which we describe in more detail in the following sections.

Chakra and relationships

Participants raised multiple concerns related to their *chakras* (Table 3). The most heavily emphasized concerns were about Kichwa people's relationships both with their traditional food and

Table 3. Main concerns of the Kichwa people regarding the Six Pillars of Food Sovereignty. A number of occurrences of each theme, and the number of respondents raising each theme, are listed here for reference, but should not be interpreted as an indicator of relative significance.

Food Sovereignty Pillar	Sub-theme (Inductive coding)	# of Occurrences	# of Respondents	Example Quote
1. Focuses on food for people	*Chakras* are becoming spaces of tensions between cash-crop farming vs. subsistence farming	17	10	"Our *chakras* used to be exclusively for producing our food. We had lots of products to eat and to share with the birds and other animals that used to come to our *chakras*. Now cacao is taking over the space in the *chakras*" (Policymaker01, Female, Kichwa, August 09, 2018).
	Changing foodways	25	14	"I know some Kichwa people that prefer to sell the food from their *chakras* and buy food from the supermarket instead. It is like buying stuff from outside gives them value as a person" (NGO05, Female, Foreigner, April 25, 2019).
2. Values food providers	Chakra Mamas' suffer discrimination when they bring their *chakra* products to urban areas	17	10	"Chakra Mamas sell their products on the sidewalks, without any shelters. (...) The customers in the city do not want to pay a fair price for these products" (Policymaker04, Female, Kichwa, April 29, 2019).
	Violence and gender inequalities affect the wellbeing of Kichwa women	13	5	"Being an entrepreneur is not easy for Kichwa women. When we started Sinchi Warmi, some of us suffered violence from our own husbands. Men in the community felt threatened by the idea that women could make more money than them (CBT08, Female, Kichwa, July 29, 2018).
3. Localizes food systems	*Chakra* products are not valued in local markets; instead, markets import food from other regions.	17	10	"The local authorities discriminate against our Chakra Mamas, because they do not give them adequate spaces to sell their products". (CBT09, Male, Kichwa, August 05, 2018).
	Non-Kichwa people lack interest in *chakra* products and the overall Kichwa gastronomy.	13	11	"If you walk in El Tena or El Coca [main cities in the upper Amazon region of Ecuador], you will find lots of fast food places, but restaurants offering Amazon cuisine will be rare" (Academia01, Male, Mestizo, August 21, 2018).
4. Democratic control over the food system	Access to arable land for Kichwa people is declining	21	9	"Collective land tenure has been our [Kichwa people] way to secure land access. But, mining and oil companies have been pushing the government to let them enter to our territories." (CBT13, Male, Kichwa, August 05, 2018).
	Lack of legal protection of traditional knowledge related to food systems.	3	3	"Our traditional knowledge is not privatized or patented; it is common to all Kichwa people. I feel that this common condition makes it easier for outsiders to profit from our knowledge" (Entrepreneur03, Male, Kichwa, April 24, 2019).
5. Builds knowledge and skills	Traditional knowledge should inform development projects in the area	11	6	"During a project of sustainable agriculture, we realized that it is crucial to include Chakra Mamas as facilitators because they are the ones that have the knowledge of traditional agriculture" (Policymaker04, Female, Kichwa, April 29, 2019).

(*continued*)

Table 3. Continued.

Food Sovereignty Pillar	Sub-theme (Inductive coding)	# of Occurrences	# of Respondents	Example Quote
	Youth interests in traditional knowledge and food practices	14	7	"My mom brought me to the *chakra* when I was a little girl. She taught me how to take care of our *chakra.* I compare my experience with my youngest son's experience; he does not know how to work in the *chakra*" (CBT08, Female, Kichwa, July 29, 2018).
6. Works with nature	Biodiversity loss	19	14	"When I was a little girl, I remember eating different species of fruits, bugs, mushrooms, and fish. Now, it is almost impossible to find these foods here. I do not know the reason, maybe it is deforestation or the climate change" (CBT07, Female, Kichwa, July 27, 2018).
	Invasive species	21	21	"Tilapia is not native from the Amazonia. I think that tilapias are the cause of the disappearance of our native fish" (Policymaker01, Female, Kichwa, August 09, 2018).
	Inadequate waste management practices	12	10	"Some communities recycle materials such as plastic bottles, but the local government does not have a recycling program to pick up these materials from the communities" (NGO1, Female, Mestizo, August 13, 2018).

land and with non-Kichwa people. For example, one of the participants described the problem of hunger as a relational challenge rather than the result of a lack of food:

> As Kichwas, we learn from our ancestors that the forest is abundant. Kichwa people that leave the land and move to urban areas are the ones who are starving and in need of food because they lose their connection to the land and to the people that live in the land. In the city, those connections are replaced by money. There, if you do not have money, you will starve. (CBT4, Male, Kichwa, July 30, 2018)

Thus, in this and other interviews, it was clear that relationships and connections, not food, are at the core of the *chakra* system's importance to local people. Similarly, one Chakra Mama explained that the disappearance of *chakra* gardens would have significant impacts on the sense of community and solidarity among Kichwa people, because seeding and harvesting are both important social activities, and the food grown in *chakras* is shared widely among members of the family and the community (CBT25, Female, Kichwa, August 05, 2018).

Several Kichwa and non-Kichwa participants also raised issues of colonialism as a threat to their relationships by explaining the legacy of discrimination against Kichwa people as well as a continuing social stigma around Indigeneity is keeping both non-Kichwa and younger Kichwa people from embracing *chakra* and *chakra* products. One of the non-Kichwa participants described her experience as follows:

> We were raised with the idea that consuming products from the *chakra* gardens was a symbol of poverty… So, most of the mestizo families prefer to buy their food from El TIA [local supermarket]. (Policymaker07, Female, Mestizo, April 25, 2019).

Finally, participants raised building relationships among groups historically separated as a key factor for the governance of the local food system. One of the policymakers argued that building relationships among Kichwa and non-Kichwa people living in Napo is essential to achieving a sustainable local food system that is desirable for both groups (Policymaker04, Female, Kichwa, April 29, 2019). Table 3 summarizes the main concerns that emerged during the conversations and echoes the Six Pillars of food sovereignty. The number of occurrences of each theme and the number of respondents raising each theme are listed here for reference but should not be interpreted as an indicator of relative significance.

Impacts of tourism on chakra/food sovereignty

In our second phase of research, we explored with participants the positive and negative impacts they are observing of Chakra Route tourism on food sovereignty in the area. Four themes emerged from this analysis: destination branding; the role of Kichwa women and their traditional knowledge; the place of Kichwa foods in tourism offerings; and the route's model of participatory governance.

Destination branding

As noted above, the name of the route recently changed from "Cacao Route" to "Chakra: Chocolate and Tourism." Regarding the new name, participants who were involved in choosing it noted that neither the route itself, nor the goals of the route, had changed. Instead, the new name was chosen to signify their explicit desire to support not just a certain kind of cropping but the entire local food system. The "Cacao Route" brand, many explained, put cacao and chocolate at the center of tourism development; the "Chakra: Chocolate and Tourism" brand centered the whole *chakra* system, which is a much more holistic representation of the Kichwa cultural and environmental values they want tourists to experience. One of the Kichwa leaders explained, "the Chakra Route restores the production and consumption of products that have ancestrally been part of our diets" (CBT10 Male, Kichwa, April 23, 2019).

Indeed, several participants discussed how the new brand better accounts for their values regarding agrobiodiversity and traditional food consumption. It was a concern among many – as noted in Table 3 under Pillar 1 – that *chakras* were becoming a site of tension between traditional and cash crop activities. One participant explained the new brand helped to address this concern, especially among Chakra Mamas, that the emphasis on cacao might drive Kichwa farmers further away from *chakra* in favor of export markets (NGO6, Female, Mestizo, April 29, 2019).

Gender empowerment and traditional knowledge

Several participants discussed how the Chakra Route is an opportunity for Kichwa women in general and Chakra Mamas specifically to bring additional income to their communities and, in so doing, to reduce the chances of discrimination and violence that they face when leaving their communities to sell their products or look for jobs in urban areas. One participant explained that the Chakra Route is inspiring Kichwa women to become entrepreneurs because they can create a tourism business in their own community using the assets (their *chakra* gardens and traditional knowledge) they already have (CBT08, Female, Kichwa, July 29, 2018). Several participants identified *Sinchi Warmi* as an emblematic case of gender empowerment in the area. One of the NGO officers described *Sinchi Warmi* as "a project that is inspiring other Kichwa women to come together and create opportunities to improve their quality of life, while staying close to their families and *chakras*" (NGO6, Female, Mestizo, April 29, 2019). Many Chakra Mamas are also using their encounters with tourists to promote respect for the traditional knowledge of Kichwa people. Chakra Mamas, generally, are the people who guide any tourism activities on the route related to *chakra* gardens and traditional food. They are using this opportunity to make their traditional knowledge and practices more visible to both youths in their community and tourists. Chakra Mamas make a ritual of educating visitors on how they must behave when entering and interacting with their *chakra* gardens. As an example, during Author 1's visit to a *chakra* garden, the Chakra Mama painted Author 1's face with a mixture made from achiote seeds (*Bixa orellana*) before entering the *chakra* garden (see Figure 3). She explained that this protocol is an act of

Figure 3. The ritual of face painting before entering a chakra garden. Author 1 stands with a Chakra Mama after having her face painted, and before entering the chakra garden of the Tamia Yura community-based tourism project in Tena, Ecuador. Photo credit: Andres Santafe (Summer 2018).

respect for the spirits that live in the *chakra* gardens and that it serves to protect the visitor from snakes and other dangers while in the forest.

Several participants lamented a general lack of interest among restaurants in urban areas in *chakra* products, which they see as a barrier to further developing tourism that supports local and sustainable foods. Some participants argued this situation is specifically the result of cultural barriers between Kichwa and non-Kichwa people. One government officer responsible for tourism in the province explained it this way:

> Most of the chefs in the urban areas do not know how to create menus that really feature local foods. It is crazy to see that almost all the restaurants in El Malecon only offer fast food. I think that these chefs assumed that all foreigners only eat fast food (Policymaker02, Male, Mestizo, August 13, 2018).

Generally, however, many people commented on how the agrobiodiversity of the *chakra gardens* is an opportunity to make restaurant menus sustainable and representative of local culture. For instance, some Kichwa restaurants are using mushrooms and other vegetables in place of tilapia fish, a common ingredient which, despite its popularity, is a concern for many participants because tilapia is an invasive species. This was an important point of contestation for many regarding the sustainability of food choices in the route: Tilapia is an efficient and inexpensive species for aquaculture and can be easily incorporated into traditional dishes. In an informal survey of the menus of 20 restaurants located in the most touristic spots along the route, we found tilapia fish included in 11 of the menus. Several people expressed concern, which research (Bittencourt et al., 2014) supports, regarding the impacts of tilapia aquaculture on native fish and ecosystems. A policymaker, for example, commented:

> Although the introduction of tilapia in the region is promoted as a strategy for food security, many Kichwa communities are requesting projects to restore native fish that they think are being vanished by the invasion of tilapias in the rivers (Policymaker01, Female, Kichwa, August 09, 2018).

Further evidence of this concern is a project to replace tilapia fish on menus by restoring native fish in ponds located close to their *chakra* gardens started in 2019 by the *Sinchi Warmi* organization.

Participatory governance and food sovereignty legislation

Another finding that we report here because we feel it warrants more research relates to how people talked about governance and legislation. As crucial to the success of the Chakra Route, some participants highlighted the diversity of groups involved in the effort (Table 1), as well as the support of provincial and national food sovereignty legislation. Leaders explained that the Chakra Route's participatory governance model creates an opportunity to connect diverse actors and empower them in broader work related to food sovereignty. One leader, for example, specifically pointed to the inclusion of non-Kichwa entrepreneurs as bringing together Kichwa and non-Kichwa people to talk about the importance of the *chakra* system for the food sovereignty of all people living in the region (Policymaker04, Female, Kichwa, April 29, 2019). Two of the stakeholders involved in this project also acknowledged that an explicit goal of the Chakra Route is to contribute to the food sovereignty in the Napo province (Policymaker04, Female, Kichwa, April 29, 2019; NGO1, Female, Mestizo, August 13, 2018). In this way, the Chakra Route is contributing positively to food sovereignty by promoting more democratic control over the food system (Table 3, Pillar 4).

Discussion

Following the premise that food sovereignty is the right that people have to define their own food and agriculture systems (Nyéléni, 2007), we started this research asking Kichwa people about their own understanding of food sovereignty in the context of Kichwa-led tourism. *Chakra* gardens emerged as a symbolic embodiment of food sovereignty for the Kichwa, which is a distinct cultural system with an ethic that closely echoes with the Six Pillars Framework. Far from being just a piece of land where food is produced, Kichwa people see *chakra* gardens as having a mix of social, economic, and ecological benefits. They are a locus for practicing and teaching culture, for building relationships, for addressing the deep historical legacies of colonialism, and for realizing a diverse mix of material as well as non-material outcomes. As such, the *chakra* provides an important counterpoint to the dominant global discourse on agriculture, which tends to focus only on export market value and the productive capability of agricultural systems to feed the world (Reganold & Wachter, 2016). For the same reasons, *chakra* gardens are also noteworthy in the context of tourism, an industry that historically has a complex relationship with commodification, poverty, and environmental degradation in Indigenous contexts (Devine, 2017; Gascón, 2015; Jamal, 2019; Whyte, 2010).

Chakra, as a concrete embodiment of the abstract notion of food sovereignty, provides locals with a constant touchstone by which they can define and design how tourism will unfold in their communities. In this sense, Kichwa people are exercising their right to self-determination and promoting justice as recognition in their territories (Jamal, 2019; Whyte, 2010). Several authors have noted that too-narrow a focus on the distributional outcomes of tourism can undermine tourism's potential contributions to improving a more pluralistic approach to justice, sustainability, and well-being for Indigenous communities (Jamal, 2019; Weaver, 2010; Whitford & Ruhanen, 2016). Here, we have tangible examples of how the mere inclusion of elements from Indigenous food systems in the tourism industry is not enough to claim that tourism is contributing positively to the well-being of host communities. On its own, cacao-oriented tourism had the potential to provide economic benefits to farmers, but early on, locals realized it was also pushing farmers to move away from growing subsistence crops. By reorienting tourism around the centerpiece of the entire traditional food system, participants in the route believe that they can now use tourism as a way to build their communities without sacrificing their traditional foods, values, and identity.

The Chakra Route also shows how connecting tourism to food sovereignty promotes empowerment and representation of Indigenous people in the governance of tourism in their territories. In this way, Indigenous people can overcome a role in the tourism industry that often renders them invisible aside from as objects of cultural difference (Jamal, 2019; Nielsen & Wilson, 2012; Whyte, 2010). The Chakra Route provides Kichwa women with opportunities to guide the development of tourism in their territories in a way that is respectful to them and their culture. This finding is similar to Lee's (2018) report on how traditional food systems make tourism work for the cultural safety and well-being of Basque women fishers. Therefore, food sovereignty in Indigenous tourism is also an opportunity to empower Indigenous women.

The Chakra Route is building relationships among Kichwa and non-Kichwa people by promoting the inclusion of non-Kichwa entrepreneurs and encouraging them to use *chakra* products and develop business partnerships with Kichwa entrepreneurs. Some of the participants in this research believe that these actions are helping to overcome legacies of racism and colonialism and to break down cultural and discriminatory barriers between Kichwa and non-Kichwa people in the province of Napo. As such, the Chakra Route may also be serving as a cross-cultural platform to promote reconciliation, solidarity, and coexistence of peoples whose histories are mired in conflict (Edelman et al., 2014; Loring, 2016).

Finally, our research evoked interesting preliminary findings regarding the role of state-based policy for food sovereignty, and how this policy can be a powerful tool for achieving sustainable and just tourism development in Indigenous and rural contexts. In the Chakra Route, participants explained that national and provincial legislation for food sovereignty provides important guidelines for how to develop tourism in the absence of tourism-specific legislation. Several studies argue that the Ecuadorian state has failed to effectively implement its food sovereignty legislation (Clark, 2016; Giunta, 2014), but what we see in our case may be an important example of non-state actors (Indigenous organizations) strategically leveraging food sovereignty legislation to promote sustainable and just tourism that promotes reconciliation. This has been seen elsewhere, as scholars observed how the *Buen Vivir* ("good living") ethos, which shapes the legislation of countries such as Ecuador and Nicaragua, has similarly inspired communities to develop tourism projects that are more socially inclusive and ecologically sustainable (Chassagne & Everingham, 2019; Fisher, 2019). Future research could explore in greater detail how food sovereignty statutes justify and become mobilized in local tourism development even without explicit intervention or regulation from the state.

Conclusion

Our research started with the oft-made observation that tourism, generally, has not been an effective form of development for Indigenous people. In the Amazon region of Ecuador, however, innovation in Indigenous-led tourism appears to be countering this trend. We show concrete examples of both positive and negative impacts Indigenous-led tourism on the social, cultural, and psychological dimensions of traditional food systems. We also see evidence of how Kichwa people, by linking tourism with their food sovereignty or *chakra* system, are working to mitigate negative impacts and promote positive ones. While time must pass before the long-term benefits and impacts of the Chakra Route can be assessed, our study suggests that tourism and Indigenous values can be reconciled constructively to facilitate Kichwa well-being, self-determination, and autonomy, and contribute to the conservation of biocultural diversity in Amazonia.

Overall, our research also suggests that the concept of food sovereignty can be a useful boundary concept for engaging multiple actors in a collaborative discussion of justice, livelihoods, well-being, and sustainability. Boundary concepts are generally those concepts that can facilitate cross-cultural collaboration and exchange despite not having clear-cut definitions. The collaborative process we engaged in with locals to explore the contextual aspects of food sovereignty was essential for building our relationships and also developing the shared understanding necessary to fully explore the symbolic and practical aspects of *chakra* in local livelihoods and culture.

The primary limitation of our research is that it focuses only on the perspectives of people in the host communities and does not explore the perspectives and values of tourists. Future research from the perspectives of the visitors will provide more insights on how to make tourism sustainable for all the actors in the tourism experience. We also encourage future research to explore the dynamics of tourism and traditional food systems in other countries where food sovereignty is recognized as a critical element of their legislation.

Acknowledgments

The authors would like to acknowledge and thank M.Sc. Ayme Tanguila and her family for their support while doing the fieldwork of this project. We also acknowledge the support, time, and trust provided by all the organizations and community members of the Chakra Route, RICANCIE, and FEPTCE. Thank you as well to Dr. Tazim Jamal and the anonymous reviewers for their constructive feedback.

Disclosure statement

No potential conflict of interest was reported by the authors.

Funding

This work was supported by the Graduate Research Fellowship from the School of Environment and Sustainability of the University of Saskatchewan

References

Adams, W., & Hutton, J. (2007). People, parks and poverty: Political ecology and biodiversity conservation. *Conservation and Society*, *5*(2), 147–183. https://www.jstor.org/stable/26392879

Agar, M. (2013). *The lively science: Remodeling human social research*. Mill City Press, Inc.

Agyeman, J. (2008). Toward a 'Just' Sustainability? *Continuum*, *22*(6), 751–756. https://doi.org/10.1080/10304310802452487

Amankwaa, L. (2016). Creating protocols for trustworthiness in qualitative research. *Journal of Cultural Diversity*, *23*(3), 121–127.

Ambelu, G., Lovelock, B., & Tucker, H. (2018). Empty bowls: Conceptualising the role of tourism in contributing to sustainable rural food security. *Journal of Sustainable Tourism*, *26*(10), 1749–1765. https://doi.org/10.1080/09669582.2018.1511719

Ashley, C., Roe, D., Goodwin, H. (2001). *Pro-poor tourism report No1. Pro-poor tourism strategies: Making tourism work for the poor*. https://www.cabdirect.org/cabdirect/abstract/20013079514

Balée, W. L. (2013). *Cultural forests of the Amazon: A historical ecology of people and their landscapes*. University of Alabama Press.

Berkes, F. (2008). *Sacred ecology*. Taylor & Francis.

Bittencourt, L. S., Silva, U. R. L., Silva, L. M. A., & Dias, M. T. (2014). Impact of the invasion from Nile tilapia on natives Cichlidae species in tributary of Amazonas River, Brazil. *Biota Amazônia*, *4*(3), 88–94. https://doi.org/10.18561/2179-5746/biotaamazonia.v4n3p88-94

Brand, F., & Jax, K. (2007). Focusing the meaning (s) of resilience: Resilience as a descriptive concept and a boundary object. *Ecology and Society*, *12*(1), 23. https://doi.org/10.5751/ES-02029-120123

Braun, V., & Clarke, V. (2006). Using thematic analysis in psychology. *Qualitative Research in Psychology*, *3*(2), 77–101. https://doi.org/10.1191/1478088706qp063oa

Castellón, M. (2015). *Género y producción de cacao en sistema chacra en la Reserva de Biósfera Sumaco en la provincia de Napo-Ecuador* [Master thesis, Facultad Latinoamericana de Ciencias Sociales]. FLACSO Andes Repository. http://hdl.handle.net/10469/7649

Chassagne, N., & Everingham, P. (2019). Buen Vivir: Degrowing extractivism and growing wellbeing through tourism. *Journal of Sustainable Tourism*, *27*(12), 1909–1925. https://doi.org/10.1080/09669582.2019.1660668

Chellaney, B. (2019, January 11). *Indigenous groups are the world's endangered environmental guardians*. The Globe and Mail. https://www.theglobeandmail.com/opinion/article-indigenous-groups-are-the-worlds-endangered-environmental-guardians/

Clark, P. (2016). Can the state foster food sovereignty? Insights from the case of Ecuador. *Journal of Agrarian Change*, *16*(2), 183–205. https://doi.org/10.1111/joac.12094

Clifford, J. (1983). On ethnographic authority. *Representations*, *2*(1), 118–146. https://doi.org/10.2307/2928386

Coca Pérez, A. (2016). El turismo comunitario en el mundo de la globalización. El caso de los quichuas amazónicos. *Gazeta de Antropología*, *32*(1), artículo 08. http://hdl.handle.net/10481/42873

Coq-Huelva, D., Higuchi, A., Alfalla-Luque, R., Burgos-Morán, R., & Arias-Gutiérrez, R. (2017). Co-evolution and bio-social construction: The Kichwa agroforestry systems (chakras) in the Ecuadorian Amazonia. *Sustainability*, *9*(10), 1920. https://doi.org/10.3390/su9101920

Coq-Huelva, D., Torres-Navarrete, B., & Bueno-Suárez, C. (2018). Indigenous worldviews and Western conventions: Sumak Kawsay and cocoa production in Ecuadorian Amazonia. *Agriculture and Human Values*, *35*(1), 163–179. https://doi.org/10.1007/s10460-017-9812-x

Coté, C. (2016). "Indigenizing" food sovereignty. Revitalizing Indigenous food practices and ecological knowledges in Canada and the United States. *Humanities*, *5*(3), 57. https://doi.org/10.3390/h5030057

Devine, J. A. (2017). Colonizing space and commodifying place: Tourism's violent geographies. *Journal of Sustainable Tourism*, *25*(5), 634–650. https://doi.org/10.1080/09669582.2016.1226849

Du Rand, G., Heath, E., & Alberts, N. (2003). The Role of local and regional food in destination marketing. *Journal of Travel & Tourism Marketing*, *14*(3–4), 97–112. https://doi.org/10.1300/J073v14n03_06

Edelman, M., Weis, T., Baviskar, A., Borras, S. M., Jr, Holt-Giménez, E., Kandiyoti, D., & Wolford, W. (2014). Introduction: Critical perspectives on food sovereignty. *The* Journal of Peasant Studies, *41*(6), 911–931. https://doi.org/10.1080/03066150.2014.963568

Fisher, J. (2019). Nicaragua's Buen Vivir: A strategy for tourism development? *Journal of Sustainable Tourism*, *27*(4), 452–471. https://doi.org/10.1080/09669582.2018.1457035

Foley, J. A., Asner, G. P., Costa, M. H., Coe, M. T., DeFries, R., Gibbs, H. K., Howard, E. A., Olson, S., Patz, J., Ramankutty, N., & Snyder, P. (2007). Amazonia revealed: Forest degradation and loss of ecosystem goods and services in the Amazon Basin. *Frontiers in Ecology and the Environment*, *5*(1), 25–32. https://doi.org/10.1890/1540-9295(2007)5[25:ARFDAL]2.0.CO;2 https://doi.org/10.1890/1540-9295(2007)5[25:ARFDAL]2.0.CO;2

Fondo Ecuatoriano de Cooperación para el Desarrollo. (2017). Ruta ancestral del cacao.

Gascón, J. (2015). Pro-poor tourism as a strategy to fight rural poverty: A critique. *Journal of Agrarian Change*, *15*(4), 499–518. https://doi.org/10.1111/joac.12087

Gascón, J., & Cañada, E. (2012). Turismo y soberanía alimentaria. In J. Buades, E. Cañada & J. Gascón (Eds.), *El turismo en el inicio del nuevo milenio* (pp. 98–103). Foro de Turismo Responsable.]

Giunta, I. (2014). Food sovereignty in Ecuador: Peasant struggles and the challenge of institutionalization. *The* Journal of Peasant Studies, *41*(6), 1201–1224. https://doi.org/10.1080/03066150.2014.938057

Grey, S., & Newman, L. (2018). Beyond culinary colonialism: indigenous food sovereignty, liberal multiculturalism, and the control of gastronomic capital. *Agriculture and Human Values*, *35*(3), 717–730. https://doi.org/10.1007/s10460-018-9868-2

Grey, S., & Patel, R. (2015). Food sovereignty as decolonization: Some contributions from Indigenous movements to food system and development politics. *Agriculture and Human Values*, *32*(3), 431–444. https://doi.org/10.1007/s10460-014-9548-9

Hall, C. M. (2007). Pro-poor tourism: Do 'tourism exchanges benefit primarily the countries of the South'? *Current Issues in Tourism*, *10*(2–3), 111–118. https://doi.org/10.1080/13683500708668426

Hays, D. G., & Singh, A. A. (2012). *Qualitative inquiry in clinical and educational settings*. Guilford Press.

Henderson, J. C. (2009). Food tourism reviewed. *British Food Journal*, *111*(4), 317–326. https://doi.org/10.1108/00070700910951470

Higgins-Desbiolles, F. (2010). Justifying tourism: Justice through tourism. In S. Cole, & N. Morgan (Eds.), *Tourism and inequality: problems and prospects* (pp. 194–214). CABI International.

Houck, K., Sorensen, M. V., Lu, F., Alban, D., Alvarez, K., Hidobro, D., Doljanin, C., & Ona, A. I. (2013). The effects of market integration on childhood growth and nutritional status: The dual burden of under-and over-nutrition in the Northern Ecuadorian Amazon. *American Journal of Human Biology : The Official Journal of the Human Biology Council, 25*(4), 524–533. https://doi.org/10.1002/ajhb.22404

Irvine, D. (1989). Succession management and resource distribution in an Amazonian rain forest. *Advances in Economic Botany, 7*, 223–237.

Jamal, T. (2019). *Justice and ethics in tourism*. Routledge.

Jamal, T., & Camargo, B. A. (2014). Sustainable tourism, justice and an ethic of care: Toward the just destination. *Journal of Sustainable Tourism, 22*(1), 11–30. https://doi.org/10.1080/09669582.2013.786084

Johnston, A. (2006). *Is the sacred for sale? Tourism and Indigenous peoples*. Earthscan.

Kuhnlein, H. V., & Receveur, O. (1996). Dietary change and traditional food systems of indigenous peoples. *Annual Review of Nutrition, 16*(1), 417–442. https://doi.org/10.1146/annurev.nu.16.070196.002221

Kuhnlein, H. V., Erasmus, B., & Spigelski, D. (2009). *Indigenous peoples' food systems: The many dimensions of culture, diversity and environment for nutrition and health*. FAO & CINE.

Leatherman, T. L., & Goodman, A. (2005). Coca-colonization of diets in the Yucatan. *Social Science & Medicine (1982), 61*(4), 833–846. https://doi.org/10.1016/j.socscimed.2004.08.047

Leatherman, T. L., Goodman, A. H., & Stillman, T. (2010). Changes in stature, weight, and nutritional status with tourism-based economic development in the Yucatan. *Economics and Human Biology, 8*(2), 153–158. https://doi.org/10.1016/j.ehb.2010.05.008

Lee, E. (2018). Black female cultural safety in Tebrakunna country. In B. Grimwood, H. Mair, K. Caton, & M. Muldoon (Eds.), *Tourism and wellness: Travel for the good of all?* (pp. 1–20). Lexington Books.

Loring, P. A. (2016). Toward a theory of coexistence in shared social-ecological systems: The case of Cook inlet salmon fisheries. *Human Ecology: An Interdisciplinary Journal, 44*(2), 153–165. https://doi.org/10.1007/s10745-016-9806-0

Loring, P. A. (2017). Food (in)security and food sovereignty in the North. *Northern Public Affairs, 5*(1), 56–59.

Loring, P. A., & Gerlach, S. C. (2009). Food, culture, and human health in Alaska: An integrative health approach to food security. *Environmental Science & Policy, 12*(4), 466–478. https://doi.org/10.1016/j.envsci.2008.10.006

Maffi, L. (2001). *On biocultural diversity: Linking language, knowledge, and the environment*. Smithsonian Institution Press.

Malhi, Y., Roberts, J. T., Betts, R. A., Killeen, T. J., Li, W., & Nobre, C. A. (2008). Climate change, deforestation, and the fate of the Amazon. *Science (New York, N.Y.), 319*(5860), 169–172. https://doi.org/10.1126/science.1146961

Moon, K., & Blackman, D. (2014). A guide to understanding social science research for natural scientists. *Conservation Biology: The Journal of the Society for Conservation Biology, 28*(5), 1167–1177. https://doi.org/10.1111/cobi.12326

Nielsen, N., & Wilson, E. (2012). From invisible to indigenous-driven: A critical typology of research in indigenous tourism. *Journal of Hospitality and Tourism Management, 19*(1), 67–75. https://doi.org/10.1017/jht.2012.6

Nyéléni. (2007). *Declaration of Nyéléni*. http://www.nyeleni.org/spip.php?article290

Padoch, C., & Pinedo-Vasquez, M. (2010). Saving slash-and-burn to save biodiversity. *Biotropica, 42*(5), 550–552. https://doi.org/10.1111/j.1744-7429.2010.00681.x

Patel, R. C. (2012). Food sovereignty: Power, gender, and the right to food. *PLoS Medicine, 9*(6), e1001223–4. https://doi.org/10.1371/journal.pmed.1001223

Perreault, T. (2005). Why chacras (swidden gardens) persist: Agrobiodiversity, food security, and cultural identity in the Ecuadorian Amazon. *Human Organization, 64*(4), 327–339. https://doi.org/10.17730/humo.64.4.e6tymmka388rmybt

Popkin, B. M., & Gordon-Larsen, P. (2004). The nutrition transition: Worldwide obesity dynamics and their determinants. *International Journal of Obesity, 28*(S3), S2–S9. https://doi.org/10.1038/sj.ijo.0802804

Quan, S., & Wang, N. (2004). Towards a structural model of the tourist experience: An illustration from food experiences in tourism. *Tourism Management, 25*(3), 297–305. https://doi.org/10.1016/S0261-5177(03)00130-4

Regan, P. (2010). *Unsettling the settler within: Indian residential schools, truth telling, and reconciliation in Canada*. UBC Press.

Reganold, J. P., & Wachter, J. (2016). Organic agriculture in the twenty-first century. *Nature Plants, 2*, 15221. https://doi.org/10.1038/nplants.2015.221

Richardson, R. B. (2010). *The contribution of tourism to economic growth and food security*. https://DOI: 10.22004/ag.econ.97140

Rogerson, C. M. (2012). Tourism–agriculture linkages in rural South Africa: Evidence from the accommodation sector. *Journal of Sustainable Tourism, 20*(3), 477–495. https://doi.org/10.1080/09669582.2011.617825

Rosenberg, R. (2018). *Tourism and land grabbing in Bali: A research brief*. https://www.tni.org/en/publication/tourism-and-land-grabbing-in-bali

Rudel, T. K., Bates, D., & Machinguiashi, R. (2002). A tropical forest transition? Agricultural change, out-migration, and secondary forests in the Ecuadorian Amazon. *Annals of the Association of American Geographers, 92*(1), 87–102. https://doi.org/10.1111/1467-8306.00281

Schiavoni, C. M. (2017). The contested terrain of food sovereignty construction: Toward a historical, relational and interactive approach. *The Journal of Peasant Studies, 44*(1), 1–32. https://doi.org/10.1080/03066150.2016.1234455

Smith, L. T. (2012). *Decolonizing methodologies: Research and indigenous peoples*. Zed Books Ltd.
Toledo, V., & Barrera-Bassols, N. (2008). *Memoria biocultural: La importancia ecológica de las sabidurías tradicionales*. Icaria Editorial.
Torres, R. (2003). Linkages between tourism and agriculture in Mexico. *Annals of Tourism Research*, *30*(3), 546–566. https://doi.org/10.1016/S0160-7383(02)00103-2
Trosper, R. L. (2002). Northwest coast indigenous institutions that supported resilience and sustainability. *Ecological Economics*, *41*(2), 329–344. https://doi.org/10.1016/S0921-8009(02)00041-1
Ulloa, A. (2014). Geopolíticas del desarrollo y la confrontación extractivista minera: elementos para el análisis en territorios indígenas en América Latina. In B. Gobel, & A. Ulloa (Eds.), *Extractivismo minero en Colombia y América Latina* (pp. 425–458). Universidad Nacional de Colombia & Bero-Amerikanisches Institut.
World Tourism Organization -UNWTO. (2019). *Recommendations on sustainable development of Indigenous tourism*. https://www.e-unwto.org/doi/pdf/10.18111/9789284421299
Uzendoski, M. (2005). *The Napo Runa of Amazonian Ecuador* (Vol. 57). University of Illinois Press.
Weaver, D. (2010). Indigenous tourism stages and their implications for sustainability. *Journal of Sustainable Tourism*, *18*(1), 43–60. https://doi.org/10.1080/09669580903072001
Weiler, A. M., Hergesheimer, C., Brisbois, B., Wittman, H., Yassi, A., & Spiegel, J. M. (2015). Food sovereignty, food security and health equity: A meta-narrative mapping exercise. *Health Policy and Planning*, *30*(8), 1078–1092. https://doi.org/10.1093/heapol/czu109
Whitford, M., & Ruhanen, L. (2016). Indigenous tourism research, past and present: Where to from here? *Journal of Sustainable Tourism*, *24*(8–9), 1080–1099. https://doi.org/10.1080/09669582.2016.1189925
Whyte, K. (2010). An environmental justice framework for indigenous tourism. *Environmental Philosophy*, *7*(2), 75–92. https://www.jstor.org/stable/26168043
Whyte, K. (2018). Food Sovereignty, Justice, and Indigenous Peoples. In A. Barnhill, M. Budolfson, & T. Doggett (Eds.), *The Oxford handbook of food ethics*. (pp. 345–366). Oxford University Press.
Williams, L. K., & Gonzalez, V. V. (2017). Indigeneity, sovereignty, sustainability and cultural tourism: Hosts and hostages at 'Iolani Palace, Hawai'i. *Journal of Sustainable Tourism*, *25*(5), 668–683. https://doi.org/10.1080/09669582.2016.1226850
Wilson, S. (2008). *Research is ceremony: Indigenous research methods*. Fernwood Publishing.
World Bank. (2019, September 24). *Indigenous people*. https://www.worldbank.org/en/topic/indigenouspeoples
Yin, R. (2009). *Case study research: Design and methods*. SAGE Publications, Inc.

Roots tourism: a second wave of Double Consciousness for African Americans

Alana Dillette

ABSTRACT
The African diaspora's 're-migration' to presumed homelands in Africa has been steadily gaining popularity. In fact, with the recent emergence of the Black Travel Movement, including organizations and companies that cater specifically towards Black travelers, the industry has seen a surge in travelers wanting to '*go back home*'. This idea of returning to the homeland is permeated with symbolism and meaning, including a longing to regain a sense of social and cultural repatriation that was lost hundreds of years ago during the initial slave trade from Africa's West coast, one of the most notable countries, being Ghana. Ghana has received quite a bit of attention in academic literature, however, few have analyzed the extent to which this type of roots travel can be connected to the identities of African-Americans and their search for social justice. In order to address this gap, this study will explore the lived experiences of African-American roots travelers on a trip Ghana. Using an ethnographic case study approach, data was collected before, during and after the trip. Findings reveal a form of '*Double Consciousness*' and a sense of heightened civic commitment to social justice as a result of the travel experience.

> Just as a tree without roots is dead, a people without history of cultural roots also becomes a dead people. You take a tree, you can tell what kind of tree it is by looking at the leaves. If the leaves are gone, you can look at the bark. But, when you find a tree with the leaves gone, and the bark gone, everything gone, you call that what? A stump. And you can't identify a stump as easily as you can identify a tree. Malcolm X (1967)

Introduction

The practice of tracing one's roots to an imagined or real ancestral homeland is not a new phenomenon. In the case of the African continent, these pilgrimages can be traced back to the 1830's and have been a source of great socio-cultural, political and economic discussion amongst scholars (Drachler, 1975; Edmondson, 1986; Mensah, 2015; Mowatt & Chancellor, 2011; Schramm, 2004; Yankholmes & Timothy, 2017). From a sustainability perspective, literature highlights three critical dimensions, economic efficiency, environmental integrity and social equity and justice (Burton, 1987; Edwards, 2005; Klein-Vielhauer, 2009). A 2018 report by the United Nations World

Tourism Organization (UNWTO), *Tourism for Development*, highlights the links between tourism and sustainable development goals, including the goals of equality, peace and justice (UNWTO, 2018). Ancestral pilgrimages may contribute to sustainable tourism through social equity and justice by revealing greater meaning behind one's ethnic identity and/or supporting transformation through encountering stories that lead to changing ones world view of social injustice, spurring tangible action to reduce inequities (Barton & Leonard, 2010). This movement of people has resulted in the formation of a niche tourism market known as 'roots tourism' (Timothy, 1997). Roots tourism includes those tourists who travel to a destination, specifically chosen for its connection to ancestry. In the case of many African-Americans, these pilgrimages are a result of one of the most significant spreads of African Diasporas, the transatlantic slave trade.

In the tourism sphere, we have seen an uptake in diaspora Africans from around the globe begin to travel throughout Africa. Many African-Americans have been led to travel to West Africa, and in particular, many travel to Ghana, one of the leading roots tourism destinations for African Diasporas (Schramm, 2004; Timothy & Teye, 2004). This uptake in African diaspora roots tourism is rooted in a 'Back to Africa' movement started during the post-reconstruction era by Chief Alfred Sam. Chief Sam's ambitious plan inspired a future 'Back to Africa' movement started by the Universal Negro Improvement Association in 1922 (Benton & Shabazz, 2009). This movement continued with the concept of 'Africa interest', a time in which African-Americans were encouraged to relocate back to their homelands, and in particular Ghana, to re-settle and influence socio-economic development (Alex-Assensoh, 2010). This movement is still experiencing development in the tourism realm. In 2013, the UN declared the *International Decade for People of African Descent*. More recently, Ghanaian President Akufo-Addo, declared 2019 *The Year of the Return* as the slogan for the country, marking 400 years since the beginning of the transatlantic slave trade. The slogan served as a marketing tool to welcome African Diasporas back to Ghana, and has been supported by a myriad of festivals, homecomings and special events. In tandem with this push from the Ghana Tourism Authority, momentum in the United States has also been growing in the form of the 'Black Travel Movement' (BTM) (Dillette, Benjamin, & Carpenter, 2018). Birthed from counter narrative stories shared through social media, BTM has provided a platform for companies like Nomadness Travel Tribe, Travel Noir and Tastemakers Africa to flourish. Each of these organizations, in addition to a plethora of others, have dedicated themselves to promoting and providing travel opportunities for people of color around the globe, and in many cases, opportunities to travel back to Africa. Over the years, though the movement has shifted names and leaders, the underlying motivation has remained the same – creating a safe space for Diasporic Africans to explore the roots of their identity through travel. This continuous exploration of identity links to W.E.B. Dubois' sociological theory of Double Consciousness, a "phenomenological description of self-formation under conditions of racialization" (Itzigsohn & Brown, 2015, p. 232). Pilgrimages back to Africa have been around for some time now and have received quite a bit of attention in the academic literature (Bruner, 1996; MacGonagle, 2006; Mensah, 2015; Mowatt & Chancellor, 2011; Reed, 2004, 2010; Shackley, 2001; Yankholmes & Akyeampong, 2010; Yankholmes & Timothy, 2017). However, amongst the literature, few studies have explored how this type of tourism affects the diaspora's identity and their propensity to display a greater commitment to the sustainable tourism development goals of equality and social justice. Pritchard, Morgan, and Ateljevic (2011) call for tourism research to embrace and commit to work that uncovers the power of tourism to impact social justice and universal equality for human rights. Additionally, Tribe (2008) argues for the continued critical assessment of tourism, suggesting that research should be informed by the dominant narratives and power structures in society. Therefore, the goal of this study is gain a deeper understanding of the lived experiences of Black travellers on a roots tourism trip to Ghana through the lens of Double Consciousness. Double Consciousness suggests that self-identity formation is the result of social processes, built through ongoing interaction, with the caveat that the 'colour line' creates different processes of self-formation among racialized groups limiting communication and mutual recognition for these groups (Itzigsohn & Brown, 2015; Rawls, 2000). More specifically, this study explored how this roots tourism trip affected these travellers' identity formation process and its consequence on their action and commitment towards social justice initiatives. This work extends previous research on roots tourism in Africa by adding depth and breadth to understanding the impact that this type of travel has on identity formation and links to equality and social justice.

Literature review

Roots tourism: a journey for the Black Diaspora

An exploration of roots tourism for the Black Diaspora must begin with a discussion of the horrific and brutal Trans-Atlantic slave trade. During this slave trade, Black Africans were captured, trafficked and sold across the Atlantic Ocean to both Europe and the Americas (Teye, Turk, & Sönmez, 2011). Though the numbers are difficult to pinpoint, scholars estimate that 11 to 12 million Africans were packed onto European vessels headed for a life of slavery (Timothy & Teye, 2004). Approximately 9 to 10 million captured slaves survived the horrific conditions of the middle passage, after which they landed and were subsequently sold to White plantation owners and slave traders in the Caribbean, Latin and North America. This slave trade has had extraordinarily enduring effects on the global dispersal of Black[1] Africans who are now spread across the world and referred to today as the Black Diaspora. Those Africans that ended up in the United States went on to endure slavery, and the eventual breeding of slaves up until 1865 when slavery was abolished. Though 1865 marked the date of emancipation for slaves in the United States, these now relabelled 'African-Americans' had a long road to recovery ahead.

Most ex-slaves became sharecroppers[2] or took on other low-skilled jobs where they still faced significant discrimination up until 1964 when Jim Crow[3] segregation was outlawed. These practices not only ingrained attitudes of racism amongst American society, but they also made it difficult for Black people to experience leisure, travel and tourism in the same way their White counterparts did, resulting in an invisibility of Black people in the travel sphere (Buzinde, Santos, & Smith, 2006). This evident marginalization of Black people in the tourism realm is not altogether surprising given the historical tenets of normalized White supremacy in American society. Many of the cautious travel behaviours of Black people (often travelling in familial groups & avoiding unfamiliar places) can be traced back to the collective memory linked to slavery (Johnson & Smith, 1999).

Despite the adversity Blacks faced, many still managed to establish themselves as successful. Small groups of prosperous Blacks began to participate in leisure and travel activities as early as the late 1800's (Foster, 1999). Determined Blacks, including one of the most respected ex-slaves and abolitionists Frederick Douglas, created leisure lifestyles, gallivanting in significant numbers along the railroads in the United States, some venturing off to other countries in Europe and Africa (Foster, 1999). Those who were able to travel abroad, recount finally feeling a release from the racism ever present in their home country. Though determined, this group of Black Americans did not travel without experiencing discrimination. As a result, many Black people sought to self-segregate themselves from others to achieve total relaxation, a classic example of how Black people dealt with the consistent warring identities of Double Consciousness. This came in the form of entertaining in their own homes, travelling in all Black groups, patronizing Black businesses and travelling back to the African continent seeking a connection to their roots (Yankholmes & Timothy, 2017). Many of these practices are still within the comfort zone of Black travellers and led to the creation of BTM (Dillette et al., 2018) as a safe space for people of colour to experience the freedoms and transformation of travel. One way these organizations provide this is by offering roots tourism experiences to the African continent.

Roots tourism emphasizes travel to places where visitors have a direct connection through individual lineage or family history (Benjamin, Kline, Alderman, & Hoggard, 2016). Within the literature, topics explored have included motivations for traveling to ancestral homelands (Duval, 2002; Timothy & Teye, 2004), roots travel as a form of pilgrimage (Esman, 1984), homeland tourism and ethnic identification (Garrido, 2011; Taylor, Levi, & Dinovitzer, 2012), migrant tourism (Skrbiš, 2007), the ethics of roots tourism (Austin, 2000), the impact of roots tourism on the economy (Lew & Wong, 2005), packaging roots tourism (Austin, 2000; Maddern, 2004) and the relationship between locals and visitors (Stefansson, 2004). Within the broader roots tourism literature, motivations for roots tourism included the desire to visit relatives (Duval, 2002) while some have come to perceive roots tourism as a form of pilgrimage with the goal of "reasserting, reaffirming or perpetuating their heritage" (Timothy & Teye, 2004, p. 112). Other studies focused on homeland and migrant travel revealed experiences of competing emotions as a result of searching for connection (Garrido, 2011; Skrbiš, 2007; Taylor et al., 2012). Travellers experienced a sense of closeness in tandem with a sense of distance, making them acutely aware that they are tourists in places they consider to be their homeland (Taylor et al., 2012). More specifically, Ghana has been the subject of numerous studies including an exploration of African-American expatriate experiences (Yankholmes & Timothy, 2017), dark tourism experiences at slave dungeons (Bruner, 1996; MacGonagle, 2006; Mowatt & Chancellor, 2011; Reed, 2004; Timothy & Teye, 2004; Yankholmes & Akyeampong, 2010) how residents perceive African-American vs. White tourists, (Teye et al., 2011), an ethnographic comparison of Ghana and Sierra Leone (Benton & Shabazz, 2009) and the factors that impact roots tourism experiences (Mensah, 2015). Within this literature on roots tourism in Ghana, Benton and Shabazz (2009) set out to understand the Pan-African phenomenon of packaging culture to the African Diaspora. Results revealed opposing forces between the government and locals with regards to the authenticity of packaging. In 2011, Mowatt and Chancellor conducted interviews with 14 African-Americans about their experiences visiting the Cape Coast and Elmina castles and found three emergent themes, a desire to reclaim, emotional attachment and the images will never leave me. Also in 2011, Teye et al., set out to understand how African-American and White tourists were perceived by locals. Findings revealed significant differences between the perceptions of African-American and White tourists, namely – Black tourists were viewed more favourably in general, while White tourists were viewed more favourably from an economic standpoint. Mensah (2015) found four factors that significantly influence a roots tourism experience in Ghana, host-guest relationships, authenticity, emotion and appearance of slave castles. Most recently, in 2017, Yankholmes and Timothy found that a form of social distance between African-American expatriates and local Ghanaians exists. Although Ghana has been the subject of number of roots tourism studies, little work has been devoted to exploring the complexities of roots tourism and its impacts on identity and social justice involvement. In an attempt to reveal stronger links between sustainable tourism, diasporic identity formation and social equity, this study delves deeply into the experiences of roots tourists, utilizing the sociological theory of Double Consciousness as the lens.

W.E.B. Du Bois' theory of Double Consciousness

The main argument of sociological theories of the self is that "the self is the result of social processes, constructed and reconstructed through ongoing social interaction" (Itzigsohn & Brown, 2015, p. 232). Fundamental to this argument is equally reciprocated recognition and communication between different members of society, foundations upon which classical sociological theorists William James (James, Burkhardt, Bowers, & Skrupskelis, 1890) and George Herbert Mead (Mead, 1964) base their theories. Unaddressed by these classical theorists, is the argument that there are significant boundaries to communication and mutual recognition for racialized groups in society (Du Bois, 1903). Steeped in the tenets of Critical Race Theory (CRT) (Delgado &

Stefancic, 2017) Du Bois' theory of Double Consciousness fills an unresolved gap in the theorizing of the self by discussing the presence of an impalpable boundary between dominant society and racialized groups, often erasing the presence of multiculturalism and transnationalism. For diaspora Africans in the United States, this societal push towards 'sameness' revealed itself as Double Consciousness, a theory originally coined in the 1897 Atlantic Magazine essay, *"Strivings of the Negro People"* and later republished with revisions in the revolutionary book *The Souls of Black Folk* (Du Bois, 1903). In this book, Du Bois describes the phenomenological experiences of the racialization of identity:

> ... the Negro is a sort of seventh son, born with a veil, and gifted with second-sight in this American world, - a world which yields him no true self-consciousness, but only lets him see himself through the revelation of the other world. It is a peculiar sensation, this double-consciousness, this sense of always looking at one's self through the eyes of others, of measuring one's soul by the tape of a world that looks on in amused contempt and pity. One ever feels his two-ness,— an American, a Negro; two souls, two thoughts, two unreconciled strivings; two warring ideals in one dark body, whose dogged strength alone keeps it from being torn asunder (Du Bois, 1903, p. 4)

An examination of this passage reveals a complexity to Double Consciousness that introduces us to the idea that as "an American" and "a Negro", one must consider two identities, two souls, two thoughts – all in competition with one another. In essence, the passage exposes three dimensions to the theory of Double Consciousness, the veil, twoness and second sight (Bruce, 1992; Itzigsohn & Brown, 2015).

1. *The veil.* The first element is 'the veil', which could also be referred to as 'the colour line'. The experience of the veil refers to the power of structural racism embedded in dominant society and its influence in Black life, thought and identity (Delgado & Stefancic, 2017). For example, in order to maintain dominance in society, negative stereotypes about Black life and thought are spread (Bruce, 1992). These stereotypes in turn seep into the fabric of everyday Black life, influencing their vision of their own identity. Due to these structural inequalities, the Black experience in America is shrouded with a veil and heavily influences how they experience their social world. For example, Whites project their constructions of Black life and identity onto the veil, and Blacks internalize these projections. However, because Whites maintain dominance in society, the veil acts as a one way mirror, preventing full acknowledgement of the humanity of Blacks as a racialized group (Itzigsohn & Brown, 2015).
2. *Twoness.* Twoness refers to the internal process of the external gaze (the veil) leading to internal conflict within the African-American posing the question, what does it truly mean to be an African and an American, hundreds of years removed from ones original place of familial lineage (Bruce, 1992)? The consistent structural influences of everyday racism positions the African-American in two distinctly different communities, the dominant community which denies their humanity and shrouds them with a veil and the Black community that provides them with a safe space for support and growth (Delgado & Stefancic, 2017; Itzigsohn & Brown, 2015). Because of the history of slavery and segregation in the United States, African-Americans have always had to deal with these two sides of society, having to learn the art of 'code switching' to find acceptance and success.
3. *Second sight.* The third element of Double Consciousness, second sight, Du Bois' refers to as 'the gift'. Second sight refers to racialized groups becoming aware of their position in society, leading to the potential for them to see the world on the other side of the veil (Bruce, 1992). This awareness can turn into a gift as it forces the racialized person to learn how to move beyond the boundaries of the veil and live in a world of opportunity and possibility. This second sight gives the Black person a view into the dominant world, creating the possibility for neutralization and racial reconciliation between the two groups (Itzigsohn & Brown, 2015).

Du Bois' theory of Double Consciousness is a complex phenomenon and asks us to consider the long lasting impacts the practice of slavery has had on human identity. In its linkages to CRT, the theory of Double Consciousness is committed to continuously seeking social justice through challenging structural racism in society and transforming the identity of racialized groups. Unfortunately, in the current climate, this struggle between two warring ideals still abounds.

Although Double Consciousness has been used to explore the impacts of slavery and the diaspora experience in the field of sociology (Bruce, 1992; Dayal, 1996; Itzigsohn & Brown, 2015; Moreiras, 1999), little has been done to apply Double Consciousness in the travel and tourism literature. There is some mention of Double Consciousness in sport tourism, revealing that Black Norwegian athletes feel disconnected from Norwegian society (Andersson, 2007) and as a brief footnote to explain diaspora dispersal similarities and differences between Jewish Americans and African-Americans in their experiences with homeland tourism (Powers, 2011). However, though the theory is mentioned, none of these studies have used Double Consciousness as a central theoretical lens. As it pertains to roots tourism, many studies are lacking any theoretical underpinnings altogether, while others use different theoretical lenses including the concept of stranger-hood (Simmel, 2008; Yankholmes & Timothy, 2017) transnational identity (Mowatt & Chancellor, 2011) the sociological concept of ambivalence (Taylor et al., 2012) and critical theory (Benjamin et al., 2016). Though under-utilized, the theory of Double Consciousness has the potential to expose how dominant narratives have been used to manipulate and racialize the African-American identity and its use is timely and in line with a call for more critical methodologies in tourism studies (Gibson, 2019). Therefore, Double Consciousness is used as a theoretical lens to answer the research question: What are the impacts of roots tourism on the identity formation process and its subsequent consequence towards social justice initiatives?

Methodology

Ghana has a long history of being at the forefront of the 'Back to Africa' movement, receiving more Black visitors than any other African nation annually (Benton & Shabazz, 2009). In fact, 2019 was marked the *Year of Return*, a tourism campaign to commemorate the 400th anniversary of the first slave ship landing. In addition to this movement, there has also been an increase of Black travel groups dedicated to promoting and providing enriching travel experiences for people of colour (Dillette et al., 2018). An organization in this movement sets the stage for this study. The trip was entitled *Experience Ghana: Journey to a Real Life Wakanda*[1] and occurred over 10 days during August 2018. The group included ten African-Americans (5 men and 5 women) between the ages of 23-60 and involved experiences in the cities of Accra, Cape Coast and Kumasi and included visits to slave dungeons, historical sites, local villages, markets, nature reserves and beaches.

In order to explore the lived experiences of these African-American travellers in Ghana, a qualitative ethnographic case study design was employed. At the base of this blended methodological approach is ethnography. Ethnography is "the description and interpretation of a culture or social group" (Holloway, Brown, & Shipway, 2010, p. 76). The goal of conducting ethnography is to explore the behavior of participants including their cultural interactions and how these interactions bring meaning to their lives (Barbour, 2010). Qualitative case study design takes a more outward looking approach, specifically bounded by time and space (Yin, 2017). In combination, an ethnographic case study design allows the blending of concepts from each method and lessens the limitations of each. In the case of this study, an ethnographic approach to the research was used bounded within a case study protocol.

In qualitative field research, the researcher plays the role of a participant and an observer, maneuvering back and forth between multiple worlds – the cultural world that they are studying, and the perspective of their own world, which they cannot escape or deny (Dibley, 2011).

Due to the complexity of this role, it is important to understand the researchers identity and positionality in the field (Creswell & Creswell, 2017). The primary researcher identifies as a Black multi-cultural woman, which helped her considerably in gaining the trust of the group travelling to Ghana. She is from a multi-cultural family, with Caribbean and Canadian roots, but has lived in the United States for over a decade. She has travelled to Africa on numerous occasions, however, this trip marked her first visit to Ghana. This information is pertinent because it helped her to understand the varied experiences of participants. Prior to the trip, the primary researcher knew one participant personally, but had not met any of the other participants on the trip. This could be considered a disadvantage as it may have hindered some participants from being open with her. However, the use of multiple forms of data, prolonged access and interactions and a significant focus on member checking and trustworthiness lessened the effects of this limitation.

Data collection and analysis

Data collected for this study consisted of direct observation, a focus group and semi-structured interviews. Multiple forms of data collection allowed for triangulation to contribute to trustworthiness (Denzin, 2017). Data was analyzed utilizing thematic analysis (Braun & Clarke, 2012). While Du Bois' theory of Double Consciousness was used as an overarching framework, this theory does not prescribe a-priori themes; therefore, an inductive approach to the thematic analysis was used (Saldaña, 2015). Audio-recorded data was transcribed and coded along with observational field notes. Each data set was coded separately exploring latent meaning. A search for disconfirming evidence throughout the fieldwork process provided deeper insight into opposing themes (Crabtree & Miller, 1999). To strengthen the trustworthiness of the data analysis process, criterion including transferability, member checking, and confirmability were utilized (Nowell, Norris, White, & Moules, 2017). Raw data were stored and saved separately at the start of the analysis process as a benchmark for evaluating transferability (Nowell et al., 2017). Additionally, member checking of transcripts and codebooks was used throughout the analysis process as a means to authenticate meaning (Carlson, 2010) and to allow participants the opportunity for clarification and expansion when desired (Houghton, Casey, Shaw, & Murphy, 2013). Codebook comparison between free codes, themes, major themes and direct quotes was reviewed by an outside researcher to strengthen the confirmability of the findings.

Pre-trip

Semi-structured interviews. Two weeks prior to the trip, individual phone interviews were conducted with each participant. These interviews were conducted with the goal of providing each participant a chance to speak with me individually before the trip began to make sure they felt comfortable with the study. Information from these interviews provided a basis for understanding each participants background and their decision to travel to Ghana.

During-trip

Direct observation. Direct observation is used in both ethnography and case study research. Using myself as the lens, I observed and interacted with all ten participants throughout their trip to Ghana. The observations were unstructured and occurred on each day of the trip. I observed how trip participants interacted with each other and with local Ghanaians as well as their reactions to different sites and experiences. Each day, I recorded these observations through detailed field notes and self-audio recordings. Taking detailed notes provided a strong basis for coding themes (Onwuegbuzie, Leech, & Collins, 2010). The direct observations throughout the trip allowed me to more deeply understand the culture and experiences of the group as they were developing, and triangulate this understanding with other data (Marshall & Rossman, 2014).

The focus group. Focus groups are used in both ethnography and case study research, though they are more typical in case studies (Fusch & Ness, 2015). One semi-structured focus group was conducted on the last afternoon of the trip to elicit the thoughts, perspectives, feelings and experiences of the group. The focus group allowed for a flexible dialogue between group members (Brockman, Nunez, & Basu, 2010). The focus group was audio-recorded with participant consent accompanied by researcher notes.

Post trip

Semi-structured interviews. Individual semi-structured interviews were conducted with each participant three months after the trip. Interviews were conducted over the phone, audio-recorded and lasted between 45 – 75 minutes. The interviews were designed to focus on how the trip impacted each individual, including their identity and propensity to participate in social justice, as well as to gain clarification on observations and field notes (Onwuegbuzie et al., 2010).

Findings

Experiencing (dis) connection to place, space & people

Roots travelers have many expectations when taking a trip to their homeland, especially when it is their first time. In line with Mowatt and Chancellor (2011), many travelers expressed their hope of feeling a sense of home, connection and community while in Ghana. As a major theme, this is represented through the term AKWABBA, which translates to "You are Welcome", a term used to welcome us around the country resulting in experiences of connection, community and authenticity. In stark contrast to these experiences, many participants also felt a sense of disconnection and inauthenticity, or, as Du Bois put it, a sense of 'twoness' (Du Bois, 1903). Interviews prior to the trip revealed a shared sense of hope to find inspiration and authentic connections while in Ghana. Elijah, who is a songwriter shared that he hoped to *"broaden my perspective on African people, their cultures and traditions and I want to connect this with myself, with being African-American".* Another participant, Roxanne, went even further and discussed how DNA testing helped her to decide which African country to visit, and how this helped to shape her expectations. During her pre-trip interview, she shared:

> My results directly link me to the Gold Coast (Ghana) and so that is why I am going. Obviously, I don't expect to meet any actual family members, but to be in a place where people look like me, are shaped like me, have the same hair as me… these are the small things I am looking forward to, to for once, feel at home.

In line with previous research (Duval, 2002; Mowatt & Chancellor, 2011; Timothy & Teye, 2004), this motivation to travel 'home' was spread across all participants in the group. Additionally, participants chose to travel in a group to experience community and connection (Lee & Scott, 2017). This reoccurring theme of *'travelling Black'* dates back to the late 1800's (Foster, 1999) and highlights connections to Du Bois' theory of Double Consciousness. Due to structural inequalities, the Black experience is shrouded with stereotypes that seep into the everyday experiences of Black life (the veil). In turn, this 'veil', motivates Black people travel to a place where people *look like them*, searching for a sense of 'oneness' as opposed to 'twoness' (Du Bois, 1903). This sheds light on an ongoing issue that should not be ignored – the travel and tourism industry is still missing the mark on creating inclusive spaces. In the particular case of roots travel to African countries, this trend of segregated travel should not be ignored. Observations from the researchers field notes highlight this even more:

> There was a clear and distinct line between groups touring the dungeon that were White and ours that was all Black. At first, I didn't think much of this – but then I thought, wait a minute, this is really weird. It felt like I was living in the Jim Crow era. They (White people) were getting a much rosier version of the tour.

> For example, tour guides highlighting how the British (slave) masters would interact humanely with enslaved Africans – while, on our side, we were hearing detailed and sometimes graphic descriptions of what it would have been like to have been kept in the dungeon as an enslaved man awaiting the middle passage.

In addition to my own observations, Roxanne also shared feelings during our focus group about her choice to travel in an all Black group.

> Prior to coming here, there was a certain anger I felt towards White people. But, after being here, I realize the importance of opening my mind a bit more. I felt like it seemed like "oh my ancestors were taken away, forced away from their families and I deserve to be angry about it. Versus the way they (Africans) see it, their ancestors just disappeared one day, they just vanished, like the rapture, or something. It's all just really dark, some of which we (Black people) have to take responsibility for. Yes, it was the Europeans - but also, we had a role to play. The indigenous enslaved others through tribal wars - and eventually, the Europeans took over and used Christianity as a front".

This quote represents an opportunity for the creation of peace and reconciliation through travel. This type of roots tourism could greatly benefit from a more educational conversation between both Black and White travelers about the collective emotional trauma resultant from slavery, the misinformation about slavery and how we can discuss and reconcile these issues. The most important question here being, how can we, as a travel industry, create more peace through tourism rather than a wider divide?

Though the major underlying motivation for this trip was linked to community and connection, a search for disconfirming evidence within the data revealed that many participants also experienced a form of disconnection, misunderstanding and inauthenticity. Similar to previous findings in homeland tourism studies, (Garrido, 2011; Skrbiš, 2007; Taylor et al., 2012), participants shared that, although their goal was to find a sense of home and community in Ghana, many felt a distinct difference that separated them from local Ghanaians.

> It's become so clear to me that they're (Ghanaians) so much a part of what is within us. It helped me to understand, "Well, yes, I do feel this connection here, but on the other hand, I am distinctly different as well. I see now where these influences that I didn't recognize come from, where they derive from, 'cause they're not really a part of American culture, and they're not accepted in American culture either" And I am not African, that is clear, I am an American – but I have these identifiable characteristics about me that clearly link me to this place. And so I hope to be able to go home and be able to exude that like a light off me, without being afraid or ashamed, we shouldn't be fearful of who we really are.

Again, the sense of Du Bois' 'twoness' appears here, beginning to show hints that James is experiencing a sort of 'second sight' when he says "I am not African, that is clear, I am an American'. During the post-trip interviews, similar feelings of inauthenticity and disconnection were shared. As an example, India shared an experience that shocked and disheartened her, leading her to feel that though she came to Ghana looking for a sense of community and connection, the reality is, she was just another tourist.

> One of the experiences that I did not expect is to feel a certain level of inauthenticity while traveling around Ghana. I expected to feel a stronger connection, but I realized, to a certain extent, Ghanaians view all of us (tourists) the same, we are 'abroni'. For example, in certain places, it felt like we were just a speck on the wall, watching locals live their lives and in others - it was like a performance, put on just for us (tourists). There was one particular day that we got approached by a man on the street who offered to tour us around his neighbourhood. We accepted, but once we got down there - I realized - oh my goodness, this is slum tourism.

This realization brings to light the power distance that exists between western tourists and locals. As Teye et al. (2011) found, Black and White tourists are viewed differently in some respects, but, the bottom line remains the same, they are viewed as a means to support the economy. Black Americans in particular, often come to the realization that while they travel to Africa in search of community and connection, their day-to-day existence is strikingly different than most locals they encounter while traveling. This realization laid the groundwork for a deeper discussion about identity that would reveal itself more in the post trip interviews.

Am I Black enough? Identity reconstruction on being African-American through the lens of Double Consciousness

Travel of all kinds has the ability to promote cross-cultural understanding and encourage global perspectives. Roots tourism has the added layer of travelers choosing their destination based on a self-identification with the culture and traditions of their chosen geographical region. As a major theme, this is represented as identity reconstruction through the lens of W.E.B. Du Bois' theory of Double Consciousness. Throughout all methods of data collection, participants questioned their identities as African-Americans, grappled with understanding the disparities between Africa and America while linking Ghanaian culture to the African-American spirit. During the focus group on the final evening of the trip, an unprompted discussion around identity was brought up by one participant and led to an interesting conversation amongst the group. Kalel begins:

> After this trip, I can confidently say that I identify as an African in America now. I don't identify as African-American anymore, 'cause I know that I am displaced. Hypothetically, if slavery didn't happen, I would still be in Africa. Maybe I would have made the choice to go to America, but it would have been my choice.

In response Ramona jumps in:

> You may feel this way, but I definitely don't. To be honest, at this moment, I feel more confused than I did at the start of the trip. I mean, we are tourists in a place that we're supposed to call home. I'm never going to get to know it from growing up here or from learning the language here as my first language or eating the food, or wearing the clothes. But at the same time, I do not feel this sense of home in Oakland ... it's like we are the orphans of the world.

Finally, Kevin, acting as a middle ground responds saying:

> From my point of view, I get where both of you are coming from, I really do. For me is, this trip has given me a greater understanding of why I am the way I am, and why Black (American) culture is the way it is. Like I said earlier, I cannot claim being an African, because I am not. But I also realize that I am not fully American in the way that America tries to prescribe my identity to me. I am cut from many cloths, some of which I now know is culturally African. I know that my loud and boisterous ways, my dancing and attraction to colourful clothing can be linked to this land, and now, I can go back home and feel less apologetic about this ... we all have to realize that, we are who we are, and yes, this is rooted in the horrific past of our enslaved ancestors.

From this conversation, it is clear that many participants felt a strong sense of Du Bois' 'veil' when considering their identity. For example, Kevin's description reveals a connection to Du Bois' argument that Whites project their constructions of Black life and identity onto the veil, and Blacks internalize these projections. However, this statement highlights Kevin breaking away from this veil and reconstructing his identity based on the "many cloths" that Ghana helped him to identify. Follow up interviews were also rich with stories about how participants had worked through this discussion about identity in the final focus group discussion. Desmond shared a new philosophy he had come to in the months after the trip.

> I'm still trying to wrap my mind around it. It gave me more insight into how much of a concept race truly is, and how it doesn't exist we just project it out into the world. Really, I've come to realize that race is just a myth, we are all human, I wish that we lived in a world that could really and truly embrace that, but we don't. So, for now, I just continue trying to spread the love, and be unashamed about who I am, which is a mixture of so many things, some of which do originate in Africa, and I am proud of that now.

In contrast to Desmond, other participants shared that they are still struggling with navigating what it means to have a multi-faceted identity. Amanda, the Jamaican American participant, shares her own internal struggles with 'twoness' and how the trip to Ghana has caused her to question her identity of being both mixed-race and multi-cultural.

> For me, this has been a lifelong internal struggle, because I am two halves representing two races, but society only recognizes me as one. In the states, if you have a drop of Black blood, you're Black and that's how people view you. But, in Ghana, there are not a lot of mixed-race people. So, I am not Black, I am not White. I am not African, and I am not fully American, I have been questioning, who am I?

Though this example is specific to Amanda's mixed background, this internal tug-of-war between multiple identities (Du Bois, 1903) was also acknowledged by other participants in the group. In some cases, participants shared how travelling to Ghana helped them to more deeply understand and embrace aspects of their culture that they have previously been ashamed of. As an example, James reflected on how Ghana has allowed him to confidently reconstruct his identity to include small facets of Ghanaian culture, such as wearing bright colors in a professional setting in his life. As another example Andrea shares a similar experience with contradicting feelings.

> Well, there was one thing I remember, when we were in Jamestown on a Sunday afternoon and just walking through a local village, some people, they just started dancing - and it wasn't a party per say, someone just heard a song they liked and started dancing. It was so beautiful, they were just dancing for the joy of it. And this made me think about what we've lost, You know, in my family, we do this, but in public, its not acceptable, it's sadly, a loss of our cultural roots.

These experiences reveal the beginnings of identity reconstruction along similar lines of the dilemma that Du Bois argued over a century ago. Much of the internal conflict that we can see represented in these quotes ride the fine line of balance that African-Americans face between two warring identities. However, it is also evident that, as a result of these warring identities, participants began to experience a 'second sight', a 'gift' (Du Bois, 1903) that they took responsibility for upon returning home from Ghana.

Travelling Black. An activator for social justice

Travel and tourism are not often linked to social justice. However, as Barton and Leonard (2010) argued "tourism can serve as a vehicle for sustainable community development by contributing to equity and social justice" (pg. 298). In the case of this roots tourism trip to Ghana, it was important to explore the impacts the trip had on participants social circles, and if these impacts resulted in any form of action on the part of the individual. As a major theme, this is represented through understanding roots tourism as an activator for sustainability through social justice (UNWTO, 2018). This theme is embodied within the overarching theoretical lens of Double Consciousness through the activation of social justice as a result of participants experiences with their identity in a majority Black country. The acute awareness of 'twoness' felt at home, led travelers on this roots tourism trip to Ghana in search of acceptance and a form of 'oneness'. Ironically, their experiences led to a different sort of 'twoness' (identifying cultural roots while also recognizing a stark power distance), and eventually to a 'second sight' which revealed itself in the form of social justice activism. Participants shared their experiences participating in social justice activism, engaging in counter-narrative storytelling, embracing leadership roles in giving back to their communities as well as simply spreading the word about the importance of travelling to Africa. To begin, India shares her reflections during her interview after returning from Ghana:

> The colors and designs of the fabric represent everything I feel about my trip to Ghana, everything that Ghana has imprinted within me. I am more vibrant, colorful, and unapologetic about who I am and what I believe in than ever before. Before taking this trip, I was afraid of what people who think, of my colleagues thinking I was too 'Black centric', or my White friends thinking that I don't respect them. I now realize that my silence was simply a veil that I was hiding behind, a way for me to push the injustices I see day in and day out under the rug. We still have significant issues with racism in this country. And I will no longer be quiet. Recently, I hosted the first meeting in my company for my Black colleagues and I to address issues specific to us that we face in the workplace. Ghana helped me find the courage to do this.

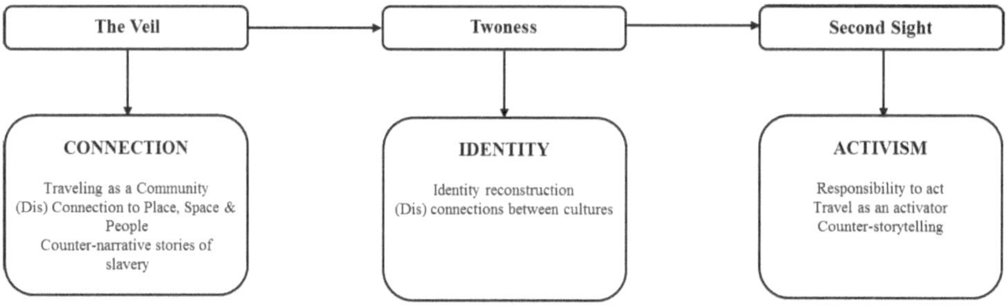

Figure 1. Graphical representation of findings.

In the example above, India reflects on a specific initiative she was able to start, revealing how Ghana tangibly impacted her civic commitment to social justice. The internal dilemma she describes here of being stereotyped as too 'Black centric' is supported by one of the significant issues with Double Consciousness – the power of White stereotypes in Black life and thought (Delgado & Stefancic, 2017). However, the resultant action from the dilemma also reveals a tangible shift towards 'second sight' in the actionable form of a social justice initiative focused on racial equity (Du Bois, 1903). Other participants also discussed steps they are taking to either begin or continue their commitment to social justice work in and outside their home communities. Ramona shares how she has begun a tangible effort to give back to micro-lending for female entrepreneurs in Ghana:

> I started putting money away in an account to contribute to micro-lending. Prosper was talking a lot about micro-lending in Ghana and how a lot of the businesses that have been started in the past five years have been from women. A lot of the barriers for them to enter into business is them not being able to secure funding. At home, I am already involved in organizations that impact Black women and girls - but I will say that the trip has solidified this work, and emboldened me to do more.

Another participant, Andrea, shares how she is taking active steps to support other Black musicians in her community:

> I have seen a change in how I interact with my community. For example, I am a music teacher and I recently sponsored a young Black musician to come to Atlanta and take a course, a very important course. This is how I am making a difference. I also speak up more, to my friends, my family, my White friends, my conservative friends. Silence is no longer my go to.

These stories of actionable change and expanded mindsets give us hope that travel truly can make a difference in the worlds and minds of those traditionally underserved and overlooked in society. They also shine a light on the continued everyday racism that is faced as a significant trickle-down effect of slavery. Participants shared candidly how societal racism and underrepresentation for African-Americans has impacted them in their lives, and how, travelling to Ghana has emboldened them to continue in their search for clarity of identity and self, as well as equality for societal injustices. Findings expand the applicability of Du Bois' theory by revealing links between the roots tourism experience and the three dimensions of Double Consciousness (Figure 1). In the case of social justice, findings highlight that the shifting of the African-American identity led to a tangible sense of responsibility to get involved in equitable social justice initiatives. Finally, from a sustainability perspective, findings confirm that roots tourism does in fact have the power to impact the sustainable tourism goals of social justice and universal equality for human rights (UNWTO, 2018).

Discussion and conclusion

The act of travelling home has long been a pilgrimage for people around the globe. However, today, over 400 years after the beginning of the transatlantic slave trade, travelling home has taken on a different meaning for African diasporas. As exemplified in this case study, roots tourism experiences involve a search for connection, identity and social justice. Unfortunately, this study also confirmed that the continued marginalization of Black travelers impacts their motivation to travel in homogenous groups and, that Black travelers continue to struggle with the construction of their identity (Du Bois, 1903). This finding supports the notion of 'the veil' argued by Du Bois' that structural racism is embedded in dominant society and therefore has a significant influence in Black life, thought and identity. This veil ultimately led participants to travel to Ghana in search of a 'oneness'.

Major themes in the data also suggest that Black travelers who make the journey to Africa do in fact have the opportunity to gain a deeper understanding of how African culture is embedded in their identity and communities. This expanded depth of understanding is an important outcome for roots tourism travelers, especially those of the African Diaspora. Having been disconnected from their origins for so long, many Black travelers are now seeking truth about their past, in hopes of forming a stronger connection to the African continent. These experiences are possible, in part, due to the rising popularity of social media, allowing Black travelers to have a platform to share their counter-narrative stories (Dillette et al., 2018). This mechanism is what allowed the BTM group 'Travelling Black' to thrive, and lead people on this trip to Ghana. Another important, and underexplored area to note is the connection between roots tourism and the now popular DNA testing services such as Ancestry and 23andMe.

Disconfirming evidence also suggested that with the search for connection comes dis-connection, revealing a sense of inauthenticity and power distance. This finding further illustrates Du Bois' concept of 'twoness'. Though participants were led to Ghana in search of community and connection, many experienced a different kind of twoness, realizing that although they came in search of home, they are in fact just tourists looking from the outside inwards. Though this is in line with previous research that has suggested issues with disconnection and inauthenticity for this type of tourism (Benton & Shabazz, 2009; Garrido, 2011), this case study discovered a link between experiences of (dis)connection with identity (re)construction. Findings support the notion argued by W.E.B. Du Bois' theory of Double Consciousness, that the African-American identity is significantly marred by continued issues of everyday racism, stereotypes and internal conflict between two significantly different identities. In a more positive direction, study participants became liberated to activate ideals of identity reconstruction and social justice responsibility pointing towards the third dimension of Double Consciousness, second sight. Through the experiences participants gained travelling to Ghana, they became more acutely aware of their position of privilege in the global society. Essentially, travelling to Ghana allowed participants to gain a sense of rootedness in an identity that they otherwise would not have experienced, while glimpsing a view into the dominant world in which they live. Occurrences of becoming emboldened to take action on issues of inequality and social justice provide evidence that tourism truly does have the ability to be hopeful.

This study, though innovative in its use of Double Consciousness in the context of tourism, is not without its limitations. It is limited in that it takes a very American-centric viewpoint of roots tourism through the eyes of African-Americans. Findings from different demographic groups within the African diaspora (i.e., Afro-Caribbean or Afro-Latino) may vary and are worth exploring. It is also limited in that the data only represents a snapshot in time and lacks a longitudinal context. It would be interesting to follow up with this group of roots travelers in future years to see if and how the impact of this trip has continued. Though limitations do exist, findings from this study support the call from Pritchard et al. (2011) for research that uncovers the power of tourism to impact social justice and the universal equality for human rights. A continued search for awareness into how inequalities in tourism are perpetuated is important for creating the authentic and inclusive spaces for traditionally marginalized groups. Moreover, Tribe (2008) has argued for the continued critical assessment of tourism, positing that our research should be informed by the "current configurations of power and the operation of dominant ideological practices" (p.

253). This work certainly adds to the critical assessment of tourism and brings accountability to the power structures that have marginalized underrepresented groups for too long.

Notes

1. The term Black and African-American is used interchangeably to refer to people with African ancestry.
2. Sharecropping is a form of agriculture where landowners allow farmers to farm their land in exchange for a share of the crop.
3. Jim Crow laws were laws to enforce legal racial segregation in the United States of America.

Disclosure statement

No potential conflict of interest was reported by the author(s).

Data availability statement

Data for this case study is primary qualitative data and is available in the raw format of transcribed focus groups and interviews if requested.

References

Alex-Assensoh, Y. M. (2010). African Americans, African immigrants and homeland-diaspora development in Africa. *African Diaspora*, *3*(2), 207–234. doi:10.1163/187254610X526922

Andersson, M. (2007). The relevance of the Black Atlantic in contemporary sport: Racial imaginaries in Norway. *International Review for the Sociology of Sport*, *42*(1), 65–81. doi:10.1177/1012690207081829

Austin, N. (2000). Tourism and the transatlantic slave trade: Some issues and reflections. *The Political Economy of Tourism Development in Africa*, 208–216.

Barbour, A. (2010). Exploring some ethical dilemmas and obligations of the ethnographer. *Ethnography and Education*, *5*(2), 159–173. doi:10.1080/17457823.2010.493399

Barton, A. W., & Leonard, S. J. (2010). Incorporating social justice in tourism planning: Racial reconciliation and sustainable community development in the Deep South. *Community Development*, *41*(3), 298–322. doi:10.1080/15575330903444051

Benjamin, S., Kline, C., Alderman, D., & Hoggard, W. (2016). Heritage site visitation and attitudes toward African American heritage preservation: An investigation of North Carolina residents. *Journal of Travel Research*, *55*(7), 919–933. doi:10.1177/0047287515605931

Benton, A., & Shabazz, K. Z. (2009). "Find their Level". African American roots tourism in Sierra Leone and Ghana. *Cahiers D'études Africaines*, *49*(193-194), 477–511. doi:10.4000/etudesafricaines.18791

Braun, V., & Clarke, V. (2012). Thematic analysis. In H. Cooper, P. M. Camic, D. L. Long, A. T. Panter, D. Rindskopf, & K. J. Sher (Eds.), *APA handbooks in psychology®. APA handbook of research methods in psychology, Vol. 2. Research designs: Quantitative, qualitative, neuropsychological, and biological* (p. 57–71). American Psychological Association.

Brockman, J. L., Nunez, A. A., & Basu, A. (2010). Effectiveness of a conflict resolution training program in changing graduate students style of managing conflict with their faculty advisors. *Innovative Higher Education*, *35*(4), 277–293. doi:10.1007/s10755-010-9142-z

Bruce, D. D. (1992). WEB Du Bois and the idea of double consciousness. *American Literature*, *64*(2), 299–309.

Bruner, E. M. (1996). Tourism in Ghana: The representation of slavery and the return of the black diaspora. *American Anthropologist*, *98*(2), 290–304. doi:10.1525/aa.1996.98.2.02a00060

Burton, I. (1987). Report on reports: Our common future: The world commission on environment and development. *Environment: Science and Policy for Sustainable Development, 29*(5), 25–29. doi:10.1080/00139157.1987.9928891

Buzinde, C. N., Santos, C. A., & Smith, S. L. (2006). Ethnic representations: Destination imagery. *Annals of Tourism Research, 33*(3), 707–728. doi:10.1016/j.annals.2006.03.008

Carlson, J. A. (2010). Avoiding traps in member checking. *The Qualitative Report, 15*(5), 1102–1113.

Crabtree, B. F., & Miller, W. L. (1999). *Doing qualitative research*. Sage Publications. Thousand Oaks, CA, USA.

Creswell, J. W., & Creswell, J. D. (2017). *Research design: Qualitative, quantitative, and mixed methods approaches*: Sage publications. Thousand Oaks, CA, USA.

Dayal, S. (1996). Diaspora and double consciousness. *The Journal of the Midwest Modern Language Association, 29*(1), 46–62. doi:10.2307/1315257

Delgado, R., & Stefancic, J. (2017). *Critical race theory: An introduction* (Vol. 20). NewYork University Press: New York, NY, USA.

Denzin, N. K. (2017). *The research act: A theoretical introduction to sociological methods*. Routledge. New York, NY, USA.

Dibley, L. (2011). Analysing narrative data using McCormack's Lenses. *Nurse Researcher, 18*(3), 13–19. doi:10.7748/nr2011.04.18.3.13.c8458

Dillette, A. K., Benjamin, S., & Carpenter, C. (2018). Tweeting the black travel experience: Social media counternarrative stories as innovative insight on# TravelingWhileBlack. *Journal of Travel Research, 58*(8), 1–16. doi:10.1177/0047287518802087

Drachler, J. (1975). *Black homeland/black diaspora: Cross-currents of the African relationship*. Associated Faculty PressInc.

Du Bois, W. E. B. (1903). *The souls of Black Folk*. London: Longmans, Green and Co. Ltd.

Duval, D. T. (2002). The return visit-return migration connection. In M. Hall & A. M. Williams,*Tourism and migration* (pp. 257–276). Springer: New York, NY, USA.

Edmondson, L. (1986). Black America as a mobilizing diaspora: Some international implications. In G. Sheffer, *Modern diasporas in international politics* (pp. 164–211). Croom Helm: Kent, UK.

Edwards, A. R. (2005). *The sustainability revolution: Portrait of a paradigm shift*. New Society Publishers. Gabriola Island: BC, Canada.

Esman, M. R. (1984). Tourism as ethnic preservation: The Cajuns of Louisiana. *Annals of Tourism Research, 11*(3), 451–467. doi:10.1016/0160-7383(84)90031-8

Foster, M. S. (1999). In the face of" Jim Crow": Prosperous blacks and vacations, travel and outdoor leisure, 1890-1945. *The Journal of Negro History, 84*(2), 130–149. doi:10.2307/2649043

Fusch, P. I., & Ness, L. R. (2015). Are we there yet? Data saturation in qualitative research. *The qualitative report, 20*(9), 1408.

Garrido, M. (2011). Home is another country: Ethnic identification in Philippine homeland tours. *Qualitative Sociology, 34*(1), 177–199. doi:10.1007/s11133-010-9186-5

Gibson, C. (2019). Critical tourism studies: New directions for volatile times. *Tourism Geographies*, 1–19. doi:10.1080/14616688.2019.1647453

Holloway, I., Brown, L., & Shipway, R. (2010). Meaning not measurement: Using ethnography to bring a deeper understanding to the participant experience of festivals and events. *International Journal of Event and Festival Management, 1*(1), 74–85. doi:10.1108/17852951011029315

Houghton, C., Casey, D., Shaw, D., & Murphy, K. (2013). Rigour in qualitative case-study research. *Nurse Researcher, 20*(4), 12–17. doi:10.7748/nr2013.03.20.4.12.e326

Itzigsohn, J., & Brown, K. (2015). Sociology and the theory of double consciousness: WEB Du Bois's phenomenology of racialized subjectivity. *Du Bois Review: Social Science Research on Race, 12*(2), 231–248. doi:10.1017/S1742058X15000107

James, W., Burkhardt, F., Bowers, F., & Skrupskelis, I. K. (1890). *The principles of psychology* (Vol. 1). London: Macmillan.

Johnson, C., & Smith, P. (1999). *Africans in America: America's journey through slavery*. Houghton Mifflin Harcourt. Boston, MA, USA.

Klein-Vielhauer, S. (2009). Framework model to assess leisure and tourism sustainability. *Journal of Cleaner Production, 17*(4), 447–454. doi:10.1016/j.jclepro.2008.07.006

Lee, K. J., & Scott, D. (2017). Racial discrimination and African Americans' travel behavior: The utility of habitus and vignette technique. *Journal of Travel Research, 56*(3), 381–392. doi:10.1177/0047287516643184

Lew, A., & Wong, A. (2005). Existential tourism and the homeland: The overseas Chinese experience. In C. Cartier & A. A. Law, *Seductions of place: Geographical perspectives on globalization and touristed landscapes* (pp. 286–300). Routledge. New York, NY, USA.

MacGonagle, E. (2006). From dungeons to dance parties: Contested histories of Ghana's slave forts. *Journal of Contemporary African Studies, 24*(2), 249–260. doi:10.1080/02589000600770007

Maddern, J. (2004). The 'Isle of Home'is always on your mind: Subjectivity and space at Ellis Island Immigration Museum. In T. Coles & D. J. Timothy, *Tourism, diasporas and space* (pp. 167–185). Routledge. New York, NY, USA.

Marshall, C., & Rossman, G. B. (2014). *Designing qualitative research*. Sage publications. Thousand Oaks, CA, USA.
Mead, G. H. (1964). On social psychology. Chicago: University of Chicago Press
Mensah, I. (2015). The roots tourism experience of diaspora Africans: A focus on the Cape Coast and Elmina Castles. *Journal of Heritage Tourism*, *10*(3), 213–232. doi:10.1080/1743873X.2014.990974
Moreiras, A. (1999). Hybridity and double consciousness. *Cultural Studies*, *13*(3), 373–407. doi:10.1080/095023899335149
Mowatt, R. A., & Chancellor, C. H. (2011). Visiting death and life: Dark tourism and slave castles. *Annals of Tourism Research*, *38*(4), 1410–1434. doi:10.1016/j.annals.2011.03.012
Nowell, L. S., Norris, J. M., White, D. E., & Moules, N. J. (2017). Thematic analysis: Striving to meet the trustworthiness criteria. *International Journal of Qualitative Methods*, *16*(1), 160940691773384. doi:10.1177/1609406917733847
Onwuegbuzie, A. J., Leech, N. L., & Collins, K. M. (2010). Innovative data collection strategies in qualitative research. *The Qualitative Report*, *15*(3), 696–726.
Powers, J. L. (2011). Reimaging the imagined community: Homeland tourism and the role of place. *American Behavioral Scientist*, *55*(10), 1362–1378. doi:10.1177/0002764211409380
Pritchard, A., Morgan, N., & Ateljevic, I. (2011). Hopeful tourism: A new transformative perspective. *Annals of Tourism Research*, *38*(3), 941–963. doi:10.1016/j.annals.2011.01.004
Rawls, A. W. (2000). "Race" as an interaction order phenomenon: WEB Du Bois's "Double Consciousness" thesis revisited. *Sociological Theory*, *18*(2), 241–274. doi:10.1111/0735-2751.00097
Reed, A. (2004). Sankↄ fa site: Cape Coast Castle and its museum as markers of memory. *Museum Anthropology*, *27*(1-2), 13–24. doi:10.1525/mua.2004.27.1-2.13
Reed, A. (2010). Gateway to Africa: The pilgrimage tourism of diaspora Africans to Ghan Doctoral Dissertation: Department of Anthropology, Indiana University, Bloomington.
Saldaña, J. (2015). *The coding manual for qualitative researchers*. Sage Publications. Thousand Oaks, CA, USA.
Schramm, K. (2004). Coming home to the motherland: Pilgrimage tourism in Ghana. In S. Coleman & J. Eade, *Reframing pilgrimage* (pp. 143–159). Routledge. New Fetter Lane, London, UK.
Shackley, M. (2001). *Managing sacred sites: Service provision and visitor experience*. Cengage Learning EMEA. Bedford Row, London, UK.
Simmel, G. (2008). The stranger. In T. Oakes & P. L. Price, *The cultural geography reader* (pp. 323–327). Routledge. Milton Park, OX, UK.
Skrbiš, Z. (2007). From migrants to pilgrim tourists: Diasporic imagining and visits to Medjugorje. *Journal of Ethnic and Migration Studies*, *33*(2), 313–329. doi:10.1080/13691830601154294
Stefansson, A. H. (2004). Sarajevo suffering: Homecoming and the hierarchy of homeland hardship. In F. Markowitz & A. H. Stefansson, *Homecomings: Unsettling Paths of Return* (pp. 54–75).Lexington Books. Lanham, Maryland, USA.
Taylor, J., Levi, R., & Dinovitzer, R. (2012). Homeland tourism, emotion, and identity labor. *Du Bois Review: Social Science Research on Race*, *9*(1), 67–85. doi:10.1017/S1742058X12000069
Teye, V., Turk, E., & Sönmez, S. (2011). Heritage tourism in Africa: Residents' perceptions of African-American and white tourists. *Tourism Analysis*, *16*(2), 169–185. doi:10.3727/108354211X13014081270404
Timothy, D. J. (1997). Tourism and the personal heritage experience. *Annals of Tourism Research*, *24*(3), 751–754. doi:10.1016/S0160-7383(97)00006-6
Timothy, D. J., & Teye, V. B. (2004). American children of the African diaspora: Journeys to the motherland. In T. Coles & D. J. Timothy, *Tourism, diasporas and space* (pp. 125–137). Routledge. Milton Park, OX, UK.
Tribe, J. (2008). Tourism: A critical business. *Journal of Travel Research*, *46*(3), 245–255. doi:10.1177/0047287507304051
UNWTO. (2018). Tourism for development – Volume I: Key areas for action. Retrieved from https://www.e-unwto.org/doi/pdf/10.18111/9789284419722
Yankholmes, A., & Timothy, D. (2017). Social distance between local residents and African–American expatriates in the context of Ghana's slavery-based heritage tourism. *International Journal of Tourism Research*, *19*(5), 486–495. doi:10.1002/jtr.2121
Yankholmes, A. K., & Akyeampong, O. A. (2010). Tourists' perceptions of heritage tourism development in Danish-Osu. *International Journal of Tourism Research*, *12*(5), 603–616. doi:10.1002/jtr.781
Yin, R. K. (2017). *Case study research and applications: Design and methods*. Sage Publications. Thousand Oaks, CA, USA.

Heritage tourism, historic roadside markers and "just representation" in Tennessee, USA

Candace Forbes Bright , Kelly N. Foster, Andrew Joyner and Oceane Tanny

ABSTRACT

The American landscape is increasingly populated with memorial tourist sites showing a devotion to the past. In the last published statewide study of Tennessee historical roadside markers, Jones (1988) analyzed the 1,170 roadside markers across the state. In his analysis, markers devoted to black history or white women accounted for only 0.7% ($n = 8$) and 0.8% ($n = 9$), respectively, of all markers. At the time of this study there were more roadside markers solely dedicated to David Crocket (0.9%, $n = 11$) than to either of these groups. Additionally, Native American people merely accounted for 3.0% ($n = 35$), while white men accounted for 11.7% ($n = 137$) – including 3% dedicated to Klu Klux Klan Founder Nathan Bedford Forrest (Jones, 1988). This particularly visible expression of public history serves primarily to preserve a white, Protestant, male history of the area (Jones, 1988). Using data on the 313 historical roadside markers erected since 1988, we (1) update the analysis of Jones (1988) and (2) discuss the lack of a "just representation" of non-white male history in these markers that has carried forward to 2019. The study is informed by social representation and critical race theory.

Introduction

Public history is a critical component of the ways that contemporary society memorializes the past, and the American landscape is increasingly populated with tourist sites showing a devotion to iterations of public history (Doss, 2010; Alderman & Inwood, 2013). These representations of the past in the American South, however, often focus on white hegemonic notions of the past, either overtly or covertly through racist messages. The controversy surrounding Civil War monuments, most of which were erected during the Jim Crow era, serves as just one example of the tensions running through the American South. Historical roadside markers are another way in which public history is memorialized across the United States. Both their content and intent have been increasingly scrutinized by both the communities in which they are erected and by the communities they represent, as well as public historians and geographers. Despite this attention to historical roadside markers in recent scholarship, they have been relatively underscrutinized in the state of Tennessee and their role in representation as cultural and heritage tourism sites deserves greater attention.

Silberberg (1995, p. 361) defined cultural tourism as "visits by persons from outside the host community motivated wholly or in part by interest in historical, artistic, scientific or lifestyle/heritage offerings of a community, region, group, or institution." Within cultural tourism, heritage tourism occurs when visitors engage with museums, festivals, humanities, and historic preservation sites such as memorial sites (Seba, 2011). Smaller memorial sites, however, are erected and visited in towns, large and small, across the United States; historical roadside markers are among these. Memorial sites are often indicated by historical markers that bear texts or images meant to commemorate people, events, and places associated with these sites. In fact, Jones (1988) argued that, "perhaps two of the most visible and easily recognized components of what has been called public history are found in historic roadside markers and listings in the National Register of Historic Places" (p. 19). Historical markers are commonly placed along major highways in conjunction with tourist spots and are also referred to as "roadside markers."

The practice of using historical markers to commemorate the past originates from England, where William Ewart, a member of parliament, proposed the first commemorative marker program in 1863. Following Ewart, the Society of Arts constructed two plaques, celebrating the lives of poet Lord Byron and Napoleon III in 1867 ("The History of Blue Plaques," n.d.). The first historical marker program in the US began in Virginia in 1927 when William E. Carson, the first chairman of the Virginia Conservation Commission (Bardet, 2012), erected a handful of markers along U.S. 1 between Richmond and Mount Vernon (Virginia Department of Historic Resources, 2019). Other states, such as North Carolina, followed a few years after by starting their own historical marker programs (North Carolina Department of Natural and Cultural Resources, n.d.). Despite this long history, or perhaps because of this long history, there is a need to reevaluate historical roadside markers through the lens of sustainable and socially responsible tourism. As cultural and heritage tourism sites, these markers are dominated by sectional interests that inherently represent those who hold power in the community (Hall, 2003).

As much of American history is seen through historical markers, it is essential that they reflect all American history or what we refer to herein as providing a "just representation." That is, markers should include and represent *all* aspects of American culture, both tangible and intangible, equitably and fairly. In just representation, the intangible cultural heritage—defined as "the practices, representations, expressions, as well as the knowledge and skills, that communities, groups and in some cases individuals recognize as part of their cultural heritage" (Intangible Heritage Convention as cited in Ruggles & Silverman, 2009)—is incorporated into the tangible heritage, such as historical roadside markers, for all social groups. Just representation as used in our study entails a more equitable representation of minority groups and gender representation in historical roadside markers as tangible heritage sites.

Historical markers are important for the value they add to cultural and heritage tourism (Otterstrom & Davis, 2016). Moreover, they accentuate places, people, and events that may have otherwise remained unknown to the public. They also direct the tourist gaze to select national and local history. History and events of the past are crucial parts of social life as they give communities a sense of self-identity; the memorialization and representation of place is inherently a social process rendering public memory a social construct (Alderman, 2012). However, historical markers usually represent what is deemed important to remember by a subset of the broader community (Alderman, 2010). As access to power often determines who makes decisions about historical markers, it is imperative from a sustainable tourism perspective to address important questions about the role of stakeholders, their aims, and the power and politics present in tourism (Hall, 2003). The purpose of this paper is to assess the extent of "just representation" of non-white male history in Tennessee historical markers.

The paper is organized as follows: In the next section, we introduce the literature used to conceptualize just representation, including as overview of social representation theory and of extant research on historical marker research in Tennessee. This is then followed by the details of the study approach, outlining the research questions, the dataset, and the data coding analysis. We then present the results, followed by a discussion of these results and directions forward.

Conceptualizing "just representation"

An emphasis on sustainable tourism includes considering social impacts, including internal divisions within communities (Buckley, 2012). Although community-based planning is promoted in both bureaucratic and academic approaches to tourism development (Murphy, 1988; Hall, 2003), it is one among many components of promoting sustainable tourism. To understand the role of the community, we must first address power dynamics—politics determine power and who gets what (Lasswell, 1936); thus, public representation is a form of power. Decisions affecting tourism are representations of power and the values of those with control over scarce resources, rendering representation in tourism as political in nature. Thereby, "any satisfactory account of politics and the political must contain the element of human conflict; of groups of human beings in constant struggle with each other over resources and ideas about the distribution of resources" (Thirft & Forbes, 1983, p. 247). Lukes (1974; 2004) presented power as the successful control of one group over another group in which the group in power secures the compliance of the group not in power. In Lukes' conceptualization of power, like Bachrach and Baratz (1970) before him, power is intrinsically related to authority, coercion, force, influence, and manipulation. Although the role of community in tourism and tourism planning in increasingly recognized (Hall, 2003), regarding tourism development, it is well established that power is not distributed evenly within the community (Hall & Jenkins, 1995; Hall, 2003), which is in part related to issues of public participation in tourism planning, such as difficulty in attaining representativeness in the planning stages, as well as addressing the general apathy of citizens (Jenkins, 1993).

Heritage tourism is a large component of tourism worldwide, and its intricate link to culture and society, bringing its historical past to the attention of present and future generations of visitors it is imperative to consider its sustainability from a social and cultural perspective (Garrod & Fyall, 2000). The concept of sustainable development, defined by Pearce (1993, p. 4) as the desire to "pass onto the next generation a stock in capital assets no less than the stock we have now", is also applicable in the study of representation of historical markers. Here, both the tangible and intangible aspects of culture and heritage must be considered. The Intangible Heritage Convention describes the long-term aspect of intangible culture as that which is transmitted across generations and part of community identity (Ruggles & Silverman, 2009). Du Cros (2001) argued that there are two objectives to sustainable heritage tourism planning: (1) to determine the most appropriate places for heritage tourism development and (2) to determine the most appropriate way to manage these sites for sustainability. However, as she notes the "ways to manage" the sites are power laden and do not equally incorporate or represent the community in which these sites are situated. Intangible heritage such as cultural rituals and traditions passed down orally is particularly difficult to ensure fairness representation, for unlike physical built sites, they leave little to no material traces. We therefore argue that from a sustainability perspective, just representation of cultural heritage entails fair consideration and incorporation of intangible heritage along with tangible cultural heritage in heritage tourism development.

Both heritage tourism and sustainability overlap in the concept of preservation. They both seek to conserve sites, but from a tourism development standpoint, also to commodify them for visitor experience (Du Cros, 2001). From a societal perspective, however, many groups in society are not given the same consideration in heritage tourism planning. Among these, in Tennessee and in much of the United States, are Blacks, Native Americans, and White women. There is a large body of literature on the intersection of landscape representation and race in the American

South (Inwood & Martin, 2008; Dwyer et al., 2013), but limited studies that have included representation and the symbolic-discourse of multiple social groups, providing a comparative perspective on the social inclusion of non-White male history. Thus, we introduce "just representation" through the lens of social representation theory to contribute to the scholarship and praxis on the social sustainability of historical roadside markers. We also draw from critical race theory to inform our framework of social justice and inclusiveness for "just representation."

Social representation theory

Moscovici's (1961, 1981, 1983, 1984, 1988) social representation theory reformulates Durkheim's (1898) conceptualization of collective representation in which social representations perform as interpretive framings for understanding society and culture (Doise, 1989; Bright & Carter, 2018). In other words, representations are part of the social world and form socially constructed ways of knowing (Fredline & Faulkner, 2000), including the narrativized world (Holland et al., 1998). Social representations are anchored and objectified through situations, objects, and narratives that contribute to an individual or collective realm of knowledge. Thus, *objectification* is a tool of meaning making that varies from group to group and interacts with past experiences to make sense of the world. It is through objectification that social representations are acted out collectively and create their object (Wagner et al., 1995). In the context of historical markers, this theory provides that the signs do not hold value alone, but provide larger interpretations about the values of society. They reflect beliefs about the world in both what select groups seek to memorialize and in the convictions of those groups about what is of social importance. As representations and their respective objectifications are formed by social units, the narratives they construct favor some social conditions over others (Wagner et al., 1995). Roadside markers are in this sense objectification devices.

To adequately assess just representation in Tennessee roadside markers, social representation theory is coupled with critical race theory. Recognizing roadside markers as objectification devices, it is important to understand the values that are represented and the social constructions that are implied in these representations, including those related to race, ethnicity, and sex. Together, these theories inform the framework that guides this research. As such, the messages presented on historical markers as objectification devices can be analyzed to better understand past and present constructions of power and race relations.

Following the discussion above, it can be argued that historical roadside markers perform as interpretive framings for understanding the values of society; they narrate a version of history through the anchoring of these events and sites of public history into our cognitive mappings. The historical roadside markers are also products of socialization that provide individuals with cues about the world around them (Fredline & Faulkner, 2000)— the representations that are constructed present a constructed historical past that continues to structurally reinforce specific ideologic meanings, power, and race relations. In the American South, the dominant social representation is a "nostalgic social representation" that idolizes a simpler, distant, and more peaceful time—one that did not actually exist (McKittrick, 2016; Bright & Carter, 2018). Critical race theory (Delgado & others, 1989; Delgado & Stefancic, 2017), like social representation theory, focuses on knowledge shared and development of representations, central to which is race and power. Thus, following social representation and critical race theory as described above, we use the following theoretical framework to guide our study: (1) we identify roadside markers as objectification devices that serve as a conceptual tool (story-telling device); (2) they represent a history that is deeply culturally engrained; (3) this history embodies structural and historical racism; and as such, (4) the narratives can be analyzed to assess the inequity in representation in the stories memorialized.

Historical roadside marker research in Tennessee

Like much of the American South, the state of Tennessee has a long and shameful history of slavery, racism, and segregation. Native Americans, Blacks, and women have been unjustly treated and have lacked representation and power within the state. There is evidence of Native American presence in Tennessee dating back to the Archaic Period. In the late 1600s, these tribes (Cherokee, Yuchi, Shawnee, Creek, Chickasaw, among others) came into contact with Euro-Americans and were forcefully removed within the century, violating the Proclamation of 1763, which reserved the Tennessee land to the Native Americans. As all of their land became ceded to Tennessee by the American government under the Revolutionary War, the Native Americans were forcefully removed under President Andrew Jackson's Indian Removal Act in the 1830s. Those seeking to remain were forced into internment camps. Seven tribes remained in colonial Tennessee and today, two tribes are federally recognized in the state: Choctaw and the Muscogee "Creek" Nation Citizens. There are an additional eight tribes currently seeking recognition in the state (Central Band of Cherokee, Cherokee Wolf Clan, Remnant Yuchi Nation, Chikamaka Cherokee, Far Away Cherokee, Upper Cumberland Cherokee, American Indian Association of Millington, and Native Cultural Circle) (House Bill 3299/Senate Bill 3123).

Black history in the state is equally problematic. Despite the abolition of slavery in 1865, gross abuse of blacks and other minority groups continued unabated for decades (Tischauser, 2013). There were laws that protected racists actively until the mid-1960s, when the civil rights movement gained in popularity. These laws were referred to as Jim Crow laws, a suite of rules and legislation designed to enforce racial segregation (Tischauser, 2013). Black people were not allowed to enjoy the same benefits as whites in public and private places; they were not allowed to live in the same neighborhoods as whites, eat at the same restaurants, or even attend the same schools. It was not until 1956 that the first public high school in Tennessee was desegregated, in the small town of Clinton (West, 2018). Blacks were not allowed to vote and even though slavery was abolished, blacks were often forced into sharecropping and other exploitative situations (Tischauser, 2013). This segregation and racism has impacted the way history was recorded in Tennessee as the efforts and contributions of blacks and other minorities to the history of the state were downsized at the advantage of the contributions of white males, who at the time had and still have more power in lawmaking and society shaping. For instance, regarding gender, Tennessee's first female senator, Marsha Blackburn, was not elected until 2018.

The history of these groups within the state is not given equitable visibility in memorialization. American history is bound by accounts and contributions of multiple people, with different backgrounds. It is rich with diversity, and reflective of its current society. Still, the commemorative landscape excludes various dimensions of American history and American people, and seems to focus on a specific group: white men (Alderman, 2010). According to Alderman in *Surrogation and the politics of remembering slavery in Savannah, Georgia (USA)* (2010), African Americans used traditional, often oral, methods to conserve memory. Unfortunately, African Americans did not have the opportunity to share their memory with a larger audience because their commemorative practices were contained within their own community (Alderman, 2010). Additionally, they had limited access to public spaces (Alderman, 2010). For instance, "when African Americans occupied city streets and parks to celebrate holidays, such as Emancipation Day or the birthdays of Abraham Lincoln and Frederick Douglass, they exercised only temporary use of these public spaces before they reverted back to white control" (Alderman, 2010, p. 92). From heritage sites to schools, minorities are deculturized in that their cultures and history are not given representation and often erased or minimalized (Spring, 2016). Similarly, women are often invisible in American history, which indicates the relative importance they were given in society compared to men (Clark et al., 2004).

To date, only two scholarly works have explored roadside historical markers in Tennessee since the inception of the program in 1950, following the design of Virginia and North Carolina:

Jones' (1988) statewide study, and more recently, but limited to one county, Abdallah (2015). In *More to the story: Historical narratives and the African American past in Maury County, Tennessee,* Jaryn Abdallah explored historical production in Maury County, Tennessee (2015). In an attempt to gain insight about the way Maury County's history is portrayed, Abdallah examined historical publications, surveying historical markers (n = 27) and monuments (n = 22) (Abdallah, 2015). Out of the 27 historical markers of Maury County, 22 were dedicated to the history of white residents, while only 5 mentioned African Americans (Abdallah, 2015). Out of the five, two recognized the contribution of African Americans and just one focused on slavery (Abdallah, 2015). These historical markers are erected in places highly visited by the African American community (Abdallah, 2015). Again, some of these markers were cumbersome to obtain and the African American Heritage Society of Maury County had to take major efforts to have them erected (Abdallah, 2015). No similar studies exist focusing on the coverage of Women in History or Native Americans in Tennessee.

In the only published statewide study of Tennessee historical roadside markers, Jones (1988) analyzed the 1,170 roadside markers erected across the state. In his analysis, markers devoted to Black History or Famous White Women accounted for only 0.7% (n = 8) and 0.8% (n = 9), respectively, of all markers. In context, at the time of this study there were more roadside markers solely dedicated to David Crocket (0.9%, n = 11) than to either of these groups. Additionally, Native American markers only accounted for 3.0% (n = 35), while Famous White Men accounted for 11.7% (n = 137) – including 3% that is dedicated to Klu Klux Klan Founder Nathan Bedford Forrest (Jones, 1988). This particularly visible expression of public history serves a white, Protestant, male history of the state (Jones, 1988).

Social importance

Jones' (1988) findings are thus echoed in Abdullah's (2015) findings. Just as Abdallah (2015) argued that for visitors (or even locals) unfamiliar with Maury County's history, seeing the topics commemorated by their historical roadside markers can be misleading, making people assume that white residents majorly shaped Maury County's history. These representations ultimately exclude African American views and experiences from the overall narrative (Abdallah, 2015). Jones (1988), nearly three decades earlier, found that African Americans and other minorities are often omitted from these commemorative sites. The lack of representation not only threatens public memory but also historical accuracy, and accuracy is latently important when trying to celebrate the past (Bardet, 2012). Moreover, lack of representation of African Americans from Maury County's historical production prevents the county from making economic gains from their history (Abdallah, 2015). This is particularly relevant as Maury County's economy considerably benefits from tourism, and "the exclusion of people of color from the representation of the past may also impact their access to power in the present" (Abdallah, 2015, p. 37). These issues speak to the relationship between power and the production of history through tourism (Norkunas, 1993) where "the exclusion of people of Mexican, Native American and Chinese descent from the representation of the past" has negatively impacted "their access to power in the present" (Abdallah, 2015, p. 37). This exclusion is important for a number of reasons; for example, presentations of the past shape the public's perception and knowledge of culture and heritage (Jones, 1988), as well as wider social discourse around race and gender.

Similarly, Hanna and Hodder (2015) shed light upon the disparities between historical accuracy and the landscape commemorative markers is Fredericksburg, Virginia (Hanna & Hodder, 2015). After crucially studying Fredericksburg's commemorative landscape, paying close attention to markers commemorating slavery and emancipation, the authors concluded that the city needs more markers dedicated to these events, as they are central to the city's history but are not reflected adequately in the commemorative landscape (Hanna & Hodder, 2015).

Alderman (2010) adds that the ability to erect more of these markers is often held by bodies of individuals that do not reflect the population. More specifically, in the U.S., state governments, as well as local organizations and individuals, can fund and therefore influence what historical markers to erect. The content of commemorative site memorialization is indicative of the cultural value or political importance of a limited group's perception of the past that may not reflect the larger population. "If the values held by those responsible for deciding what is worthy of remembrance reflect the common heritage or a more selective past, because markers and listing are examples of what perspicaciously could be termed 'client,' or 'constituency' history" (Jones, 1988, p. 18). Looking at the profile of such groups in the examples shown illustrates an overwhelming majority of white males, especially in the South. The ability to fund and make decisions about historical markers gives white males control over memory (Alderman, 2010), and helps sustain their power over American history, enabling the distortion of historical facts and omission of contributions made by racial minorities and women.

Study approach and research questions

In his analysis of national register listings and roadside historic markers in Tennessee, Jones (1988, p. 28) asserted:

> Most recently, in 1987, as the result of newly conscious and organized constituency, Ida B. Wells, a late nineteenth-century cvil rights activist and newspaper editor in Memphis, was memorialized with a marker. It is tempting to say that this represents the beginning of an increase in marker and listing behavior from the black or black-female constituency, but it is too early to tell.

In our study, we assess the accuracy of this statement and critically reflect on its meaning. Using data on the 313 historical roadside markers erected since 1988, we (1) update Jones's (1988) analysis and (2) use the lens of socially just inclusion, from the perspective of social representation theory and critical race theory, to discuss the significance of the lack of inclusion of non-white male history in these markers. Southern sites of past reflection—those sites marking landscapes of public history and encouraging visitors to reflect on this history– are fertile for critical reflection on both the past and how it is portrayed in the present (Cook, 2018). To this effect, two research questions guide this study:

1. How has representation of people of color and women changed since Jones (1988)? In this question we explore not just the frequencies by category, but the content of the signs and how they speak to what society values and whose values are commemorated.
2. How well do the historical roadside markers represent the demographics of the communities in which they are erected? We explore this question through a focus on county-level demographics, as well as the funding source.

Data set

We analysed secondary data provided by The Historical Marker Database (HMDB; www.hmdb.org) that included all markers in the state of Tennessee that have been erected since 1988 (n = 313). The dataset included: a unique Marker ID; the title of the marker; the year erected; the latitude and longitude; the town, county, and state; and a link to see a picture of the marker to view its content and the funder. We only selected markers erected since 1988 because those prior to that were included in Jones's previous analysis. Detailed information on methods follows.

Data coding and analysis

In our analysis we first replicate Jones's (1988) methods of marker content analysis using his codes and we then respond to his call that, "... comprehensive and detailed content analysis of marker texts... can provide further information leading to more accurately supported qualitative judgements" (p. 27). Recognizing Jones' findings that non-white male social groups were largely underrepresented prior to 1988, as present not only in the frequency of representation of African Americans, Women, and Native Americans, we sought to reassess inclusion of these groups specifically. To this effect, we used both content analysis and GIS mapping. In all, analysis took place in six steps: (1) Assessing presentation across the state; (2) Comparing thematic preferences in presentations of the past and how they relate to Jones' analysis; (3) Expanding the analysis of the markers by expanded themes albeit with our expressed interest of assessing the change in emphasis on the Native American, Black History, and Famous White Women themes; (4) Expanding upon Jones' analysis to better understand the content of the signs and their funders by analyzing the distribution of the markers, focusing on urban/non-urban, Famous White Women, Black History, and Native American, across the state; (5) Assessing the content of the markers that were coded as Native American, Famous White Women, and Black History. Recognizing that by coding any reference to these groups as a marker representative of these groups that the content might not actually commemorate them or highlight something significant about them, we coded markers as directly related to the group, about a white person in history (but mentioning the marker group), or ancillary (being only related to the space or place, but not commemorating the group); and (6) Finally, we assessed the funders of the 66 markers designated (whether previously coded as ancillary or direct commemorating) as Native American, Famous White Women, and Black History.

Content analysis

This research employs a qualitative approach to providing additional meaning and insight to the content of Tennessee historic roadside markers. Here, we not only approach our content analysis from a qualitative perspective but seek to further delve into what the text may indicate about the cultural value of minority groups. In approaching our analysis this way, we further Bengtsson's (2016) stance that one of the strengths of content analysis is that it "is more than a counting process, as the goal is to link the results to their context or to the environment in which they were produced" (p. 9). For our study, we conducted a content analysis of the text inscribed on historical markers post-1988 in Tennessee using the aforementioned HMDB database. We did this by first coding all historical markers in the same 18 categories that were present in the Jones, 1988 article. Current categories were selected *a priori* based on categories present in the aforementioned research. Because we did not have information as to the criteria Jones used to code markers into categories, we applied a very liberal threshold. For example, if the marker mentioned African Americans, slaves, or black people, we coded that marker as Black History, meaning it had some association with African American culture in Tennessee. We did this for all 313 markers present in the post-1988 data set. After this first stage analysis, we then examined all those that were coded as mentioning Black History, Famous White Women, or Native Americans. In this second stage of content analysis we were looking for deeper meaning behind the text inscribed on the marker. At this stage, we shifted to emergent coding to allow for the most flexibility in analysis. From this, we developed three themes in how the originally coded person/group was mentioned in the historical marker text: 1) the marker was culturally relevant or spoke to the individual/subgroup itself; 2) the marker mentioned the individual or group in relation primarily to a white or male historical figure; 3) the marker mentioned primarily a significant geographical space or event. Simply put, we noted whether the marker was truly about the person or culture or whether the person/culture mentioned was simply a passing note

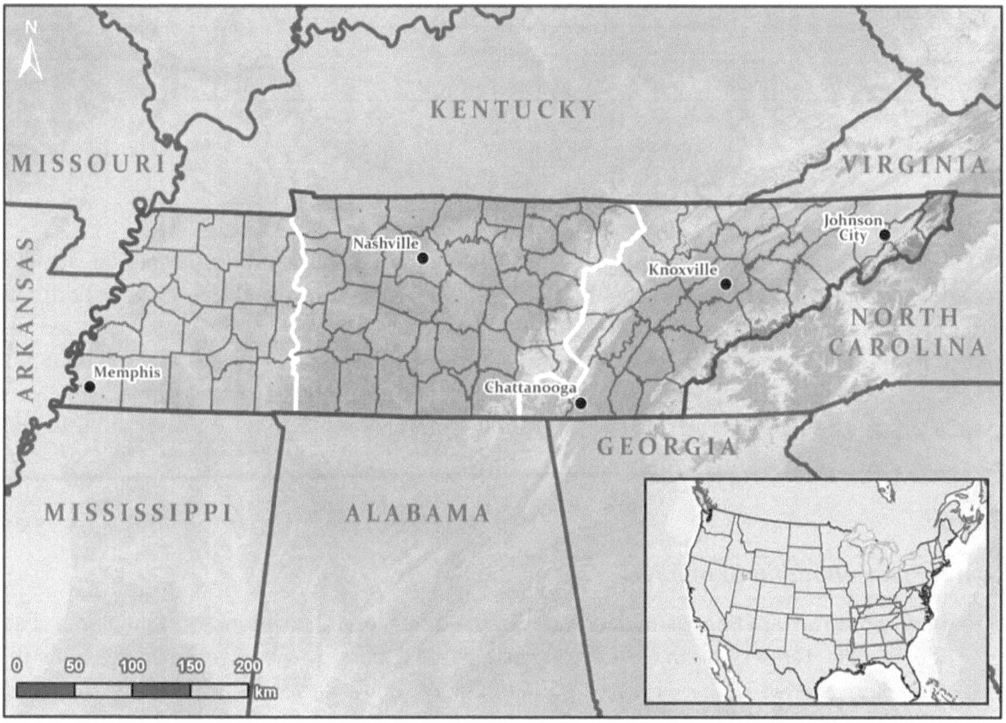

Figure 1. Study area showing all 95 Tennessee counties with white lines representing the three major regions of the state (West, Middle, East). Memphis is the principle metro area in West Tennessee and Nashville is the principle metro area in Middle Tennessee. East Tennessee has three principle metro areas: Chattanooga, Knoxville, and the Tri-Cities (Johnson City, Kingsport, Bristol). Image credit: Authors, 2020.

in another narrative. Utilizing content analysis in this particular way allows us to look at trends and patterns in the data in a unique way (Stemler, 2001).

GIS mapping

The historical marker database contained latitude and longitude coordinates, allowing them to be mapped as points using a geographic information system (GIS: ArcGIS Pro 2.3). Once mapped, multiple partitions were created: 1) markers in urban locations and non-urban locations (based on US Census-designated urban boundaries), 2) markers representative of white female history, 3) markers representative of black history, and 4) markers representative of Native American history. Additionally, demographic data from the 2010 US Census were obtained at both the county and Census tract levels in GIS format (US Census Bureau, 2010). Percent of population was mapped at the county level in Tennessee for female, black, and Native American populations. Marker locations were spatially joined to the Census tracts, which included population total and totals by race (i.e., black and Native American), allowing for more localized analysis. Traditionally, Tennessee is divided into three geographic regions: West Tennessee, Middle Tennessee, and East Tennessee (The Three Grand Divisions, n.d.). These regions were mapped and markers were queried by region to examine spatial trends since 1988.

Reflexivity and positionality of the researchers

The authors of this research come from different backgrounds, but share an interest in race relations and social equity. Two authors identify as white females, one author identifies as a white male, and one author identifies as a black female. Three of the authors are American born and

Table 1. Marker distribution by the three grand divisions of Tennessee.

	Jones		Since Jones	
Region	Number	Percent	Number	Percent
West	167	19	90	28.8
Middle	367	41	146	46.6
East	359	40	77	24.6
TOTAL	893	100%	313	100%

one immigrated to the United States within the last five years. Of the authors, three are sociologists and one is a geographer, but all authors share the perception that there is the opportunity to improve diversity in nearly all social representations, past and present. This perception led the researchers to analyze the depth or extent of diversity using roadside markers as the medium of study. It is acknowledged that the researchers' personal and professional positionalities impact their individual views, as well as collective views presented herein, regarding the research topic.

Results

Geographic distribution of markers

Comparing Jones' distribution of markers across the three grand divisions of Tennessee—East Tennessee, Middle Tennessee, and West Tennessee (see Figure 1)—we see an increase in the share of markers erected in the west and middle regions, but a decrease in the share erected in the east region (see Table 1). Like Jones, we find that in the sampling of markers across the regions, there exists a disproportionate consciousness of history across the state. However, we find that the regional disparity has actually lessened since 1988. Finally, regarding geographic distribution, in alignment with Jones, we find distribution across the state to be somewhat reflective of population centers in Tennessee, with Shelby County (Memphis) standing out in the west and Davidson County (Nashville) standing out in the middle (see Figure 2(a)).

Thematic distribution of markers

Regarding thematic preferences in presentations of the past and how they relate to Jones' analysis (see Table 2) using the thematic categories of the National Register nomination form (as presented in Jones' coding scheme), we found significant differences across the categories, with the exceptions of religion, government, and Native American. More specifically, there was an increase in the share of markers for historic districts, business/industry, religion, education, transportation, Native American, military, and miscellaneous, while there was a decrease in the share of markers for houses, and government[1].

Racial distribution of markers

While we find substantial increases for Black History and Famous White Women, we find minimal, if any change in the emphasis on Native American history (see Table 3). Like Jones, we find that Military and Famous White Men hold the highest shares of markers, but Border Markers are no longer accounting for a large share of the markers. Furthermore, the increase of the share of markers related to Famous White Men overshadows the increase we found in our categories of interest. This is likely due to Jones' coding of markers into a single category, while we accounted for plurality of markers. Thus, what we are capturing as an increase could simply be Jones' observation:

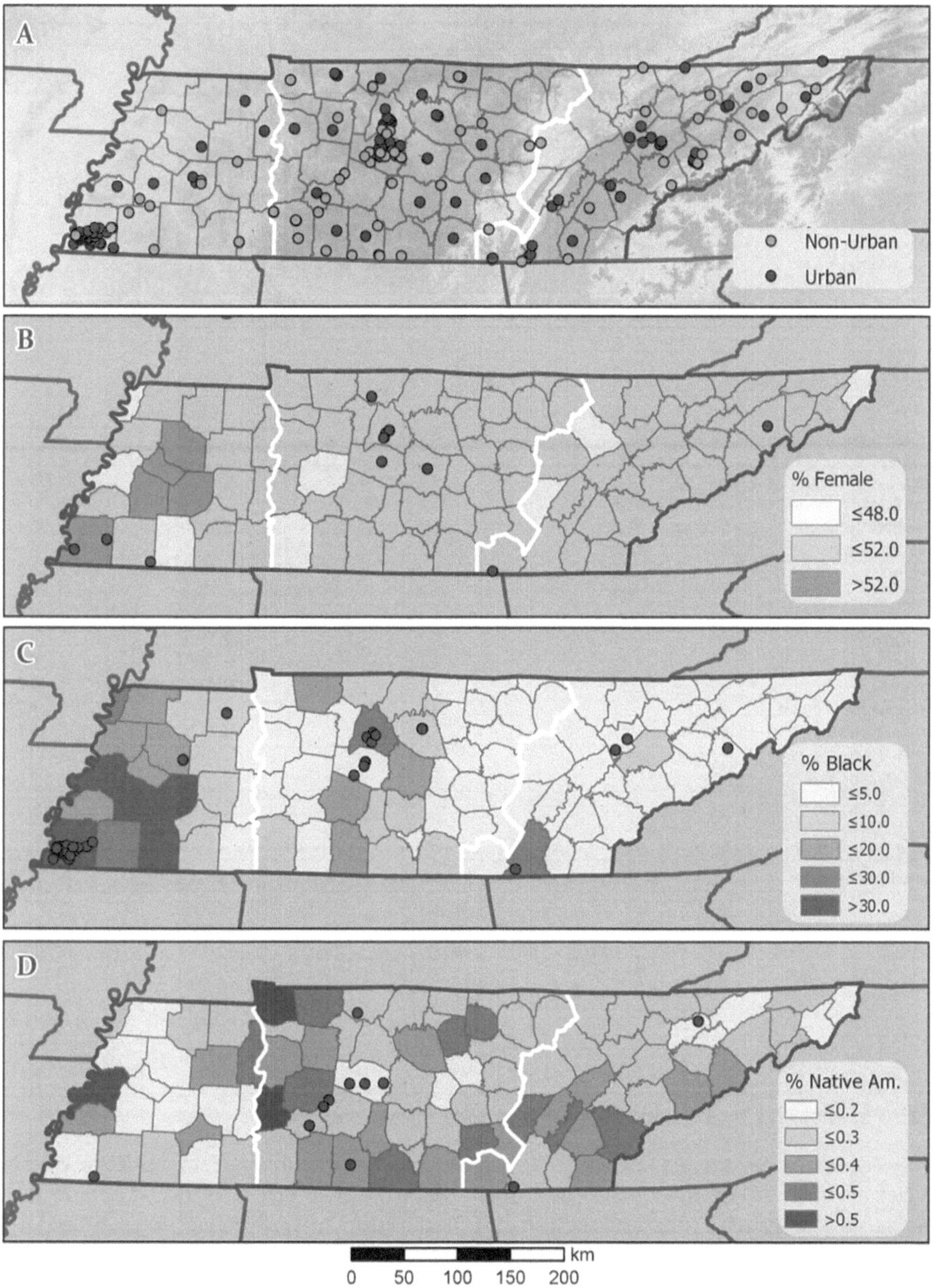

Figure 2. Tennessee historical markers located in urban/non-urban locations (A), representative of white female history (B), representative of black history (C), and representative of Native American history (D). B, C, and D also show the percent of the concurrent minority population by county. Image credit: Authors, 2020.

The fact that so many markers and listings were found to be either directly or indirectly related to famous white men, their 'firsts,' their wars, their politics, and their religion, betrays a previously unexamined dimension to these two public history programs: they almost exclusively represent a well-to-do white male Protestant constituency and reflect what it values in the past (1988, p. 26).

Table 2. Marker distribution by thematic categories of the National Register nomination form.

	Jones (N = 893)		Since Jones (N = 313)	
Category	Number	Percent	Number	Percent
Historic Districts	91	10	122	39.0
Houses	400	45	28	8.9
Business, Industry	98	11	52	16.6
Religion	86	10	40	12.8
Education	47	5	49	15.7
Government	47	5	12	3.8
Transportation	42	4	33	10.5
Native American	27	3	11	3.5
Military	19	2	112	35.7
Miscellaneous	36	5	71	22.7

Table 3. Marker distribution by enhanced thematic categories.

	Jones (N = 1170)		Since Jones (N = 313)	
Theme	Number	Percent	Number	Percent
Military	227	19.4	112	35.8
Border Markers	215	18.4	1	0.3
Famous White Men	137	11.7	113	36.1
Native American	35	3	11	3.5
Governors	45	4	7	2.2
"Firsts"	168	14	21	6.7
Education	70	6	49	15.6
Black History	8	0.7	42	13.4
Animals	4	0.3	0	0
Religion	56	5	40	12.8
Famous White Women	9	0.8	14	4.5
Business, Industry	53	4.5	52	16.6
Miscellaneous	132	11.3	71	22.7
David Crockett	11	0.9	2	0.6

Table 4. Marker distribution across the grand divisions of Tennessee based on urban boundaries, race, and gender.

Historical Markers	West TN	Middle TN	East TN	All of TN
Total	90 (28.8%)	146 (46.6%)	77 (24.6%)	313
Urban	75 (32.5%)	100 (43.3%)	56 (24.2%)	231 (73.8%)
Non-Urban	15 (18.3%)	46 (56.1%)	21 (25.6%)	82 (26.2%)
Famous White Women	3 (21.4%)	8 (57.1%)	3 (21.4%)	14 (4.5%)
Black History	28 (66.7%)	10 (23.8%)	4 (9.5%)	42 (13.4%)
Native American	1 (9.1%)	8 (72.7%)	2 (18.2%)	11 (3.5%)

Density of markers by population and demographics

Middle Tennessee has the highest share of markers coded as non-urban, Famous White Women, and Native American, relative to their total share (i.e., more than 46.6% of the markers in these categories are in Middle Tennessee; see Table 4). Black History markers, however, have been predominantly erected in West Tennessee. Acknowledging that demographics are not equally distributed across the state and that markers represent local history, we then mapped these select themes to county-level demographics in Figures 2(a–d). We further assessed representation in Figures 3(a,b) by looking at Tennessee historical markers representative of Black History by percent of black population residing in the census track and representative of Native American history by percent Native American history residing in the census tract, respectively. Visual analysis of these maps reveals that (1) markers are denser in urban population areas, (2) there is a greater density of Black History makers in counties with a higher percentage of black residents, (3) there are not distinct patterns in Native American or Famous White Women markers, and most

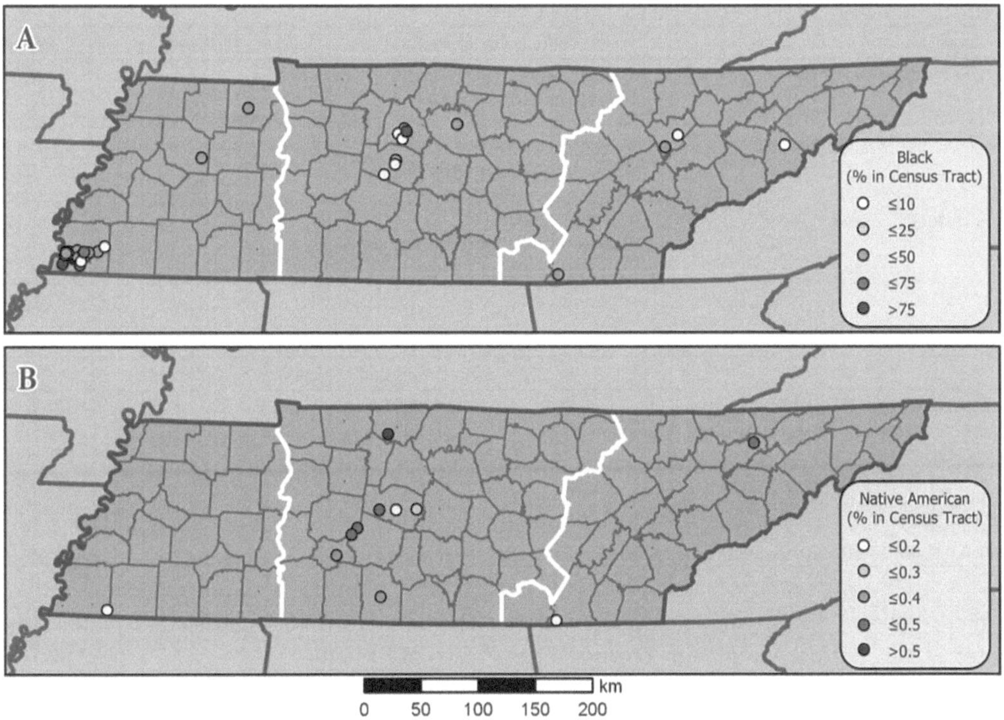

Figure 3. Tennessee historical markers representative of black history and shaded by percent of the black population residing in the concurrent Census tract (A). Tennessee historical markers representative of Native American history and shaded by percent of the Native American population residing in the concurrent Census tract. Image credit: Authors, 2020.

importantly, (4) there is ample opportunity to promote a more "just representation" and inclusive historical portrayal through Tennessee historical roadside markers.

Analysis of markers designated as native american, women's and black history

After coding markers as directly related to the group, about a white person in history (but mentioning the marker group), or ancillary (being only related to the space or place, but not commemorating the group), the numbers of sign designated as representing a minority group significantly decreased. Specifically, only 27.27% (n = 3) of the Native American markers were considered to commemorate Native Americans, 46.15% (n = 6) of the Famous White Women markers were considered to commemorate white women, and 88.10% (n = 37) of the Black History markers were considered to commemorate black persons (see Table 5). For example, the "Normal Depot" which is coded as an African American roadside marker actually does not commemorate anything that is culturally significant to African Americans. Rather, the inscription on the marker notes that the Normal Depot is "A brick Craftsman-style building with a tiled hip roof, it was a commuter station with separate white and 'colored' waiting rooms." (https://www.hmdb.org/marker.asp?MarkerID=84112) and goes on to discuss how students arrived at the building and when it was decommissioned and sold. The mention of separate waiting rooms by race is hardly unique given the year, 1912, and certainly does contribute to a cultural narrative for the African American community - rather the African American community is mentioned as a byproduct of the operations of the building. Another example is Dorothy Harrison Eustis who is mentioned in a historical marker but not for her achievement of opening The Seeing Eye, the famous dog guide training school, but she is instead mentioned in relation to a male

Table 5. Content designation based on emphasis on key themes.

Category	Content Designation	Number	Percent
Native American (N = 11)	Directly related	3	27.27
	Relation to white person	2	18.18
	Relation to space or place	6	54.55
Famous White Women (N = 13)	Directly related	6	46.15
	Relation to white person or male	2	15.38
	Relation to space or place	5	38.46
Black History (N = 42)	Directly related	37	88.10
	Relation to white person or male	0	0.00
	Relation to space or place	6	14.32

Table 6. Funding sources of key themes.

	Native American	Famous White Women	Black History
Tennessee Historical Commission	1 (9.09%)	1 (7.69%)	6 (14.29%)
Local or Government Historical Group (other than TN HC)	7 (63.64%)	7 (53.85%)	12 (28.57%)
University	0	0	2 (4.76%)
Private Citizen or another group	0	2 (15.38%)	17 (40.48%)
Family of person on sign	0	0	2 (4.76%)
Government agency	0	0	1 (2.38%)
Funder not indicated	0	0	2 (4.76%)

partner - "It was Frank who persuaded Dorothy Harrison Eustis to establish a school in the United States." (https://www.hmdb.org/marker.asp?MarkerID=61940)

Analysis of funding

Seven funding types emerged in this analysis (see Table 6). While Native American and Famous White Women markers were predominantly funded by local and government groups other than the Tennessee Historical Commission, Black History markers were predominantly funded by private citizens or related groups.

Discussion and directions forward

Southern sites of past reflection are fertile for critical reflection on both the past and how it is portrayed (Cook, 2018) and our landscape is increasing populated with historical tourist sites (Doss, 2010; Alderman & Inwood, 2013). This research has examined the representation of Famous White Women, Native Americans, and Black History in the historical marker memorial landscapes or "memoryscape" of Tennessee. Like all states in the American South, Tennessee has extensive racial history riddled with white privilege and supremacy that is still reflected in current relations and processes of social representation; thus, the findings presented herein are not surprising to the researchers. We have not only sought to assess if Jones' prediction that the placement of the Ida B. Wells marker in 1987 possibly marked "the beginning of an increase in marker and listing behavior from the black or black-female constituency" (p. 28), but we have conducted and presented a critical reading of content of the markers erected over the past three decades to identify the presence and absence of minorities in Tennessee's memorial landscape as presented through historical roadside markers.

While we began this research with the purpose of assessing these sites across Tennessee and evaluating Jones' prediction, we find that little progress has been made in promoting a more equitable or just representation through historical roadside markers in the last three decades. Carlson's (1980) analysis of all listings on the National Register of Historic Places as of 1976 revealed an over-emphasis on urban, influential citizens and reproduced a biased view of the past. In regards to Tennessee historical roadside markers, we concur, as did Jones (1988) and

Abdullah (2015), that we have failed to reach a just representation of history as a state. While we are unable to capture societal values in our research, our thematic analysis is revealing of perceptions of what or whose history is important in Tennessee. Results indicate that even today, in 2019, that there are more historical markers dedicated to Conference General Nathan Bedford Forrest, founder of the Ku Klux Klan, than all white women and Native Americans combined. While we see an increase in markers dedicated to Black History, if we only account for those directly related (removing those identified as ancillary), then we have only five more African American markers across the whole state than those dedicated solely to Forrest. Therefore, we conclude that past events in Tennessee remain to be framed from a white, Protestant perspective and that little progress has been made over the past three decades. Following social representation theory, this entails that this version of history is anchored in the historical roadside markers as objects and in their content as narratives. This history constructs meaning for social context in the past and present—a reality that is socially constructed and narrated through roadside markers as objectification devices. These narratives then provide social cues or ways of knowing. However, from a sustainability perspective, our analysis reveals that the social cues that they are providing are not equitable or socially just.

This finding is easily situated within a larger body of literature documenting divergent, segregated narratives for black and white America (Shipler, 1997). From a social representation theory and more specifically, from a critical race theory lens—the view that race is central to American culture and intersects class and gender—it is important to understand the potential impact of this bias in the memorialization of history on minorities. In 1988, Jones wrote,

> In Tennessee the orientation mentioned earlier is one-sided and value-biased and robs entire cultural and ethnic groups of the sense of identity and continuity that history is legitimately supposed to provide… It has made the public's view of the past a monolithic, avocational, and sentimental hallucination that ignores the variegated richness of our collective heritage. The granting of historical legitimacy through markers and listings might even be interpreted as an exercise in social control, an attempt to maintain the values of a secular order confronted by the dynamics of social change (p. 29).

Our analysis confirms Jones' observation as we conclude that historical memorialization is a form of power. In the case of Tennessee, this power has yet to fully be used to challenge the dominant historical narrative (see Delgado & & others, 1989) and hence we have both failed to achieve a just representation to date and have ample opportunity to rectify this in the future. We thus find that this lack of power presents opportunity in the state in that the narrativized social representations are of a sentimentalized South (c.f., Bright & Carter, 2018). The stories represented by these sites do not present a just representation in that they convey a biased social representation and can be considered a medium for negotiation of this version of history. Through the lens of critical race theory, the voice of black America must be expressed to offer competing versions of history and to challenge the white-centric comfortable versions of history (Delgado & Stefancic, 2017). Through efforts such as historical roadside marker programs, consciously or unconsciously, the story of black history, of Native American history, and of white women in history has been edited from the perspective of white males, thus altering the American collective memory (see Delgado & Stefancic, 2017).

We conclude that the representation of non-white men is limited and unjust. Tennessee was built upon more than a white-male past and Tennessee has more than a white-male present. We have an opportunity to highlight and commemorate differences and uniqueness. In this sense, we find meaning in the signs we assessed, but also in the ones that are absent. Through analyzing the spatial patterns of historic marker locations since 1988, we see that there are many counties across Tennessee that have proportionally higher minority populations, but few, if any, makers have been erected in these areas during the study period. For example, 69 historic markers were established in Shelby County (where Memphis is located), with 26 of those representing black history, two representing white female history, and one representing Native American history. However, four nearby counties with very high black populations (>30%) in

West Tennessee (Lauderdale, Haywood, Madison, and Hardeman) have no historic markers representing non-white males. These areas may represent future opportunities for new historic markers to better reflect their communities. Historical markers provide educational information, but they have also become icons of popular culture that demonstrate how we invest importance in past people, events, and sites. In this sense, aligning with social representation theory, historical roadside markers provide an identity to the community, but this identity is not always fair or just; in fact we might argue that it is rarely representative of the actual community. In other words, as markers of public history, these sites a narrativizing or anchoring a version of history that is not representation of all social groups. As such, these sites either consciously or subconsciously contribute to the social discourse in these communities in giving them this unjust identity. Thus, memoryscapes reproduce and construct history, but as we have found, this is not a representative history nor is it socially just. Rather, it is a mythical version of history curated to legitimate popular, comfortable versions of history.

Alderman and Campbell (2008, p. 338) assertion that sites across the American South are facing symbolic excavation or "the unearthing of difficult and long suppressed (and repressed) historical narratives… through memory work, the construction and representation of the past." This is also perhaps present in the initiative taken by non-government bodies to erect Black History markers. To see change in the representation of history through historical markers and in heritage tourism more broadly, we have to take a closer look at the funders of history and how this impacts representation. When funders are distorted in their representation, a distorted view of the past may be replicated; they reproduce value-biased views of the past that do not reflect their constituency. The funders are often responsible to constituencies that are not reflective of the broader population. Thus, Jones (1988, p. 26-27) noted, "Other constituencies have found no way to express their history within the confines of the criteria utilized in these two processes, because one constituency has dominated the means of expression." While we find instances of "memorial entrepreneurship" (c.f., Dwyer et al., 2012; Bright et al., 2018) or efforts to impact these memorial sites, we find more opportunity for engagement by communities to promote just presentation.

Aligning with sustainable tourism related to public history, a greater emphasis should be placed on community-based planning and inclusion. Although a challenging task (Jamal & Getz, 1995), coordination between those erecting objects narrativizing the past and creating tourism sites of them and the communities they represent is imperative. As noted by Hall (2003, p. 99), community-based planning "does not automatically lead to either sustainable tourism development or even a reduction in the amount of conflict surrounding tourism development." In other words, communities are also complex entities with their own politics and conflicts. The results presented herein reflect unequal power and politics within tourism at the community level. Hall and Jenkins (1995) presented that tourism development often entails the suppression of certain issues reflective of power dynamics between stakeholders. While many traditional models of sustainable tourism focusing on community development, such as Murphy (1985) fail to account for representation of the community and power-politics, we recognize that there are politically imposed limitations on decision-making in tourism that are reflected in decision-making and thus reflected in representation (Crensen, 1971; Hall, 1994, 2003). The communities of interest in this study are largely faced with invisibility and unjust representations and are up against a long history of suppressing their cultural identity.

Just as Jenkins (1993) listed the apathy of citizens as one of the impediments to promoting more sustainable tourism, we note that there is perhaps an evolving interest, and thereby relevance, in roadside historical markers. Just as many heritage managers would not even consider themselves in the tourism business (Croft, 1994), there are many individuals who would drive by roadside markers without noticing or giving them consideration. Although historical markers are often thought of as an archaic way to learn about the history of a place or region, in today's digital age, they are increasingly accessible through mobile and online apps (Georgia Historical

Society, n.1994d.). For instance, it is possible today to plan whole trips revolving around roadside markers based on routes generated online. Moreover, nearly every roadside historical marker can be easily viewed online. This digital cataloguing, perhaps, presents a new body of stakeholders as these heritage tourism sites can be accessed across the world without travel. Although cultural heritage tourism is defined by travel directed to experience these sites (Bright & Carter, 2016), it is conceivable that digital access presents a new body of consumers that have social significance in the development in the future of historical roadside markers.

For praxis, the results herein should be of interest to public historians, geographers, sociologists, and other academics, as well as community leaders, not only in Tennessee, but in other states and countries as well. While there are numerous methodological challenges to assessing how well markers reflect their constituency and if inclusive history is portrayed and protected, this research advances the literature by both enhancing the methods and data provided by Jones (1988), but also by taking a more critical stance on the social importance of a more just, representative history. It is important that scholars, as well as communities themselves, consciously advocate for a more socially just memorialization of history that challenges the comfortable myths that are currently portrayed. To do so, we have to first be aware of the social and historical ideologies that are present in Tennessee's presentation of history, which is not unique to other regions in the United States. This research has identified white privilege in historical memorialization and sought to speak to the opportunity to counter the mythical, nearly hegemonic, portrayals of our past.

Note

1. While percent change could be calculated to compare Jones' thematic category shares to our analysis, this would be misleading because Jones' placed each sign into a single theme, while we found numerous markers fell under more than one category. More specifically, 217 of our 313 markers were coded as two themes.

Disclosure statement

No potential conflict of interest was reported by the author(s).

Data availability statement

Raw data were generated at www.hmdb.org. Derived data supporting the findings of this study are available from the corresponding author, Candace Forbes Bright, on request.

ORCID

Candace Forbes Bright http://orcid.org/0000-0003-1205-2492

References

Abdallah, J. E. (2015). *More to the story: Historical narratives and the African American past in Maury County, Tennessee* [Doctoral dissertation], Middle Tennessee State University.
Alderman, D. H. (2010). Surrogation and the politics of remembering slavery in Savannah, Georgia (USA). *Journal of Historical Geography, 36*(1), 90–101. https://doi.org/10.1016/j.jhg.2009.08.001
Alderman, D. H. (2012). History by the Spoonful" in North Carolina: The textual politics of state highway historical markers. *Southeastern Geographer, 52*(4), 355–373. https://doi.org/10.1353/sgo.2012.0035
Alderman, D. H., & Campbell, R. M. (2008). Symbolic excavation and the artifact politics of remembering slavery in the American South: Observations from Walterboro, South Carolina. *Southeastern Geographer, 48*(3), 338–355.
Alderman, D. H., & Inwood, J. F. (2013). Landscapes of memory and socially just futures. In N. C. Johnson, R. H. Schein, & J. Winders (Eds.) *The Wiley-Blackwell companion to cultural geography* (pp. 186–197). Wiley. https://doi.org/10.1002/9781118384466.ch18
Bachrach, P., & Baratz, M. S. (1970). *Power and poverty: Theory and practice.* Oxford University Press.
Bardet, P. (2012). Demarcating territory: Historical markers in the United States. *Miranda, 6*(6), 1–9. https://doi.org/104000/miranda.2920
Bengtsson, M. (2016). How to plan and perform a qualitative study using content analysis. *NursingPlus Open, 2*, 8–14. https://doi.org/10.1016/j.npls.2016.01.001
Bright, C. F., & Carter, P. (2016). Who are they? Visitors to Louisiana's River Road plantations. *Journal of Heritage Tourism, 11*(3), 262–274. https://doi.org/10.1080/1743873X.2015.1100627
Bright, C. F., & Carter, P. (2018). Social representational communities and the imagined Antebellum South. *Sociological Spectrum, 38*(1), 24–38. https://doi.org/10.1080/02732173.2017.1396944
Bright, C. F., Alderman, D. H., & Butler, D. L. (2018). Tourist plantation owners and slavery: A complex relationship. *Current Issues in Tourism, 21*(15), 1743–1760. https://doi.org/10.1080/13683500.2016.1190692
Buckley, R. (2012). Sustainable tourism: Research and reality. *Annals of Tourism Research, 39*(2), 528–546. https://doi.org/10.1016/j.annals.2012.02.003
Carlson, A. W. (1980). An analysis of historic preservation in the United States as reflected by the National Register of Historic Places. *The Journal of American Culture, 3*(2), 245–267. https://doi.org/10.1111/j.1542-734X.1980.0302_245.x
Clark, R., Allard, J., & Mahoney, T. (2004). How much of the sky? Women in American high school history textbooks from the 1960s, 1980s and 1990s. *Social Education, 68*(1), 57–63.
Cook, M. R. (2018). Textual politics of Alabama's historical markers: Slavery, emancipation, and civil rights. In Stanley D. Brunn and Roland Kehrein (Ed.), *Handbook of the changing world language map* (pp. 1–31).
Crensen, M. A. (1971). *The un politics of air pollution: A study of non-decisionmaking in the cities.* Johns Hopkins University Press.
Croft, T. (1994). What price access? Visitor impact on heritage in trust. In J. M. Fladmurk (Ed.), *Cultural tourism* (pp. 169–178). Donhead Publishing.
Delgado, R. (1989). Storytelling for oppositionists and others: a plea for narrative. (Legal Storytelling). *Michigan Law Review, 87*(8), 2411–2441. https://doi.org/10.2307/1289308
Delgado, R., & Stefancic, J. (2017). *Critical race theory: An introduction* (Vol. 20). NYU Press.
Doise, W. (1989). Attitudes et repreasentations sociales. In D. Jodelet (Ed.), *Les repreasentations sociales.* PUF.
Doss, E. (2010). *Memorial mania: Public feeling in America.* University of Chicago Press.
Du Cros, H. (2001). A new model to assist in planning for sustainable cultural heritage tourism. *International Journal of Tourism Research, 3*(2), 165–170. https://doi.org/10.1002/jtr.297
Durkheim, É. (1898). Représentations individuelles et représentations collectives. *Revue de Métaphysique et de Morale, 6*(3), 273–302.
Dwyer, O., Butler, D., & Carter, P. (2013). Commemorative surrogation and the American South's changing heritage landscape. *Tourism Geographies* , *15*(3), 424–443. https://doi.org/10.1080/14616688.2012.699091

Fredline, E., & Faulkner, B. (2000). Host community reactions: a cluster analysis. *Annals of Tourism Research, 27*(3), 763–784. https://doi.org/10.1016/S0160-7383(99)00103-6

Georgia Historical Society. (n.d.). "Historical Markers." *Georgia Historical Society*, https://georgiahistory.com/education-outreach/historical-markers/

Garrod, B., & Fyall, A. (2000). Managing heritage tourism. *Annals of Tourism Research, 27*(3), 682–708. https://doi.org/10.1016/S0160-7383(99)00094-8

Hall, C. M. (1994). *Tourism and Politics: Power, Policy, and Place*. John Wiley & Sons.

Hall, C. M. (2003). Politics and place: An analysis of power in tourism communities. *Tourism in Destination Communities*, 99–114.

Hall, M., & Jenkins, J. (1995). *Tourism and public policy*. Routledge.

Hanna, S. P., & Hodder, E. F. (2015). Reading the signs: Using a qualitative Geographic Information System to examine the commemoration of slavery and emancipation on historical markers in Fredericksburg, Virginia. *Cultural Geographies, 22*(3), 509–529. https://doi.org/10.1177/1474474014548161

Holland, D., Skinner, D., Lachicotte, W., & Cain, C. (1998). *Identity and Agency in Cultural Worlds*. Harvard University Press.

Inwood, J., & Martin, D. (2008). Whitewash: White privilege and racialized landscapes at the University of Georgia. *Social & Cultural Geography, 9*(4), 373–395. https://doi.org/10.1080/14649360802033882

Jamal, T. B., & Getz, D. (1995). Collaboration theory and community tourism planning. *Annals of Tourism Research, 22*(1), 186–204. https://doi.org/10.1016/0160-7383(94)00067-3

Jenkins, J. (1993). Tourism policy in rural New South Wales-policy and research priorities. *GeoJournal, 29*(3), 281–290. https://doi.org/10.1007/BF00807047

Jones, J. B. (1988). An analysis of national register listings and roadside historic markers in Tennessee: A study of two public history programs. *The Public Historian, 10*(3), 19–30. https://doi.org/10.2307/3378527

Lasswell, H. D. (1936). *Politics: Who gets what, when, how*. Whittlesey House.

Lukes, S. (1974). *Power: A radical view*. Macmillan International Higher Education.

Lukes, S. (2004). *Power: A radical view*. Macmillan International Higher Education.

McKittrick, K. (2016). Diachronic loops/deadweight tonnage/bad made measure. *Cultural Geographies, 23*(1), 3–18. https://doi.org/10.1177/1474474015612716

Moscovici, S. (1961). *La Psychanalyse, Son Image et Son Public*. Presses Universitaire de France.

Moscovici, S. (1981). On social representations. In J. P. Forgas (Ed.), *Social cognition: Perspectives on everyday understanding* (pp. 181–209). Academic Press.

Moscovici, S. (1983). The coming era of social representation. In J. P. Codol & J. P. Leyens (Eds.), *Cognitive approaches to social behavior* (pp. 3–69). Nijhoff.

Moscovici, S. (1984). The phenomenon of social representations. In R. M. Farr & S. Moscovici (Eds.), *Social representation* (pp. 3–69). Cambridge University Press.

Moscovici, S. (1988). Notes toward a description of social representation. *European Journal of Social Psychology, 18*(3), 211–250. https://doi.org/10.1002/ejsp.2420180303

Murphy, P. (1985). *Tourism: A community approach*. Methuen.

Murphy, P. E. (1988). Community driven tourism planning. *Tourism Management, 9*(2), 96–104. https://doi.org/10.1016/0261-5177(88)90019-2

Norkunas, M. K. (1993). *The politics of public memory: Tourism, history, and ethnicity in*. Suny Press.

North Carolina Department of Natural and Cultural Resources. (n.d.). About the Historical Marker Program. https://www.ncdcr.gov/node/5947?fbclid=IwAr0zJFT4WgailR4ZexuMAsp1pnFMOYw55wzUs6fg-zcLnnsjY6ML6bn_Wq4

Otterstrom, S. M., & Davis, J. A. (2016). The uneven landscape of California's historical landmarks. *Geographical Review , 106*(1), 28–53. https://doi.org/10.1111/j.1931-0846.2015.12133.x

Pearce, P. L. (1993). Fundamentals of Tourist Motivation. In D. G. Pearce and R. W. Butler (Eds.), *Tourism research: Critiques and challenges* (pp. 113–134). Routledge.

Ruggles, D. F., & Silverman, H. (2009). From tangible to intangible heritage. In *Intangible heritage embodied* (pp. 1–14). Springer.

Seba, J. (2011). A position paper on cultural and heritage tourism. In *Tourism and hospitality* (pp. 130–143). https://doi.org/10.1201/b12231-13

Shipler, D. (1997). *A country of strangers: Blacks and whites in America*. Alfred A. Knopf, Inc.

Silberberg, T. (1995). Cultural tourism and business opportunities for museums and heritage places. *Tourism Management, 16*(5), 361–365. https://doi.org/10.1016/0261-5177(95)00039-Q

Spring, J. (2016). *Deculturalization and the struggle for equality: A brief history of the education of dominated cultures in the United States*. Routledge.

Stemler, S. (2001). An overview of content analysis. *Practical Assessment, Research & Evaluation, 7*(17), 1–6.

The History of Blue Plaques. (n.d.). https://www.english-heritage.org.uk/visit/blue-plaques/about-blue-plaques/history-of-blue-plaques/

The Three Grand Divisions. (n.d.). Tennessee Department of State. https://sos.tn.gov/sites/default/files/Pg.%20639%20Three%20Grand%20Divisions.pdf

Tischauser, L. V. (2013). *Jim crow laws*. Greenwood.
Thirft, N., & Forbes, D. (1983). A landscape of figures: political geography with human conflict. *Political Geography Quarterly*, *2*, 247–263.
US Census Bureau. (2010). 2010 Census. U.S. Census Bureau. Retrieved August 12, 2019, from https://www.census.gov/programs-surveys/decennial-census/data/datasets.2010.html
Virginia Department of Historic Resources. (2019). Historical Highway Markers. https://www.dhr.virginia.gov/highway-markers/
Wagner, W., Elejabarrieta, F., & Lahnsteiner, I. (1995). How the sperm dominates the ovum—Objectification by metaphor in the social representation of conception. *European Journal of Social Psychology*, *25*(6), 671–688. https://doi.org/10.1002/ejsp.2420250606
West, C. V. (2018, March 1). "Clinton Desegregation Crisis." *Tennessee Encyclopedia*, Tennessee Historical Society. https://tennesseeencyclopedia.net/entries/clinton-desegregation-crisis/.

Resisting marginalisation and reconstituting space through LGBTQI+ events

Oscar Vorobjovas-Pinta and Anne Hardy

ABSTRACT

The field of event studies has attracted a breadth of research on the triple-bottom line of economic efficiency, environmental integrity and social equity. The focus of many studies related to event tourism, however, has fallen upon the economic and environmental dimensions of events with far less attention on "social equity." The potential of events tourism to facilitate justice and equity for marginalised and minority groups has been especially overlooked. LGBTQI+ communities utilise gay events, such as pride parades, as mediums to communicate their identities and seek support from broader society. This paper examines a unique festival space where LGBTQI+ communities can resist marginalisation and exclusion, counteract stereotypical images and representations, and reconstitute space to fully embrace their identity and their communities. This study draws on neo-tribal theoretical insights to examine the case of the Broken Heel drag queen festival, held in the rural Australian town of Broken Hill. The study analyses Instagram posts using a netnographic approach to explore the spatiality of this festival and participant practices on the journey and at the festival site. Results points to the importance of LGBTQI+ events such as the Broken Heel festival to counter marginalisation and promote social justice and sociality for LGBTQI+ people through individual and collective expression of emotional connectedness and non-heterosexual identities.

Introduction

Every year in September, around 6500 festival attendees, many dressed in drag, will travel from their homes (many of which are located in urban locations such as Sydney and Melbourne) to the remote outback town of Broken Hill, Australia, to celebrate drag, queerness and the LGBTQI+ communities in a festival with the name Broken Heel (Baird, 2019; see also https://www.bhfestival.com/). The location for the festival is quite unusual. Broken Hill is affectionately known as the capital of the outback, or Silver City, due its former success as a silver ore mining town. Located on the border of two states, New South Wales and South Australia, Broken Hill is inland and located near the desert. It is 1100 km by road from Sydney and 900 km from Melbourne. In 2018, Broken Hill's population was just under 18,000 (Australian Government, 2019). When one *googles* images of Broken Hill and the outback, they are delivered a variety of images related to the Australian outback such as mines, photographs of red earth, remote, dry and arid landscapes, and sun worn farmers wearing iconic Australian hats (Akubras) and flannel (plaid) shirts. The Australian outback has traditionally been depicted as a rugged and predominantly masculine landscape (Little & Panelli, 2007; Scott & Biron, 2010). Broken Hill is a traditionally conservative and heteronormative space and the Broken Heel festival provides a humorous and provocative challenge to this image. Festival attendees delight in the irony of the festival location, its name and the challenge that it faces them.

This study examines the Broken Heel drag queen festival in outback Australia through a social justice and neo-tribal lens. In doing so, it examines how marginalised people, albeit those who are gathering temporarily, can coalesce and challenge the heteronormative imagining of space through these types of events. Public everyday spaces are charged with meaning and values (Malpas, 2012), but overwhelmingly, such spaces are heteronormative and even heterosexually masculine (Pielichaty, 2015; Vorobjovas-Pinta & Hardy, 2016). Gay spaces are heterogenous in their nature and may vary in terms of diversity, openness and exclusivity. Walters and Jepson (2019) suggest that event spaces not only cause and perpetuate social marginalisation, but also, can help to overcome it. They can serve as an important tool for building stronger community bonds, identity, and self-development. Hassanli, Walters and Friedman (2020) assert that events can act as counterspaces for individuals by facilitating their adaptive response to marginalisation. With this in mind, the paper aims to respond to the following two research questions:

- How do LGBTQI+ events like the Broken Heel festival reduce marginalisation and promote greater social justice?
- How useful is neo-tribal theory to examine the role that festivals play in advancing social justice and sociality?

The paper is structured as follows. We commence with an appraisal of the concepts of social justice, marginalisation and exclusion, and then go on to assess neo-tribal theory, spatialisation, resistance and transformation. We then set the context for the study, the Broken Heel Festival. Following the data analysis that uses a netnographic approach, the paper concludes with discussion and conclusions.

Social justice, marginalisation and exclusion

The notion of equity and justice has tended to focus on economic issues, drawing on cost-benefit analyses, visitor expenditures etc., but even those in relation to marginalised and minority groups appear to be strongly overlooked. These oversights have been noted; a study by Jamal and Camargo (2014) in Quintana Roo, Mexico indicated that profitability and return on investment is often favoured in lieu of local Mayan residents' cultural and economic wellbeing. Other studies have addressed the need for a greater inclusion of different community groups in tourism but have tended to focus on economic benefits (Gustavsson et al., 2014; Schellhorn, 2010) which are often supported by the state through policy means and/or investment in tourism infrastructure (Blake, 2001; Bojer, 2003). A range of studies has focused on issues of exclusion and dislocation, such as dispossession from Native lands, and exclusion from protected areas and local commons (Devine & Ojeda, 2017; Higgins-Desbiolles, 2008; 2009). Environmental justice issues related to exclusion and physical health have also been examined, particularly, in relation to diverse populations and communities of colour (Dangi & Jamal, 2016; Lee & Jamal, 2008). However, few studies have provided in-depth exploration of such justice issues related to LGBTQI+ communities (Dworkin & Yi, 2003; Ginwright & James, 2002).

LGBTQI+ communities can be seen as examples of where marginalised people have established new platforms themselves to engage in a direct dialogue with wider society and the state, and, as such, contribute to individual as well as collective wellbeing of those LGBTQI+ communities. Enclavic spaces like LGBTQI+-friendly destinations and gay cruises offer spaces to build social networks, friends and supportive *communitas* (Vorobjovas-Pinta, 2018a). There are also other more ephemeral or temporary spaces where opportunities arise to resist and change dominant discourses, prejudices and injustice. These include public events such as pride parades, rainbow weeks or similar events. Whilst born from radical protests, pride parades or similar LGBTQI+ events celebrate social and self-acceptance and achievements that are enabled by legal and policy frameworks such as marriage equality laws. LGBTQI+ communities around the world are known to use gay events, such as pride parades, as mediums to resist discrimination, marginalisation and invisibility, communicate their identity and to seek support from broader society for the right to belong as equal members of society (Ong & Goh, 2018).

It is important to note that LGBTQI+ communities within the various spaces mentioned above are also highly diverse. Drag performers, for instance, are well established within LGBTQI+ cultures (Drysdale, 2019). Their performance spaces are usually confined to places (such as gay clubs or bars) or LGBTQI+ events (such as parades and festivals) (Drysdale, 2019). Drag shows can range from amateur acts at local gay bars to extravagant theatrical performances. Drag queen acts imply a theatrical performance of gender where often serious societal issues are showcased through sarcasm and humour.

Marginalisation and discrimination

Marginalisation has been referred to as a 'product of the cultural politics' that plays out differently in the context of locally-specific social relations (Hubbard, 1998, p. 56). LGBTQI+ voices remain largely marginalised (Verrelli et al., 2019). The contemporary marginalisation of various individuals and community groups is not always an objective fact, but rather a power play bound by their placement on the margins of social and geographical imagination (Hubbard, 1998; Smith & Pitts, 2007). Marginalised individuals and community groups are often subject to covert and overt discrimination, bullying and animosity. Stereotypes, prejudice, and discrimination also prevail in social arrangements, creating cultural, religious or sexual marginalisation, social exclusion and severe impacts on health and well-being (Kaniuka et al., 2019; Verrelli et al., 2019). The consequences of such marginalisation result in higher rates of psychological distress and depression (Vorobjovas-Pinta & Hardy, 2016; Walters & Jepson, 2019). Dworkin and Yi (2003) argue that there is a distinct correlation between such social injustice and mental health among diverse groups. LGBTQI+ people, for instance, continue to experience poor mental health and elevated suicide rates, in addition to discrimination, harassment and violence even in Australia or other countries in the Global North (Kaniuka et al., 2019; Vorobjovas-Pinta & Robards, 2017). In fact, the mental health of gay, lesbian and bisexual people is amongst the poorest in Australia (Verrelli et al., 2019). As a consequence, members of LGBTQI+ communities often seek spaces that are inclusive, supportive and free of discrimination. In the tourism literature, such spaces are referred to as 'gay space' where members of the LGBTQI+ communities can address the social and emotional nature of their identity and relate to the identities of others (Vorobjovas-Pinta, 2018a). Such spaces qualify not only as leisure spaces (Blichfeldt et al., 2013), but also as the catalysts for political advocacy and human rights movements (Caudwell, 2018; Ong & Goh, 2018).

Neo-tribal theory

In recent years, LGBTQI+ events have spread beyond the cities into more regional and remote areas. Examples from around the world include ChillOut in Daylesford (Victoria, Australia), Pride on the Plains in Auburn (Alabama, United States), Agrogay Festival in Monterroso (Galicia, Spain) and even a Pride Parade in McMurdo Station (Antarctica). However, unlike fixed enclavic destinations and holiday spaces, or mobile but still enclavic LGBTQI+-friendly cruises, the spaces in which such events occur are temporary. One of the lenses through which events have been explored is neo-tribal theory. This section explores the term and some of the theoretical constructs employed, with the view to further understand its potential use in examining LGBTQI+ events from the perspective of social justice and well-being.

The concept of neo-tribes emanated from sociology. The term was coined by Shields (1992) and later defined by Michel Maffesoli's *Le Temps des Tribus* (1988; later translated as *The Time of the Tribes*, 1996) as being: 'without the rigidity of the forms of organisation with which we are familiar, [neo-tribe] refers more to a certain ambience, a state of mind, and is preferably to be expressed through lifestyles that favour appearance and form' (Maffesoli, 1996, p. 98). The uptake of the concept in sociology and cultural studies was largely due to its challenge to notions such as individualisation and social fragmentation, that were argued to have been exacerbated by the demise of bonds created by class and industrial society (Beck, 1992; Giddens, 1991). Since its conceptualisation, the neo-tribal theory has been applied within marketing, consumer behaviour, health, tourism and leisure contexts.

The rationale for the formation of a neo-tribe is the desire to coalesce because of a shared sentiment or passion. This is built upon the notion of *communitas* (Turner, 1974) and it has been suggested that 'the link is more important than the thing' (Cova & Cova, 2002, p. 595). The individual may be a member of multiple groups (Cova & Cova, 2002; Kozinets, 1999) and groups may consist of those with shared personality traits or values, and may act in a collective manner (Cova et al., 2007). The tribe is often heterogeneous in terms of age, sex, income etc., and involves individuals who are linked by a shared passion or emotion towards an activity or lifestyle. Using Mafessoli's concept of *tribus* (tribes), Bennett (1999) argued in early research that the identity of participants in the British dance music scenes he studied were fluid and constructed rather than fixed and were not indicative of a particular social class, but rather denoted a form of sociality (hence they formed neo-tribes).

Research by Hardy and Robards (2015) reinforced such sociality. They argued that neo-tribal 'tourists may be fruitfully segmented by commonalities of intangible aspects, such as 'a shared sense of sentiment,' 'tourist ritual,' 'collective bonding,' and 'belonging'' (Hardy & Robards, 2015, p. 443). They discussed Rvers (recreational vehicle travellers) as a neo-tribal culture (with sub-cultures within) and suggested explorations of their symbolic and behavioural characteristics to better identify such intangible characteristics of sociality (a suggestion we take up in our case study). Hardy et al. (2013) proposed two aspects of neo-tribal existence: symbolic characteristics such as group membership, and behavioural characteristics manifested through language, rituals and interactions at performance sites. Tourism research has demonstrated the important role that one of these aspects, namely ritual, can play in developing a space within which rituals and co-creation of identity and belonging may occur (Goulding & Shankar, 2011; see also Hardy & Robards, 2015; Kriwoken & Hardy, 2018).

Spatiality, resistance and transformation

Neo-tribes were once defined as being fluid and unstable (Maffesoli, 1996), being formed by ephemeral groupings of people, and as groupings that lacked rigidity in their organisation (Maffesoli, 1996). Ephemerality as a constant has since been challenged, most notably by Robards and Bennett (2011), who suggested that neo-tribes could exist as enduring groupings of people. Hardy and Robards (2015) and Hardy et al. (2013) built upon this and suggested membership ranged from long-lasting groupings with members who had prolonged involvement, to members with fleeting involvement in highly fluid groupings. Kriwoken and Hardy (2018) suggested that neo-tribal formation could be either spontaneous, where groups develop from people with similar passions coming together; or, alternatively, they may be facilitated, such as being created by a company who wishes to promote their products or services. While the neo-tribe was originally conceptualised as a group of people who come together in physical space (according to Maffesoli, 1996), the advent of the internet has led to the inclusion of virtual gatherings (explored by Robards & Bennett, 2011 and Cova et al., 2007).

The role of space as a conduit for the formation and well-being of these various groupings of neo-tribes is an important consideration (Vorobjovas-Pinta, 2018a, 2018b). Vorobjovas-Pinta (2018a, p. 9) work indicated that space must be tangible and equally shared by its members to maintain neo-tribal centrality: 'space is the connective thread, as visible in its consistent interplay with fluidity in membership, shared sentiment, and rituals and symbols.' The spatial element is a significant fulcrum of neo-tribal coalescence and provides a symbolic stage for neo-tribal performance as well resistance (see Armour, 2018; Canosa, 2018; Hart, 2018). Neo-tribes can be transformative, playful and empowered regarding product and service development (Cova et al., 2007). Maffesoli's Maffesoli (1996) portrayal of postmodern structures of society questions the traditional perceptions of politics and, in particular, morality, and supports an argument that neo-tribal social justice is informed by pragmatic morality, that is built upon the collective and proximate mores, aesthetics and experiences.

Neo-tribal sociality has also been understood to be a highly emotional matter driven by the collective manifestations of activism and a desire to challenge institutional and societal stasis (Lippens & Kearon, 2005). Insights can also be gained from the study of reactions to anomie and moral disorientation during the transition from pre-industrial to industrial society, as most famously contained in Durkheim's (1893/2014) *The Division of Labour in Society* (Vorobjovas-Pinta, 2017). Neo-tribes have been demonstrated as being able to re-define roles, challenge stereotypes and resist entrenched norms, creating their own personas and products (Cova et al., 2007, p. 4). We argue that the Broken Heel festival is a neo-tribal event that, while fleeting and temporary, can work to reduce marginalisation and facilitate social justice through a range of practices that resist and transform entrenched norms, prejudices and conceptions set in the rural aesthetic of outback Australia.

Case study: The Broken Heel drag queen festival

This case is not intended to explore the emergence and the meaning of the 'drag' in the binary gender system. It engages with the drag queen culture as a performance and as a theme that attracts people from different walks of life to attend the Broken Heel Festival in Broken Hill, New South Wales. The concept of 'drag' is rather fluid but generally refers to cross-dressing practices, where people of the (usually) opposite sex impersonate an uber-masculine (drag king) or an uber-feminine (drag queen) character to engage in the political work of drag (Rupp et al., 2010). Indeed, 'drag queens often rely on the 'real' male body beneath the performance of femininity' (Drysdale, 2019, p. 25). It is a common misconception that drag queens are mocking women and femininity (Bishop et al., 2014). Rather, they are challenging traditional binary gender and sexual identities (Rupp, Taylor & Shapiro, 2010).

Background and context

Broken Hill, the 'city in the desert', is a mining town in the far west of outback New South Wales, Australia (Australian Government, 2019). The town is best known for being one of Australia's oldest mining towns. It is also known as a birthplace of global mining giant BHP (Broken Hill Proprietary). The population of Broken Hill is just under 18,000 with a median age of 45 (Australian Bureau of Statistics, 2016). It should be noted that being a mining town, a large proportion of people (not included in the statistics) are the so called FIFO (fly-in fly-out) mine workers, who come to work at the mines from all around Australia for extended periods of time. Broken Hill's cityscape offers a unique mix of architecture: 'historic government buildings reflect the New South Wales colonial government style of architecture, the residential buildings reflecting those of South Australia, and, commercial buildings, such as hotels, display characteristics similar to Melbourne's late Victorian architectural period' (Australian Government, 2019).

In 1994, an Australian movie was released titled *Priscilla, Queen of the Desert*. Directed by Stephan Elliot, the movie followed the story of Tick (whose drag character was Mitzi) and their friends Adam (Felicia) and transgender Bernadette, who travelled from Sydney to Alice Springs in order to perform at a hotel where Tick's ex-wife worked. Their journey encountered many hurdles including homophobic abuse, violence, and racism, but became a transformative journey of self-discovery and reconnection, as Tick was able to reconnect and come out to his son. The movie received significant acclaim, grossing over $18 million in Australia and winning many awards, including an Oscar for Best Costume design. While it was criticised by some for creating humour that appealed to homophobic and sexist-stereotypes (Cook, 2010), it has been heralded for its positive portrayal of LGBTQI+ themes and its challenge to the notion of masculinity that has been inherently linked to the notion of the Australian outback. It has since become a cult classic, resulting in a highly successful musical and significantly for this paper, provided the inspiration for the Broken Heel Festival, located in Broken Hill which is regarded as the 'spiritual home' of the movie (Broken Heel Festival, 2019). The impact of the movie has been studied from within the film research genre; it has been lauded as marking a new genre of 'ocker' (typical Australian) comedies (Kunze, 2013) and referred to as the 'glitter cycle', whereby the central characters attempt to deviate from the norms of society (Rustin, 2001).

The first Gay Mardi Gras or Broken Heel Festival was organised in 2014 as a tribute to Priscilla, Queen of The Desert. The three-day festival grew to become a major highlight in Broken Hill's event calendar attracting over 6500, of whom 70% do not live in the town. Indeed, many festival goers will travel for several days from cities such as Sydney (a 1100 km drive) and Melbourne (a 900 km drive) in order to attend the festival (Baird, 2019). The festival now runs over three days and is hosted by Philmah Bocks and Ester Le Rovere. One of the most visible aspects of the festival is the parade, called the Main Drag in Drag, held on day two in the centre of Broken Hill.

Method: a netnography approach

For this study, an observational netnographic approach (Kozinets, 2019) was used to explore the role of the festival in reducing marginalisation and promoting social justice, as well as evidence of neo-tribal sociality. Netnography uses ethnographic techniques to explore communication, attitudes and behaviour online (Kozinets, 2015). In this case social media in the form of Instagram was used to access the data, via the use of search tools such as hashtags and handles in the form of the symbol @ (Gretzel, 2017; Gretzel & Hardy, 2019). A third technique, the use of location enabled tags for Broken Heel was also facilitated – this is where the Instagram user enables the location of the photograph to be visualised. The use of these three tags enabled an understanding of the festival, as we were able to use two types of data –text and visuals. The richness of this process facilitated the discovery of emergent themes and patterns as well as the discovery of 'black swans', defined as data points that challenge assumptions (Kozinets, 2015).

Data gathering and analysis

Data was gathered and collated one week prior to the festival (6th September 2018) until one month after (14th October 2018). The decision to conduct two layers of data collection was taken in order to assess festival goers' online sentiment as they drove through the outback, and also to explore their online sentiment while at the festival. Several parameters underpinned our approach. Only posts whose account was public were used. The second requirement was to only collect posts that contained handles directed towards public sites, such as (@) or hashtags (#), or who were geo-located. We collected posts which were tagged with the following handles and hashtags: #brokenheel, #brokenheelfestival and @brokenheelfestival. The focus on hashtags has been justified in previous research as they are regarded as marking a post's contribution to a conversation on Instagram, thus acting as a rich semiotic indexing system (Djerf-Pierre et al., 2016). The process of tagging may be regarded as an intent to publicise posts and in the absence of consent (which is not possible when collecting large online data sets). The requirement to only use these posts provided some assurance that the account holder was seeking to maximise their exposure. Further requirements when analysing the data were that the posts be de-identified, either visually or by deleting the username, and only the captions of the posts were explored; comments were not included in this analysis. This process resulted in the collation of 546 posts related to the festival. First, all posts were screened – posts with groups of friends and no landscape or township vistas were deleted as this research was focussed on the depiction of drag and/or queerness in space. This process resulted in 61 posts being retained as they contained images of the festival goers travelling to, from, or taking part in the festival, and space was clearly visible within it.

The visuals captured for this study contained people in drag or dressed in a costume, in an outback setting or in the township of Broken Hill. Our analysis involved a concurrent assessment of both text and the imagery used in the Instagram posts. This approach was taken as the research team found that the text was often written to be a juxtaposition to the photographs, thus is would not have been appropriate to temporally separate phases of analysis. The analysis of the extracted data involved hermeneutic circles of reading, re-reading and discussing. Focus was given to the content of the visuals, rather than their composition (Albers & James, 1988; Gretzel & Hardy, 2019). The assessment used a metaphoric rather than metonymic perspective (Stepchenkova & Zhan, 2013, p. 591), so that the images were not assessed at face value, but rather were assessed as 'symbols that collectively allude to a meaning that lies outside of the particular picture'. The coding was done by one coder and cross-checked by a second. Neo-tribal theory helped the coding and analysis by keeping in mind the major themes presented in the previous section, including marginalisation and exclusion, discrimination, resistance and transformation, as well as spatiality and sociality as construed by neo-tribes.

Reflexivity and positionality of the researchers

Assessing the entire posts (visual plus text in the Instagram posts) not only facilitated a rich qualitative assessment, it also allowed the researchers to situate themselves within the research. The lead author of this paper is a member of the LGBTQI+ community. This allowed for an insider positionality, whereby the author was already familiar with the vernacular of gay culture and the experience of being a gay man in a predominantly heteronormative society. The insider approach also yielded an emic perspective which conveys the meaning of the internal language, symbols and a defined culture (Vorobjovas-Pinta & Robards, 2017). An emic perspective adopts a stance as if looking 'at things through the eyes of members of the culture being studied' (Willis, 2007, p. 100) and endeavours to capture particular groups' 'indigenous meanings of real-world events' (Yin, 2016, p. 16). The second author, whilst not identifying as LGBTQI+, has a background in social aspects of sustainable tourism and experience in ethnographic research. During the data collection process, the second researcher kept a reflexive journal to document their initial thoughts prior to data collection and then the changes in opinion with regards to the function and role of the road trip and outcomes of the festival.

Results

Results from the analysis are discussed under three major thematic topics that emerged from this analysis: 1) Challenging the outback imagery; 2) Appropriating the sociality and heteronormativity of the road trip; and 3) Reclaiming gay space in an outback town.

Challenging the outback imagery

The pictures and the accompanying captions of the 61 posts that were examined appeared to purposely challenge and disrupt the traditional view of the outback. For instance, the text accompanying one post stated: 'Life outback is never a drag, darlings. Dig it? [...]'. The accompanying image contained a photograph of two drag queens posing against an old, rusted, iron structure with the red sand of outback Australia and large blue skies in the background, thus challenging the stereotypical notion of the outback (Figure 1).

Irony and sarcasm often became apparent when both the text and the images from posts were viewed at the same time. Numerous posts described how their drag outfits, wigs, feathers, 'outrageously high heels', and their presence as 'wild women', etc. enabled the festival participants to disrupt the 'soft edges' of outback ideal. For example:

> *Just what this country needs. Another cock in a frock on a rock #queenofthedesert #brokenheelfestival*

> *Broken hill it's been real, we came, we saw, we wore many wigs! 👯❤ 💋 Here's me and some wild women [drag queens] at Silverton 😍 thanks for the memories [...] @brokenheelfestival and everyone we met*

> *There are no soft edges here in Broken Hill, the historic mining town in Outback New South Wales. The outfit of the day #ootd around here is a Visi-Vest, except for this past weekend when the Broken Heel festival was on and feathers, glitter and outrageously high heels were de rigeur. They even had a special train for the event - the Stiletto Express 👠 #brokenhill #mining #visitnsw #outback #australianoutback #outdoors #brokenheelfestival #explore #seeaustralia*

The images that were assessed featured drag queens and the flamboyantly dressed people in the rough Australian outback, often with sweeping, red earthed landscapes as a backdrop. The flamboyant 'dress-code' that is so unique to the metropolitan LGBTQI+ scene created a stark contrast to the Australian bushland. It could be interpreted as a challenge to the traditional portrayal of the outback as it is being 'queered'. Clothing and appearance codes for the LGBTQI+ communities have been documented as serving as semiotic markers and non-verbal means of communication of one's identity and stance (Schofield & Schmidt, 2005). Indeed, Schofield and Schmidt (2005, p. 319) further elaborate that clothes and appearance provide LGBTQI+ communities with the potential to act out their possible identities and pertain to the 'theatrical self': 'if you're a man in drag, you don't sit there quietly, you make a scene about everything'. This is indeed what the Instagram images were depicting. For example, one image captioned 'a cock in a frock on a rock [...]', depicted a bearded person dressed in a sparkly dress with over-the-top make-up and a glittery wig, climbing clumsily over a red rock in the picturesque outback (Figure 2).

Roadkill is a problematic image of outback roads in Australia and this issue was also tackled by some festival goers. A picture captioned 'found a desiccated Skippy 😂😂 Drag in the Desert for The Broken Heel festival! [...]', featured a high heeled person dressed in a pink sparkly dress, in a stereotypical feminine pose, holding a tail of a kangaroo carcass. The imagery of the highly colourful person in a dress challenged the stereotypical notion of a farmer having to deal with dead livestock and wildlife. At the same time, it also challenged the stereotypical imagery of red and khaki colours used to depict the outback aesthetic.

The posts examined under this theme challenged the traditional idyll of the outback and resisted stereotyping through irony, sarcasm, posturing and commentary on how their drag costumes and identity challenged traditional views and perceptions of the outback. The spatiality of the outback setting, from the road trip to the festival site, provided an opportunity to actively challenge stereotypes and 'perform' their resistance, as noted in posts such as: 'if you're a man in drag, you don't sit there quietly, you make a scene about everything'. The festival goers were challenging, resisting and appropriating the outback for themselves, resisting marginalisation

Figure 1. Life outback is never a drag, darlings. Dig it? [...] . Image source: Public domain (Instagram).

and dispossession in the process. These performative acts of resistance and re-possession occurred not only at the site, but also during the road trip itself.

Appropriating the sociality and heteronormativity of the road trip

The act of the road trip has been documented as contributing towards 'the alleviation of drudgery, loneliness and isolation' (Laderman, 1996, p. 41). From a neo-tribal perspective, road

Figure 2. A cock in a frock on a rock […]. Image source: Public domain (Instagram).

trips have been documented as an activity that enhances sociality. Hardy et al. (2013) recognised the highly mobile and social worlds that exist amongst tourists travelling on the open road. They also documented the significance of self-actualisation as a motivation and/or outcome of road trips; for many travellers, the road trip offers an escape from the routines of daily life and offers an opportunity for new beginnings and transitional journeys (Hardy et al., 2013). We saw evidence of this in our data. For example, see Figure 3:

However, while the road trip offers the opportunity for self-actualisation, escape from the routine of daily life and new beginnings, the field of tourism studies and the notion of tourism have been critiqued for being inherently heteronormative (Waitt et al., 2008). Furthermore, in Australia, road trips tend to also be associated with danger and masculinity in the Australian imagery. Wolff (1993, p. 224) in her seminal work on the metaphors of travel argued that there

Figure 3. PRISCILLA - Dream Achieved ✓ Somewhere between Broken Hill and Sydney I stood atop of the Mundi Mundi Plains and watched the sunset […] Ever since I first saw the film Priscilla Queen Of The Desert I have admired the courage and strength of the topics explored throughout the narrative […] My life will be for ever changed and I cannot thank all the sisters that made it possible for a group of people to come together and find others just like them […]. Image source: Public domain (Instagram).

is an 'intrinsic relationship between masculinity and travel'. Biber (2001, p. 27) situated the road trip within the context of Australian cinematography as being 'within the fantasies of violence, freedom and liberation; fantasies that are performed in opposition to competing fantasies of family, law and control.' These themes were disrupted by the Broken Heel participants who could be regarded as a temporary neo-tribal group of LGBTQI+ travellers who, each year, reconstitute the outback idyll and resist stereotypic images of masculinity and traditional family sociality. The journey itself is highly symbolic and relevant to reclaiming space and infusing it with LGBTQI+ qualities. For this study, the Instagram posts pertaining to the road trip to Broken Hill depicted non-masculine travellers on the road, heading to a particular destination. For example:

178klms to Broken Hill #brokenheelfestival #brokenheel

Quick stop at Orange and we're back on our way to Broken Heel Festival. #brokenheelfestival @brokenheelfestival [...]

The notion of self-discovery could be seen in almost all of the posts. Images of drag queens with a backdrop of the outback offers not only a notion of resistance, but also hope for acceptance. There was a clear indication that festival goers sought a destination where they could be themselves – Broken Hill. The journey formed a seminal role in reclaiming their identity and resisting stereotypes such as machoism and masculinity evident in traditional representations of the outback. For example, an image captioned 'what should we do in Broken Hill? #roadtrip #dramaqueen #brokenheelfestival', depicts a young well-built man wearing very short shorts with a naked torso and a jacket with shoulders covered in black feathers. The pose and the hashtag 'dramaqueen', immediately destroyed the notion of a road trip as being 'macho' and 'violent'. Similarly, another picture captioned 'off we go! #brokenheelfestival #pink #ontheroad', depicted two women in pink wearing boas and pink cowboy hats holding hands and posing next to their ute (Figure 4).

Figure 4. Off we go! #brokenheelfestival #pink #ontheroad. Image source: Public domain (Instagram).

Reclaiming gay space in an outback town

The Instagram images and text that we explored depicted drag queens and kings and their supporters, flamboyantly dressed and carrying rainbow and transgender community flags through one of the main streets of Broken Hill. Some festival attendees carried posters or were dressed as characters from the movie. Some captions read:

> *Believe it or not, they're all waiting for a bus called Pricilla that's full of fabulous Queens #brokenheel*
>
> *Serving Fem Queen Realness #dragqueen #trans #transsister #brokenhill #bhfestival #brokenheelfestival2018 #silverstilletos #silvercitystiletto #empowerment #pride #Aboriginal #trans girl #queenoftheoutback #desertdiva*
>
> *About to sashay down the main drag of Broken Hill [...] @brokenheelfestival parade*

Generally, pride parades and similar events temporarily occupy the traditionally heteronormative spaces and create LGBTQI+ pockets of public visibility and presence of sexual otherness (Browne, 2007). The Broken Heel Festival is no different in this regard. Its participants resisted marginalisation and actively appropriated and transformed the outback town to a heterogenous LGBTQI+ space where they could 'sashay down the main drag of Broken Hill' and situate themselves as part of the outback community. For example, Figure 5 illustrates the attendees marching in the parade, called the Main Drag in Drag, on one of the main streets in the centre of Broken Hill. This involves an element of being oneself amongst the like-minded people, whilst, at the same time depicting queering of a public – heteronormative – space. Interestingly, the post justifies the value of the event by emphasising the number of tourists this event brought to town and the economic value of the event.

Discussion

The nexus between neo tribal festival behaviour and social justice

By seeking to challenge the utopian imagination of queerness, the subjects of this study intrinsically call for greater social justice and inclusivity for people who were, and still are, unable to be their true selves. Gorman-Murray and Waitt (2009) approach this issue through the social cohesion perspective by emphasising the need to look beyond economic and public order and focus on interrelationships, tolerance, acceptance, and trust at the local community level. The visual depiction of drag queens coalescing in on outback town illustrated how events can be used to achieve social justice, through challenging space, re-imagining of imagery, re-imagining the reputation of a township, and ultimately challenging the notion of marginalisation and the potential achievement of social justice.

Broken Heel festival attendees may be seen as a neo-tribe brought together by mutual and proximate mores. Social justice is sought through an emotive and rather pragmatic morality built upon shared values, aesthetics and experiences. Furthermore, it is driven by the collective conquest of space and is imbued with the collective manifestations of activism that challenges the heteronormative and masculine status quo of outback Australia. Neo-tribal membership has been defined as fluid and ephemeral in its nature: members 'form, dissolve and reform alongside fluctuating consumer values' (Vorobjovas-Pinta, 2018a, p. 5). Indeed, whilst the majority of the festival-related visuals feature people in drag and pride parades, a number of posts depict arrival and departure moments of the festival. Furthermore, the preparation and transformation of an individual into drag can be understood as straddling this inside/outside boundary. This observation builds upon the argument that space plays a fundamental role for the neo-tribal existence (Vorobjovas-Pinta, 2018a, 2018b). Neo-tribes have been said to coalesce around spaces that have been pre-designed for them. These are comfortable, inherently familiar and convenient spaces.

Figure 5. Broken Heel Festival [...]. What a weekend it has shaped up to be, can't believe how many tourists we have in town [...]. Image source: Public domain (Instagram).

However, this research found that a neo-tribe might also occupy a space that is unusual, uncomfortable and non-conforming to the traditional 'values' of a neo-tribe.

The festival examined in this study is unusual in that events are generally developed by communities within which the event will be held (Duffy & Mair, 2018); but in the case of the Broken Heel festival, a community that lives externally and remotely has facilitated the event and travels to the event location. Consequently, the notion of the neo-tribe is especially pertinent for this festival. Neo-tribes are a fleeting gathering of people who come together for a shared passion and purpose and then return to their everyday realities. Such attendance is socially constructed by the festival attendees themselves as a rite of passage and as an expression of the collective values. For Hetherington (1997), liminality is linked to the notion of 'heterotopia', a site, where normative assumptions are brought into question through the performance of alternative identities.

In the case of the Broken Heel festival, drag performances occur in a traditionally heterosexual space. Consequently, the coalescence of a neo-tribe of drag queens and their supporters in the rough 'macho' Australian outback is both transformative and resistant. This aligns with research that has recognised that festivals and events can act as spaces where a community can express itself and in doing so, can contribute to social justice (Derrett, 2003; Duffy & Mair, 2018; Ong & Goh, 2018). It also concurs with Markwell and Waitt (2009, p. 148) who suggested that the sexualised LGBTQI+ festival spaces 'can disrupt, subvert or 'queer' conventional boundaries of sexuality imposed by the dualism of Western thought (straight/gay), and at the same time reinscribe them'.

Challenging the rural stereotype

The landscape of the Australian outback has been dominated by the heteronormative traditions of living and have been predominantly constructed and negotiated around the concept of hypermasculinity (Coad, 2002; Little & Panelli, 2007). Little and Panelli (2007, p. 180) elaborate that sexual identities and aspirations in the Australian outback are 'in tune with nature, and conventional constructions of masculinity and femininity are supported'. The flamboyantly dressed festival attendees stand in a stark juxtaposition with the traditional depiction of the rough Australian outback, thus disrupting the 'outback idyll'. In a similar vein, Gorman-Murray et al. (2012) discussed the disruption of the Australian 'rural idyll' and positioned it as an outcome or a symptom of a metropolitan lifestyle. The captions in the Results section illustrate the sentiment of the disruption and embrace the queering of the 'outback idyll'. The outback/metropolitan juxtaposition can act as a catalyst for social justice to spread in the rural/outback settings. Furthermore, it could be understood that the subjects themselves are aware of their disruptive nature. They promote the awareness of difference and offer a view that Australian outback should not be perceived in the traditional terms of masculinity and/or femininity. Such an emphasis might serve as a reflection of social change in the community. Indeed, the depiction of drag in the Instagram posts not only queers the existent perceptions of the uber-masculine outback Australia, but also contributes towards the normalisation of such visual representations. By challenging the rural outback imagery, we can destigmatise the prevalent perceptions of non-heterosexual people and deconstruct the gender role stereotypes that outback Australia currently imbues. Fundamentally, the inner confusion stemming from the incongruous collage of 'drag' and 'outback' can (re)shape our social justice value stance.

Redefining the road trip

In their work on Route 66, Caton and Santos (2007) suggested that the journey was viewed by road trippers as an evolving corridor, and that meaning came from participants' experiences of driving the road and viewing the landscape. In particular, Caton and Santos (2007, p. 381) found their participants 'believed that particular elements of value from the past could be retained alongside or combined with newer developments to achieve optimal utility.' Similarly, in the case of this research, the juxtaposition of drag against the backdrop of the outback acts as a symbol of hope for acceptance in the future. This has synergies with the work of Noy (2004) who suggested that tourism can be illustrated as a 'springboard' for tourists to construct their identity through the images and stories that they share with others. In the case of those travelling to Broken Heel, their Instagram posts acted not only at the individual level of identity construction, but also as a springboard to encourage broader social acceptance of the LGBTQI+ communities.

Our findings also have synergies with Wang's (1999) concept of existential authenticity – defined as a state where tourists feel more connected to their real selves while travelling, than they normally would in their everyday lives. In the case of those traveling to Broken Hill, two opportunities arose for this to take place. The first was during the liminal experience of the 500-1000 km road trip, which afforded the opportunity for festival goers to be their true selves, that were, in turn, portrayed through their Instagram posts. Secondly, once festival attendees arrived at Broken Hill, the festival itself offered a further opportunity for existential authenticity to occur. The gathering of the festival attendees was a symbol of liberation from the heteronormative strictures where neo-tribal sociality was enabled by the celebration of common identities, traditions and interactions.

Conclusions

The highly specific nature of the study described in this paper provides valuable insights into how event spaces like the Broken Heel drag queen festival can enhance social justice and well-being of LGBTQI+ people. Events such as pride festivals usually aim to contribute to a wider cultural cohesion by challenging and actively opposing the notions of marginalisation and exclusion, discrimination and xenophobia, resistance and transformation, as well as spatiality and sociality. This research illustrated a further point – that LGBTQI+ events enable neo-tribes, formed by people from disparate walks of life, to be mobilised and collectively express their emotional connectedness and non-heterosexual identities. In this case the event goes one step further; it disrupts the heteronormative image of the outback Australia. For this neo-tribe, the event is the catalyst for reimagining and reconstituting the heteronormative space (Ammaturo, 2016; Markwell & Waitt, 2009). Neo-tribalism thus plays a significant role not only in terms of solidifying community bonds through membership, shared sentiment, rituals and symbols, but also in terms of challenging stereotypes and resisting entrenched norms.

The study contains several limitations. Firstly, we need to fully understand the spectrum of opinions that exist beyond the social media platform of Instagram. The analysis only evaluated posts with predetermined hashtags and handles and contained posts that described the event in a positive light, thus the research might be skewed towards one general opinion. The hashtags we explored may not have captured the negative connotations associated with the event. While we only used posts that were expressly outwardly directed (e.g. non-private accounts with hashtags to link the post to public fora), we were unable to use private photographs in this research due to the sensitive nature of the research.

Future research should also assess the spectrum of stakeholders involved in the event. For example, this study does not capture the views of the locals, both those that attend and those that do not attend the event. It is unclear how they feel about the influx of non-heteronormative event goers. Future research could seek to understand how the Broken Hill community perceives the event and whether its fleeting group of neo-tribal members have contributed towards the achievement of social justice. And lastly, this study did not address how the road trip and festival constitute a form of 'secular pilgrimage' and belonging in a liminal space for a short period of time, annually. Future research is needed to advance understandings of the importance of annual LGBTQI+ events in addressing social injustice and facilitating the legitimacy, well-being and recognition of diverse LGBTQI+ groups (i.e., that LGBTQI+ people are heterogeneous with particular cultural groups and orientations, rather than the stereotyped notions that often exist of this marginalised population).

Disclosure statement

No potential conflict of interest was reported by the author(s).

ORCID

Oscar Vorobjovas-Pinta http://orcid.org/0000-0002-2666-9453
Anne Hardy http://orcid.org/0000-0003-1461-2967

References

Albers, P. C., & James, W. R. (1988). Travel photography: A methodological approach. *Annals of Tourism Research*, *15*(1), 134–158. https://doi.org/10.1016/0160-7383(88)90076-X

Ammaturo, F. R. (2016). Spaces of pride: A visual ethnography of gay pride parades in Italy and the United Kingdom. *Social Movement Studies*, *15*(1), 19–40. https://doi.org/10.1080/14742837.2015.1060156

Armour, Z. (2018). Dedicated followers of PaSSion: The neo-tribal participation of UK clubbers and the 're-memory' of gift-exchange online. In A. Hardy, A. Bennett, & B. Robards (Eds.), *Neo-Tribes: Consumption, leisure and tourism* (pp. 137–152). Palgrave Macmillan.

Australian Bureau of Statistics. (2016). 2016 Census QuickStats: Broken Hill. https://quickstats.censusdata.abs.gov.au/census_services/getproduct/census/2016/quickstat/SSC10592?opendocument

Australian Government. (2019). National heritage places - City of Broken Hill. https://www.environment.gov.au/heritage/places/national/city-broken-hill

Baird, L. (2019, March 1). Gay pride delivers tourism boom to regional centres. *Australian Financial Review*. https://www.afr.com/politics/gay-pride-delivers-tourism-boom-to-regional-centres-20190301-h1bv02

Beck, U. (1992). *Risk society: Towards a new modernity*. Sage.

Bennett, A. (1999). Subcultures or neo-tribes? Rethinking the relationship between youth, style and musical taste.Sociology, 33(3), 599–617. https://doi.org/10.1177/S0038038599000371

Biber, K. (2001). The threshold moment: Masculinity at home and on the road in Australian cinema. *Limina: A Journal of Historical and Cultural Studies*, *7*, 26–46.

Bishop, C. J., Kiss, M., Morrison, T. G., Rushe, D. M., & Specht, J. (2014). The association between gay men's stereotypic beliefs about drag queens and their endorsement of hypermasculinity. *Journal of homosexuality*, *61*(4), 554–567. https://doi.org/10.1080/00918369.2014.865464

Blake, M. (2001). Distributive justice, state coercion, and autonomy. *Philosophy & Public Affairs*, *30*(3), 257–296. https://doi.org/10.1111/j.1088-4963.2001.00257.x

Blichfeldt, B. S., Chor, J., & Milan, N. B. (2013). Zoos, sanctuaries and turfs: Enactments and uses of gay spaces during the holidays. *International Journal of Tourism Research*, *15*(5), 473–483. https://doi.org/10.1002/jtr.1890

Bojer, H. (2003). *Distributional justice: Theory and measurement*. Routledge.

Broken Heel Festival. (2019). Your guide to the 2019 Broken Heel Festival: Programme. https://static1.squarespace.com/static/56c939c359827e3fd85160cc/t/5d6d200a70b556000174dc13/1567432831246/BHF_programme_2019.pdf

Browne, K. (2007). A party with politics? (Re)making LGBTQ pride spaces in Dublin and Brighton. *Social & Cultural Geography*, *8*(1), 63–87. https://doi.org/10.1080/14649360701251817

Canosa, A. (2018). Consumption, leisure and the "doof" neo-tribes in the Bryon Shire. In A. Hardy, A. Bennett, & B. Robards (Eds.), *Neo-Tribes: Consumption, leisure and tourism* (pp. 169–184). Palgrave Macmillan.

Caton, K., & Santos, C. A. (2007). Heritage tourism on Route 66: Deconstructing nostalgia. *Journal of Travel Research*, *45*(4), 371–386. https://doi.org/10.1177/0047287507299572

Caudwell, J. (2018). Configuring human rights at EuroPride 2015. *Leisure Studies*, *37*(1), 49–63. https://doi.org/10.1080/02614367.2017.1383505

Coad, D. (2002). *Gender trouble down under: Australian masculinities*. Presses Universitaires de Valenciennes.

Cook, A. (2010). More than just a laugh: Assessing the politics of camp in The Adventures of Priscilla, Queen of the Desert. In C. Lin & J. McSweeney (Eds.), *Representation and contestation: Cultural politics in a political century* (pp. 3–26). Rodopi.

Cova, B., & Cova, V. (2002). Tribal marketing: The tribalisation of society and its impact on the conduct of marketing. *European Journal of Marketing*, *36*(5/6), 595–620. https://doi.org/10.1108/03090560210423023

Cova, B., Kozinets, R. V., & Shankar, A. (2007). *Consumer tribes*. Routledge.

Dangi, T., & Jamal, T. (2016). An integrated approach to 'sustainable community-based tourism. *Sustainability*, *8*(5), 475. https://doi.org/10.3390/su8050475

Derrett, R. (2003). Making sense of how festivals demonstrate a community's sense of place. *Event Management*, *8*(1), 49–58. https://doi.org/10.3727/152599503108751694

Devine, J., & Ojeda, D. (2017). Violence and dispossession in tourism development: A critical geographical approach. *Journal of Sustainable Tourism*, *25*(5), 605–617. https://doi.org/10.1080/09669582.2017.1293401

Djerf-Pierre, M., Ghersetti, M., & Hedman, U. (2016). Appropriating social media. *Digital Journalism*, *4*(7), 849–860. https://doi.org/10.1080/21670811.2016.1152557

Drysdale, K. (2019). *Intimate investments in drag king cultures: The rise and fall of a lesbian social scene.* Palgrave Macmillan.

Duffy, M., & Mair, J. (2018). Engaging the senses to explore community events. *Event Management, 22*(1), 49–63. https://doi.org/10.3727/152599517X15111988553991

Durkheim, E. (1893/2014). *The division of labor in society* (W. D. Halls, Trans.). Palgrave Macmillan.

Dworkin, S. H., & Yi, H. (2003). LGBT identity, violence, and social justice: The psychological is political. *International Journal for the Advancement of Counselling, 25*(4), 269–279. https://doi.org/10.1023/B:ADCO.0000005526.87218.9f

Giddens, A. (1991). *Modernity and self-identity: Self and society in the late modem age.* Polity Press.

Ginwright, S., & James, T. (2002). From assets to agents of change: Social justice, organizing, and youth development. *New Directions for Youth Development, 2002*(96), 27–46. https://doi.org/10.1002/yd.25

Gorman-Murray, A., & Waitt, G. (2009). Queer-friendly neighbourhoods: Interrogating social cohesion across sexual difference in two Australian neighbourhoods. *Environment and Planning A: Economy and Space, 41*(12), 2855–2873. https://doi.org/10.1068/a41356

Gorman-Murray, A., Waitt, G., & Gibson, C. (2012). Chilling out in 'cosmopolitan country': Urban/rural hybridity and the construction of Daylesford as a 'lesbian and gay rural idyll. *Journal of Rural Studies, 28*(1), 69–79. https://doi.org/10.1016/j.jrurstud.2011.07.001

Goulding, C., & Shankar, A. (2011). Club culture, neotribalism and ritualised behaviour. *Annals of Tourism Research, 38*(4), 1435–1453. https://doi.org/10.1016/j.annals.2011.03.013

Gretzel, U. (2017). Social media activism in tourism. *Journal of Hospitality and Tourism, 15*(2), 1–14.

Gretzel, U., & Hardy, A. (2019). #VanLife: Materiality, makeovers and mobility amongst digital nomads. *E-review of Tourism Research, 16*(2/3), 1–9.

Gustavsson, M., Lindström, L., Jiddawi, N. S., & de la Torre-Castro, M. (2014). Procedural and distributive justice in a community-based managed Marine Protected Area in Zanzibar. *Tanzania. Marine Policy, 46*, 91–100. https://doi.org/10.1016/j.marpol.2014.01.005

Hardy, A., Gretzel, U., & Hanson, D. (2013). Travelling neo-tribes: Conceptualising recreational vehicle users. *Journal of Tourism and Cultural Change, 11*(1-2), 48–60. https://doi.org/10.1080/14766825.2013.783584

Hardy, A., & Robards, B. (2015). The ties that bind: Exploring neo-tribal theory's relevance to tourism. *Tourism Analysis, 20*(4), 443–454. https://doi.org/10.3727/108354215X14400815080686

Hart, M., (2018). #Topless Tuesdays and, #w. & Wednesdays, Digitally mediated neo-tribalism and NSFW selfies on Tumblr. In A. Hardy, A. Bennett, & B. Robards (Eds.), *Neo-tribes: Consumption, leisure and tourism* (pp. 207–220). Palgrave Macmillan.

Hassanli, N., Walters, T., & Friedmann, R. (2020). Can cultural festivals function as counterspaces for migrants and refugees? The case of the New Beginnings Festival in Sydney. *Leisure Studies, 39*(2), 165–180. https://doi.org/10.1080/02614367.2019.1666296

Hetherington, K. (1997). *The badlands of modernity: Heterotopia and social ordering.* Routledge.

Higgins-Desbiolles, F. (2008). Justice tourism and alternative globalisation. *Journal of Sustainable Tourism, 16*(3), 345–364. https://doi.org/10.1080/09669580802154132

Higgins-Desbiolles, F. (2009). Indigenous ecotourism's role in transforming ecological consciousness. *Journal of Ecotourism, 8*(2), 144–160. https://doi.org/10.1080/14724040802696031

Hubbard, P. (1998). Sexuality, immorality and the city: Red-light districts and the marginalisation of female street prostitutes. *Gender, Place and Culture: A Journal of Feminist Geography, 5*(1), 55–76. https://doi.org/10.1080/09663699825322

Jamal, T., & Camargo, B. A. (2014). Sustainable tourism, justice and an ethic of care: Toward the just destination. *Journal of Sustainable Tourism, 22*(1), 11–30. https://doi.org/10.1080/09669582.2013.786084

Kaniuka, A., Pugh, K. C., Jordan, M., Brooks, B., Dodd, J., Mann, A. K., Williams, S. L., & Hirsch, J. K. (2019). Stigma and suicide risk among the LGBTQ population: Are anxiety and depression to blame and can connectedness to the LGBTQ community help? *Journal of Gay & Lesbian Mental Health, 23*(2), 205–220. https://doi.org/10.1080/19359705.2018.1560385

Kozinets, R. (1999). E-tribalized marketing?: The strategic implications of virtual communities of consumption. *European Management Journal, 17*(3), 252–164. https://doi.org/10.1016/S0263-2373(99)00004-3

Kozinets, R. (2015). *Netnography: Redefined.* Sage.

Kozinets, R. (2019). *Netnography: The essential guide to qualitative social media research.* (3rd ed.). Routledge.

Kriwoken, L., & Hardy, A. (2018). Neo-tribes and Antarctic expedition cruise ship tourists. *Annals of Leisure Research, 21*(2), 161–177. https://doi.org/10.1080/11745398.2017.1286512

Kunze, P. (2013). Out in the outback: Queering nationalism in Australian film comedy. *Studies in Australasian Cinema, 7*(1), 49–59. https://doi.org/10.1386/sac.7.1.49_1

Laderman, D. (1996). What a trip: The road film and American culture. *Journal of Film and Video, 48*(1-2), 41–57.

Lee, S., & Jamal, T. (2008). Environmental justice and environmental equity in tourism: Missing links to sustainability. *Journal of Ecotourism, 7*(1), 44–67. https://doi.org/10.2167/joe191.0

Lippens, R., & Kearon, T. (2005). Introduction and editorial overview: Emerging imaginaries of regulation, control, and oppression. *Social Justice, 32*(1), 1–4.

Little, J., & Panelli, R. (2007). Outback' romance? A reading of nature and heterosexuality in rural Australia. *Sociologia Ruralis*, *47*(3), 173–188. https://doi.org/10.1111/j.1467-9523.2007.00434.x
Maffesoli, M. (1996). *The Time of the tribes: The decline of individualism in mass society* (D. Smith, Trans.). Sage.
Malpas, J. (2012). Putting space in place: Philosophical topography and relational geography. *Environment and Planning D: Society and Space*, *30*(2), 226–242. https://doi.org/10.1068/d20810
Markwell, K., & Waitt, G. (2009). Festivals, space and sexuality: Gay pride in Australia. *Tourism Geographies*, *11*(2), 143–168. https://doi.org/10.1080/14616680902827092
Noy, C. (2004). This trip really changed me': Backpackers' narratives of self-change. *Annals of Tourism Research*, *31*(1), 78–102. https://doi.org/10.1016/j.annals.2003.08.004
Ong, F., & Goh, S. (2018). Pink is the new gray: Events as agents of social change. *Event Management*, *22*(6), 965–979. https://doi.org/10.3727/152599518X15346132863292
Pielichaty, H. (2015). Festival space: Gender, liminality and the carnivalesque. *International Journal of Event and Festival Management*, *6*(3), 235–250. https://doi.org/10.1108/IJEFM-02-2015-0009
Robards, B., & Bennett, A. (2011). My tribe: Postsubcultural manifestations of belonging on social network sites. *Sociology*, *45*(2), 303–317. https://doi.org/10.1177/0038038510394025
Rupp, L. J., Taylor, V., & Shapiro, E. I. (2010). Drag queens and drag kings: the difference gender makes. *Sexualities*, *13*(3), 275–294. https://doi.org/10.1177/1363460709352725
Rustin, E. (2001). Romance and sensation in the "Glitter" cycle. In I. Craven (Ed.), *Australian cinema in the 1990s* (pp. 133–148). Frank Cass.
Schellhorn, M. (2010). Development for whom? Social justice and the business of ecotourism. *Journal of Sustainable Tourism*, *18*(1), 115–135. https://doi.org/10.1080/09669580903367229
Schofield, K., & Schmidt, R. Ä. (2005). Fashion and clothing: the construction and communication of gay identities. *International Journal of Retail & Distribution Management*, *33*(4), 310–323. https://doi.org/10.1108/09590550510593239
Scott, J., & Biron, D. (2010). Wolf Creek, rurality and the Australian gothic. *Continuum: Journal of Media & Cultural Studies*, *24*(2), 307–322. https://doi.org/10.1080/10304310903576358
Smith, A., & Pitts, M. (2007). Researching the margins: An introduction. In M. Pitts & A. Smith (Eds.), *Researching the margins: Strategies for ethical and rigorous research with marginalised communities* (pp. 3–41). Palgrave Macmillan.
Shields, R. (1992). The individual, consumption cultures and the fate of community. In R. Shields (Ed.) *Lifestyle shopping: The subject of consumption* (pp. 99–114). Routledge.
Stepchenkova, S., & Zhan, F. (2013). Visual destination images of Peru: Comparative content analysis of DMO and user generated photography. *Tourism Management*, *36*, 590–601. https://doi.org/10.1016/j.tourman.2012.08.006
Turner, V. (1974). *Dramas, fields, and metaphors: Symbolic action in human society*. Cornell University Press.
Verrelli, S., White, F. A., Harvey, L. J., & Pulciani, M. R. (2019). Minority stress, social support, and the mental health of lesbian, gay, and bisexual Australians during the Australian Marriage Law Postal Survey. *Australian Psychologist*, *54*(4), 336–346. https://doi.org/10.1111/ap.12380
Vorobjovas-Pinta, O. (2017). Gay neo-tribes: An exploration of space and travel behaviour [Doctoral dissertation]. University of Tasmania.
Vorobjovas-Pinta, O. (2018a). Gay neo-tribes: Exploration of travel behaviour and space. *Annals of Tourism Research*, *72*, 1–10. https://doi.org/10.1016/j.annals.2018.05.008
Vorobjovas-Pinta, O. (2018b). It's been nice, but we we're going back to our lives: Neo-tribalism and the role of space in a gay resort. In A. Hardy, A. Bennett, & B. Robards (Eds.), *Neo-tribes: Consumption, leisure and tourism* (pp. 71–87). Palgrave Macmillan.
Vorobjovas-Pinta, O., & Hardy, A. (2016). The evolution of gay travel research. *International Journal of Tourism Research*, *18*(4), 409–416. https://doi.org/10.1002/jtr.2059
Vorobjovas-Pinta, O., & Robards, B. (2017). The shared An insider ethnographic account of a gay resort. *Tourist Studies*, *17*(4), 146879761668756–146879761668387. https://doi.org/10.1177/1468797616687561
Waitt, G., Markwell, K., & Gorman-Murray, A. (2008). Challenging heteronormativity in tourism studies: Locating progress. *Progress in Human Geography*, *32*(6), 781–800. https://doi.org/10.1177/0309132508089827
Walters, T., & Jepson, A. S. (2019). *Marginalisation and events*. Routledge.
Wang, N. (1999). Rethinking authenticity in tourism experience. *Annals of Tourism Research*, *26*(2), 349–370. https://doi.org/10.1016/S0160-7383(98)00103-0
Willis, J. W. (2007). *Foundations of qualitative research: Interpretive and critical approaches*. Sage.
Wolff, J. (1993). On the road again: Metaphors of travel in cultural criticism. *Cultural Studies*, *7*(2), 224–239. https://doi.org/10.1080/09502389300490151
Yin, R. K. (2016). *Qualitative research from start to finish*. The Guilford Press.

Slow food justice and tourism: tracing Karakılçık bread in Seferihisar, Turkey

İlkay Taş Gürsoy

ABSTRACT

Using actor-network theory and the concept of the sociological imagination, this study follows the traces of an ancestral food, namely Karakılçık bread, in the Cittaslow destination of Seferihisar in İzmir, Turkey. Actor-network theory is used to explore economic, social, and cultural relationships and everyday practices in the making and selling of this ancestral bread. Tracing the bread in the networks of human and non-human relations helps towards understanding how slow food justice may be enacted in the hybrid socio-ecological-technological network of slow food tourism. The study shows the importance of local spaces like local farmers' markets that enable the sharing and conservation of food heritage through the social ordering of traditional Karakılçık bread. The festivalization of bread and the co-created experiences in local food places express values like sociality and conviviality that contribute to the well-being of residents and tourists alike. As interpreted in this study, slow food justice is a holistic concept that considers not simply principles like fairness and equity, but also resident and tourist well-being, cultural flourishing, and heritage conservation.

Introduction

> "This bread takes me back to my childhood. I am happy to find it here. I will continue to visit the farmers' market."
>
> Participant at the Ancestral Bread and Armola Cheese Festival, Ulamış village, İzmir, Turkey

This study traces the linkages and relationships in local and touristic practices related to the preparation and consumption of Karakılçık bread. This is an ancestral bread that is locally produced at Ulamış village in Seferihisar, a municipality within the metropolitan area of İzmir, Turkey. As an important staple in many cultural traditions, bread takes different forms. In Turkey, bread symbolizes human effort for survival, as reflected in an idiom known to the locals: "The bread is in the mouth of the lion" [ekmek aslanın ağzında]. Another idiom, "for the sake of salt and bread" [tuz-ekmek hakkı] refers to showing gratitude to people who have provided food and help to you. In Turkey, bread is considered an essential food. According to a study of the Turkish Grain Board in 2016, the annual consumption of bread per capita was 121 kg (Toprak Mahsülleri Ofisi, 2016). The paper traces the making and sharing of Karakılçık bread in the Cittaslow destination of Seferihisar, where the slow movement and slow food practices are being enacted. Seferihisar became Turkey's first Cittaslow member in 2009 and has pioneered the dissemination of the slow city movement nationwide (Cittaslow Türkiye, 2014; see also Go Turkey Tourism, 2019). Interrelated economic, social and cultural implications arise around the production and consumption of this traditional food as part of collective and convivial food activities in local spaces visited by tourists, such as various village markets (see Seferihisar Municipality, 2017). Relational insights emerge for a *slow food justice* paradigm for tourism that emphasizes not just fairness and equity but also resident and tourist well-being, cultural flourishing, and heritage conservation.

Inspired by Bennett (2010), the study follows the traces of bread as a vital, vibrant material enmeshed in networks and relationships in everyday and touristic spaces in Seferihisar. It is theoretically and methodologically informed by actor-network theory (ANT). To better understand the interconnectedness of humans and non-humans in this approach, I also draw upon the socio-ecological imagination proposed by White et al. (2016) and the "vital materiality" described by Bennett (2010, p. vii). The socio-ecological imagination emphasizes the hybrid nature of social phenomena that combine various elements of society, nature, materiality, etc. and the "tangled connections" between "lively and active agencies of many types", both human and non-human (White et al., 2016, p. 116, 121). Bennett (2010, p. 23) claims a vibrant agency for things as an "efficacy or effectivity" distributed across an "ontologically heterogenous field". She argues that "the figure of an intrinsically inanimate matter may be one of the impediments to the emergence of more ecological and more materially sustainable modes of production and consumption" (2010, p. ix). This theoretical framework is relevant for our study of traditional Karakılçık bread as it enables thinking beyond dualisms, accepting the relationality and heterogeneity of the social world which is co-shaped by contingent, unstable, multiple hybrid actor-networks. Neuman (2019, p. 91) similarly suggests studying food practices "as tied to each other in bundles and nexuses."

The research undertaken below is structured as follows. The next section discusses slow food justice and tourism. This is followed by a section explaining the theoretical and methodological framework of socio-ecological imagination and ANT. The case study is then presented, including the research design, setting, methods, study insights and discussion. The paper closes with conclusions and directions for future research.

Slow food justice and tourism

Food is more than just a nutritional or economic good; it is "a multidimensional cultural artefact capable of linking issues regarding the relationships between place and identity, and the material and symbolic" (Everett, 2008, p. 338). Food becomes a medium of communication as an everyday social practice (Neuman, 2019, p. 80), where it is also subject to power politics. Herring (2015, p. 18) notes the "the right to food" as an increasingly important component of global social movements for food security, driven by concerns about the unsustainability of industrialized food systems and the urgency for affordable, reliable, good-quality food that is available to all populations without discrimination of race and class. It brings to the fore issues of equity, and social justice (Jung, 2016). Justice cannot be found "in situations in which those who produce the food cannot avoid hunger and insecurity" (Borras & Franco, 2015, p. 255). Bioregional and local foodways offer environmentally and socially sustainable and equitable alternatives to globalized and industrialized food systems, and the slow food movement is one such hopeful avenue as described below.

The slow movement

The Slow Food Movement was initiated by Carlo Petrini and a group of activists in 1986 in Italy and officially founded in 1989 (Slow Food, 2018). It is part of the global slow movement that involves initiatives such as slow cities, slow tourism, and slow travel to tackle the problems created by growth and speed-oriented production and consumption systems (Forné & Jamal, 2020). The movement's principles aim at enabling access to *good, clean,* and *fair* food for all (for more on these, see: Slow Food, 2015a). Similarly, the Slow City Movement (Cittaslow) is a form of urban activism which was started in 1999 by a network of municipalities in Italy to disseminate the philosophy of slow food (see Pink, 2008). To participate in the global network of slow cities, municipalities are required to implement projects that intersect across the economy, community, and environment (Cittaslow International, 2016).

The slow movement emphasizes the connectedness of the individual with both society and nature. It prioritizes clean production that does not harm the environment and also helps to put ecologically sustainable gastronomy into practice. Eco-gastronomy refers to "a recognition of the strong connections between plate and planet, and the fact that our food choices have a major impact on the health of the environment and society" (Slow Food, 2015b). A slow food is a just food: it should be available at "accessible prices for consumers" while providing "fair conditions and pay for small-scale producers" and it should be "good, "healthy", "quality and flavorsome" (Slow Food, 2015b). It challenges and resists the fast foods and plastic packaged global consumer culture by providing an alternative, meaningful way to engage socially and joyfully in everyday food experiences. Sociality plays an important role in facilitating slow food justice by sustaining local producers, food entrepreneurs and service providers. "Small-scale agricultural production, well-rooted local economies and food artisans" are seen as the leading players of conviviality (Petrini et al., 2012, p.16).

Fair food and slow food tourism

Fairness is an important component of slow food and food justice. Fair food refers to food that is "respectful of social justice, meaning fair pay and conditions for all concerned - from production to commercialization to consumption" (Slow Food, n.d). Food justice holistically embraces production and consumption, and emphasizes fairness in the above sense as well as in "equitable access not only to healthy, culturally appropriate food, but also to the benefits of food production and distribution for all communities" (Purifoy, 2014, p. 376). It includes residents, tourists, and citizens in general with respect to quality of life and public health and safety, watchful of environmental justice in relation to adverse environmental impact of industrial food system (Purifoy, 2014). Also inherent in food justice is food security and self-determination, enabling food provisioning and self-sustenance through food sovereignty. Food sovereignty refers to community engagement in determining the priorities and characteristics of the local food system (Chappell, 2015, p. 727). Alkon and Agyeman (2011, p. 8) note its political and justice dimensions, saying that "like procedural justice, food sovereignty moves beyond the distribution of benefits and burdens to call for a greater distribution of power in the management of food and environmental systems". The slow food movement helps to create the context for food sovereignty and food justice building relationships among residents by promoting their participation in projects, activities, committees, etc. Such social relationships and "socialities" are the sources and locations of agency for the movement, according to Pink (2008, p. 171).

Slow Food Tourism plays a valuable role here as part of the slow food movement. Forné and Jamal (2020) consider slow food tourism as "an ethical microtrend for the Anthropocene", associated with guiding principles based on slow, sustainable, secure, and democratic participative processes. Alongside these are the key principles of the slow movement noted earlier: fair, clean and good food for all concerned, empowering and engaging communities for taking action for their livelihoods while cherishing conviviality, sociality, mutuality (co-creation), responsibility, the senses, and mindfulness. Being slow points to a cultural turn emphasizing the importance of mindfulness in consumption, production, and everyday choices (Köstem, 2017). Jamal and

Camargo (2014, p. 17) use the term *ecocultural justice* to refer to "the active involvement of a destination's resident stakeholders (including indigenous, low income, diverse and minority groups) in the development and marketing of their ecocultural goods for tourism purposes". Food is, of course, an important cultural activity that sustains everyday life and cultural heritage, and the practices offers a common space to engage with visitors.

Tourists are part of slow food justice as co-producers of conviviality, responsibility and sociality by actively seeking knowledge, establishing rapport with producers, behaving responsibly, paying a fair price for the food while appreciating the efforts of producers, and promoting the slow food philosophy to other people. They are active participants of farmers' markets and consumers of community-supported agricultural products (Petrini et al., 2012, p. 18). Thus, the slow movement emphasizes creating connections between place, products, and tourists. However, knowledge does not always result in ethical decisions because habits, routines, and economic constraints also affect the behavior of tourists (Cairns & Johnston, 2018). Moreover, the sustainability of practices related to slow travel and tourism is not merely environmental but also social and cultural (see Fuentes & Svingstedt, 2018). A just and sustainable tourism destination, according to Jamal and Camargo (2014, p. 27), is the one which "facilitates economic as well as social-cultural equity, respect and well-being, rather than merely material benefits for the 'industry', the promotional agendas of destination marketers and the utilitarian calculus of government stakeholders" (see also Jamal, 2019).

Drawing on the above, slow food justice can be seen to be a similarly holistic concept that focuses not only on aspects like equity, but also on the common good for both humans and non-humans. It includes taking the vitality and agency of non-humans into account (Bennett, 2010). After all, food production and consumption result from collaborative acts of both human and non-humans (seeds, living organisms in the soil, water, animals, climate, markets, etc.). Recognition and respect for non-humans involved in food production encourages alternative relationships and approaches to tourism, such as a "degrowth" paradigm that counters the unsustainable mass consumption and commodification of ecological, social, and cultural goods and relationships. However, degrowth is not "in exact opposition to economic growth" (Martinez-Alier et al. 2010, p. 1741, 1742); rather, it encourages the "re-examination of the dominant economic values of affluent societies" to "liberate societies from the growth imperative and open up space for a more critical approach to development and well-being" (D'Alisa et al., 2013, p. 213). Slow food tourism reinstates the respect for non-human others by focusing on practices like ethical eating and by creating sociality, learning, celebration, and sharing (Forné & Jamal, 2020). This study uses ANT and the socio-ecological imagination to better understand the inclusion of non-human entities. These approaches provide a valuable framework for taking a holistic, integrated, and culturally situated approach to slow food justice.

Actor-network theory and the socio-ecological imagination

Actor-Network theory

Actor-network theory is a "a conceptual frame for exploring collective sociotechnical processes" that was primarily developed by Bruno Latour, John Law and Michel Callon (Crawford, 2004, p. 1). ANT gives "equal analytical status and attention to human and nonhuman actors" (Van der Duim et al., 2017, p.141; Latour, 2005) and aims "to understand complex social situations by following actors and how they associate" (Van der Duim et al., 2014, p. 589). It offers "a disparate family of material-semiotic tools, sensibilities, and methods of analysis that treat everything in the social and natural worlds as a continuously generated effect of the webs of relations within which they are located" (Law, 2009, p. 141). It reflects anti-essentialist movements that reject predefined divisions among nature, society, agent, structure, etc. (Latour, 1999, p. 20). Instead, ANT emphasizes relational materiality, "which presupposes that all entities achieve significance in relation to others (Crawford, 2004, p. 1). An actor-network is an "emergent effect" of moving actants within an entangled socioecological network: "people, organizations, objects, technologies, and spaces are all concurrently brought together in the performance," explains Van der Duim (2007, p. 965, 961). Actants refer to "acting agents" and "interveners" that include human and non-human components, and both human and non-human actants have the same capacity to influence relations and other actors in the actor-network (Latour, 2004, p. 75).

Bennett (2010, p. 20) suggests considering "thing power" in terms of assemblages. An assemblage is an "open ended collective" of elements working together (Bennett, 2010, p. 24). In an assemblage, the influencing power of a thing is not absolute or individual, but emerges in relation to other human and non-human beings. The context and landscape in which tourism takes place can therefore be viewed as tourismscapes consisting of actor-networks with human and non-human assemblages. Van der Duim et al. (2017, p. 139.) conceptualize tourismscapes as "contingently assembled and interdependent socio-material configurations", which are "economic, physical, technological, political, spatial and social at the same time". According to Van der Duim, Ampumuza and Ahebwa, "this basically means that we cannot take anything as given, as everything is an effect of relational practices" (2014, p. 590). Food in tourismscapes can thus be viewed as an actant in the assemblage of tourism, where the agency of the food as an actant is determined in relation to the other entities in the actor-network.

The relational aspect of ANT is therefore very important to keep in mind in this ANT research on bread as a vital ancestral food. Mares and Peña (2011, p. 199) suggest that we "stretch our understanding of what it means to grow and eat food justly" by thinking of food in a broader context involving social relations and cultural practices. Domaneschi (2012, p. 319) proposes studying the "human/non-human networks of social agents involved and the shifting balance in the governance of the food system". For example, Ren (2011) investigated Oscypek Cheese to understand the role of tourism objects in place-making, studying the actor-networks, translations, and orderings of the local cheese to unraveling various discourses, relations, and practices related to tourism. Beard et al. (2016, p. 99) use ANT to trace "food" and "food ideas" as key actants in Llangollen, a slow city tourism destination in the UK. Holifeld (2009) uses ANT for an environmental justice case study of Cass Lake, Minnesota. He recommends ANT as a relevant approach for justice related studies because ANT can be used to trace the "multiple, competing accounts" of what creates inequalities rather than providing a priori causes of inequality in the society (Holifeld, 2009, p. 655).

Key concepts from ANT that guide the approach of the present study include ordering, translation, materiality, and multiplicity. Ordering simply refers to "attempts at control or management" (Kendall & Wickham, 2001, p. 27), which involve "materially heterogenous set of arrangements and processes" (Franklin, 2008, p. 30). Ordering relates to the ways dynamic networks are ordered into "more or less stabilized networks" (Van der Duim et al., 2013, p. 5). Kendall and Wickham note that ordering, which is "about acting on the world", is never complete (2001, p. 28, 32). Each ordering can link to other orderings that can translate to become something different with a different impact (Franklin, 2008, p. 31). Translation is "the process of establishing communication or making connections ... dependent on more or less material intermediaries" (Jóhannesson, 2005, p. 139). Materiality relates to "the role of the material and nonhuman elements ... in the processes of ordering" (Van der Duim et al., 2013, p. 5). Multiplicity is an important characteristic of ANT, and is "a property of things, not of humans interpreting things" (Latour, 2005, p. 116, see note 165). In ANT thinking, "an actor is what is made to act by many others" (Latour, 2005, p. 46) and any action mobilizes multiple objects on its way (Latour, 2005, p. 72). "Multiple relational orderings" and "relational materiality of the social world" are inherently part of the network building and consolidation process (Jóhannesson, 2005, p. 134). The next section builds on the above to introduce the "socio-ecological imagination" which is important for the study presented further below as it enhances the vitality and agency of bread as a socio-political actant.

The socio-ecological imagination

The socio-ecological imagination is derived from C. Wright Mills' idea of sociological imagination, which established a link between the individual and the society by connecting "biography" to "history", and "private troubles" to "public issues" (1959, cited in White et al., 2016, p. 1). Besides, it focuses on the mutual transformation capacity of nature and society to create possibilities and constraints for each other (White et al., 2016, p. 2). It is inspired by Marx's theoretical work on metabolic rift, the dual alienation of land and labor from means of production, and the mutually transformative capacity of nature and society to create possibilities for, and impose constraints on, each other (Marx, 1973 [1857-1858] cited in White et al., 2016).

According to Goodman and Goodman (2001, p.98), food networks are "metabolic relations: between agricultural nature and social labour in production and the corporeal and symbolic act of human food consumption". The socio-ecological imagination adopts "an entangled hybrid worldview" of social, natural, cultural, political, and material co-creations (White et al., 2016, p. 15). Slow food justice fits into the relational approach to socio-natures by linking gastronomy to "among other things, politics, agriculture and the environment" (Slow Food, n.d., p. 4). Slow food justice conceptualizes value as relational, in that "the 'economic value' of things change from a quantitative absolute to a qualitative value, where 'one's full unfolding depends on the unfolding of others and vice versa'" (Salleh et al., 2015, p. 97). The socio-ecological relations of slow food assign a kind of "transformative agency" to food in contrast to conventional counterparts (Goodman & Goodman, 2001, p. 112).

Power differences in the society based on gender, class, etc. affect how and by whom environmental problems are narrated, experienced, and solved. The socio-ecological imagination acknowledges that "environmental problems are inescapably social problems" (White et al., 2016, p. xviii). This brings to the fore the issue of hybrid politics that comprise not only market, state, and technologies but also civil society, social movements, and active citizens (White et al., 2016). Slow food justice is enacted by civil engagement through networks connecting producers with co-producers and by building local food communities, educating consumers on food, and collectively protecting food biodiversity (Slow Food, n.d.). These networks provide the examples of alternative production relations based on inclusion rather than the "social indifference of competitive markets" (Salleh et al., 2015, p. 97). The socio-ecological imagination emphasizes relations "between the ecological, personal and communal conditions of production and life" and enables a local-global perspective, connecting global food justice problems to everyday life in the local, such as in work, play, and eating (White et al., 2016, p. 214). Taken together, ANT and the socio-ecological imagination offer a valuable theoretical and methodological approach to study fair food and slow food relationships in the context of *slow food justice* and Karakılçık Bread, in Seferihisar, İzmir (Turkey).

Case study: Karakılçık bread in Seferihisar, İzmir, Turkey

Research approach

The socio-ecological imagination and key concepts from ANT guided the research questions and study of Karakılçık Bread described below, including ordering, translation, materiality, and multiplicity. The following research questions are related to local and touristic practices of Karakılçık Bread: (i) How is the local Karakılçık ancestral bread ordered? Specifically, (ii) how does translation or assembling take place (Jóhannesson, 2005) in this actor-network? (iii) How is the materiality of human and non-human elements co-constitutive of the collective and social ordering? (iv) How is the multiplicity of this bread enacted in the heterogeneous assemblages of the Seferihisar tourismscape?

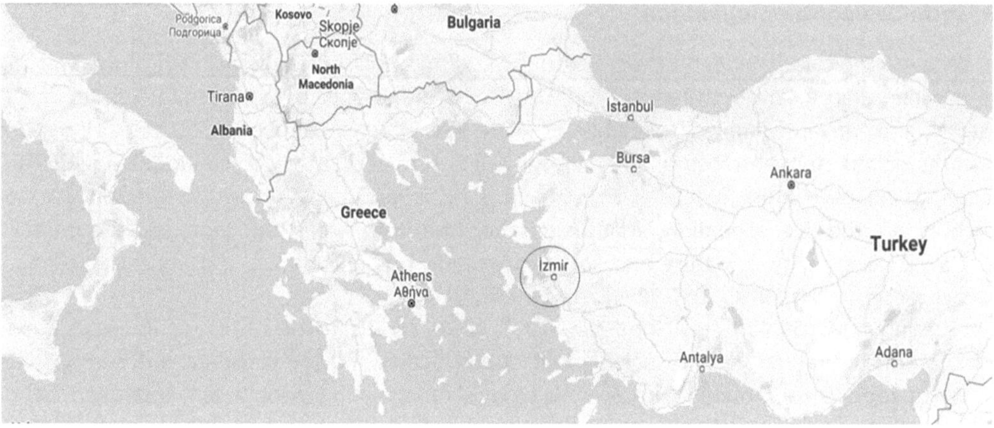

Figure 1. Location map of İzmir in Turkey.
Source: Google Maps (n.d.). [İzmir]. Retrieved 2019, December 8 from https://www.google.com/maps/@38.386738,26.8271683,9.4z.

Interview questions were also guided by ANT concepts. For example, to find a starting point to enter the actor-network of Karakılçık bread, I visited the local administrator, the muhtar of Ulamış village, to ask about the activities associated with slow food production and tourism. I then contacted local producers to ask about production and marketing practices, the availability of fair income and working conditions, their participation in slow city activities (sociality), and the tourists' relationship with them. During the field study, purposive sampling was used to identify these actors to trace the relationships and practices related to the bread. Similar, qualitative research and ANT approaches have been used successfully in previous food related and tourism studies (e.g. Cook, 2004; Ren, 2011; Beard et al., 2016).

Understanding the notions and principles of slow food tourism justice described earlier helped to inform the overall research questions, interview questions, and the use of ANT. However, the study was not constrained by specific justice parameters. Rather, they influenced my thinking during both data analysis and discussion in an emergent and iterative manner by going back and forth between the data and the literature on food tourism and justice described earlier. This approach stays true to ANT in that it does not aim to "reveal" or "show" justice, but rather to understand what it means by tracing the relationships and practices of ancestral bread making and consumption in the heterogeneous assemblages of everyday life and touristic spaces.

Research setting and background

Seferihisar

Seferihisar is one of 30 towns within the metropolitan area of İzmir, located on Turkey's western Aegean Sea coast, approximately 47 km from İzmir city center (see Figure 1). Seferihisar's 21 neighborhoods include several villages that were recategorized in 2012 as neighborhoods with the same name under Law No. 6360 on metropolitan municipalities (Seferihisar Municipality, 2019a). Despite this change in legal status, they are nevertheless still commonly referred to as villages. In this study, they will also be referred to as villages to reflect how they are named in daily language. Ulamış neighbourhood, where the ancestral bread is produced, is one of these villages. Agricultural production (particularly olives and citrus fruit) plays a major role in the local economy of Seferihisar. Once known only as a small coastal town with tourism facilities, it has emerged as a well-known tourism destination due to its membership of the international Cittaslow (slow city) network and its adoption of a degrowth development model (one the principles of the slow movement). Degrowth means "democratic and redistributive downscaling of production and consumption" (Joutsenvirta, 2016, p. 23) for the sake of environmental sustainability, social justice, and well-being (Demaria et al., 2013, p. 209).

Seferihisar's highway connection with İzmir makes it easy for urbanites to visit, especially for weekend getaways to enjoy festivities and food in the many villages contained within the township (see: Go Turkey Tourism, 2019).

Research methods

I was active in the field doing participant observation, interviews, and gathering relevant documents. Following ANT, it is the actants in the network that show the researcher where to look, and helps to guide purposive sampling (Beard et al., 2016). An important actant, for instance, was the former mayor of Seferihisar (M), who pioneered the adoption of the Cittaslow label in the town and in Turkey (20 February 2019). I also participated in the annual farmers' festival of Ulamış village (3 August, 2019), and visited the village during several weekends and weekdays (31 May, 17 August 2019, 14 September 2019, 18 October in 2019), when I undertook direct observations of the farmers' market and everyday life. This helped to guide further purposive sampling and facilitated conceptualizations based on "intimate empirical engagements" (White et al., 2016, p. 122).

While the field study was originally planned to involve only Ulamış village, the Karakılçık bread transcended its boundaries. Indeed, ANT does not easily conform to "the strict confines of a pre-given spatial field of study" (Ren, 2011). For example, although Orhanlı village (a neighbourhood of Seferihisar) had not been selected to visit, the network expanded and this new site emerged as a part of the assemblage while tracing Karakılçık Bread. I therefore visited Orhanlı village as well as part of my data gathering exploration (5 December 2019 and 5 January 2020). Here, the municipality had collaborated with the Doğa [Nature] Association to transform a former school building into the Doğa[Nature] School, and partnered with BirdLife International Turkey to help establish this nature conservation and education facility (BirdLife, 2020). Doğa School is relevant for the study in that it operates Yavaş Dükkan [Slow Store] which is a non-profit organization aiming to support biodiversity and indigenous production by connecting local producers with the consumers of food products (Yavaş Dükkan, n.d.) After finding several small shops selling Karakılçık bread in central İzmir, I also interviewed the shopkeepers in that area, thereby expanding the network of assemblages further in tracing the bread.

Interviews were conducted in Turkish as all participants were Turkish residents. The interviews with visitors and local producers took place on farmers' market days and during various festivals and were recorded if the participants consented. For those participants who did not approve of recording, the researcher took notes in the field. The interviews took between 15 and 60 minutes. Participant details are presented in Table 1. As a person interested in slow living who tries to avoid industrialized food and mass producers, I was reflexive and aware of my position as a Turkish researcher and the potential influence of my values on the many "small decisions which build up and shape data collection and narrative output" (Beard et al., 2016, p. 103).

In order to follow the material and communicative traces as well as the related practices beyond geographical boundaries, audio-visual materials (videos, photographs) and documents were included in the analysis, as Ren (2011) did while following the food. I transcribed the interview recordings and notes myself (and translated as needed in writing this article). Aided by an independent researcher, I coded and generated themes inductively from the data by using qualitative analysis software (QSR Nvivo). While the literature review provided some background, the analysis was an iterative analytical process with a continuous movement between the existing background ANT framework and the emerging insights from the data that further informed notions of fair food and slow food justice. The audio-visual and textual materials were analyzed to identify how human and non-human connections are generated; how such connections create

Table 1. Profile of study participants.

L1	Local administrator (muhtar) of Ulamış village, male, age 53
L2	Local inhabitant, not involved in farmers' market, Ulamış, 50
L3	Local inhabitant, not involved in farmers' market, Ulamış, 80
PU1	Producer/seller of bread in Ulamış, female, age 50
PU2	Producer/seller of bread in Ulamış, female, age 25
PU3	Seller at the farmers' market in Ulamış, female, age 46
PU4	Seller at the farmers' market in Ulamış, male, age 60
V1	Visitor, farmers' market, Ulamış, female, age 42
V2	Visitor, farmers' market, Ulamış, male, age 43
V3	Visitor, farmers' market, Ulamış, male, age 30
V4	Visitor, farmers' market, Ulamış, female, age 28
V5	Visitor, farmers' market, Ulamış, female, age 52
D1	Doğa School, civil society reprensentative, Orhanlı, male, age 25
D2	Doğa School, civil society reprensentative, Orhanlı, male, age 28
D3	Doğa School, civil society reprensentative, Orhanlı, female, age 28
D4	Doğa School, civil society reprensentative, Orhanlı, male, age 30
S1	Urban food shopkeeper, male, age 40
S2	Urban food shopkeeper, female, age 25

new entities and practices. The focus was on actions, relationships and practices, for ANT instructs the researcher to look for action and "action is *always* interaction" (Sayes, 2014, p. 140, emphasis in original).

Tracing the bread: Study insights

Based on the data gathering and analysis, this section traces and deconstructs Karakılçık bread to understand what is happening within the assemblages and to identify the interactions that make "just" food (i.e. fair, clean and good food, following slow movement principles) visible in the tourism network, i.e. in the touristic practices and tourismscapes, as described earlier.

Discursive practices: Naming and enacting the bread

The ancestral bread of Ulamış village is called Karakılçık bread. Discursive acts associate the bread with its basic ingredient, Karakılçık wheat, which is considered to be healthy and unmodified in contrast to industrial wheat. The name Karakılçık bread differentiates it from other types of local bread by denoting the kind of wheat used and the processes involved in making this traditional break; it is an actant in a heterogeneous network of slow food tourism and an appealing draw for visitors at local farmers' markets that serves both residents and tourists. Figure 2 shows how the bread is displayed and becomes a theme and attraction.

The name of the ancestral bread is often used by the local municipality and the metropolitan municipality to highlight the local food heritage in tourism marketing materials and advertising, where they emphasize the untouched, ungenetically modified bread that is made from patrimonial seeds. Within each assemblage of tourism and non-tourism actors and relationships, bread is transformed or translated into an actant that connects with other actors in the assemblage and influences various relationships. While the materiality (shape, texture, smell, weight, taste, etc.) of bread stays the same in each assemblage, the wheat is an important actant with the power to create dynamic change in the assemblage of bread. The name Karakılçık associates the bread to a local food heritage through both the ancestral Karakılçık wheat seeds still being used as well as the traditional bread-making craft. Here, wheat and bread are ordered in association with the labor of women and the conduct of everyday village life. Figure 3, for example, shows the role of women in mural paintings that depict their integral relationship to wheat and ancestral bread. One participant emphasized that she makes the bread in the same way she learnt from her mother (PU3). Women's labor is at the forefront of the farmers' market, which provides

Figure 2. An elderly peasant woman sells Karakılçık bread at the farmers' market.
Photo credit: Photo taken by the author.

opportunities for sociality and empowerment of women in economic, social, and cultural life while selling homemade food, bread, other agricultural products, and ornaments. "I am very happy to earn money by myself" said a participant (PU1).

Heritage creation and conviviality

> It all started with a handful of ancestral seeds of [Karakılçık] wheat that an old peasant farmer found at the bottom of an old chest in his house (M).

The quote above belongs to the former mayor of Seferihisar who recalls how the patrimonial wheat seeds were discovered and revitalized for use in the present day. When asked about Karakılçık wheat, most participants from Ulamış village, both locals and visitors, mentioned the same archetype (PU1, PU4, PU2, V1, V2, V3): the discovery of something precious or inherited from ancestors. The patrimonial seeds, which are agricultural entities, were reclaimed as a lost heritage from a sacred treasure chest (M) (see: Seferihisar Municipality, 2016). They may look like ordinary seeds but they are not– this particular seed lays claim for the bread's ancestral heritage through discursive and situated acts of heritage creation and conservation.

> This is our heritage. I am happy that we have found it. It is our task to protect it for our children (L2).

In the tourism context, the patrimonial seeds create and re-create heritage through discursive acts as well as performances and practices, such as represented by the festivalization of seeds and bread. For example, the slow city movement launched a "Seed Exchange Festival" aimed at protecting, celebrating, and sharing native seeds. As one participant explained, the ancestral

Figure 3. Mural paintings in Ulamış village.
Photo credit: Photo taken by the author.

seeds were "discovered" and more widely understood for its ancestral and health values through this festival (L1). Sociality is present in the assemblage of patrimonial seed and the festival while the farmers' market is also a primary source of conviviality and sociality for the participants, including residents, producers, and tourists. As one participant put it, "I try to buy from the farmers' market. The products are fresh and I just like the ambiance there" (V2). During a report about Karakılçık on the public television channel's main news program, a peasant farmer in Ulamış village said:

> The seeds are precious for us. They survived for hundreds of years. If we protect them, they will be our legacy to our children (TRT Haber, 2019).

As another participant explained, "The seeds were about to go extinct" (L3). Here, the restoration of the patrimonial heritage of native seeds is closely linked to long-term sustainability and inter-generational heritage, being passed from the past to the present and future (children) as a "legacy."

Festivalization: from seed to bread and cheese

The Seed Exchange Festival, which has been held in Seferihisar since 2011, has aroused interest in native seeds and enabled Turkey's diverse native seeds to be recognized as historically valuable and part of its natural and cultural heritage. As the festival's name indicates, it involves the festivalization of traditional seeds – translating the native seeds into a tourist attraction and making them available to the visiting public. However, it is also related to bureaucratization, in that only those seeds registered in national records with proper certification can be traded in Turkey since 2006 (Law No: 5553). This festivalization occurs through the voluntary exchange of native unregistered seeds in a hybrid context of assembling agriculture with tourism in the actor-network.

Other food actors: mutuality (co-creation)

Karakılçık Bread has a festival dedicated to its name in the Ulamış: Ancestral Bread and Armola Cheese Festival. Here, bread is enacted alongside another food, Armola cheese, as part of the local food heritage. Bread and cheese appear together on the festival logo and promotional materials, thereby connecting two local foods. The production of Armola cheese is quite difficult. "It is like a performance" said a participant while showing me on his mobile his performance at the festival, which involved stuffing an animal skin (L1). He described in detail how Armola cheese should be produced:

> I remember how they did it when I was a kid. Particularly at Eid al-Adha, they used to peel the skin of the goat or sheep without tearing it apart. Then they used to stuff the skin with yogurt, milk, and salt. They used to let it dry. It would take many days. Armola cheese requires skill, first of all (L1).

Increasing the importance of local foodscapes and food heritage in tourism can create direct earning opportunities for small-scale local businesses, particularly those involved in agriculture, food, and crafts. "I saw the festival on TV last year. This year, I am here at the festival" (V4) said a visitor. At the festival, the posters, music, visitors, dances, and performances of food heritage produced and reproduced the existence of the village. Tourism and the festival have re-ordered the habitat of peasants as a tourism destination, with nearly 50,000 people visiting the village during the festival in 2019 (Seferihisar Municipality, 2019b) and travel agencies organizing day trips from İzmir (see: Martı Tur, 2019).

However, except for the farmers' market and the festival, Karakılçık is missing from other touristic spaces and practices in Seferihisar. For instance, when asked about the direct contact of visitors with Karakılçık bread at the local hotels or restaurants, some participants noted that most hotels and restaurants do not serve local bread due to its higher price (PU2) , hard texture (V4), habits (V1), or because they just do not see this food in the context of sharing and sustaining heritage through tourism (V5). The absence of such food heritage in the hospitality sector co-creates an actor-network where economic concerns take over and market-based solutions tend to sidestep sustainability issues and opportunities related to cultural conservation and resident involvement. Slow food justice is entailed by the collective empowerment of local communities to engage in practices related to the making and sharing of traditional foods like the ancestral Karakılçık bread.

Socio-ecological-technological assemblages

While moving in the network by following the traces of the wheat, I discovered through my conversations and interviews another version of the seed in production and use that I call the *replica seed*. The ancestral seeds used today are re-productions of the original seeds. It took almost five years to start fully-fledged production of Karakılçık wheat using replica seeds generated from the original patrimonial seeds (L1). Although they are called ancestral seeds and have similar physical characteristics to the originals, they are different because the network of relations around them and the technology used to produce them today are different. The current Karakılçık seeds are socio-technological assemblages involving the methods applied at the seed protection centers, agricultural technology, and the experience of the farmers. The *socio-technological seed* is one where technology is a new actor in the network that creates new possibilities for producing and re-producing the ancestral seed, for instance, for cultivation and consumption at a larger scale. Jóhannesson emphasizes that such a translation is a socio-material process: "A translated actor is not the same as the original, or better, it is both similar and different" (2005, p. 140). Figure 4 below shows the billboard located in İzmir city center to announce the re-production of Karakılçık wheat.

Karakılçık has been touching and altering many lives during its journey through food and tourism networks, particularly the lives of the peasant farmers. The mayor of İzmir (who is the former mayor of Seferhisar slow city) initiated a new agricultural project in November 2019 to increase production of Karakılçık wheat to meet the demands of the city and the country;

Figure 4. Karakılçık wheat at the street as part of urban assemblage.
Photo credit: Photo taken by the author.

making Karakılçık based food products, such as flour and bread, available to a larger group of people. According to the mayor, this project is an effort to reverse the industrial agricultural model based on unhealthy modern seeds (Haber Türk, 2019). In this version, discursively and politically, the Karakılçık wheat is ordered as an actant of urban activism for accessible food and clean agriculture.

Providing fair, clean, good food to all

The village administrator says that people from all around Turkey call him to ask for the bread. "They call me and say, I heard about Karakılçık bread. Could you send some to me by cargo? I ask the producer friends here. They send and earn money" (L1). In addition, as part of its slow city activities, the municipality launched a website for the rural development cooperative of Doğanbey village to sell local products online, including Karakılçık bread (Seferibakkal, 2019). The municipality sets a standard selling price to eliminate price competition. Karakılçık bread used to be contextualized in situ because it was sold in the farmers' markets in Ulamış and Sığacık, and at several sales points of Seferihisar Municipality. However, its reputation has since spread beyond the town's geographical boundaries. It has recently started to be sold in small shops in İzmir, illustrating the fluid nature of actor-networks. Participant shopkeepers suggested several reasons for the increasing demand:

> The municipality's announcements about production of Karakılçık wheat in İzmir influenced people. They are curious about Karakılçık wheat and bread (S1).

> Because people want to access healthy food and Karakılçık wheat is not genetically modified (S2).

Action and performance for slow food justice

One of the mottos of the slow movement is "From farm to fork", which underlines the connection of urban and rural lives (Slow Food, 2019). "You cannot become Cittaslow by saying I am

Figure 5. An excerpt from the brochure of Yavaş Dükkan showing operating principles.
Photo credit: B. Özkırlı & M. Koyaş. Source: Yavaş Dükkan [Slow Store] [Brochure]. (n.d). Izmir, Turkey: Doğa Association.

Cittaslow. You have to take actions, to do projects" said the former mayor (M). Everyday practices matter. "We don't know who produces our food. We don't see their face. It is as if the large retail shops produced them for us" (V3) said one participant, highlighting the absence of human connection in daily food practices. In the tourism assemblage, direct contact between producers and consumers is valued, as shown in the festivalization of bread.

Apart from the local administrator, none of the interviewees were participating in the slow city committees. However, the village residents are engaged in theater as one was founded in the village in 2018. "From fields to the theatre scene" was the way this sociality type was discursively enacted in the media (Hürriyet, 2018). Such local interactions open room for *performative justice*. Jamal and Hales (2016) note the link between the acts people perform and what they identify as morally acceptable or unacceptable. Justice here is not only based on a moral or ethical stance; it is an act, a performance of "affective, embodied participants engaged in performative struggles and resistance to injustices in everyday life" (Jamal & Hales, 2016, p. 179).

In this regard, international normative frameworks can provide the impetus. The Cittaslow transnational movement reinforces the agency of small places (Pink, 2009, p. 453) and "the role of human sociality in the production of activist-led processes of change" (Pink, 2008, p. 163). The agricultural products and food by themselves matter in the creation and maintenance of connections to others. Yavaş Dükkan is both an actant of fair food and an assemblage of various food culture practices. "The products are discovered and collected from various locations in the country" (D2).

Figure 5 shows the operating principles of Yavaş Dükkan on the left-hand side. It starts by stating that "Yavaş Dükkan is a local goods network". It emphasizes that, in this network, "the distinction between producer and consumer disappears" while food products result from "the collective work done by all living-beings", not only humans. "The cycle of production is transparent and all steps are a result of relationships based on kindness"; "no living being is harmed", and no ingredient used is harmful for humans. On the right-hand side, a peasant woman teaches young people about a local craft. As Danermark and Gellerstedt (2004, p. 343) argue, social patterns of representation, interpretation, and communication are important performative tools for facilitating fair foods and slow food justice.

Fair foods, gender equity, and fairness

"Local food systems can help to create social justice in the form of stable, meaningful jobs in communities", say Alkon and Agyeman (2011, p. 6). In line with the local development model of the slow movement, Seferihisar municipality has opened farmers' markets where peasant farmers can sell their products without paying rent to the municipality. The municipality also established women's cooperatives to increase their labor participation. Besides rural development, cooperatives have been opened that specialize in one or more agricultural products. These cooperatives aim at civil participation in decision making as well as economic gains (M). Such local practices help to facilitate *gender equity, women's empowerment* and socio-economic *fairness.*

There is also evidence of government interventions to sustain the cultural heritage of bread as a cultural good. To support Karakılçık wheat production, the municipality of Seferihisar and İzmir metropolitan municipality make ancestral seeds available to local farmers and extends a guarantee to purchase the wheat for twice the price of modern wheats. This is necessary because Karakılçık wheat is less productive than other forms, including genetically modified imports used in industrial agriculture. As a peasant farmer explains, "If Karakılçık gives one unit of output, the other wheats give three units" (L1). Here, the "powers of the different elements in the assemblage shift and interact in different ways" (Woodward, 2019, p. 77), as economics, equity, dignity, and sustainability intersect. One participant argued that "the new development model should honor the jobs that people do in agriculture" (D1) while another emphasized the importance of socially acknowledging the value of rural workers (D4).

The socio-ecological imagination

As the mayor of Izmir metropolitan municipality explained, "this project, which may look simple, is a step for sustaining our food" (Haber Türk, 2019). Tracing food (bread) to sustainability is in line with the socio-ecological imagination. Participants from Doğa School in Orhanlı Village, for instance, emphasize activating the public space between the individual and the market by creating new understandings. Doğa School is a non-governmental, non-profit organization. The new understandings arise from new assemblages whereby food is directly connected with people, such as when residents "emotionally connect with their food and the towns where they live" (D3). When asked about why they engage in food heritage, a couple of participants responded:

> Because nature is us. There is a circular balance in nature. If a local craft or food goes extinct, there are many others, animals or plants, that are influenced by it. Birds, animals, plants, humans are all parts of the vulnerable eco-system (D2).

> In the places we visited for bird watching, we witnessed the local food heritage that was about to get forgotten. We aim to protect the bio-diversity (D1).

Figure 6 below shows how actants such as birds and flowers assemble with human beings in an actor-network of local food, craft, and cultural heritage. Other examples of actants in this network are traditional soap, sesame seed oil, olives, and olive oil.

In the assemblage of the bread, the simplicity of the bread-making technique and the ingredients, the absence of artificial flavors, additives, and mass production technology facilitate the

Figure 6. An excerpt from the brochure of Yavaş Dükkan showing the variety of actors in the food network.
Photo credit: M. Koyaş, Y. İlemin, V. Eroğlu & Ö. Sözen. Source: Yavaş Dükkan [Slow Store] [Brochure]. (n.d.), İzmir, Turkey: Doğa Association.

socio-ecological imagination and enable the "thing power" of the bread in Bennett's (2010) terms:

Thing-power gestures toward the strange ability of ordinary, man-made items to exceed their status as objects and to manifest traces of independence or aliveness, constituting the outside of our own experience (p. xvi).

As one participant put it, "The texture of the bread is hard, and it smells like the ones my grandma used to make. It reminds me of my childhood" (V4).

Discussion

To understand how slow food is ordered in a tourism network (and its implications for food heritage and slow food justice), this study followed an ordinary non-human actor, namely a local ancestral bread in a Cittaslow destination (Seferihisar). In line with the ANT methodological framework, the bread represents "the multiplicity of associations between humans and non-humans" (Barry, 2006, p. 5). That is, bread is not an individual or a static agent with a given set of characteristics "but rather an entity" whose characteristics are "formed through its shifting network of relations with other actors" (Barry, 2006, p. 4). Slow Food Tourism as being enacted through the production and consumption of Karakılçık bread corroborates slow food principles. It emphasizes the decommodification of food and providing "fair, clean and good" bread for residents and visitors while empowering and engaging rural peasant communities, women, and small-scale producers towards sustainable livelihoods and long-term sustainability, cherishing conviviality, sociality, and mutuality (co-creation).

In ANT thinking, "agency is a relational effect", not a certain capacity nor a priori superiority (Van der Duim et al., 2017, p. 140, 141). This study traced how human and non-human actors create associations and relationships, working together in creating and sharing the bread. The case study helps to understand multiple orderings of the bread as a vital, vibrant actant within assemblage of slow food tourismscapes in the slow city destination. Production of Karakılçık bread in the village is part of the town's slow city activities, and bread interacts with visitors in local festivals and village markets, at the women's cooperative and other small retail outlets. The labor of the producers, and the social, ecological, and cultural aspects of food production are made visible to consumers (residents and visitors) through textual and visual materials, and by the personal communication, practices and performance of the producers that are enacted in relation to other actants. This is absent in other food products on store shelves (commodified and static).

Karakılçık bread, a non-human actant, is enacted by multiple discursive and situated practices, by emphasizing its ancestral heritage, by performing it, and by reproducing it in socio-ecological-technological assemblages. Both ways of ordering showed the agency of Karakılçık wheat in network building. The agency of Karakılçık wheat is enabled by its vitality and its material characteristics, such as being non-genetically modified, which relates to its agency as a "fair, clean and good" food (echoing slow food principles here). Karakılçık wheat is ordered as an actant within the assemblage of local food heritage, which includes the women's cooperative, and as an actant within the assemblage of rural and urban agricultural production. While neither the bread nor the other actants "reveal" justice, glimmers of insight for slow food tourism justice arise by tracing the entangled relationships and agencies in the various assemblages. As Hitchings (2003, p. 106) notes, "the way entities are understood is a product of the relationship in which they are located".

The slow food movement aims to provide fair, clean, and good food to all. According to the metropolitan municipality, growing ancient wheat in İzmir's agricultural plains is a step towards improving access to healthy wheat and healthy food. Ancestral bread is translated into an urban actant of good agricultural practices, fair and equitable to the farmers and to the women in the agricultural cooperatives who gain recognition, empowerment, and are the bearers of the traditional knowledge and practice of making the bread. Cultivating Karakılçık wheat for the bread contributes to food security. However, political struggles are evident in countering economic commodification and market constraints. Karakılçık bread was markedly absent from the local hospitality network, e.g., being served in local hotels. Yet, due to increasing awareness of the health benefits of Karakılçık wheat aided by İzmir municipality promotional campaigns, demand for the bread is increasing in urban areas and the municipality actively encourages (and subsidizes) local farmers to cultivate patrimonial wheat seeds in the urban agriculture assemblage. Since Karakılçık wheat is less productive than modern wheats, to encourage farmers to produce ancestral wheats, the municipality guarantees farmers twice the average price for modern wheats (an insight for redistributive justice).

The act of translating rural Karakılçık wheat to an urban setting via the agricultural strategy of the metropolitan municipality and the emergence of new urban assemblages for the city reflect the agency of slow cities like Seferihisar for engaging in sustainable and fair agriculture practices that benefited the local populace. The women's cooperative model was developed in Seferihisar within the Cittaslow framework. The seed exchange festival was inspired by the international slow movement for protecting native seeds against standardization and genetic modification by global companies. The study also shows assemblages that facilitate sociality. Touristic spaces like the farmers' market and local festivals were the main sources of sociality and conviviality. The vitality and material agency of bread or the "thing power" of bread in an assemblage with nature, agriculture, and social practices can create meaningful relationships between the consumer, food, and the producer (see also the example of food products exhibited and sold by the Doğa School shown in the previous section).

According to the *socio-ecological imagination*, the transformation of the individual is inter-connected with the transformation of the society. This raises a concern regarding the micropolitics of personal transformation. Micropolitics increases "receptivity to a politics of becoming" whereby "new forms of subjectivity might be cultivated" (Cameron & Gibson, 2005, p. 320). The slow movement, for example, offers a micropolitical possibility of being: that is, being slow. According to Cittaslow International (2016), becoming slow as part of the degrowth movement is not related to the speed of living. It is related to quality of life, well-being, praise for anti-utilitarianism, and critiques of development, democracy, justice, bio-economics, and ecology (Demaria et al., 2013). Slow bread practices and relationships in the actor-network enact a "performance" in everyday life and in the entangled touristic spaces (Jamal & Hales, 2016), giving rise to mutuality (co-creation), equity and women's empowerment, sociality and conviviality in the tourism network.

Conclusions

The actor-networks of relationships shaping the vibrant agency of Karakılçık bread is heterogenous. It involves both human and non-human actors that are ordered by the constant circulation and interrelations of network entities at various levels, ranging from individual to local, national, and international. Tourism is intimately interwoven in these relational networks. The present study uses ANT to follow the traces of ancestral Karakılçık bread in the Cittaslow of Seferihisar (part of the international slow movement). Besides being an essential food and an ancestral heritage, bread is part of the tourismscape in Ulamış village and other local villages in the district. The festivalization of food is evident in tracing the bread in Ulamış. This vital, vibrant actant is continuously connected with everyday and touristic places (e.g., local festivals and farmers' markets) through relational practices engaging residents and visitors. Discursive and symbolic acts abound. The walls of the houses are hand-painted with figures of women farmers collecting wheat and making bread. The actor-networks and assemblages that cross-cut agriculture, environment, and tourism demonstrate how ancestral wheat, bread, and people interact in this slow food destination to sustain the place and creatively conserve and celebrate cultural heritage.

Tracing the bread using ANT doesn't "show" justice but in the slow city desination of Seferihisar, the relationality of practices and participants offer preliminary insights for slow food justice and tourism. Slow food justice entails fairness and equity at the individual level and in macro level policies, as well as heritage conservation of cultural goods like ancestral Karakılçık bread, and communal well-being through sociality, conviviality, and safe, clean, and good food. Access to safe and nutritious food should be a right for everyone, residents and visitors alike. This can be facilitated by increasing the awareness of the civic public and visiting tourists about mutual dependency and mutual responsibility between producer and consumer, and enabling opportunities to engage in slow encounters, conviviality, appreciation, and enjoyment of local foods and cultural goods, such as local gastronomy and ancestral foods like Karakılçık bread.

The study has some limitations. Its geographical scope is limited to one destination and its analytical scope is limited to one type of fair food, bread. Further studies to broaden the scope to new objects and destinations may help to understand alternative assemblages with different actants. The study is also limited with respect to the importance of international alliances in the enactment and sustainability of fair food and conservation of food heritage in socio-ecological-technological systems (technology is an actant in this translation). Future research can focus on the macropolitics of food security, bioregionalism and sustaining cultural foodways, and the micropolitics of just eating, eco-gastronomy and co-creating slow food experiences, aided by the socio-ecological imagination.

Disclosure statement

No potential conflict of interest was reported by the author(s).

References

Alkon, A. H., & Agyeman, J. (2011). Introduction: The food movement as polyculture. In A. H. Alkon & J. Agyeman (Eds.), *Cultivating food justice* (pp. 1–21). MIT Press.

Barry, A. (2006). Actor-network theory. In A. Harrington, B. I. Marshall & H. P. Müller (Eds.), *Encyclopedia of social theory* (pp. 4–5). Routledge.

Beard, L., Scarles, C., & Tribe, J. (2016). Mess and method: Using ANT in tourism research. *Annals of Tourism Research*, *60*, 97–110. https://doi.org/10.1016/j.annals.2016.06.005

Bennett, J. (2010). *Vibrant matter: a political life of things*. Duke University Press.

Borras, S. M., & Franco, J. C. (2015). Food, justice and land. In R. J. Herring (Ed.), *The Oxford handbook of food, politics, and society* (pp. 253–272). Oxford University Press.

Cairns, K., & Johnston, J. (2018). On (not) knowing where your food comes from: meat, mothering and ethical eating. *Agriculture and Human Values*, *35*(3), 569–580. https://doi.org/10.1007/s10460-018-9849-5

Chappell, M. J. (2015). Global movements for food justice. In R. J. Herring (Ed.), *The Oxford handbook of food, politics, and society* (pp. 717–738). Oxford University Press.

Cittaslow International. (2016). How to become. http://www.cittaslow.org/content/how-become.

Cittaslow Türkiye C (2014). Seferihisar. https://cittaslowturkiye.org/cittaslow-seferihisar/.

Cook, I. (2004). Follow the thing: Papaya. *Antipode*, *36*(4), 642–664. https://doi.org/10.1111/j.1467-8330.2004.00441.x

Cameron, J., & Gibson, K. (2005). Participatory action research in a poststructualist vein. *Geoforum*, *36*(3), 315–331. https://doi.org/10.1016/j.geoforum.2004.06.006

Crawford, C. (2004). Actor-network theory. In G. Ritzer (Ed.), *Encyclopedia of social theory* (pp.1–4). Sage.

D'Alisa, G., Demaria, F., & Cattaneo, C. (2013). Civil and uncivil actors for degrowth society. *Journal of Civil Society*, *9*(2), 212–224. https://doi.org/10.1080/17448689.2013.788935

Demaria, F., Schneider, F., Sekulova, F., & Martinez-Alier, J. (2013). What is degrowth? From an activist slogan to a social movement. *Environmental Values*, *22*(2), 191–215. https://doi.org/10.3197/096327113X13581561725194

Danermark, B., & Gellerstedt, L. C. (2004). Social justice: Redistribution and recognition-a non-reductionist perspective on disability. *Disability & Society*, *19*(4), 339–353. https://doi.org/10.1080/09687590410001689458

Domaneschi, L. (2012). Food social practices: Theory of practice and the new battlefield of food quality. *Journal of Consumer Culture*, *12*(3), 306–322. https://doi.org/10.1177/1469540512456919

BirdLife. (2020). Doğa [Nature]. https://www.birdlife.org/europe-and-central-asia/partners/turkey-doga.

Everett, S. (2008). Beyond the visual gaze?: The pursuit of an embodied experience through food tourism. *Tourist Studies*, *8*(3), 337–358. https://doi.org/10.1177/1468797608100594

Forné, F., & Jamal, T. (2020). Slow Food Tourism: An ethical microtrend for the Anthropocene. *Journal of Tourism Futures*, ahead-of-print (ahead-of-print). https://www.emerald.com/insight/content/doi/10.1108/JTF-10-2019-0120/full/html. https://doi.org/10.1108/JTF-10-2019-0120

Franklin, A. (2008). The tourism ordering: taking tourism more seriously as a globalising ordering. *Civilisations*, *57*(57), 25–39. https://doi.org/10.4000/civilisations.1288

Fuentes, C., & Svingstedt, A. (2018). The practice of slow travel: Understanding practitioners' recruitment, careers and defections. In L. James, C. Ren, & H. Halkier (Eds.), *Theories of practice in tourism* (pp. 10–28). Routledge.

Goodman, D., & Goodman, M. K. (2001). Sustaining foods: organic consumption and the socio-ecological imaginary. In M. Cohen, and J. Murphy (Eds.), *Exploring sustainable consumption: environmental policy and the social sciences* (pp. 97–119). Elsevier.

Google Maps. (n.d.). [İzmir]. Retrieved 2019, December 8 from https://www.google.com/maps/@38.386738,26.8271683,9.4z

Go Turkey Tourism. (2019). Slow city Seferihisar. https://www.goturkeytourism.com/destinations-turkey/seferihisar-first-cittaslow-of-turkey.html

Haber Türk (2019, November 5). Karakılçık buğdayı İzmir ovalarına geri dönüyor [Karakılçık wheat is back to the plains of İzmir]. [News channel]. https://www.haberturk.com/izmir-haberleri/73163506-karakilcik-bugdayi-izmirin-ovalarina-geri-donuyorkarakilcik-bugdayinin-uretimi-menemen

Herring, R. J. (2015). How is food political? Market, state, and knowledge. In R. J. Herring (Ed.), *The Oxford handbook of food, politics, and society* (pp. 3–42). Oxford University Press.

Hitchings, R. (2003). People, plants and performance: On Actor-Network Theory and the material pleasures of the private garden. *Social & Cultural Geography*, *4*(1), 99–114. https://doi.org/10.1080/146493603200049333

Holifeld, R. (2009). Actor-network theory as a critical approach to environmental justice: a case against synthesis with urban political ecology. *Antipode, 41*(4), 637–658.
Hürriyet. (2018, January 18). Tarladan sahneye [From field to stage]. [Newspaper]. https://www.hurriyet.com.tr/tarla-dan-sahneye-40701711
Jamal, T. (2019). *Justice and ethics in tourism*. Routledge.
Jamal, T., & Camargo, B. A. (2014). Sustainable tourism, justice and an ethic of care: Toward the Just Destination. *Journal of Sustainable Tourism, 22*(1), 11–30. https://doi.org/10.1080/09669582.2013.786084
Jamal, T., & Hales, R. (2016). Performative justice: New directions in environmental and social justice. *Geoforum, 76*, 176–180. https://doi.org/10.1016/j.geoforum.2016.09.014
Joutsenvirta, M. (2016). A practice approach to the institutionalisation of economic growth. *Ecological Economics, 128*, 23–32. https://doi.org/10.1016/j.ecolecon.2016.04.006
Jóhannesson, G. T. (2005). Tourism translations: Actor–network theory and tourism research. *Tourist Studies, 5*(2), 133–150.
Jung, Y. (2016). Food provisioning and foodways in postsocialist societies: Food as medium for social trust and global belonging. In J.A. Klein & J. L. Watson (Eds.), *The handbook of food and anthropology* (pp. 289–307). Bloomsbury.
Kendall, G., & Wickham, G. (2001). *Understanding culture: Cultural studies, order, ordering*. Sage.
Köstem, B. (2017, April 6). Yavaş Yaşa [Live slowly]. [Video] TEDxBursa. https://www.youtube.com/watch?v=yyYQfz7mViE
Latour, B. (2004). *Politics of nature: How to bring the sciences into democracy*. Harvard University Press.
Latour, B. (2005). *Reassembling the social*. Oxford University Press.
Latour, B. (1999). On recalling ANT. In J. Law & J. Hassard (Eds.), *Actor network theory and after* (pp. 15–25). Blackwell.
Law, J. (2009). Actor-network theory and material semiotics. In B. S. Turner (Ed.), *The new Blackwell companion to social theory*. (pp. 141–158). Wiley-Blackwell.
Mares, T. M., & Peña, D. G. (2011). Environmental and food justice: Toward local, slow, and deep food systems. In A. H. Alkon & J. Agyeman (Eds.), *Cultivating food justice*. (pp.197–219). MIT Press.
Martı Tur [Martı Tour]. (2019). Ata ekmeği turu. [Ancestral Bread Tour]. https://www.martitur.com/gunluk-194/ata-ekmek-festivali-almola-senligi-turu-izmir-cikisli-seferihisar-ata-ekmek-festivali-turu.html
Martinez-Alier, J., Pascual, U., Vivien, F.-D., & Zaccai, E. (2010). Sustainable de-growth: mapping the context, criticisms and future prospects of an emergent paradigm. *Ecological Economics, 69*(9), 1741–1747. https://doi.org/10.1016/j.ecolecon.2010.04.017
Marx, K. (1973 [1857-1858]). *Grundrisse*. Vintage.
Mills, C. W. (1959). *The sociological imagination*. Oxford University Press.
Neuman, N. (2019). On the engagement with social theory in food studies: cultural symbols and social practices. *Food, Culture & Society, 22*(1), 78–94. https://doi.org/10.1080/15528014.2018.1547069
Petrini, C., Bogliotti, C., Rava, R., Scaffidi, C. (2012). The central role of food [Congress Paper]. https://slowfood.com/filemanager/official_docs/SFCONGRESS2012__Central_role_of_food.pdf.
Pink, S. (2008). Re-thinking contemporary activism: From community to emplaced sociality. *Ethnos, 73*(2), 163–188. https://doi.org/10.1080/00141840802180355
Pink, S. (2009). Urban social movements and small places. *City, 13*(4), 451–465. https://doi.org/10.1080/13604810903298557
Purifoy, M. (2014). Food policy councils: Integrating food justice and environmental justice. *Duke Environmental Law & Policy Forum, 24*(2), 375–398.
Ren, C. (2011). Non-human agency, radical ontology and tourism realities. *Annals of Tourism Research, 38*(3), 858–881. https://doi.org/10.1016/j.annals.2010.12.007
Salleh, A., Goodman, J., & Hosseini, S. A. H. (2015). From sociological imagination to 'ecological imagination': Another future is possible. In J. P. Marshall, & L. H. Connor (Eds.), *Environmental change and the world's futures. Ecologies, ontologies and mythologies* (pp. 96–109). Routledge.
Sayes, E. (2014). Actor-network theory and methodology: Just what does it mean to say that nonhumans have agency? *Social Studies of Science, 44*(1), 134–149. https://doi.org/10.1177/0306312713511867
Seferibakkal. (2019). Karakılçık Ekmeği. [Karakılçık bread]. https://seferibakkal.com/urun/karakilcik-bugday-ekmegi/.
Seferihisar Municipality. (2019a). Muhtarlıklar [Neighbourhoods]. http://seferihisar.bel.tr/en/mahalle-muhtarlari/.
Seferihisar Municipality. (2019b, August 5). Ancient bread and armola cheese festival. Retrieved from: http://seferihisar.bel.tr/en/ancient-bread-and-armola-cheese-festival/
Seferihisar Municipality. (2016, August 17). Ancient bread of seferihisar. http://seferihisar.bel.tr/en/ancient-bread-of-seferihisar/
Seferihisar Municipality. (2017, June 8). Ancient bread and cheese festival. http://seferihisar.bel.tr/en/ancient-bread-and-cheese-festival/
Slow Food. (n.d.). Welcome to our world. https://www.slowfood.com/about_us/img_sito/pdf/Companion08_ENG.pdf
Slow Food. (2015b). Slow food terminology. https://www.slowfood.com/about-us/slow-food-terminology/

Slow Food. (2018). Our history. https://www.slowfood.com/about-us/our-history/
Slow Food. (2015a). Our philosophy. https://www.slowfood.com/about-us/our-philosophy/
Slow Food. (2019, December 13). European NGOs urge to improve the farm to fork strategy. https://www.slowfood.com/sloweurope/en/european-ngos-urge-to-improve-the-farm-to-fork-strategy-2/
Toprak Mahsülleri Ofisi [Turkish Grain Board]. (2016). Ekmek israfı ve tüketici alışkanlıkları [Bread wastage and consumer habits]. http://www.tmo.gov.tr/Upload/Document/ekmek/tmobrosuryeni2.pdf.
TRT Haber. (2019, May 31). Ata yadigarı tohumlar gen bankalarında korunuyor [Seeds inherited from our ancestors are protected in the seed banks]. [News video]. *TRT Haber.* https://www.trthaber.com/haber/turkiye/ata-yadigari-tohumlar-gen-bankalarinda-korunuyor-417728.html.
Van der Duim, R., Ampumuza, C., & Ahebwa, W. M. (2014). Gorilla tourism in Bwindi Impenetrable National Park, Uganda: An actor-network perspective. *Society & Natural Resources, 27*(6), 588–601. https://doi.org/10.1080/08941920.2014.901459
Van der Duim, R., Ren, C., & Jóhannesson, G. T. (2013). Ordering, materiality and multiplicity: enacting actor-network theory in tourism. *Tourist Studies, 13*(1), 3–20. https://doi.org/10.1177/1468797613476397
Van der Duim, R., Ren, C., & Jóhannesson, G. T. (2017). ANT: A decade of interfering with tourism. *Annals of Tourism Research, 64*, 139–149. https://doi.org/10.1016/j.annals.2017.03.006
Van der Duim, R. (2007). Tourismscapes an actor-network perspective. *Annals of Tourism Research, 34*(4), 961–976. https://doi.org/10.1016/j.annals.2007.05.008
White, D., Rudy, S., & Gareau, B. (2016). *Environments natures and social theory*. Palgrave.
Woodward, S. (2019). Material cultures. In V. May & P. Nordqvist (Eds.). *Sociology of personal life* (pp. 74–86). Red Globe Press.
Yavaş Dükkan [Slow Store]. (n.d.) Hakkımızda [About us]. https://www.yavasdukkan.net/hakkimizda/

The dialogic negotiation of justice

Becky Shelley, Can-Seng Ooi and Lisa Denny

ABSTRACT

The purpose of the paper is to reveal the complexities of negotiating justice. We present the case of workers' experiences during a long-running industrial dispute at Australia's first legal casino. First, we consider the concept of justice, drawing on discussions from tourism studies. This notion is considered in relation to an industrial dispute at the casino. Second, we use Bakhtin's dialogic imagination to explore the issues, primarily through the voices of the workers. The concepts of heteroglossia, polyphony and carnivalesque facilitate layering the voices and concerns of the workers. Formal and informal strategies to exert influence are present. Within the carnivalesque, there is order and disorder. Different parties present a cacophony of voices, asserting the just-ness of their position. We discuss how, as boundaries of social action are negotiated, the parties are effectively negotiating justice. Third, we suggest how, from our Bakhtinian perspective, justice is a social activity, doing justice is a social process, and achieving justice is hard. Finally, we share some broader suggestions and reflections on the complexities and contradictions.

Introduction

"It's like, I guess it's like - sounds awful - putting a frog into a saucepan of water. It'll just sit there if it's cold or if you put him in boiling water, he'll jump out. If you slowly bring it to boil, he'll just sit there and die. I don't want that". Analogy given by a union member and casino worker to understand his employer's proposal to reduce his working conditions.

In Australia, industrial rights are enshrined in legislation and embodied in cultural norms, such as the notion of a 'fair go'. There are few people in the world who do not believe in justice and fairness (Thomas, 1993; Newton, 2019). Yet, as values, they are not fixed (Jamal, 2019). Workers' ideas of justice and that of their employers might be presumed to differ (Schwartz, 2005). Research on this paper started in 2018 during a dispute between workers and management regarding a new enterprise bargaining agreement at Wrest Point Hotel and Casino (Wrest Point) in Hobart. Wrest Point is Australia's oldest legal casino.

From the many different theories and approaches to justice and fairness, it seems that any call for justice and fairness is an assertion of how-things-ought-to-be and assumes a selective perspective and agenda. Most assume their stance is the 'fair' or 'just' one. Few would acknowledge starting from a position of unreasonableness. Bakhtin's (1981) heuristic of the carnivalesque is used to reveal the complexities of the situation and to consider the broader lessons from our case. Industrialised democracies maintain an economic system that reduces industrial justice to a process of negotiation; reaching an agreement. Given the uneven distribution of benefits in the current capitalist and industrial relations systems, one can argue that the systems are inherently unfair. A semblance of justice is basically just a feeling of fairness in the exchange. Our participants frequently described how it felt being in negotiation and in dispute. Asserting their rights was empowering, conflicting, demotivating and stressful. The casino management and owners were vulnerable in the process of negotiated justice. Wrest Point management did not achieve their desired outcome. After twenty months Wrest Point capitulated, and the dispute was settled just before the peak summer tourism season. Was the agreed-upon settlement between employer and the workers fair or just?

The authors invited Wrest Point management as well as union members to participate in the research. The authors also considered media statements, web-pages, and media reports. Wrest Point declined to be interviewed because the dispute was unresolved at the time of the research project. This is their right, participation in research must be voluntary. Paradoxically, silence is also part of the cacophony. Workers felt that pliant acquiescence would lead to a deterioration of their working conditions. Workers expressed their discontent by speaking with family, friends and customers, as well as through symbolic collective actions. Our case illuminates the difficult reality of the workers' pursuit of justice in a tourism-dependent casino.

We approach the analysis in the following way. In a brief exploration of the notion of 'justice' we consider the concept of procedural justice and the opportunities and constraints afforded within the Australian industrial relations system. Building on this situational understanding, we introduce Bakhtin's idea of the dialogic imagination which characterises the social world as dynamic and entangled (Bakhtin, 1981). The concepts of heteroglossia, polyphony and the carnivalesque enable layering of the voices and concerns of the workers and the stance of the casino. We note Bakhtin's influence within the social sciences broadly and within tourism studies. In this section we connect the concept of the carnivalesque with the ideas of polyphony and heteroglossia. Dialogism is a heuristic in which to consider ambiguity and incongruence during the dispute. We present a road-map to guide the reader through the arguments. Our analysis makes clear that the social activity *within* the dispute process introduces new situations, behaviours and actions. This leads to our concluding section in which we argue that the findings resonate more broadly and highlight polyphony and the carnivalesque within the notions of justice and fairness. There is organisation and disorganisation. Different parties presented discordant voices, trying hard to be more persuasive, more indignant. Justice and perceived injustice are present.

Justice

Following Jamal (2019), this article acknowledges that justice is an ambiguous and vague concept. Applying the idea of justice requires context for instance, distributive justice addresses concerns at the societal level and retributive justice focuses on blame and punishment (Jamal, 2019, p. 31). The case we are presenting of workers grievances in a gambling-dependent tourism business deals with one form of "participatory and procedural justice" (Jamal, 2019, p. 31). In our case, despite labour negotiation and industrial relations laws, the drawn-out process demonstrates a grubby reality of navigating participatory processes and procedures. No one would disagree with the need for justice, but just acts occur within and are framed by social contexts, resulting in multiple interpretations even if the procedures are clear. The social world is necessarily ambivalent and nuanced. We use the dialogic imagination in this study because it enables exploration of the complexities behind the negotiation of fair working conditions within a set of established procedures.

Jamal (2019) points out that freedom and equality are fundamental principles of justice. Workers, for instance, have freedom to choose where they labour or whether to engage in industrial action. These freedoms are, however, constrained by the realities of competition in the labour market and the degree to which industrial laws protect worker rights and interests. In the context of this study we are particularly concerned with 'justice as fairness'. While this article seems to be looking at how the profits from a gambling and tourism business are distributed to workers, there are wider issues about how fairness is negotiated in an emerging and dynamic social context. At the personal level, workers lamented the emotional costs they bore in pleasing customers and in being flexible with their work arrangements. At the business level, it was about processes and organisational structures that extract profit for owners. Workers disagreed on why they should consent to losing existing employment conditions when the business was highly profitable.

In Australia, the Fair Work Act (2009) stipulates the processes to be followed to negotiate enterprise level agreements. The enterprise bargaining system is intended to provide workers with fair wages and conditions in exchange for productivity improvements. The 'bargained' agreement is subject to a 'no disadvantage test', which requires all employees covered by the agreement to be 'better off' overall than they would be under the award. The crux of the dispute at Wrest Point Casino related to the roster. Management was seeking to shorten the notice period given to workers of the hours they would be required to attend work. The United Voice (Australia) union members voted against the offer multiple times during a protracted period of negotiations from March 2017 until December 2018. Under Australian law there is provision for protected industrial action so that employees or employers can support or advance their claims during the bargaining period. Our study focuses on this period. A dialogical approach – through the concepts of heteroglossia, polyphony and carnivalesque – allows us to systematically layer and situate the voices and concerns of the workers in a gambling-dependent tourism business during an industrial dispute.

Dialogism

Many social scientists have adopted Bakhtin's ideas into research. In 2010 Hollinshead exhorted those who work in tourism studies or management to be trained "in the dialogic forms of knowing and in the multiple-voice cognitions" of key theorists of societal and cultural identification (Hollinshead, 2010, p. 509). Bakhtin's concepts have been deployed by some researchers interested in tourism. For example, Rickly-Boyd (2009) consideration of personal narratives references Bakhtin's concept of the 'chronotope' to consider temporal and spatial dimensions of the tourism experience. Others have pointed to Bakhtin's notion of the carnival and the presence of carnivalesque texts within medical tourism destination promotional practices that challenge dominant norms of what their nation is capable of doing (Buzinde & Yarnal, 2012, p. 787). They do not take Bakhtin's ideas literally and instead use it as a tool to imagine and analyse society (Ooi, 2013).

Bakhtin introduced complexity into the reading of texts, so similarly dialogism imagines and characterises the social world as heterogeneous and dynamically intertwined. This relational reading of society unpacks the many inter-locking forces in society. A dialogic social science sees society as complex: there is consistency and order, as well as turmoil and conflict (Ooi, et al., 2014). Unlike the post-modernist perspective of deconstructing and then questioning its own analysis, a dialogic approach deconstructs the social phenomenon and reconstructs the order that is observed. Unlike the functionalist ontology, however, a dialogically-imagined social world is not naturally integrated (Ooi, et al., 2014). Instead, the order of the world is seen as maintained, negotiated and engineered. As a result, the politics of social control as found in critical theories is part of the dialogic imagination. It is thus difficult to talk of dialogism as an independent paradigm. Scholars using dialogism have learned from and responded to other social theories. The radical perspectives in critical theories, the integrationist neatness of functionalism and the playful poetics of post-modernism are accepted and confronted in the dialogic imagination. Bakhtin was not clear in his definition of a dialogic relationship. In this study we focus on three dimensions. The first is on-going inter-relations between social phenomena. As Gardiner and Bell (1998, pp. 4–6) argue, socio-cultural phenomena are on-going, interlinked between individuals and groups, involving a multiplicity of different contexts, discourses and circumstances. The second dimension is mutual animation of social phenomena. This means that social phenomena and the socially constructed meanings must be understood in relation to each other, and not be seen in isolation. The third related element in the dialogic imagination is the holistic understanding of all social manifestations in relation to their complementing and conflicting contexts and circumstances.

Carnivalesque alludes to the seeming disorder and yet orderly proceedings of the carnival (Bakhtin, 1981; Stallybrass & White, 1986). The concept accentuates multiple cultures and spheres of activities in the casino business, which cannot be totally managed, suppressed or controlled. The concept of carnivalesque brings together heteroglossia and polyphony. Multiple contexts and voices characterise social reality. A person holds different, even contrasting views and carries a repertoire of social behaviour. Because a person can behave differently within the range of appropriate behaviour, the outcome of the interaction remains uncertain. For example, workers in our study were permitted to wear badges to indicate their dissatisfaction with their employer while serving in restaurants and casinos. The badges drew attention to the dispute but, at the same time, the services to customers were not disrupted. There were rules and unruliness. Ambivalence and ambiguity define the carnivalesque social situation, and, like in the on-goings of a carnival, can take off in different directions as parties negotiate, persuade and interact. Individuals involved may feel happy and frustrated as the dynamic carnivalesque unfolds. Figure 1 sets out the structure of the argument. In this paper, we identify the myriad of contrasting and contradicting activities in the workplace to accentuate the seeming order and disorder, and by doing so, we consider the voices and the context of actions within the sections on polyphony and heteroglossia.

Closely related to carnivalesque is the concept of polyphony. Polyphony highlights multiple voices (Bakhtin, 1981, pp. 331–336; Vice, 1997, pp. 112–148). When a point of view is being articulated, there is also an overarching voice or the voice of the narrator. Just like in this journal article, our voice is omnipresent even though we do not use the personal pronoun in most instances. In social interactions, individuals may marginalise their own voices and instead invoke other voices in their arguments and opinions. For example, our research participants spoke of public support for them and how customers rallied behind them. Each worker constructs an argument and narrative. The concept of polyphony reminds us who constructs arguments, whose voices are invoked and whose are hidden during social exchanges. Voices that are marginalized are also important. Taken together, the voices reveal the politics of the conversation. The dialogue amongst different parties, the cacophony of voices, illuminates the social complexity and ambiguity. So, this study looks at how people attempt to make sense of their own situation by speaking for and about 'justice' and 'fairness'. It demonstrates how these values become a cohering focal point in the cacophony of voices. The multiple meanings of justice and fairness emerge, yet there is no clear understanding of these ideas. Access to procedural justice does not solve the problem of what people feel is just. Instead the cacophony of voices reveals the layers of arguments that constitute people's understanding of justice and fairness.

Following from polyphony, this paper focuses on how individuals take on their situated identities and roles such as being a union member, a colleague, a manager, a friend and a family member when they communicate. Heteroglossia points to the multiple contexts existing in social situations that individuals navigate (Bakhtin, 1981, pp. 325–326; Vice, 1997, pp. 18–44). Heteroglossia links closely to polyphony and the carnivalesque. Heteroglossia or multiple contexts are embedded in the work situation, for instance, and the workplace entails a common understanding of what constitutes appropriate behaviour and how social actions should be

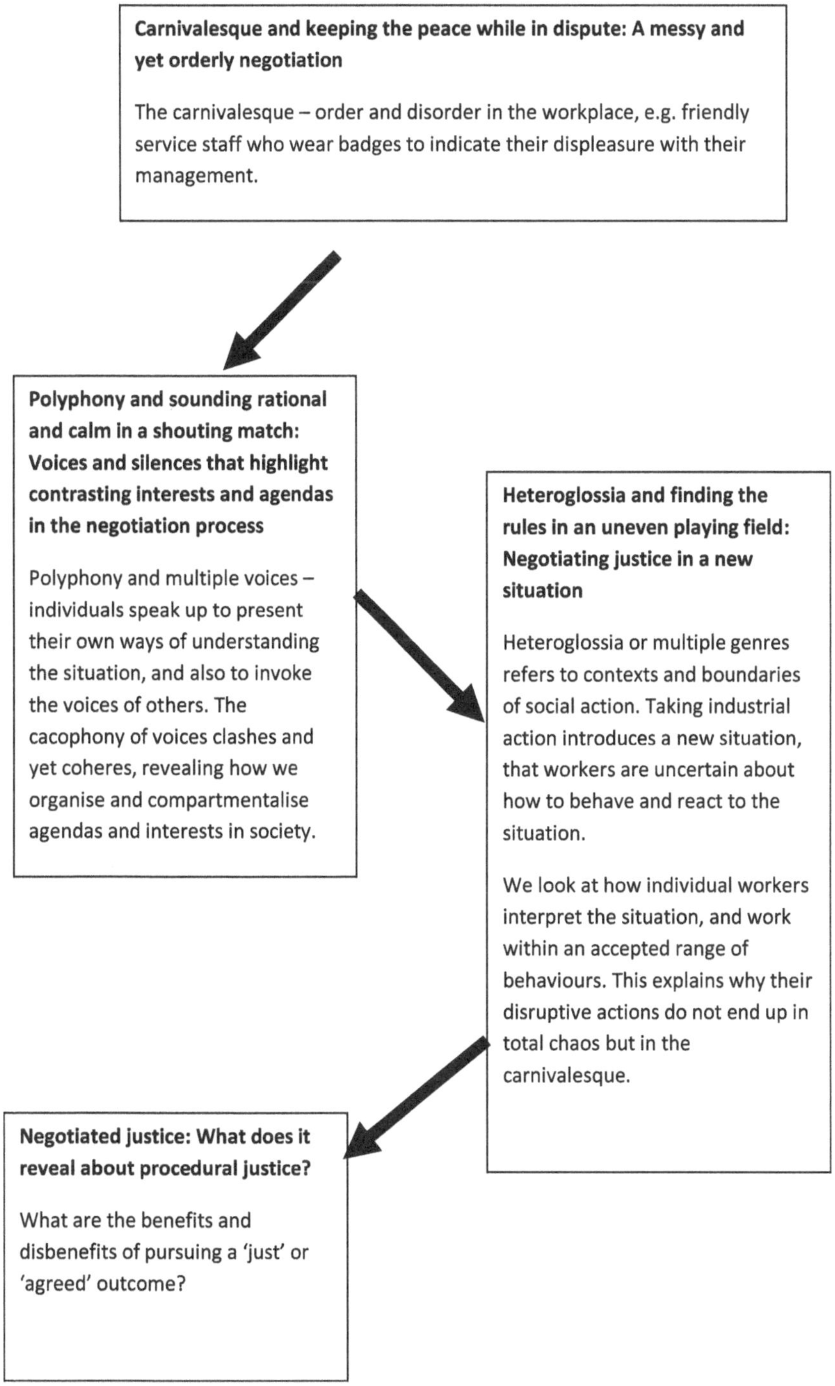

Figure 1. Structure of the argument.

interpreted. Disagreements are common but there are established industrial processes to handle them. In this study, industrial action creates a new set of contexts that workers respond to. For instance, at the broad level, workers are to serve their employers and in more ordinary circumstances, they also express support and loyalty to the company. But when workers are in dispute with their employers, the situation changes. Relations are reframed and straddle professional behaviour during work and frustration about work when not working. In the context of this study, we look at the negotiation of social boundaries between employers, employees and customers. It explains why there is order in the carnivalesque, as individuals have tacitly agreed to what is appropriate behaviour despite their differences. Procedures help facilitate the negotiation of their work conditions.

We present the multi-faceted and entwined dynamics of claiming justice. Indignation and fairness were voiced in the workers' fight for justice, as management sought to change their working conditions. The union members understood Wrest Point management's justification but challenged it. Their loyalty to their employer was tested. Their concept of justice as fairness was at stake. One worker observed that the dispute was "about our wages and conditions. It's about respecting your employees. It's about treating people with dignity" (I3). Union members attempted to mobilise public support as well as support from customers and punters of the casino-hotel complex. Voicing their position as it evolved from a supportive stakeholder group within the industry to an activist one required reframing the issues. Dialogism as a heuristic device illustrates this.

Data collection

Before we present the case, and in the spirit of the dialogic imagination, we present some background stories in this study. Part of the genesis for the project was the experience of one of the researchers – Becky Shelley – driving past protesting workers waving hand-made placards on the side of the road outside Wrest Point Casino, eliciting honks of support from the passing traffic. Meanwhile, protestors campaigning against a proposed tourism development on the mountain that frames the capital city of Hobart were mobilising thousands of community members. The contrast in scale of media attention was notable and suggested the challenge for workers in such industries to generate popular support for their position. Becky contacted the union offices and sought permission to photograph the placards and began a discussion with Can-Seng about how to make sense of two tourism industry related campaigns that were occurring simultaneously. A research design took shape, conceived as an exploration of the growing pains of the tourism industry in Tasmania.

The methodology for the study was informed by inductive qualitative design (Crotty, 1998). It was also influenced by Can-Seng's appreciation of Bakhtinian scholarship within the social sciences (Ooi, et al., 2014). Ethics approval for the study was granted by the Human Research Ethics Committee (Tasmanian) Network (H0017640). Information about the study was sent to the United Voice union for distribution and to senior management at Wrest Point. Ultimately, four participants volunteered to be interviewed from the union. We were not seeking to find a general voice for or of the union but instead wanted to note the layers of interpretations of the situation by some workers. This includes how they see fairness and justice in their cases. Our participants worked in different departments at Wrest Point and have had different levels of engagement with the industrial action. As will be presented later, we noticed that the workers shared similar sentiments and articulated similar views in different ways. Our goal however is to present a slice of the uncertainty and chaos within the stability and lawfulness of a dynamic situation. We have not provided detail about the context of each worker's employment out of an ethical commitment to ensure that research participants are not able to be re-identified.

Interviews were audio-recorded, transcribed in full and given to the participants. Their interviews reshaped our interest into questions of justice and justice as fairness. Becky Shelley and Can-Seng Ooi conducted in-depth interviews over two days. Two participants continued contact with the researchers via email correspondence to further expand on their interviews or provide additional information, such as the images of the 'Fed Up' badges. After the authors compared transcripts a process of reflexive review occurred where we read, commented and challenged each other's analyses. In our interpretive inquiry we situate the qualitative process as fragile and interactive (Jamal & Hollinshead, 2001). Our methodological considerations on the dialogic imagination incorporate Can-Seng Ooi's oeuvre (Ooi & Ek, 2010; Ooi & Strandgaard Pedersen, 2010; Ooi, et al., 2014).

There were many issues involved and the workers felt that the casino wanted to 'take this, and this and this and this and we're going to offer you nothing in return' (I4). The final unresolved issue in the dispute hinged on a proposed reduction in the notice period given to workers about when they would be required – rostered – to work. One worker summarised their concern:

> (I2) While you're starving people for work and saying, "well you can't work anywhere else, you must be available to us 24/7", how is that helping your business? It's not…

The proposed rostering arrangements made it difficult for workers to know in advance when they would have days off. This also meant that workers would not receive the compensation they would get by stepping in at the last minute. Another worker recounted that the offer had made them feel as if they had the "shit scared out of me" and that was why they got involved. We asked why they were scared, and they responded:

> (I3) Because I stood to lose so much. Every year pay conditions whittle down a little worse and worse. Until this time when it was a massive cut to pay and conditions and that was too bitter a pill to swallow for me, so I started becoming very active.

In sum, the two main points of contention of the workers are first, the lack of respect and consultation with them, resulting in an increased sense of helplessness; and second, the new rostering rules and pay offer would reduce their working conditions. There were other changes proposed by the employer and the workers that had been resolved.

The carnivalesque and keeping the peace while in dispute: a messy and yet orderly negotiation situation

During the industrial dispute, Wrest Point continued its operations and it would seem it was business as usual. The prolonged collective negotiation escalated into actions that might embarrass Wrest Point, such as demonstrations at the boundaries of the premises and media reports on the dispute. There were contrasting activities; workers continued to serve their employer and the same workers were expressing their unhappiness with their employer while at work. Similarly, for the employer, the managers maintained that they care about employees and yet ensuring that industrial actions did not disrupt their business. Customers had to make sense of the contrasting and contradictory messages when they encountered different customer service staff. It was like a dance hall where people were doing an awkward tango.

In the service-intensive hospitality business, customer-facing workers play an important role in producing a quality service experience. Pretending to be happy is part of the emotional labour (Ooi & Ek, 2010; Hochschild, 2012). That pretension created an orderliness which did not disrupt customer services. But, the dispute did spill over to customers as many workers at Wrest Point were wearing protest badges at work (Figure 2). While serving customers they maintained professionalism, but their badges communicated their discontent. One worker plastered their car with posters and left it in a prominent position in the car park. Another reported on his experience:

> (I4) Customers ask, "what's that about?" Mostly they're sympathetic when you explain …it's 18 months since we have had a pay rise. People are sympathetic to that and especially when you say, "and the only pay rise they're proposing is only 2% and then only if we give up conditions." Yeah, people are generally sympathetic to us.

Figure 2. Campaign badges: I'm Voting No;1.8% ISN'T A FAIR DEAL; Their profits are UP, our wages are DOWN; We are Fed Up; and We are Federal Group.
Photograph: anonymous United Voice member.

As will be discussed in polyphony, the badges appropriated the messages used by the employer. As has been observed in other protest movements, symbolic acts such as wearing badges were contingent and transitory, and served multiple audiences (Shelley, 2001, 2005). One of the badges featured the words "Fed Up" written over the top of the company slogan "We are Federal Group". Wrest Point is part of the Federal Group.

The "Fed Up" badge was a parody of the badges given to workers by Wrest Point during the State election in 2018. The Tasmanian Labor Party campaigned for the tightening of gambling regulations and the Federal Group (owners of Wrest Point) rallied its employees to support the incumbent Liberal Party policy in order to protect its interests in casino and electronic gaming machine across pubs and clubs in Tasmania. Wrest Point workers were asked to demonstrate support for the company by wearing badges declaring "We are Federal Group". The company's stance during the enterprise bargaining process was a let-down for the workers who anticipated reciprocal support from Federal Group after participating in the electoral campaign to protect the Federal Group's economic interests.

(I4) In particular, after that was resolved [the Liberal Party was re-elected] and they still wouldn't give us a better offer, a lot of people were really disappointed and upset about that after we'd supported them. We collectively had supported them during the election. They weren't now willing to support us. And someone came up with the idea of creating a badge that looks substantially just like the one that they had given us to wear but where it says "Federal", it says "Fed Up". So, it's a kind of call back to those badges. What's been interesting though, within the workplace, obviously there are going to be some people that are very, just very pro-company, so some of them in reaction to our badges, they've taken to wearing their, "We are Federal Group" badges.

So, there were two sets of badges being worn by staff members during the bargaining period: the official and the unofficial ones. It was not a coherent picture. There was an abandoned plan to distribute stickers for customers to show their support for workers. But because of concern about legal threats, that strategy was not pursued by the union. Actions were bounded by the legislative framework, so that the options available were constrained.

Nonetheless, established social and cultural norms were affected, creating chaotic situations. One research participant recalled

> (I4) There are managers within Wrest Point who understand the right to withdraw your labour, to go on strike as a human right. It's a human right recognised by the UN and a lot of people sort of understand that you don't take it personally when your workers walk off the job. There are other managers who have taken it very personally and have been willing to sort of target people who have done it. Or there's the bars, in particular, whenever casuals have participated in strikes, it's been not uncommon that as soon as the strike would end, every casual would just get sent home. And being casual there's no finish time and you can just do that. So, it's been really nasty in a lot of departments. There's been a lot of managers who have been sort of seen to be almost storming around the building and yelling, "where are the fucking staff", that's a direct quote from one of the senior managers at Wrest Point. It hasn't been conducted in a respectful manner and I think that's the thing that really sticks out, just how nasty some elements within management have gotten.

The uncertainty and chaos in the carnivalesque are also felt outside the workplace:

> (I3) I'm stressed out. I'm snappy at home. I'm irritable. I've lost some weight, got dark circles under my eyes. It's really stressful. I want it to end and lots of people want it to end and I think the company's counting on that, and I think that's been a, I think that's just, that's their game plan. Just wear them down until they won't fight back anymore.

During the negotiations, the situation was messy. The outcomes were uncertain, but the organisation and the union continued to work and to confront each other. The disorder arose from the disagreements and the frustrations felt by the workers and managers, as well as the contrasting messages presented by service staff. But there was also order which stemmed from a need to respect the rules of industrial action, workers acting professionally and maintaining respectful relationships with co-workers and management. The next two sections will dissect the social processes that created and maintained order and disorder.

Polyphony and sounding rational and calm in a shouting match: Voices and silences that highlight contrasting interests and agendas in the negotiation process

The previous section focused on the carnivalesque during the industrial action period. The dynamic of the situation was also embedded in the interests and agendas of the different negotiating parties. Polyphony is a concept that helps unpack these interests and agendas considering how voices are layered and articulated in the situation. A person speaks for oneself, and at the same time, speaks for others as they imagine them to be. The dialogic imagination brings these voices into focus to highlight the interests and agendas of what the speakers say, do not say, and how they speak for themselves and for others. This is part of the negotiation in furthering their own cause within a procedural justice framework.

Wrest Point supports and engages with the local community. It pays its workers above the standard award rate in the hospitality industry. It has a series of corporate social responsibility activities:

> Wrest Point has proudly been a supporter and active member of our local community for more than 40 years. In addition to entertaining many thousands of Tasmanians and being a major employer, we've formed lasting partnerships with local organisations to help shape our community. This support takes many forms – financial, in-kind, and employee participation (Wrest Point, 2018).

Wrest Point fosters its image as a responsible corporate citizen that serves not only the community but also treats its workers well. During the 2018 State election campaign, the Federal

Group mobilised its workers through a campaign focused on protecting jobs in local pubs and clubs. Also as mentioned earlier, during the election campaign, the union suspended negotiations. I1 recalled what happened then:

> During that time management sent out personal letters: "Dear staff member, you need to vote Liberal if you want to have a chance of keeping your job, otherwise Labour is going to close us down", so they were posted out to our personal addresses.

I1, together with many other workers, expressed a sense of betrayal. United Voice had agreed to suspend negotiations and had kept silent on worker grievances indicating that it too had an overarching interest; job security. The United Voice union's fight for better working conditions should not jeopardise the jobs of union members. And thus paradoxically, union members campaigned against the gambling policy of the union affiliated Australian Labor Party. The Labor Party lost the election. The workers and the union refrained from making their industrial grievance into an electoral issue. While it is debatable if that was a strategic mistake, what happened demonstrated the selectiveness of when the workers want to speak.

One interviewee reported that they communicated with Greg Farrell, Founder and Managing Director of Federal Group:

> (I4) I recently contacted Greg Farrell to, partly, make the point that, "if you want customers to have a good experience in your business, it's the staff that are going to give it to them and for that to happen you need staff to be coming in to work feeling good about the workplace. Feeling like they're being looked after. Feeling like, I'm really happy to work here and I'm cared about. So, that they're happy coming into work, feeling good, that's reflected in the sort of service you give to a customer."

The worker was trying to convince his employer that they would be more efficient and effective if they were happier. But that strategy was not useful as he did not get any response and he found the silence deafening. In polyphony, the multiple voices are not ranked equally and may not be equally loud. The goal was not just to be heard but to realise their goal of keeping their jobs, as well as to maintain their working conditions. Just as they compartmentalised when and what to say, their employers did the same.

One effective way of speaking up was through the protest badges. The workers wanted to shift customer narratives and perceptions about Wrest Point. But after months of stagnation in the negotiation, worker voices of support were receding. A participant lamented

> (I1) I think that it's hard because there's a little bit of everything. I've had a few problems with the badges. I wear the badges. Everyone knows me wearing badges. When we first started the badges, everyone was wearing them, and everyone was wearing five and in photos everyone's wearing badges. It was great. And then now, there's me, wearing badges.
>
> I ask, "where are your badges?" and some staff go "oh are we still doing that", and I say, "yeah we're still doing that, it's okay, wear your badges", "oh I don't know where mine is now". So, there's a little bit of (I get a bit frustrated), with a little bit of apathy. But also, I have to take into account that I'm different to other people and that I have to understand that that doesn't mean that they're not feeling it or they're not in with us. It's just that they don't necessarily feel comfortable wearing a badge or they're not as, they don't feel secure in either themselves or in their position or maybe their manager is more of a bully than mine.

Workers would be stronger if they stood together with a single voice. Our participants were active members of the union, but many workers were not. The workers and union members seemed divided over time. The same people even rationalised their employer's actions and spoke for Wrest Point, despite being in a derogatory manner.

> (I1) I can understand their position. They're a business they want to make money. The way they portray themselves is on their old family, "we're all a family, we all work together and isn't it great" which is how it was seen in the 70s and 80s ... It is a different atmosphere now, and again like I said earlier, they're just not interested in the worker. They're not interested in your point of view. They're not interested if you've got, you're not seeing your family. None of those. It's, "you've got a job, if you don't like it, see you later",

Figure 3. Honk to support us. Permission to use image from United Voice, taken by Jess Saunders.

> and that's all it is. And "we're not interested in what benefits you've had, what benefits you want, what conditions you want, do the job or go away".

Workers staged protests at the busy thoroughfare outside the casino, holding up signs to draw attention to their issue (Figure 3). Photos were posted online to gather further support. Our participants stated that the public was generally supportive. They were communicating with the public to put reputational pressure on Wrest Point. Wrest Point, on the other hand, publicly kept silent and privately met with union negotiators. The company communicated with employees via text messages and polls to measure support for the deal it was offering; if 50% of workers agreed, the deal would come into effect. Our participants were furious because the same deal was offered repeatedly, as if the union negotiations were not occurring. During the time we interviewed them, the participants observed that more and more workers were agreeing to that deal. However loudly they shouted, members were losing momentum and were fearful that they may lose their jobs. While they were told that they need to stand together, assert what is just and fair, and to continue to speak loudly, some members were choosing to stay silent.

But I4 talked about the silver lining and how the dispute rallied people:

> (I4) These sorts of bruising disputes can draw people out who would like to be more involved or who would be willing to be more involved but haven't felt the need to be. So, I think in that sense there's been a few of us who it's been like that for. Through this there can be some positives to come out and to look towards future negotiations and think, "how can we get ourselves into a position where we can negotiate from a better position of strength." Whether that's getting more delegates, getting more members, or, and things around the rules [industrial relations laws] themselves being broken. Will we potentially have a different set of rules to negotiate on next time?

I3 mentioned that the impact of receiving support "through the union, through the patrons was uplifting. You know, when we are striking to have our angriest patrons giving an earful to management, "Pay Staff Properly!", that's good. I want that".

Despite the cacophony of voices, uneven loudness, the ruckus that union members tried to create, Wrest Point maintained a more procedural and silent approach. It went through the process of seeking support for the deal directly from workers via text and other modes of communication. Wrest Point did not dispute that it wanted its workers to be happy and to be a responsible employer. And during the State elections, the Federal Group emphasised that it wanted to help small pubs and clubs and valued job security. The negotiation was not about it being a good employer per se but more about managing a company to be even more profitable. For workers and the employer, moral arguments about fairness were presented only in support of their own agendas, interests and causes.

Heteroglossia and finding the rules in an uneven playing field: Negotiating justice in a new context and situation

The sections on carnivalesque and polyphony present the complexity and messiness of the situation in an industrial dispute, albeit from worker perspectives. Despite the multiple angry voices, things have not gotten 'out of hand'. As there are centripetal forces, there are also centrifugal ones. The idea of justice and fairness is vague but universal. And in a democratic society like Australia, where human rights are lauded, people react to others who exert their rights respectfully even though that may result in inconveniences. Individuals behave in ways that are socially defined, and they reflect the structures and institutions of society. Heteroglossia accentuates those structures and institutions. It points to the boundaries of appropriate social behaviour. This section will show that the boundary between fair and just exchange between employer and employee is malleable but social norms and expectations have maintained a semblance of stability.

Heteroglossia emphases contexts and the boundaries of actions. The case we have presented so far alludes to the negotiation process between the employer and the union. But the situation is marred by individual anxiety, as workers interpret the new situation. The informal relations and context are just as important as the formal. The employer and employee relationship are defined formally via the employment contract but the informal contributes to how the organisation actually functions. Workers feeling that they are part of the family is an example. Similarly, individuals fear for their employment if their bosses are unhappy with them. One research participant recalled why he did not join the union until much later in the dispute.

> (I4) I think in a lot of other department people are scared to participate in that sort of way [take part in industrial action] [... But I then realise] "hold on, the company's not being fair here and we're not going to put up with it", whereas [others may] have more problematic managers or maybe they have people working there who don't speak the best English and so they don't understand what their rights are, or people who have come in from other countries where they have it much, much, much worse, they don't have the protections that we have or they don't have the rights that we have and they don't understand that here you do have the right to walk off the job or you do have the right to assert a workplace entitlement without fear of adverse action and I think across a lot of departments that's well understood what our rights are and that, and that we are allowed to stand up for ourselves where as in some other departments that's not as well understood.

Disgruntlement can lead to each of them reacting in different ways, igniting chaos in the organisation. But industrial rights are entrenched in industrial laws in Australia. It is the respect of lawful rights that make industrial actions possible. I1 spoke of not just exercising the right to take industrial action but of the necessity of organising the process properly.

> We've also done a few protests out the front with our banners and some horn honking and that sort of stuff and did a few sorts of carpark flyers and stickers around to customers. That in itself is hard because we have shift workers. We have to be understanding that some of these people start at terrible hours and finish at terrible hours so not everyone can participate so we have to sort of try well, "who's on roster this week, how many, who's been rostered when", and it's a bit of a tricky little thing. So that's kind of what we've been up to.

The union members we interviewed have navigated the contextual constraints and identified lawfully protected opportunities to protest. I1 remembered how managers tried to tell them what was acceptable and what was not.

> (I1) They are kind of powerless if we just kind of going-through-the-motion in our job. [... But] management can say things like "if you vote yes [to industrial action] you'll lose your job" and various sort of big statements that are factually incorrect. For example, we are allowed to wear badges, and then management says, "you can only wear one badge, you can't wear all those badges". And there's been so many [false claims that] we just can't list them all. All those things are all crap. We hate them, and it's bad for them to do it and they're not allowed to do it but they still do it.

Taking industrial action is stressful for workers partly because they are in a new situation. The workers must figure out what is appropriate behaviour. They will be protected within the confines of the law, but they fear that they may have crossed the line. They want to disrupt the business activities of their employer but at the same time, they want to keep their job. From the outside, the situation is carnivalesque. But it is the ambiguous boundaries of industrial actions that must be negotiated, as they fight for justice and their rights. In the middle of the dispute, workers saw little progress, I1 was glad for one thing:

> There haven't been any big wins for us or for them. There's been small things that we've all been able to cheer together. Personally, I found some really good friends which has been good. A lot of [the good experience] is more personal. I mean I've always had a bit of a voice and I've always had some issues in the workplace but for me it's been more of a personal sort of growth. And it's, yeah, it's just, that's kind of really the best things.

Colleagues become friends. How they professionally interact with each other changes accordingly. Their joint fight for justice is fodder for not only new relations but also an affirmation that their indignation is justified. They were not alone in believing that they had been unjustly treated.

Conclusion

There are few people in the world who do not believe in justice and fairness. Those values, however, are intangible. There are different regimes and regulations around the world to define fair exchanges between workers and employers, but they may not be considered just. For instance, wage-gaps between workers because of their age, gender and ethnicity are known to exist but should not be acceptable (Baum et al., 2016). Relying on the so-called market mechanism is not enough to determine a just exchange. In using the dialogical imagination, this study has identified various informal mechanisms that workers navigate in pursuing a fair exchange with their employer, in this case, Wrest Point.

We started this study when the dispute was on-going. Only union members agreed to talk to us. Regardless, the insights we garnered from our participants and other publicly available information illuminate the difficult reality of negotiated justice. As highlighted in the carnivalesque, there is order in the disorder. Different parties presented a cacophony of voices, trying hard to be louder and claiming their right to be indignant. And, in the last section on heteroglossia, we discuss how boundaries of social action are negotiated; the parties are effectively negotiating justice. It is still debatable if the ultimate outcome expressed in the enterprise bargaining agreement between the employer and the union is 'just', but it is agreed upon. The idea of fairness is invoked when there is disagreement although what is agreed upon may still not be fair. From our Bakhtinian perspective, justice is a social activity, doing justice is a social process. There are lessons.

One, the sense of justice and ethics have an indignant tendency, as we want to do what is right. The dialogic reality accentuates the multiple intersections of agendas, interests and purposes of diverse groups of people. While almost everyone supports the fight for justice and is armed with good intentions, the actions put forward are fraught with disagreements amongst 'righteous' parties. Any position on justice and fairness assumes a perspective and agenda. And thus, any call to justice is political – parties persuade, coerce, manipulate and negotiate. This position is ongoing as circumstances change and amiable adjustments are made. But the agreed-to position remains tenuous as one change may be too much for the other party.

Two, the legal framework in many countries allows for employers and workers to negotiate at enterprise level. That framework allows for certain types of protected industrial actions and bargaining periods. Many individual workers protested with 'micro-actions' such as talking to family, friends and customers about their predicament. Many rallied around common actions, such as wearing protest badges and demonstrating outside the compound to amplify their voice. It remains unclear if such actions worked as intended, for example, the workers observed that the customers continued to return to the casino.

Three, we did approach Wrest Point management, but they were not willing to communicate with us during the dispute. We respect that. Their voice remains loud despite their silence. The cacophony of voices is made up of loud voices and conspicuous silences.

Four, it was clear to us that our participants waded through a minefield of emotional responses. There was ambivalence in the social group, and many felt trapped in the economic reality. It was fight or flight. Apparently, the flight response is the norm rather than the exception. Being amiable and avoiding feelings of conflict have arguably allowed employers to erode benefits for workers. But on the other hand, when economic well-being was affected, many workers rallied together and took on the fight. The negotiation of justice is then about the negotiation of worker's satisfaction with their work conditions.

Five, following from the last three points, the industrial regime and normative frameworks in modern societies maintain an economic system that reduces justice to a process of negotiation; reaching an agreement. With inequality and the uneven distribution of benefits in the current capitalist system, one can argue that the system is inherently unfair. A semblance of justice is also a feeling of fairness in the exchange. Our participants constantly referred to how they felt. We may wonder if the ultimate negotiated agreement between employer and the workers is fair, as it is about whether the workers are feeling satisfied with the new deal.

Finally, we have woven a narrative with our own voices in this study. There were temptations to take on the worker's position in our analysis. To a certain extent we must because some of them were willing to talk to us. Regardless, we acknowledge that we were part of the process; they wanted to tell us their stories. We listened to them. This, however, does not mean that Wrest Point was 'unjust'. There can be many valid reasons for them not to talk to us. Regardless, we have presented a cacophony of voices, situated them within the social economic structures in which the industrial dispute occurred, and demonstrated that seeking justice is a social activity. Wrest Point's participation in our study would have created even more voices and a more complex carnivalesque.

Acknowledgements

United Voice union (now the United Workers Union) and its members who generously shared their time and experiences with the researchers.

Disclosure statement

No potential conflict of interest was reported by the author(s).

ORCID

Becky Shelley http://orcid.org/0000-0002-6288-592X
Can-Seng Ooi http://orcid.org/0000-0001-9842-1556
Lisa Denny http://orcid.org/0000-0001-9842-1556

References

Bakhtin, M. M. (1981). *The dialogic imagination*, edited by M. Holquist & C. Emerson. Austin: University of Texas Press.

Baum, T., Cheung, C., Kong, H., Kralj, A., Mooney, S., Nguyễn Thị Thanh, H., … Siow, M. (2016). Sustainability and the tourism and hospitality workforce: A thematic analysis. *Sustainability (Switzerland)*, *8*(8), 809. doi:10.3390/su8080809

Buzinde, C. N., & Yarnal, C. (2012). Therapeutic landscapes and postcolonial theory: A theoretical approach to medical tourism. *Social Science & Medicine*, Elsevier Ltd, *74*(5), 783–787. doi:10.1016/j.socscimed.2011.11.016

Crotty, M. (1998). *The foundations of social research: Meaning and perspective in the research process*. London: SAGE.

Gardiner, M., & Bell, M. M. (1998). *Bakhtin and the human sciences: A brief introduction, Bakhtin and the Human Sciences*, edited by M. M. Bell & M. Gardiner. London: SAGE Publications.

Hochschild, A. R. (2012). *The managed heart: Commercialization of human feeling*. 3rd ed. Berkeley: University of California Press.

Hollinshead, K. (2010). Tourism studies and confined understanding: The call for a "new sense" postdisciplinary imaginary. *Tourism Analysis*, *15*(4), 499–512. doi:10.3727/108354210X12864727693669

Jamal, T. (2019). *Justice and* ethics in tourism. London: Routledge. doi:10.4324/9781315162942

Jamal, T., & Hollinshead, K. (2001). Tourism and the forbidden zone: The underserved power of qualitative inquiry. *Tourism Management*, *22*(1), 63–82. doi:10.1016/S0261-5177(00)00020-0

Newton, A. (2019). *The business of human rights: Best practice and the UN guiding principles*. London: Routledge. doi:10.4324/9781351131193

Ooi, C-S. (2013). *How to capture and present compexity, ambivalence and ambiguity: Applying dialogism in social science research*. Copenhagen: Copenhagen Business School.

Ooi, C. S., et al. (2014). Accentuating complexity and ambiguity: Dialogism and its methodological implications. In J. W. Meged (Ed.), *Tourism methodologies - New perspectives, practices and proceedings* (pp. 155–174). Copenhagen: Copenhagen Business School Press.

Ooi, C.-S., & Ek, R. (2010). Culture, work and emotion. *Culture Unbound: Journal of Current Cultural Research*, *2*(3), 303–310. doi:10.3384/cu.2000.1525.10217303

Ooi, C.-S., & Strandgaard Pedersen, J. (2010). City branding and film festivals: Re-evaluating stakeholders' relations. *Place Branding and Public Diplomacy*, *6*(4), 316–332. doi:10.1057/pb.2010.30

Rickly-Boyd, J. M. (2009). The tourist narrative. *Tourist Studies*, *9*(3), 259–280. doi:10.1177/1468797610382701

Schwartz, M. S. (2005). Universal moral values for corporate codes of ethics. *Journal of Business Ethics*, *59*(1–2), 27–44. doi:10.1007/s10551-005-3403-2

Shelley, B. (2001). Protest and globalization: Media, symbols and audience in the drama of democratization. *Democratization*, *8*(4), 155–174. doi:10.1080/714000231

Shelley, B. (2005). *Democratic development in East Asia, democratic development in East Asia*. London: Routledge. doi:10.4324/9780203481639

Stallybrass, P., & White, A. (1986). *The politics and poetics of transgression*. Cornell: Cornell University Press. doi:10.1086/ahr/93.4.1009-a

Thomas, G. (1993). *An introduction to ethics: Five central problems of moral judgement*. Cambridge: Hackett Publishing.

Vice, S. (1997). *Introducing Bakhtin*. Manchester: Manchester University Press.

Wrest Point. (2018). *Community - Wrest Point Hotel Casino*. Retrieved from https://www.wrestpoint.com.au/community.

Conceptualizing justice tourism and the promise of posthumanism

Jaume Guia

ABSTRACT
The past two decades of tourism research have seen a growing interest in the relationship between tourism and justice. Some of this attention has focused on the just or unjust outcomes of mainstream tourism, and how it could contribute more to justice. Other research has directed the attention to the justice outcomes of alternative forms of tourism, where their increased commodification and de-politicization has limited the potential justice benefits enormously. Yet, a clear conceptualization of justice tourism is still lacking, and its theoretical grounding is still too limited. This paper addresses these concerns and aims to clarify the concept of justice tourism and advance a conceptual framework where types of justice tourism and justice through tourism are systematically identified and classified. Moreover, from the proposed conceptual framework, posthumanism emerges as a promising ethical regime with which the commodification and depoliticization of justice tourism could be reversed, and its increasing co-optation by neoliberal capitalism curved. Posthumanism's affirmative ethics and political responsibility, along with its political forms of solidarity and advocacy, can become an effective mechanism for radical transformation and a crucial catalyst for justice in tourism and tourism research.

Introduction

Justice is a complex phenomenon, ranging from justice for individuals to justice for societies at both local and global levels. Justice not only applies to human beings, but also to non-humans, e.g., rights of animals, and even rights of nature. During the 20th century, philosopher John Rawls (1971) turned liberalism towards a concept of justice grounded in contractarian, liberal egalitarian values, yet distributive in nature. Thus, in terms of rights and the fair distribution of societal goods, justice becomes fairness in the sense that individuals have rights and should have equal opportunity to access societal goods and opportunities. Feminist, post-structural and post-humanist perspectives have since introduced new turns with alternative approaches to justice which address topics such as agency, power and performativity, identity politics and *difference*, relational ethics and care, and non-human others.

Unsurprisingly, this complexity deepens when "tourism" is added to "justice." Fennell (2006) states that tourism is an inherent issue of justice. Smith and Duffy (2003, p. 92) offer different perspectives on justice and tourism, and note that among other things, "social justice is about the fair distribution of power, goods and so on within and between societies". The recent book *Justice and Ethics in Tourism* by Jamal (2019) similarly introduces distributive, procedural and socio-environmental aspects of justice associated with tourism, and calls for new ethical directions that seek what is just and good in tourism. This ethical thread has run consistently through discussions on tourism since the 1990s. Early on, Hultsman (1995) referred to "just tourism" in relation to ethical behavior among service providers and marketers: "if service providers view tourism, first as a business, and second as an experience and ethical practice, tourism ceases to be just by becoming an "industry" (Hultsman, 1995, p. 561). Goodwin (2011, p.16) similarly incorporates ethics when he presents *responsible tourism* as an approach that "recognizes the importance of cultural integrity, ethics, equity, solidarity, and mutual respect, placing quality of life at its core". Moreover, the UNWTO's (1994) concept of *sustainable tourism* introduces equity into tourism by developing, promoting and practicing tourism in a way that carries more social and environmentally friendly prospects for the ensemble of stakeholders involved, especially host communities and the environment. Here, whether mainstream or alternative, tourism should contribute to justice more effectively; hence, the emphasis shifts to *justice through tourism*.

For others, however, the relationship between justice and tourism is not so much related to being an approach, as a particular type of tourism. *Justice tourism,* as described by Scheyvens (2002), is a type of tourism which develops solidarity, mutual recognition and equity between guests and hosts, while also providing local communities with economic, social and cultural benefits, and supporting their self-determination. Higgins-Desbiolles (2008, 2009), Isaac and Hodge (2011), and Rami and Hodge (2011) take an even more restrictive view and conceptualize *justice tourism* as a particular type of tourism which involves visiting places faced with injustice and human rights violations, where visitors can advocate for justice and use activism to counter neoliberal globalization.

It can be argued that the need for a more activist stance is due to the global proliferation of neoliberal capitalism, which until now, has fed mass tourism with the commodification and mass consumerism needed to sustain capital market expansion (Mostafanezhad, 2013; 2016; Luh Sin et al., 2015, Gascon, 2019). Moreover, tourism's potential for emancipatory praxis has been depoliticized in the process. Therefore, more must be done to reclaim the potential of justice tourism amidst the proliferation of different new forms of 'alternative' tourism and the global machinery of international tourism. Justice tourism is thus an attempt to recover the potential for justice, through the key parameters of political responsibility, activism, and action.

However, the above introduction demonstrates that a clear, precise conceptualization of justice tourism is needed in order to position such an affirmative praxis. We argue in this paper that justice tourism is a political praxis characterized by political responsibility, solidarity, advocacy and social transformation. Accordingly, as our analyzes will show, the theoretical development of *justice tourism* must be situated in political responsibility, as conceived by scholars such as Arendt (1958) and Haiki (2018), and the posthumanist affirmative ethics put forward by Nietzsche (1968), Deleuze (1990) and Braidotti, 2006a). The purpose of this paper is thus to situate an affirmative, political potential of *justice tourism* in contradistinction to the potpourri of justice outcomes evinced by the various tourism scholars above (we refer to this latter form of tourism as *justice through tourism*).

In what follows, we first clarify the concept of justice tourism as it has emerged in tourism studies, noting how commodification and depoliticization has increasingly sabotaged the potential of tourism to facilitate political praxis. Then, we introduce four ethical approaches relevant for justice and tourism and elaborate on the three main parameters of justice tourism, namely, responsibility, solidarity and advocacy. Consequently, a conceptual framework for justice tourism and justice through tourism is developed and portrayed in Figures 2 and 3. As a result, we observe how posthumanist theory emerges as a promising avenue to theoretically ground justice tourism as a de-commoditized, re-politicized and affirmative praxis, and initiate a preliminary discussion on affirmative ethics and posthumanism in order to kickstart further theorizing on justice tourism ethics.

Justice tourism as it has emerged from tourism studies

The justice tourism described by Scheyvens (2002) develops solidarity, mutual recognition and equity among guests and hosts; it also provides local communities with economic, social and cultural benefits; it supports their self-determination; and gives visitors the opportunity to feel like participants in an emancipatory process rather than agents of domination. It also entails having meaningful encounters with dissimilar people, and learning about different paths to new ways of being. This approach to justice tourism encompasses most forms of alternative tourism such as responsible tourism, pro-poor tourism, volunteer tourism and many others. A more focused perspective is employed by other scholars such as Higgins-Desbiolles (2008, 2009) and Isaac and Hodge (2011), to whom justice tourism is an overtly political form of travel that involves visits to communities facing injustices and human rights violations, and actively advocating to redress harms and advance justice. For instance, in his study of water control in Palestine, Isaac (2017, p. 140) identifies justice tourism as a mechanism by which "Palestinian host communities can experience transformations towards a higher level of personal development and growth, changing life perspective, and higher levels of self-consciousness of their rights as a sovereign nation." But there is a further dimension to justice tourism, which Isaac (2017) describes thus:

> Justice tourism … as part of alternative tourism could (or should) provide a singular model of difference, in which it seeks not only to reform the inequities and damages of contemporary tourism but also to chart a footpath to a more just global order. (p. 140)

Isaac builds directly on the work of Higgins-Desbiolles (2008, 2009), for whom justice tourism should foster alternative and more just forms of globalization, and aim for radical transformation. Higgins-Desbiolles charts different forms of alternative tourism (responsible tourism, pro-poor tourism, fair-trade tourism, volunteer tourism, reality tours, and social activism tourism) along a continuum ranging from "decreasing depth of solidarity", at one end, to "increasing depth of solidarity" at the other (Higgins-Desbiolles, 2009, p. 338). She also proposes a transformation continuum that goes from maintaining the status quo at one end, to a "radical transformation of globalization" at the other (Higgins-Desbiolles, 2008, p. 359). Here, she identifies the underlying system at each stage in the transformation continuum as four processes in the global order: capitalist globalization, corporate social responsibility, fair trade, and humanistic globalization. But what justifies these processes of global ordering? And what ethical framework underpins them in order to guide and inform the transformative potential of justice tourism?

These two frameworks are a promising start, but require close scrutiny and theoretical work to situate and justify the concept of justice tourism within them. Presently, much is missing, for instance, advocacy and activism appear to merit further consideration as relevant characteristics. There is also a strong affirmative aspect that plays out through story-telling and dialogue. Residents can tell their stories of past, or current oppression, as Scheyvens (2002) notes, for example. Isaac and Hodge (2011) provide examples of how tourists can engage in cross-cultural and critical understanding, and can "become holders of the knowledge that can eventually lead to equality, democracy, and human rights for all" (Isaac, 2017, p. 143). Additionally, further work is needed to conceptualize solidarity, which Scheyvens (2002) clearly associates with justice tourism, and which drives Higgins-Desbiolles' (2009) framework. However, although the concept of solidarity has such an important role in the framework, it is only superficially introduced, and implicitly assumed to be unidimensional and continuous from one end to the other. Moreover, in Higgins-Desbiolles' transformation framework, different forms of justice tourism are also implicitly presented as being uniform and idealized, while the practical reality of some of these forms of tourism is often far from uniform, and their transformative potential may easily be subverted or eroded by the relentless intrusion of neocolonial and neoliberal values. Related processes of commodification and de-politicization (e.g., through neoliberal capital markets that emphasize profit and consumption rather than well-being) may restrict, or even neglect, solidarity and can ignore or distort any potential advocacy (Mostafanezhad, 2013; Luh Sin et al., 2015).

This raises an important question: within the frame of neoliberal globalization (and other processes reinforcing its global capitalist order), how can justice tourism be kept within the proper sphere of political responsibility, solidarity, advocacy and activism? And is political responsibility the predominant form of responsibility needed to enable transformation of entrenched structural injustices that are often deeply rooted in colonial histories and global capital markets? To start addressing this requires an in-depth exploration of the global processes of commodification and depoliticization associated with neoliberal globalization, and which influence various other systems and processes in global transformation.

The commodification of everyday life and travel

Neoliberalism fosters the subordination of state power to the requirements of the market; thus, the state outsources its responsibility for the wellbeing of the population to transnational corporations and individual publics who are expected to take care of themselves (Peters, 2006). Habermas (1989) had already warned about the administrative and economic colonization of our society, which increasingly constrict other spheres of our lives (the *lifeworld*), converting social relationships into commodities which are reified by the quantifiable exchange value of products (Slater & Tonkiss, 2001). Therefore, commodification changes the everyday life of consumption and production, and fosters a permanent search by corporate entities for new products, markets and sources of profit.

Tourism is one victim of this neoliberal governmentality that has rapidly diffused worldwide, aided by free trade agreements and globalization. We observe how cruise ships host flags of convenience registered in some 'other' country that allows them to escape environmental and labor regulations from any single source (Wood, 2000). International aid and development have not been immune to these pressures, where quasi-governmental organizations such as the World Bank and International Monetary Fund have led structural readjustment programs and funded capacity building projects with development budgets being increasingly diverted towards multinational corporations (Sharpley & Telfer, 2014).

Tourists participate in neoliberal governmentality by means of both improving their skills as entrepreneurial and competitive citizens, and acting as caring and morally responsible individuals, e.g., being 'altruistic', responsible tourists and environmentally 'green" ecotourists/consumers (Griffith, 2015). Volunteer tourism is often criticized as a product of these new moral economies (Mostafanezhad, 2013). It is seen as one more instance of a moralistic solidarity-seeking commodity culture which attempts to effect social justice through consumption (Bryant & Goodman, 2004). Goodman (2004) notes similarly how fair-trade tourism, another ethical niche within alternative tourism, falls into a similar process by directly linking producers and consumers (economically and morally) into supporting fair labor practices in the provision of goods and experiences for tourists. But how well do the good intentions of fair-trade tourists translate into political action for structural and societal change in order to safeguard precarious low-wage workers, and resist exploitation of women and child labor?

Three main social trends are evident in the processes of social transformation that comport with tourism's transformation into a neoliberal playground domestically and globally: (i) the loss of state control, with increasing privatization of development and corporatization of traditional public services such as transport, education, housing, health services, etc. (Rieff, 2002); (ii) prioritizing the consumer in order to serve the market, in such a way that tourism development becomes a utilitarian tool for policy makers so they can use destination resources, and local people for the general satisfaction of the visiting public (Gascon, 2019); and (iii) the development of moral economies that slightly corrects these trends by introducing "morality" into the practice and consumption of tourism (Sayer, 2003). As discussed below, one way or another, all three contribute to depoliticizing tourism, affecting the solidarity and advocacy potential of its stakeholders (including residents and tourists), and severely restricting the activism needed to prevent or redress injustices.

The depoliticization of tourism

As Straume and Humphrey (2010) put it, depoliticization is a contemporary tendency where the efforts of radical democracy to uphold social value over market exchange value are jeopardized by various factors such as commodification, which reifies economic value and accumulation of capital. Similar problems ensue in the commodification of relationships. For example, commodifying an ethnic group's cultural traditions and the rituals they practice instrumentally for display and profit, with little consideration for their well-being, relegates social value to a second plane, and makes what is social a tool for economic growth. The lack of global regulation mechanisms which can govern transnational capital markets make it even harder to effect political responses to political problems, reducing them to a technical and neutral process where disagreement and contestation are limited to disparate actions that can easily be denied or delegitimized (Cuttitta, 2018).

Destination boycotts, for example, require coordinated actions in order to be effective. Consider, for instance, how calls for a destination boycott of Burma/Miyanmar due to human rights abuses (including the tourism sector) was disregarded by the Lonely Planet guidebook, but strictly adhered to by other travel marketing and booking platforms such as Responsible Travel (Henderson, 2007). Despite the well-intended efforts of sustainable tourism and responsible tourism stakeholders, the process of *depoliticization* is increasingly prevalent due to the varied kinds of global drivers and influences, for example:

i. *Ideologies of neoliberalism and consumerism* are embedded in many of the attitudes that well-intended 'alternative', 'responsible' and 'sustainable tourists' bring to a destination, advancing the same neo-liberal practices and values they are supposed to combat (Armitage & Graham, 2001). Hollinshead and Suleman (2018) discuss the powerful mechanisms of worldmaking by which consumers are co-opted into a neocolonial, materialist worldview through travel and tourism, where little control can be exercised (within or across porous borders) by governments over destination marketers and numerous other stakeholders that influence marketing images, destination offerings and tourist experiences. Moreover, in the case of volunteer tourism, as noted by McGuire (2012), even if the design of the trip favors political discourse, the mere presence of volunteers can have adverse social and political consequences (see also Hockert, 2018).
ii. *Supremacism and lack of recognition of the different "other"* are common in postcolonial and settler colonial spaces. Koensler and Papa (2011) note that both justice tourism itineraries and social activism volunteering often clash with the goals of locals and introduce unintended negative outcomes including subjection of the other through stereotyping and ethnocentrism. These are commonly attributed to factors such as lack of interpretive opportunities to understand differing lifestyles of locals and visitors, lack of visitors' knowledge of the sociopolitical space, the dominance of white savior narratives, etc. (Kelly, 2016; Hollinshead, 1992). Similarly, McGuire (2012) observed how volunteer tourism often omits effective visitor education of the historical, political and economic context of the destination, or presents a uniform and ahistorical view of emerging countries which perpetuates neo-colonialism and dependency (Escobar, 1995). Lack of quality time for dialogue and

interaction is also a factor in stereotyping and ethnocentrism. Consider the fast-track tour itinerary with stops at various ethnic and indigenous sites just long enough for the tourist 'gaze' to rest upon the exotic other, purchase a few souvenirs and take a few photographs. Not surprisingly, Hollinshead (1992) warns of Western certitudes when interpreting history or present circumstances, and argues that the visitor's own life experience needs to be brought into the interpretive effort to counter idealized narratives.

iii. A *predominance of moral engagement and humanitarian compassion over political responsibility* is manifested in the cosmopolitan moral and romanticized type of individual consumption described above, and offers a cautionary warning to well-intended narratives of social and environmental justice (Levy & Sznaider, 2004). The shift towards alternative modes of consumption such as fair-trade products and by extension, fair-trade tourism, has given notions of 'responsible tourism' and 'sustainable tourism' a new social meaning, and even a new apparent image of re-politicization where consumption itself becomes a 'form' of activism (Goodman, 2004; Mowforth et al., 2008). Thus, alternative consumer products romanticize and de-politicize development by shifting the problem of structural injustice towards one of individual consumer morality. This is evidenced by Conran (2011) in the way the narratives visitors are exposed to during their exchanges with members of the host community may be dominated by affection, familiarity, friendship, mutual understanding, goodwill and humanitarian compassion, while disguising injustices embedded institutionally or historically in colonialism, for instance, all which contribute to depoliticize the tourist experience.

It could be argued then that volunteer tourism reflects a type of modern humanitarianism from the rich in the global North towards the poor in the global South (Tester, 2010), a secular equivalent of traditional religious mission travel (Brown & Morrison, 2003) that keep expanding moralistic consumerism globally. Children perceived to be under the duress of hard labor in a destination may similarly arouse compassion, rather than the political action needed to tackle complex conditions often widespread in networks enabling such social injustices (Mathers, 2010). In the absence of political activism as a driving force for change (praxis), such humanitarian forms of travel aimed at 'helping' the destination through pro-poor tourism, volunteer tourism, etc., may be seen as a utilitarian and moralistic undertaking, and lacking the political responsibility needed to address the wider structural inequalities.

Recovering the potential and promise of tourism in order to tackle structurally engrained injustices thus requires (i) a re-politicization and reclaiming of political responsibility in a way that transcends the self-centered moralistic individual and the utilitarian calculus (as in government policies and incentives oriented to benefit the greatest number of residents and tourists), and (ii) making solidarity, advocacy and activism genuinely constitutive of political action and praxis. What is needed, among other things, is a turn away from dominant market-driven narratives, and the modernist, self-centered subject, towards a non-utilitarian and political way of being and relating. A post-humanistic affirmative ethic, we will argue later, offers a co-constitutive, relational approach to political responsibility, solidarity, advocacy and action (praxis) for justice tourism.

Four ethical approaches to tourism and justice

Before introducing the three main parameters of justice tourism, namely responsibility, solidarity and advocacy, and before moving forward to elucidate justice tourism and related forms of tourism, we need first to introduce four ethical approaches under which tourism praxis may operate, and the corresponding four ethical regimes.

Figure 1, displays utilitarian, deontological, care, and affirmative types of ethics, and their concomitant ethical regimes of neoliberalism, social liberalism, humanitarianism and posthumanism. Utilitarian and deontological ethics are both representative of classical humanism, where utility and

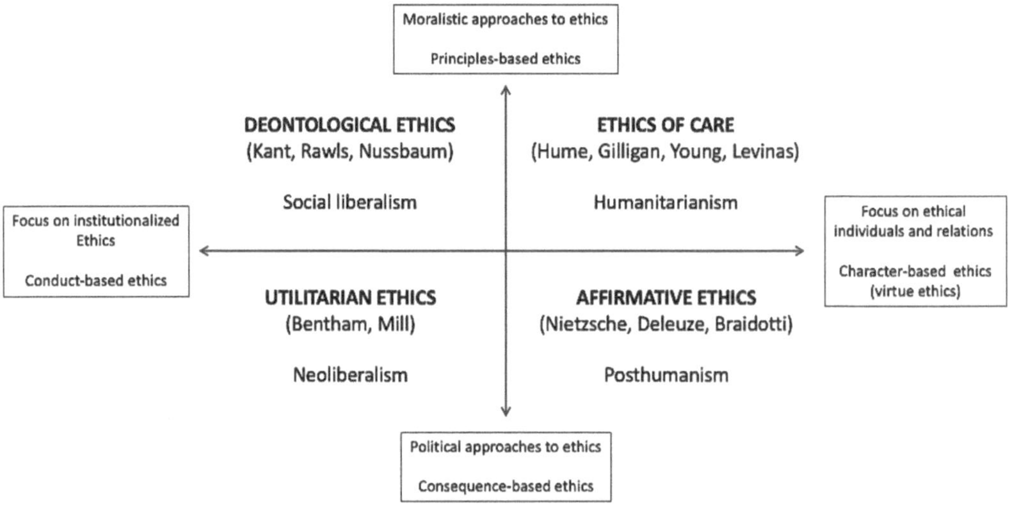

Figure 1. Ethical approaches and ethical regimes.

moral universals are framed from the centrality of humans, and for the utility and moral good of humanity. Therefore, these approaches are male-dominated, colonial and anthropocentric, leaving 'minorities' or the subaltern human and non-human in a position of inequality. In contrast, the ethics of care and affirmative ethics are approaches to ethics, which draw from feminist, postcolonial and posthumanist thought in order to try to revert this situation by putting relations and encounters with difference centerstage, rather than the dominating 'colonial' 'male' human.

Utilitarian ethics and neoliberalism

Utilitarian ethics is the dominant ethics in neo-liberal regimes and stems from the moral philosophies of Jeremy Bentham and John Stuart Mill (Quinton, 1974). Utilitarianism takes utility or the greatest happiness of the greatest number as the foundation of ethical behavior. In a business context maximizing happiness equates to maximizing profit, and even if consideration for the cost of a greater pain to other people is still considered, most of the time the only concern is private interest. Cost benefit analysis becomes the measuring tool of utilitarianism, which is a calculating approach to ethics that fosters commodification. As a result, utilitarianism is unconcerned with equity because maximizing the sum of utility is indifferent to the distribution of that sum, or because a little harm to very few individuals is acceptable if it results in a sufficiently large benefit for a sufficiently large number of individuals. All in all, utilitarianism is a type of institutional, conduct-based ethics, because it prescribes on the ethical behavior at institutional, rather than individual level; and it is political because this behavior is situational and based on the consequences or outcomes of the actions.

Deontological ethics and social liberalism

Deontological or duty ethics is the dominant ethical approach of social liberal regimes. It was mainly developed in the moral philosophies of Kant and Rawls, and opened the door to the creation of modern welfare states, liberal conceptions of social justice, and Sen's Capability approach (Dean, 2010). In deontological ethics, actions must abide to universal principles irrespective of their consequences. Actions are only morally right when carried out as a duty, rather than because a reward is expected. Moreover, the right thing to do is to act according to a moral duty, which is typically encapsulated in rules, as is the case for human rights. Therefore, like utilitarian ethics, it is conduct-based, but now this conduct responds to institutionalized moral principles, and not to outcomes. Unlike utilitarianism, social equity is now a moral percept, on the basis of which distributive justice is prescribed (Rawls, 1971).

Ethics of care and humanitarianism

Ethics of care is the dominant ethics of humanitarian and affective ethical regimes. Hume, Gilligan, Young and Levinas are representative philosophers of this approach. Like deontological ethics, the ethics of care holds on principles, particularly the moral principles of care and benevolence; not as institutionalized conduct, but rather as a moral action based on character or virtue and centered on interpersonal relations (Beasley & Bacchi, 2005). It brings the affective bases of morality to the forefront, expanding 'maternal' practices to those more distant from us, thus countering "the covert masculinist bias of mainstream notions of 'justice', given that these notions were typically legitimated by reference to universal/general/abstract and rational/disinterested moral thinking" (Beasley & Bacchi, 2005, p. 50). This is a similar ethics to Levinas' (1999) ethical interconnection with others, which focuses on the face-to-face encounter and on the relational responsibility, responsiveness and compassion, that the vulnerability of others awakens in us.

Affirmative ethics and posthumanism

Affirmative ethics is the ethics of the posthumanist ethical regime. Nietzsche, Deleuze and Braidotti are the main representative philosophers of this approach. Like the ethics of care, affirmative ethics is character-based, or virtue ethics, and centered on encounters or relations; however, it is not moralistic, nor is it based on principles. Instead, it is consequentialist and radically political. Posthumanist thinking introduces a new ethical framework for a posthumanist concept of justice, fitting for the Anthropocene. Ethical encounters with difference are mechanisms for being and becoming, and for acting with *political responsibility* in relation to the vulnerable and disempowered, whether these are human or non-human beings (a concept integral to a posthumanist ethics). This is a type of ethics which involves meeting others with response-ability and without domination; an ethics of relational virtuosity, which is neither normative, nor moralistic, but an *affirmative ethics* which wills that which occurs, not with resignation, but with affirmation (Braidotti, 2019). It is an intense and active affirmation of the self that exists in intricate relations with the other (Braidotti, 2006b). Unlike the ethics of care, such affirmative ethics enables engagement with transformative politics in order to confront the conceptual and social contradictions of our times. According to Braidotti (2006b, p. 16):

> … [A]ffirmative ethics allows us to contain the risks while pursuing the original project of transformation. This is a way to resist the dominant ethos of our conservative times that idolizes the new as a consumerist trend while thundering against those who believe in change. Cultivating the ethics of living intensely in the pursuit of change is a political act.

Three parameters of justice tourism: Responsibility, solidarity and advocacy

The concept of responsibility

To better understand the concept of responsibility and its relationship with justice tourism, three main types or responsibility are distinguished: social, relational and political responsibility. *Social responsibility* entails individuals acting in their own interest, while at the same time, having the moral obligation to act to the benefit of society, social fairness and equity (Spence & & Rutherfoord, 2001). Therefore, social responsibility resonates with an ethics of duty, i.e., a Kantian or Rawlsian ethics, and its main concern is "*acts* of wrongdoing – did she commit it or not? Did she attempt to prevent or avoid it, and if yes, with what attitude and under what circumstances?" (Zengh, 2019, p.110). In contrast, *relational responsibility* refers to the responsibility felt by individuals as humans to support and take care of those in need in order to relieve their distress and avoid contributing to their suffering. It thus coincides with the type of responsibility of care ethics. It reaches beyond norms of duty, to become a moral relationality that always puts the Other before self-interest. It is therefore an unlimited and infinite relational responsibility to the singular Other of the face-to-face relationship (Levinas, 1999).

Goodhart (2017), contends that the responsibility for systemic injustice is not some moral duty to be discharged, but rather a *political responsibility* to be taken up by those who seek to establish new conventions with others for a more just system. The concern is now on the political relationships between an individual and others: "did she publicly resist, and did she attempt to change the social circumstances that generate such wrongs?" (Zengh, 2019, p.110). The difference between the above moral types of social and relational responsibility and political responsibility lies in whether the purpose is to preserve own's subjectivity or to bring about change in the world (Arendt, 1987). Through political responsibility, institutionally and historically engrained social structural processes that have unjust consequences can be transformed (Young, 2006). This is precisely the type of responsibility characteristic of posthumanist affirmative ethics, a "political responsibility for the vulnerabilities, injustices, and hazards that our assembled life of dual being in and with the [*sic*] nature entails" (Haiki, 2018, p.170).

The concept of solidarity

Solidarity can be understood as "an obligation to act in support of particular others at a distance (or to stand ready to aid these others), with whom one shares a commitment to the achievement of justice" (Gould, 2009, p. 209). It implies a sort of social empathy and awareness of the extraordinary and difficult situation of other people, and a commitment and readiness to act in support of the vulnerable, the poor, the oppressed, or victims of violence in a way they would consider to be beneficial (Stjerno, 2005). Three types of solidarity are distinguished that are relevant for justice tourism: social, affective, and political solidarity.

Social solidarity refers to the social cohesion of a particular community; a type of natural solidarity between members of the same kinship or culture or other characteristics that members share. The shared characteristics produce social bonds among solidary members, together with positive collective moral obligations. It thus relates with the moral obligations of social responsibility outlined further above, and thus is expected to be the solidarity in deontological ethical approaches (Scholz, 2008).

Affective solidarity "is proposed as a way of focusing on modes of engagement that start from the affective dissonance that feminist politics necessarily begins from" (Hemings, 2012, p.148). It is not rooted on identity or group characteristics, but rather on mutual feelings of care and concern. It transforms the risk of disagreement into the expectation of mutual attendance to each other. It prioritizes an awareness of mutuality over one's own interpretation of difference. It is therefore directly related to the relational responsibility and the ethics of care that are characteristic of the humanitarian ethical regime.

Political solidarity, on the other hand, goes beyond individual obligations within social or affective relations. It unites people who have made a conscious commitment to challenge an unjust situation with the aim of abolishing the sources of injustice and bringing about social change. It therefore engenders political advocacy and activism. However, as Ferguson (2009) argues in order to demonstrate political solidarity, advantaged solidary members can no longer simply identify their interest with those constructed under dominant white, supremacist capitalist patriarchal systems" (p. 196), they have to necessarily transform their own identities. This requires a commitment to continually seek a diversity of perspectives in an attempt to mutually understand the unjust situation, and demands a capacity to practice radical openness, attentiveness, empathy, and love (Scholz, 2008).

These are arguments about political solidarity that resonate with the principles of posthumanist philosophy, and the Deleuzian idea that an ethical encounter "is not at the expense of difference, or even despite difference, but is absolutely the result of a dynamic drive toward difference" (Parr, 2008, p. 122). A posthumanist theory of solidarity assumes that both posthuman societies and individual subjects are constituted by, and committed to both difference and connection (Llivis, 2017, p. 42):

> Rather than erasing difference, the posthuman subject and society speak to a solidarity that thrives on differences being put in conversation with one another. … [Moreover] posthuman solidarity develops when subjects within or across societies embrace individually and work together to overcome systems that seek to hierarchize differences.

The concept of advocacy

Advocacy is described by Piccinini (2007, p. 16) as a "mix of persuasive communication and targeted actions aiming at 'pleading the cause of', 'acting on behalf of' and 'speaking out for or in support of others' … designed to change policies, positions and actions on a specific issue or cause on behalf of the voiceless". Advocacy is not only about influencing public policy, but more importantly, public opinion. Through advocacy, people can make "demands, claims, or rights of the less powerful win out over the purported interests of the more powerful" (Keck & Sikking, 1997, p.217). This is done by presenting issues in new ways, confounding expectations, and mobilizing social networks.

Three types of advocacy can be distinguished for the purpose of justice tourism: cultural, humanitarian, and political. These coincide approximately with the three characteristic elements of advocacy advanced by Barnes (1998): giving voice to vulnerable people, protecting them, and promoting their rights.

Cultural advocacy implies commitment to social inclusion and equity for a particular cultural community (Diamond, 2011). For instance, cultural tourism walking tours that rewrite postcolonial history from a subaltern perspective become instruments for the cultural advocacy of indigenous heritage. Enhancing visitors' experiential understanding of history and culture augments their recognition and respect for indigenous cultural identities, leading to the possibility of visitors becoming international cultural advocates. However, while cultural tourism can give some cultural recognition to marginalized minority groups, it has to be done critically and carefully to avoid the suppression of manifestations of injustice and oppression that would render it apolitical (Jafa, 2012).

The primary aim of *humanitarian advocacy* is to steer action "on the part of those with the power to assist, redress or enforce those in dire need" (Cotterrell, 2005, p. 113). It is grounded on the moral responsibility of humans to assist those in need and alleviate suffering, but without challenging the causes of the suffering (Gideon, 1998). Seen from this perspective, it is a form of moral advocacy which can be both secular and religious. Therefore, humanitarian advocacy uses "the language of morals and ethics rather than politics" (Chandler, 2001, p. 683), and it is a discourse of charity rather than entitlement (Ticktin, 2006).

Political advocacy is aimed at trying to influence policy-making through lobbying activities, to raise public awareness of a given cause, or to engage in street protests (Giugni & Grasso, 2018). These are channeled through a political discourse using language to challenge the causes of the oppression and injustice. However, seeking to engage governments and corporations through lobbying may prove to have limited effectiveness, as they are not the source of change so much as a reflection of it (Shellenger & Nordhaus, 2009). Similarly, raising public awareness of a cause and trying to change people's behavior has proven to have limited results (Crompton & Kasser, 2009). A more radical form of political advocacy calls for systemic change through activities that directly challenge those identities of audiences which prevent ambitious change (Gillan, 2008). Moreover, to be effective they must address underlying assumptions and wider belief structures (Groves, 2010).

Therefore, contrary to cultural and humanitarian advocacy, which are respectively related to deontological and care ethical regimes, this rather philosophical form of political advocacy conforms with posthumanism and is inspired by a posthumanist pedagogy of difference as emancipatory practice, while grounded in affirmative ethics. Literacy must be rewritten to spotlight justice and must create opportunities for people to understand otherness on their own terms (Giroux, 1992).

An elucidation of justice tourism and related forms of tourism

The conceptual and theoretical insights outlined above lead us to a much clearer elucidation of justice tourism. Figure 2 portrays the three most relevant parameters of justice tourism previously discussed: responsibility, solidarity and advocacy, and the underlying ethics through which they can be manifested: utilitarian, duty, care, and affirmative. Each row of the matrix displays the type of responsibility, solidarity and advocacy corresponding to each ethical approach; and on the extreme right side of the figure we can see the tourism types or categories of tourism forms that correspond to each of these dominant ethical approaches: unsustainable, sustainable, humanitarian (or moralistic) and justice tourism.

		JUSTICE PARAMETERS			JUSTICE-RELATED TYPES OF TOURISM
		RESPONSIBILITY	SOLIDARITY	ADVOCACY	
DOMINANT ETHICS	UTILITARIANISM *Neo-liberalism*	*Utilitarian responsibility*	*Utilitarian Solidarity*	*Utilitarian Advocacy*	UNSUSTAINABLE TOURISM *UNJUST TOURISM*
	DUTY ETHICS Social Liberalism	*Social Responsibility*	*Social Solidarity*	*Cultural Advocacy*	SUSTAINABLE TOURISM *JUSTICE THROUGH TOURISM*
	ETHICS OF CARE *Humanitarianism*	*Relational Responsibility*	*Affective Solidarity*	*Humanitarian Advocacy*	HUMANITARIAN TOURISM *JUSTICE THROUGH TOURISM*
	AFFIRMATIVE ETHICS Posthumanism	*Political Responsibility*	*Political Solidarity*	*Political Advocacy*	JUSTICE TOURISM

Figure 2. Dominant ethics and justice parameters informing (justice) tourism.

This framework resolves the limitations we identified in that of Higgins-Desbiolles (2008, 2009), and clarifies some of the conceptual confusion prevailing in the current literature surrounding tourism and justice. For example, broad tourism categories such as responsible tourism, solidarity tourism, advocacy tourism, sustainable tourism, humanitarian tourism and justice tourism are often united under the term 'alternative tourism', and seen to overlap to varying degrees; however, a clear or precise differentiation was ultimately lacking. Based on various parameters, Figures 1 and 2 offer a conceptualization that helps distinguish and situate these various types and the many forms of tourism in relation to justice tourism.

The central columns of Figure 2 show the types of responsibility, solidarity and advocacy corresponding to each ethical regime. Similarly, the column on the far right shows unsustainable, sustainable, humanitarian (or moralistic) and justice types of tourism, which directly correlate to particular types of dominant ethics.

Looking at the categories in the column on the extreme right, from top to bottom, we see first how utilitarianism lies behind all types of unsustainable tourism. The pursuit of self-interest and utility implicit in the neo-liberal conception of responsibility; the lack of solidarity manifested in its utilitarian, careless, exploitative and extractive practices; and the 'advocacy' practices characterized by utilitarian, indifferent, neo-colonial, or even antagonistic interpretations of the Other, are all well-known attributes of the most unsustainable forms of tourism. Capitalist mass tourism and other forms of detached, voyeuristic and passive sightseeing activities epitomize the utilitarianism and self-interest of neo-liberal consumption. Sex tourism, organ-transplant medical tourism, profanation of sacred rituals, poverty tourism, and other extreme types of exploitative and extractive niche tourism contribute to the destruction of land and banalization of cultures (Higgins-Desbiolles, Whyte & Tedmanson, 2013), are examples of the most unsustainable forms of tourism and illustrate the active, unethical practices these travelers adopt. To these practices, the conceptual model also adds the communicative actions of utilitarian travelers who show indifference, disdain or disseminate neo-colonial interpretations of the host communities.

The ethical framework of social liberalism introduces a deontological ethics that emphasizes duty and distributive justice. It makes a place for the forms of tourism classified within the category of sustainable tourism, which are grounded in a social conception of responsibility. This dominant social responsibility still gives pre-eminence to the interests of the traveler, but much care is taken to contribute to the host communities' well-being. Enlightened mass tourism and responsible forms of niche tourism such as slow tourism are examples of these more passive forms of sustainable and responsible tourism. Solidarity outcomes are also social, and therefore involve contributions which relate to different forms of cultural membership or identification, and which reinforce the preservation of the culture. VFR (Visiting Friends and Relatives), diaspora tourism, pilgrimage and religious tourism, and some instances of work and study immersion programmes are forms of tourism exemplifying both sustainable and solidarity tourism. Sociocultural advocacy activities entail social interaction between visitors and the members of their social networks. It is through these networks that they distribute socio-cultural knowledge of the places and communities visited and advocate for them. These communicative actions of cultural advocacy undertaken during and after the visits illustrate forms of both sustainable tourism and advocacy tourism.

Moralistic types of tourism are dominated by the ethics of care which is characteristic of humanitarianism. These tourism types obey moral conceptions of relational responsibility, where the self-interest of the visitor is now subordinated to the moral norms of care and benevolence. Pro-poor tourism and fair-trade tourism are representative forms of both responsible and humanitarian tourism. Solidarity is affective and its outcomes involve providing the vulnerable with moral and humanitarian support. Religious mission travel, celebrity humanitarianism, most types of volunteer tourism, and some academic fieldwork involve actions of affective solidarity, and can therefore be considered forms of both solidarity and humanitarian tourism. Advocacy is also a humanitarian act found in forms of tourism where the information exchanged and shared with others regarding the visitor's experience has the purpose of raising compassionate responses from audiences so that they may eventually contribute to assuaging the painful and unjust situation of the host communities. Therefore, the active contributions that these visitors may make through advocating for help to relieve the suffering of the hosts are illustrative forms of both advocacy and humanitarian tourism.

Finally, at the bottom of Figure 2, we have the case of affirmative, posthumanist ethics. Here, forms of justice tourism hinge on political forms of responsibility, where tourism is practiced affirmatively with the purpose of contributing to global transformations and eliminating systemic sources of oppression, injustice and inequality. Reality tours for global transformation are illustrative cases of facilitating political responsibility towards justice tourism. Solidarity here is also political and entails activities aimed at understanding the sources of oppression and acting in ways that contribute to transforming those sources for others in order to bring justice back. Several active forms of solidarity tourism and justice tourism are epitomized by the political activities of transnational solidary activists, as well as some cases of academic fieldwork and volunteering, when these involve political commitment and action. Advocacy is political, too, and represented by forms of tourism that engage affirmative ethics in advocacy activities; inspire listeners positively to adopt more just values, behaviors and new identities; affect policy making, or business practices; and encourage visitors to join and participate in transnational activist networks that support alternative forms of global transformation. In sum, the active contributions that these visitors make through political responsibility, political solidarity and political advocacy exemplify justice tourism.

Figure 3 offers further elucidation to situate the particular forms of tourism mentioned above, within the conceptual framework in Figure 2

		JUSTICE PARAMETERS			JUSTICE-RELATED TYPES OF TOURISM
		RESPONSIBILITY	SOLIDARITY	ADVOCACY	
ETHICAL VALUES OF PARAMETERS	UTILITARIAN	Capitalist mass tourism and other forms of careless, detached, voyeuristic sightseeing	Sex tourism, organ-transplant medical tourism, profanation of sacred rituals, poverty tourism, and other exploitative and extractive niche tourism that contribute to the destruction of land and banalization of cultures	Communicative or advocacy actions which show indifference, disdain or neo-colonial interpretations	UNSUSTAINABLE TOURISM *JUSTICE THROUGH TOURISM*
	SOCIAL	Enlightened mass tourism and responsible forms of niche tourism such as ethical slow tourism are illustrative of more detached forms of sustainable tourism	Active forms of social solidarity e.g., VFR, diaspora tourism, religious tourism, and instances of work and study immersion programmes	Communication actions of cultural advocacy	SUSTAINABLE TOURISM *JUSTICE THROUGH TOURISM*
	MORAL	Pro-poor tourism Fair-trade	Religious mission travel, celebrity humanitarianism and most forms of volunteer tourism and academic fieldwork	Actions advocating for help to relieve the suffering of the host communities	HUMANITARIAN TOURISM *JUSTICE THROUGH TOURISM*
	POLITICAL	Reality tours for ethical global transformation	Political solidarity of transnational solidary activism, and some cases of academic fieldwork and volunteering when they involve political action	Advocacy actions conducted by justice tourists with the purpose of recruiting adepts and to contributing to ethical forms of global transformation	JUSTICE TOURISM

Figure 3. Tourism forms by justice parameters and types of tourism involved.

The conceptual framework in Figure 2, and the categorization of forms of tourism in Figure 3, provide new theoretical grounding to clarify the many complexities underlying existing categorizations of tourism forms and types, and can therefore be used as a springboard for future research on justice tourism and justice through tourism. For instance, it is now clear that not all solidarity or advocacy in tourism are instances of justice tourism; there is no full correspondence between responsible tourism and sustainable tourism, as moral and political forms of responsibility do not correspond to the ethical framework of sustainable tourism; only responsibility, without solidarity or advocacy, can still bring justice outcomes although in a less direct fashion (because travelers will still be implicitly contributing to a particular type of ethical regime through their dominant ethical behavior), while solidarity and advocacy actions will make a more direct contribution to the justice outcomes of the host communities. The framework also distinguishes between justice tourism and justice through tourism, in that although sustainable tourism and humanitarian tourism differ from justice tourism, they still contribute to justice.

Discussion and implications for practice and research

A general definition of justice tourism emerges by knitting together the various nuances deconstructed above. Justice tourism can be redefined as *an explicitly political form of travel which involves visiting destinations facing some type of injustice or human rights abuse in order to* (i) *develop and promote solidarity with communities,* (ii) *become advocates of justice for these communities, and engage in activism to* (iii) *foster alternative and more just types of globalization.* This definition refers to a type of tourism that, in the current era of globalizing capitalism, engages its participants (including tour operators, residents and tourists) to take political responsibility to redress destination injustices and engage in political solidarity and political advocacy to enable praxis.

The promise of posthumanism

In the presentation of our four ethical approaches, we have seen how neoliberalism and its utilitarian ethics reify commodification, and how the consequent socialization and moralization of the neoliberal regime brought by social liberalism and humanitarianism have not delivered their promises and instead have contributed, against its initial purpose, to expand commodification to the social and moral realms. This has been so because of the inherent depoliticization of moralistic and principle-based ethical approaches, and explains the recent expansion of both commodification and depoliticization in tourism. In the case of justice tourism, this is evidenced with its increased commodification and depoliticization, which restricts or even neglects solidarity, ignores or distorts advocacy, and at best, tames these processes by introducing morality.

If justice tourism wants to move beyond the dominant neo-liberal and liberal logics of most forms of tourism (Butcher, 2002), and beyond the current moralistic and compassionate logics (Caton, 2012; Jamal & Camargo, 2014) which silently conceal underlying moralistic forms of consumerism (Mostafanezhad, 2016), it must be well aware of the underlying dominant ethics with which it is practiced, and of the complexities of radical transformation.

We have also observed how the affirmative ethics of posthumanism is the only ethical approach that is both individual and relational (as against institutional) and political (as against moralistic), and therefore is the only one that escapes utilitarianism and moralism, and embraces both inclusiveness (as against utilitarianism) and political action, i.e., political responsibility (as against moralism). Therefore, posthumanism emerges as the only ethical regime that can battle universal commodification (as against institutionalized utilitarian quantification) and bring back politicization (as again apolitical moralism). This is also the reason why posthumanism envisions subjectivities or identities that are alternative to humanism, becoming the only ethical regime that can bring radical transformation and a truly alternative form of globalization.

Therefore, for tourism to effectively contribute to justice and the expansion of justice tourism, travelers, hosts and operators alike need to become politically literate (Gale, 1994), i.e., learn to become political and act politically; learn to become-other; desire ethical encounters with difference; and embrace affirmative ethics. Political literacy and political responsibility, solidarity and advocacy prevent certain aspects of depoliticization: non-recognition, supremacism and lack of social commitment. In turn, they correct the commodification forces manifested privatizing development and privileging the consumer, thus contesting neo-liberalizing tendencies and consumerism, as well as battling the predominance of moral engagement and compassion over political responsibility and action. It is through affirmative ethics and posthumanist pedagogies of difference (Bayley, 2018) that political responsibility, solidarity and advocacy can be constructively and positively engaged to enable just tourism.

We must therefore take the promise of posthumanism seriously to further advance our understanding and research of justice tourism and its effective deployment in practice. Posthumanism, as a philosophical and reflective approach, denies human exceptionalism, advocates non-anthropocentrism, de-centers the individual subject and abolishes dualisms (Braidotti, 2013). It poses challenges to tourism because it undermines the often unquestioned ontological and ethical assumptions on which modern tourism is based as an exploitative, extractive humanist enterprise, and encourages the abolition of humanist binary divisions like host 'versus' guest, offering non-moralistic ways of engaging with tourism forms such as medical tourism, volunteer tourism or justice tourism (Cohen, 2019).

Taking posthumanism seriously means actively resisting the co-optation of tourism by the market, that is, learning to contest neoliberalism with others from within a generative, transformative mode of being (Jenkins, 2005). It paves the way for re-introducing political responsibility, solidarity and advocacy as positive world-making practices with which to subvert the current commodification and de-politicization of all forms of tourism. Therefore, posthumanist theory provides a new theoretical lens with which re-conceptualize justice tourism, introduce conceptual clarification, and open a pathway for future theoretical research.

Implications for practice and research

While the paper has sought to conceptualize justice tourism and probe into the promise of posthumanism, our findings are also useful for practitioners, managers, tourists and host communities to better discern the nuances of contributions (or lack of) offered by different types of tourism to justice.

Figure 3 provides a framework to micro-segment justice-related forms of tourism and tourism behavior; a guide for travelers to be able to self-assess their travel behavior and learn how to meaningfully contribute to justice; and a reference for host communities to help them decide what forms of tourism and types of travelers they wish to welcome (or not).

Political responsibility and solidarity can be fostered by introducing and emphasizing provocative and well-crafted narratives of shame addressed to ethically empathetic visitors. They must highlight the visitors' privileged situation (e.g., their freedom of movement and ability to advocate on behalf of the oppressed) and serve as "ornamental reminders of complicity" (Kelly, 2016, p. 741). Stirring up shame in solidarity visitors is thus particularly important if practitioners are to counteract the dominant, 'colonial' narratives of powerful oppressors that usually frame their worldview. It is not, however, the shame that negates and operates as a mechanism for social exclusion, but a shame that is transformative and that creates pedagogical possibilities (Zembylas, 2019).

Moreover, in order to encourage political advocacy, tour organizers should extend their services beyond the trip, and offer advocacy platforms and opportunities for volunteer tourists to stay engaged with the politics of injustice and active advocacy in order to educate others about this type of solidarity activism once back home.

Justice tourists must be aware of Mahrouse's (2008) findings regarding advocacy work. In her research, the accounts given by justice tourists to their fellow nationals, of injustices found and experienced in the destination after the trip, mainly drew compassion for the activists and pity for those suffering in the destination. Therefore, using sentimentalism and morality may be much less effective as a mechanism to tackle justice through humanitarian advocacy than political advocacy. The representational practices of justice tourists' advocacy actions must be affirmative and examined in terms of the audience, and of how the tourists represent themselves as the objects of empathy. Posthumanist pedagogies of difference (Bayley, 2018) must be adopted for this purpose.

In what concerns future academic research, and in light of our conceptual work here, several avenues can now be taken. Firstly, elucidating the concept of justice tourism can broaden research on various topics within areas of volunteer tourism and other justice-related types and forms of tourism. Secondly, the dilemma of whether approaches to tourism should be dominated by neo-liberal principles, social liberalism or moralistic doctrines can be finally overcome by transcending these, and further exploring the affirmative and posthumanist foundation of justice tourism. Therefore, if we want to deploy posthumanism's promising potential for advancing justice tourism and justice in tourism, we must go beyond our brief preliminary discussion on affirmative ethics and posthumanism and encourage for future research on these.

Finally, this proposed posthumanist turn also opens new avenues to further explore the implications and prospects of tourism in our current struggle to achieve the Sustainable Development Goals, and thus contribute to a more just world in the age of Anthropocene. At a time when global warming and the COVID-19 pandemic are exerting immense stress on planetary systems (including the global tourism system), and the global society as a whole is in jeopardy, new nomadic subjectivities capable of embracing affirmative forms of ethics and taking on political responsibility towards radical difference are needed more than ever if we are to secure a new and more just global order, and avoid further devastation and destruction of the Earth.

Disclosure statement

No potential conflict of interest was reported by the author(s).

References

Arendt, H. (1958). *The human condition*. University of Chicago Press.
Arendt, H. (1987). Collective responsibility. In J.W. Bernauer (Ed.), *Amor mundi* (pp. 43–50). Springer.
Armitage, J., & Graham, P. (2001). Dromoeconomics: towards a political economy of speed. *Parallax, 7*(1), 111–123. https://doi.org/10.1080/13534640010015999
Barnes, C. (1998). *Disability, disabled people, advocacy and counseling*. Jessica Kingley Pub.
Bayley, A. (2018). Posthumanism, decoloniality and reimagining pedagogy. *Parallax, 24*(3), 243–253. https://doi.org/10.1080/13534645.2018.1496576
Beasley, C., & Bacchi, C. (2005). The political limits of 'Care' in re-imagining interconnection/community and an ethical future. *Australian Feminist Studies, 20*(46), 49–64. https://doi.org/10.1080/0816464042000334537
Braidotti, R. (2006a). *Transpositions: On nomadic ethics*. Cambrige University Press.
Braidotti, R. (2006b). Affirmation versus Vulnerability: On contemporary ethical debates. *Symposium: Canadian Journal of Continental Philosophy, 10*(1), 235–254.
Braidotti, R. (2013). Posthuman relational subjectivity and the politics of affirmation. In P. Rawes (Ed.), *Relational architectural ecologies: Architecture, nature and subjectivity* (pp. 21–39). Routledge.
Braidotti, R. (2019). Affirmative ethics and generative life. *Deleuze and Guattari Studies, 13*(4), 463–481. https://doi.org/10.3366/dlgs.2019.0373
Brown, S., & Morrison, A. M. (2003). Expanding volunteer vacation participation: An exploratory study on the mini-mission concept. *Tourism Recreation Research, 28*(3), 73–82. https://doi.org/10.1080/02508281.2003.11081419
Bryant, R. L., & Goodman, M. K. (2004). Consuming narratives: the political ecology of alternative consumption. *Transactions of the Institute of British Geographers, 29*(3), 344–366. https://doi.org/10.1111/j.0020-2754.2004.00333.x
Butcher, J. (2002). *The moralisation of tourism: Sun, sand … and saving the World?*. Routledge.
Caton, K. (2012). Taking the moral turn in tourism studies. *Annals of Tourism Research, 39*(4), 1906–1928. https://doi.org/10.1016/j.annals.2012.05.021
Chandler, D. (2001). The road to military humanitarianism: how the human rights NGOs shaped a new humanitarian agenda. *Human Rights Quarterly, 23*(3), 678–700. https://doi.org/10.1353/hrq.2001.0031
Cohen, E. (2019). Posthumanism and tourism. *Tourism Review, 74*(3), 416–427. https://doi.org/10.1108/TR-06-2018-0089
Conran, M. (2011). They really love me! Intimacy in volunteer tourism. *Annals of Tourism Research, 38*(4), 1454–1473. https://doi.org/10.1016/j.annals.2011.03.014
Cotterrell, L. (2005). *Human Rights and Poverty Reduction; approaches to human rights in humanitarian crises*. ODI Working and DiscussionPapers. Available at https://www.odi.org/publications/3323-approaches-human-rights-humanitarian-crises (Accessed: 1 May 2020)
Crompton, T., & Kasser, T. (2009). *Simple and Painless? The limitations of spillover in environmental campaigning*. WWF-UK.

Cuttitta, P. (2018). Repoliticization through search and rescue? Humanitarian NGOs and migration management in the Central Mediterranean. *Geopolitics, 23*(3), 632–660. https://doi.org/10.1080/14650045.2017.1344834
Dean, H. (2010). *Understanding human need.* Policy Press.
Deleuze, G. (1990). *Expressionism in Philosophy: Spinoza.* Zone Books.
Diamond, H. A. (2011). *American Aloha: Cultural tourism and the negotiation of tradition.* University of Hawai Press.
Escobar, A. (1995). *Encountering development: The making and unmaking of the Third World.* Princenton University Press.
Fennell, D. A. (2006). *Tourism ethics.* Channel View Publications.
Ferguson, A. (2009). Iris Young, global responsibility, and solidarity. In A. Ferguson (Ed.), *Dancing with Iris: The philosophy of Iris Marion young* (pp. 199–212). Oxford University Press.
Gale, F. G. (1994). *Political literacy: Rhetoric, ideology, and the possibility of justice.* State University New York Press.
Gascon, J. (2019). History of NGDO volunteer tourism in Spain: Depoliticisation, commodification, and downturn. *Journal of Tourism History, 11*(3), 284–302.
Gideon, J. (1998). The politics of social service provision through NGOs: A study of Latin America. *Bulletin of Latin American Research, 17*(3), 303–321. https://doi.org/10.1111/j.1470-9856.1998.tb00127.x
Gillan, K. (2008). Understanding meaning in movements: A hermeneutic approach to frames and ideologies. *Social Movement Studies, 7*(3), 247–263. https://doi.org/10.1080/14742830802485643
Giroux, H. A. (1992). *Border crossings: Cultural workers and the politics of education.* Routledge.
Giugni, M., & Grasso, M. (2018). Alternative action organizations: Social solidarity or political advocacy? *The American Behavioral Scientist, 62*(6), 778–795. https://doi.org/10.1177/0002764218768855
Goodhart, M. (2017). Interpreting responsibility politically. *Journal of Political Philosophy, 25*(2), 173–195. https://doi.org/10.1111/jopp.12114
Goodman, M. K. (2004). Reading fair trade: Political ecological imaginary and the moral economy of fair trade foods. *Political Geography, 23*(7), 891–915. https://doi.org/10.1016/j.polgeo.2004.05.013
Gould, C. (2009). Varieties of global responsibility: Social connection, human rights, and transnational solidarity. In A. Ferguson (Ed.), *Dancing with Iris: The Philosophy of Iris Marion Young* (pp. 199–212). Oxford University Press.
Griffith, M. (2015). I've got goose bumps just talking about it!: Affective life on neoliberalized volunteer programmes. *Tourism Studies, 15*(2), 205–221.
Groves, A. (2010). The advocacy politics of NGOs: shaping society to respond to climate change. *E-International Relations.* https://www.e-ir.info/2010/07/13/the-advocacy-politics-of-ngos-shaping-society-to-respond-to-climate-change/
Habermas (1989). *The structural transformation of the public sphere: An inquiry into a category of bourgeois society.* MIT Press.
Haiki, J. (2018). The subject of citizenship – can there be a posthuman society? *Political Geography, 67*, 166–175.
Hemings, C. (2012). Affective solidarity: Feminist reflexivity and political transformation. *Feminist Theory, 13*(2), 147–161.
Henderson, J. C. (2007). *Tourism crises: Causes, consequences and management.* Butterworth Heinemann.
Higgins-Desbiolles, F. (2008). Justice tourism: A pathway to alternative globalisation. *Journal of Sustainable Tourism, 16*(3), 345–364. https://doi.org/10.1080/09669580802154132
Higgins-Desbiolles, F. (2009). International Solidarity Movement: A case study in volunteer tourism for justice. *Annals of Leisure Research, 12*(3-4), 333–339.
Higgins-Desbiolles, F., Whyte, K. P., & Tedmanson, D. J. (2013). Tourism and environmental justice. In K. Schwab, & D. Dustin (Eds.), *Just leisure: Things that we believe in* (pp. 91-101). Sagamore Publishing
Hockert, E. (2018). *Negotiating hospitality, ethics of tourism development in the Nicaraguan highlands.* Routledge
Hollinshead, K. (1992). White' gaze, 'red' people — Shadow visions: The disidentification of 'Indians' in cultural tourism. *Leisure Studies - Studies, 11*(1), 43–64. https://doi.org/10.1080/02614369100390301
Hollinshead, K., & Suleman, R. (2018). The everyday instillations of worldmaking: New Vistas of understanding on the declarative reach of tourism. *Tourism Analysis, 23*(2), 201–213. https://doi.org/10.3727/108354218X15210313504553
Hultsman, J. (1995). Just tourism: An ethical framework. *Annals of Tourism Research, 22*(3), 553–567. https://doi.org/10.1016/0160-7383(95)00011-T
Isaac, R. (2017). Transformational host communities: Justice tourism and the water regime in Palestine. *Tourism Culture & Communication, 17*(2), 139–158. https://doi.org/10.3727/109830417X14966810027580
Isaac, R. K., & Hodge, D. (2011). An exploratory study: Justice tourism in controversial areas. The case of Palestine. *Tourism Planning & Development, 8*(1), 101–108. https://doi.org/10.1080/21568316.2011.554048
Jafa, N. (2012). *Performing heritage: Art of exhibit walks.* Sage.
Jamal, T., & Camargo, B. A. (2014). Sustainable tourism, justice and an ethic of care: toward the Just Destination. *Journal of Sustainable Tourism, 22*(1), 11–30. https://doi.org/10.1080/09669582.2013.786084
Jamal, T. (2019). *Justice and ethics in tourism.* Routledge.
Jenkins, R. (2005). Globalization, corporate social responsibility and poverty. *International Affairs, 81*(3), 525–540. https://doi.org/10.1111/j.1468-2346.2005.00467.x

Keck, M. A., & Sikking, K. (1997). Transnational advocacy networks in the movement society. In D.S. Meyer & S. Tarrow (Eds.) *The social movement society* (pp. 217–238). Rowman and Littlefield Publishers.
Kelly, J. L. (2016). Asymmetrical itineraries: Militarism, tourism and solidarity in occupied Palestine. *American Quarterly, 68*(3), 723–745. https://doi.org/10.1353/aq.2016.0060
Koensler, A., & Papa, C. (2011). Political tourism in the Israeli-Palestinian space. *Anthropology Today, 27*(2), 13–17. https://doi.org/10.1111/j.1467-8322.2011.00796.x
Levinas, E. (1999). *Totality and Infinity*. Duquesne University Press.
Levy, D., & Sznaider, N. (2004). The institutionalization of cosmopolitan morality: The Holocaust and human rights. *Journal of Human Rights, 3*(2), 143–157. https://doi.org/10.1080/1475483042000210685
Llivis, K. (2017). *Posthuman blackness and the black female imagination*. University of Georgia Press.
Luh Sin, H., Oakes, T., & Mostafanezhad, M. (2015). Traveling for a cause: Critical examinations of volunteer tourism and social justice. *Tourist Studies, 15*(2), 119–131. https://doi.org/10.1177/1468797614563380
Mahrouse, G. (2008). Race-conscious transnational activists with cameras: Mediators of compassion. *International Journal of Cultural Studies, 11*(1), 87–105. https://doi.org/10.1177/1367877907083082
Mathers, K. (2010). *Travel, humanitarianism, and becoming American in Africa*. Palgrave Macmillan.
McGuire, M. (2012). *From volunteer tourism to a (cosmo)politics of solidarity* (Master dissertation). Honors thesis – all, Wesleyan University. https://wesscholar.wesleyan.edu/etd_hon_theses/877
Mostafanezhad, M. (2013). The politics of aesthetics in volunteer tourism. *Annals of Tourism Research, 43*, 150–169. https://doi.org/10.1016/j.annals.2013.05.002
Mostafanezhad, M. (2016). *Volunteer tourism: Popular humanitarianism in neoliberal times*. Ashgate.
Mowforth, M., Charlton, C., & Munt, I. (2008). *Tourism and responsibility: Perspectives from Latin America and the Caribbean*. Routledge.
Nietzsche, F. (1968). *The Genealogy of Morals.*, Penguin Classics.
Parr, A. (2008). *Deleuze and memorial culture*. Edinburgh University Press.
Peters, M. A. (2006). Neoliberal governmentality: Foucault on the birth of biopolitics. In S. Weber & S. Maurer (Eds.), *Gouvernementalität und Erziehungswissenschaft VS Verlag für Sozialwissenschaften* (pp. 37–49). Springer.
Piccinini, C. (2007). *Public humanitarian advocacy: Challenges, opportunities and its channeling through the celebrities* (Doctoral dissertation), University of Leeds. http://pvac-webhost2.leeds.ac.uk/ics/files/2012/05/Carlo-Piccinini-2.pdf.
Quinton, A. (1974). Utilitarian Ethics. In W.D. Hudson (Ed.), *New studies in ethics* (pp. 1–118). Palgrave.
Rami, K., & Hodge, D. (2011). An exploratory study. Justice tourism in controversial areas: the case of Palestine. *Tourism Planning & Development, 8*(1), 101–108. https://doi.org/10.1080/21568316.2011.554048
Rawls, J. (1971). *A theory of justice*. Harward University Press.
Rieff, D. (2002). *A bed for the night: Humanitarianism in crises*. Vintage.
Sayer, A. (2003). (De)commodification, consumer culture, and moral economy. *Environment and Planning D: Society and Space, 21*(3), 341–357. (https://doi.org/10.1068/d353
Scheyvens, R. (2002). *Tourism for development: Empowering communities*. Pearson.
Scholz, S. (2008). *Political solidarity*. Pen State Press.
Sharpley, R., & Telfer, E. (2014). *Tourism and development: Concepts and issues* (2nd ed.). Channel View Publications Ltd.
Shellenberger, M., & Nordhaus, T. (2009). The death of environmentalism: global warming politics in a post-environmental world. *Geopolitics, History, and International Relations, 1*(1), 121–163.
Slater, D., & Tonkiss, F. (2001). *Market society: Markets and modern social theory*. Wiley.
Smith, M., & Duffy, R. (2003). *The ethics of tourism development*. Sage Publications.
Spence, L. J., & Rutherfoord, R. (2001). Social responsibility, profit maximisation and the small firm owner-manager. *Journal of Small Business and Enterprise Development, 8*(2), 126–139. https://doi.org/10.1108/EUM0000000006818
Stjerno, S. (2005). *Solidarity in Europe: the history of an idea*. Cambridge University Press.
Straume, I. S., & Humphrey, J. F. (2010). Introduction. In I.S. Straume & J.F. Humphrey (Eds.), *Depoliticization: The political imaginary of global capitalism* (pp. 9–24). NSU Press.
Tester, K. (2010). *Humanitarianism and modern culture*. Pennsylvania State University Press.
Ticktin, M. (2006). Where ethics and politics meet: The violence of humanitarianism in France. *American Ethnologist, 33*(1), 33–49. https://doi.org/10.1525/ae.2006.33.1.33
UNWTO - United Nations World Tourism Organization (1949). *Agenda 21 for travel and tourism: Towards environmentally sustainable tourism*. WTO, WTTC and the Earth Council.
Wood, R. (2000). Caribbean cruise tourism: Globalization at sea. *Annals of Tourism Research, 27*(2), 345–370. https://doi.org/10.1016/S0160-7383(99)00073-0
Young, I. M. (2006). Responsibility and global justice: a social connection model. *Social Philosophy and Policy, 23*(01), 102–130. https://doi.org/10.1017/S0265052506060043
Zembylas, M. (2019). Shame at being human" as a transformative political concept and praxis: Pedagogical possibilities. *Feminism & Psychology, 29*(2), 303–321. https://doi.org/10.1177/0959353518754592
Zengh, R. (2019). What kind of responsibility do we have for fighting injustice? A moral-theoretic perspective on the social connections model. *Critical Horizons, 20*(2), 109–126.

World heritage and social justice: insights from the inscription of Yazd, Iran

Raymond Rastegar, Zohreh (Zara) Zarezadeh and Ulrike Gretzel

ABSTRACT

UNESCO World Heritage Sites (WHS) are known to add value to destinations. While the diverse impacts of WHS status have been investigated in relation to sustainable tourism development in various contexts, social justice perspectives remain poorly developed in the literature. We used netnographic research on social media supplemented with semi-structured interviews to investigate social justice issues arising from the WHS inscription of Yazd, Iran. Findings reveal a complexity of benefits (e.g. greater community well-being) and costs (e.g. higher prices and increased traffic) that are unequally distributed among stakeholders. We also found that cultural values represented by the WHS marginalize certain groups. Participation in decision-making processes is all but non-existent, leaving residents powerless to act upon concerns about tourism and cultural change. The research contributes to the emerging field of social justice in heritage tourism and heritage conservation. It offers theoretical and empirical insights that highlight the importance of social justice as an aspect of sustainable WHS management.

Introduction

Past studies indicate that inscription of built heritage as a UNESCO World Heritage Site (WHS) can bring about positive change for communities (e.g. Hall & Piggin, 2001). Importantly, it can support sustainable destination development (e.g. Baral et al., 2017). Yet, WHS inscription can also negatively affect a destination and its residents (e.g. Cuccia et al., 2016), and cause tensions between preservation and development (e.g. Park & Santos, 2017). Sustainable WHS management focuses on mitigating such potential negative impacts while fostering positive outcomes. This paper demonstrates that achieving sustainability requires a social justice lens to understand differential effects on diverse community and cultural groups and to reveal potential changes in their relationship with the heritage site as a result of World Heritage inscription.

Many cultural heritage sites are preserved for their "outstanding universal values" and not their specific value to diverse stakeholders. Witcomb (2012) stresses that "claims to universality inevitably rest on making the complex simple and in the process obscuring the diversity of meanings" (p. 62). Meskell (2002) links the notion of world heritage to globalization and posits that the desire for shared world heritage is not universal. Hodder (2010) argues that, rather than evaluating heritage in terms of "outstanding universal value", an evaluation with respect to social justice and community well-being is preferred, for it helps with the protection of human rights and the establishment of cultural heritage rights. A move towards a more rights-based approach to heritage management needs to consider the heritage rights of diverse groups (Logan, 2014). However, Blake (2011, p.230) argues that "cultural rights and rights relevant to culture are poorly understood and weakly implemented, if at all". This is an especially critical issue in the context of cultural heritage sites consumed by tourists (Jamal & Camargo, 2014), yet tourism scholars largely neglect issues of social justice at WHSs.

Using the WHS of Yazd as a case study, this paper seeks to illustrate how both positive and negative impacts perceived by destination residents can be unequally distributed, that the values of some resident groups can go unrecognised, and that decisions regarding WHS inscription and management can violate justice principles. Yazd was selected as the focus of our investigation because of its complexity as a WHS, its recent inscription, and its ancient emphasis on cultural diversity. The historic city of Yazd in the developing country of Iran was recognised as a WHS in 2017. The WHS encompasses the old part of the city, which is known for its unique adobe architecture. Its built structures not only represent a rare and especially diverse cultural heritage but also have the potential to significantly influence contemporary sustainable architecture. Heritage protection is a complex issue in Iran, with the "tussle between Islamic/Shiite identity and Persian ethno-linguistic identity" being a prominent theme in Iranian politics (Seyfi et al., 2019: p.311). While the WHS of Yazd includes Islamic heritage, Yazd's role as a religious melting pot and its emphasis on ancient Persian heritage is inconsistent with the values promoted by modern Iranian politics. The city is also changing due to migration, with incoming residents representing cultural heritage that differs from the one protected. The WHS inscription of Yazd therefore provides a rich opportunity to study cultural heritage rights.

Earlier studies of Yazd showed that resident perceptions before the WHS inscription were positive and supportive of future tourism development (e.g. Rastegar, 2010; Rezaei, 2017). However, since the inscription, issues such as increases in land and commodity prices have emerged. They negatively affect the well-being of some locals while others reap the gains. Additionally, a recent study by Nasrolahi et al. (2019) suggests that residents felt left out of the political process that led to the WHS inscription and have only limited understanding of the designation's meaning. This highlights potential social justice issues emerging from the WHS inscription and further warrants the selection of Yazd as a case study.

Adopting a social justice lens and a longitudinal perspective, this study investigates the WHS inscription process for Yazd and its impacts from a resident perspective. Based on social media data collected using netnographic principles (Kozinets, 2020) and supplemented with semi-structured interviews, the research uncovers positive effects on community well-being while also highlighting disparate effects on groups within the community, failures to recognize diversity in values, and missed opportunities to give community stakeholders a voice. These findings are then discussed specifically with respect to the formulation of cultural heritage rights and the need for nuanced notions of community in relation to WHS development and management.

Literature review

WHS inscription impacts

WHS inscription can generate many benefits, including universal value for humanity (e.g. Della Lucia & Franch, 2017), economic benefits and social value for residents (e.g. Bourdeau et al., 2017), and increases in tourist flows (e.g. Buckley, 2018). Cultural and heritage tourism represents one of the oldest forms of travel (Timothy & Boyd, 2006) and one of the most prosperous travel segments (Ramires, Brandao & Cristina Sousa, 2018). Many destinations promote their heritage and culture to attract both domestic and international tourists. The inscription of a site as a WHS can drive tourism development and increase local and international visibility (Drost, 1996), encourage local cultural activity (Ramires et al., 2018), preserve traditional arts and culture (Rasoolimanesh & Jaafar, 2017), and increase residents' sense of pride (Zhuang et al., 2019). Economic impacts from tourism development include job creation, increases in the standard of living, and infrastructure improvements (Choi & Sirakaya, 2006). Despite this potential to foster tourism, some research indicates little to no positive changes in visitor demand after the inscription (e.g. Patuelli et al., 2013).

A host of negative impacts on communities in or around the WHS has also been observed, such as overcrowding and increased traffic, rising property prices (Latkova &Vogt, 2012) and adverse impacts on social and cultural values within the resident community (Jaafar et al., 2017). Melubo and Lovelock (2019), for example, investigated the perspective of a Maasai community living inside the Ngorongoro Conservation Area in Tanzania and found that despite a small number of residents acknowledging some positive impacts, the majority of the community expressed antagonistic views regarding the WHS inscription and towards tourism development overall. The community felt neglected and believed that only tourists, investors, and conservationists received benefits from the WHS inscription and the resulting tourism development. Similarly, Rastegar et al. (2017) examined tourism development in the case of Mashhad, Iran, a major religious pilgrimage and cultural heritage centre in the northeast of Iran, and found that government institutions responsible for development did not seem to care about residents, thus failing to improve residents' quality of life.

Proximity to the WHS has been found to influence perceptions of positive and negative impacts (Jimura, 2011; Rasoolimanesh et al., 2017). Noor et al. (2019), in comparing two rural destinations in Malaysia, one an inscribed WHS (Lengong Valley) and one not yet inscribed (Bujang Valley), found that residents within the WHS had more positive attitudes towards tourism development than the non-WHS residents because of the direct benefits they perceived. In contrast, Rasoolimanesh et al. (2019) observed that residents living outside two WHSs in Iran (Kashan and Tabriz) had far more positive attitudes towards tourism in terms of its economic, social, and cultural returns than those living inside them because they were less directly exposed to negative tourism impacts. Thus, capturing differences in impacts is vital for understanding the effects of WHS inscription on local communities' well-being.

Sustainable WHS development and social justice

Considering residents' perceptions is important as they can significantly affect the sustainable development of WHSs (Rastegar, 2019; Caust & Vecco, 2017). However, including a social value dimension in sustainability efforts and measuring its effectiveness remain complex in practice (Díaz-Andreu, 2017). Also, ensuring positive community impacts and mitigating potential negative ones to achieve sustainable WHS development does not automatically lead to socially just outcomes (Jamal & Camargo, 2014). The way residents perceive not only the tourism impacts but also the fairness of sharing the costs and benefits can directly contribute to their quality of life (Su et al., 2019). In this vein, justice and its link to sustainability deserve special attention in the context of WHSs.

We propose that three interconnected dimensions of social justice need to be considered in relation to WHSs: 1) fair distribution of costs and benefits; 2) procedures that foster inclusiveness in planning and decision-making; and, 3) recognition of diversity (e.g. with respect to ethnicity, culture, religion, and gender) of those affected by WHS impacts. Fairness in outcomes, such as the equitable distribution of cost and benefits, demands intricate knowledge of WHS impacts and is of great practical significance, as it can enhance acceptance of development and reduce opposition by residents (Jamal & Camargo, 2014). Jamal (2019, p.28) discusses the complexity of justice in tourism by arguing that "justice is not an abstract notion for contemplation, it is closely related to people's lives, their well-being, their homes, their communities, their work, the places they travel to, and the journeys they undertake". Therefore, rather than only focusing on outcomes, sustainable tourism requires a just ongoing process by including different ideas on what, how and why things should be done. Ensuring fairness in the processes and procedures requires four elements: "(1) access to information; (2) access to and meaningful participation in decision-making; (3) lack of bias on the part of decision-makers; and (4) access to legal process for achieving redress" (Sovacool & Dworkin, 2015, p.437). This suggests that residents must be able to obtain enough information to make informed decisions and also have their voices heard. Extant heritage planning and management literature recognises the importance of participatory decision-making (e.g. Svels, 2015). However, Shrestha et al. (2019) observe that values in relation to heritage tend to mirror the interests of the most powerful groups in the community. Similarly, Smith (2006) points to an "Authorised Heritage Discourse" that gives power to professionals, politicians and public heritage management while excluding the voices of other stakeholders such as residents (see also Dans & Gonzalez, 2019). In addition, we should be concerned about diverse groups of residents having "equal participation in decision-making processes" (Venn, 2019, p 716). Therefore, inclusiveness requires recognition of diversity in residents that could be affected and respect for diversity in the values the heritage represents (Hodder, 2010).

Injustice can occur across all three of the described dimensions. The Maasai in Melubo and Lovelock (2019) research highlighted various issues such as limited access to natural resources, decline in community well-being, under-representation in decision making, unfair benefit sharing, and threat of relocation. Similarly, Camargo (2019) in her research in Palenque, Colombia found that while the UNESCO inscription brought attention to the local culture, residents, particularly women, received minimal benefits from the resulting tourism; the only income they receive is by selling fruit, sweets or posing for photos. Similar situations evolve in many WHSs where residents have minimal involvement in management and benefit sharing (Nicholas et al., 2009). Overall, there is a significant lack of understanding regarding how residents value WHSs that they live in or around, and how they feel development and increase in tourism activity affect their well-being and their relationship to the protected cultural heritage. As noted earlier, Hodder (2010) argues for closer study of social justice and well-being to better understand and define cultural heritage rights. Some studies have addressed issues of social justice in relation to natural heritage, but even these studies are limited (e.g. Della Lucia & Franch, 2017; Rastegar, 2020). Thus, despite widespread attempts to achieve sustainable tourism development in WHSs, many aspects of injustice remain ignored.

This paper highlights the complexity of social justice issues in WHSs. Our approach to better understanding social justice and well-being at individual and community levels involves studying resident perceptions and experiences related to the WHS inscription of Yazd, Iran. To do so, we use broad principles of justice to guide our research, specifically: equity and fairness, diversity and recognition, and democracy and participation, which are supported by the discussion above.

The case of world heritage inscription in Yazd, Iran

We selected Yazd as a case study for multiple reasons. First, the city is located in an ecologically fragile desert area; thus, sustainable tourism development is critical from a social and environmental point of view. Second, its earthen architecture and ancient culture are built on limited resource use and therefore serve as an inspiration for the sustainability challenges of today. Third, Yazd possesses cultural significance because of its unique historic architecture and status as a "centre of different faiths and cultures" (ICHHTO, 2016, p.15), bringing together the three religions of Islam, Judaism and Zoroastrianism. Last but not least, parts of Yazd have received WHS inscription very recently (in 2017), making it possible to study residents' perceptions as they emerge and contrast them to pre-inscription perceptions.

Setting and background

Yazd is one of the oldest cities in the world with its settlement dating back to the pre-Islamic era (3000 BC). It is a lively contemporary city in central Iran, at the crossroads of important current and traditional transportation routes (situated close to the Spice and Silk Roads) with thriving textile and ceramics industries. Its architecture has successfully resisted modern urbanisation over time, preserving many of its unique assets such as Zoroastrian temples, traditional houses and mosques, the historic garden of Dolat-abad, and the qanat system (underground irrigation canals) (UNESCO, 2019). The city is well-known for having the oldest mud-brick buildings in the Muslim world and for its windcatchers (badgirs), which are architectural elements that allow for natural ventilation in buildings (ICHHTO, 2016).

Over 600,000 residents live in the city of Yazd, with about 44,000 living within the WHS core and buffer zones (ICHHTO, 2016). The vast majority of Yazd residents (93%) is less than 65 years old (Nasrolahi et al., 2019). Yazd's population has experienced steady increases over the last 70 years (Macrotrends, 2020). Most of Yazd's inhabitants are Aryan who speak Persian with a Yazdi accent. However, recent waves of migration have brought in workers from other rural and urban areas of Iran as well as foreign nationals, mostly from Afghanistan and Iraq (Statistic Center of Iran, 2020). Currently, almost 20% of Yazd residents are immigrants from other provinces and countries, out of which 10% are foreigners (IRNA, 2019). Fathi (2016) confirms the image of Yazd as a city of religious diversity, reporting that, while Islam is the religion of 99% of Yazd residents, sizeable numbers identify as Zoroastrians, Jewish, Christian, Assyrian or "other".

Yazd has always been a main tourism destination in Iran for its culture, heritage, and handicrafts such as silk textiles and confectionaries. In the historic centre of Yazd, some of the traditional houses and buildings have become tourism infrastructure in the form of hotels and restaurants. In 2019, in the first six month of the Persian calendar (March to September 2019), the province received nearly 1.5 million national and international tourists (YJC, 2019), which represents a 180% growth in overnight stays compared to the previous year (YCHHTO, 2019). The national visitors are mainly families travelling to Yazd during occasions such as the Persian New Year or public holidays. According to the Yazd Cultural Heritage Handicrafts and Tourism Organization, a government body responsible for tourism marketing and management in Yazd, there was a 32% growth in 2019 in the number of international travellers, originating mainly from Italy, China, Germany and France (YCHHTO, 2019). This is especially noteworthy as Iran continues to suffer from a negative destination image among international tourists (Alavi & Yasin, 2000).

Due to its cultural and historical value, the old part of the city was listed as a national monument in 2005. Iran's Cultural Heritage Handicrafts and Tourism Organization (ICHHTO), the national body responsible for World Heritage properties in Iran, preserves and manages the monument. It is required to coordinate its efforts with the local community, non-governmental organizations, and other national authorities (UNESCO, 2019). In April 2016, a nomination document was prepared by ICHHTO and submitted to UNESCO. In July 2017, the city of Yazd was inscribed as a World Heritage City at the World Heritage Committee meeting in Krakow, Poland. It is important to acknowledge that the touristic development of the old part of the city had started a few years before the inscription and included the addition of wider streets to ensure transportation access, the establishment of squares as a symbol of modern urban development and the repurposing of old residential buildings into touristic infrastructure (hotels, restaurants, etc.). It is also useful to note that tourism development in general in Iran lacks behind other countries in the region (e.g., Turkey and the United Arab Emirates) and overall remains limited, even in Yazd.

Data collection and analysis

Data was collected through netnographic research on social media platforms to garner unsolicited opinions and reactions from a broad range of residents. The netnographic data was supplemented with rich insights from semi-structured interviews to ensure that the voices of those who might not post on social media would be included, and to allow for further probing into why certain perceptions existed.

Netnography

A netnographic research approach was used to capture some of the general discourse about the historic city of Yazd before and after the WHS inscription. Netnography comprises a set of guidelines and practices used to conduct qualitative social media research (Kozinets, 2020). It has been adopted in tourism and hospitality research to gather a variety of online opinions across a range of social media platforms (Tavakoli & Mura, 2018). Its advantage lies in being able to sample broadly and tap into archival material as well as ongoing discussions before developing cultural understandings of a phenomenon.

This study took a passive "lurker" approach (Kozinets, 2020) which precludes researcher participation in online communications. The passive netnographer observes the communicative activities of users and extracts information he or she believes relevant to the specified research questions, with the advantage of collecting data in natural discourse settings and without prompts by the researcher. Data collection from social media platforms involved archival research of public posts and videos regarding Yazd from two years before (2015) to two years after (2019) the inscription. Data sources were identified through English and Persian keyword searches on Aparat (a popular Iranian social media platform), Instagram, YouTube, and TripAdvisor, using terms "Yazd", "Historical city of Yazd", "Yazd, Heritage city", and "Heritage city of Yazd".

Social media research requires a deep, culturally-sensitive understanding of the users and uses of different platforms (Gretzel, 2018). Following established netnographic sampling criteria (Kozinets, 2020), the four platforms were selected because of their relevance to tourism contents as well as to Iranian social media users. Instagram is the most popular global social media app in Iran given that Facebook and Twitter are officially blocked. Instagram is widely used by different age groups. Many Iranians use Instagram not only for photo-sharing but also to write mini blogs or news (Jafari, 2015). Aparat is an Iranian video sharing platform. Video files on Aparat are accessible to all users without having to subscribe. However, to leave a comment on a video, users must create an account. Some videos in Aparat are not available to the public and payment is required; they were not considered for analysis for ethics reasons. While YouTube has been banned a few times over the years, it remains a popular video sharing platform that Iranians access via VPNs. TripAdvisor is fully accessible in Iran. Since it is not tied to transaction records, it allows residents and tourists alike to post reviews about the WHS attractions and leave comments in its destination forums.

The data collection procedure was based on Kozinets's (2020) recommendations, with the second author immersing herself in the social media data, searching and scanning for relevant documents over a period of one year. Following an initial perusal of the hundreds of results pages, the most relevant contents (posted by residents and pertaining to social justice aspects) were selected for further analysis. Links to videos in Aparat and YouTube were saved (the final data set included a total of 64 videos posted by residents or government officials) and posts and comments from all four platforms were captured through screenshots. Consent was not sought from social media users as only existing public contents stripped of any personal identifiers were used for the analysis.

Semi-structured interviews

Interviewees included ten residents of Yazd. They represent residents who derive their livelihood from tourism (e.g. as shopkeepers or restaurant owners) and residents who are not directly engaged in tourism. The sample also includes both residents living inside and outside the WHS zone (Table 1). This diversity in the sample permitted us to investigate potential differences in

Table 1. Interviewee profiles.

Pseudonym	Gender	Age	Living area*	Engaged in the tourism industry
Arman	Male	39	Outside the WHS zone	No
Mehran	Male	41	Outside the WHS zone	No
Ava	Female	56	Outside the WHS zone	No
Baran	Female	43	Inside the WHS zone	No
Saman	Male	64	Outside the WHS zone	No
Ehsan	Male	47	Outside the WHS zone	No
Farzad	Male	44	Outside the WHS zone	Yes
Neda	Female	59	Inside the WHS zone	Yes
Mahtab	Female	44	Inside the WHS zone	Yes
Jafar	Male	75	Inside the WHS zone	No

*Inside the WHS zone refers to the historic part of Yazd, while outside of the WHS zone indicates residency in the new areas of Yazd that were not subject to the WHS inscription.

the impacts of the inscription on different groups of stakeholders. Interviewees had to be living in Yazd since at least 2015 to capture perceptions before and after the WHS inscription. They were recruited through snowball sampling starting from personal contacts of the researchers to ensure that trust could be effectively established. Sampling targeted residents 35 years old and above to capture a segment of the population that is less likely to post on social media.

The data was collected over four months in 2019. The sample consisted of four females and six males aged between 39 and 75 years. The interview guide is presented in Table 2. The interview questions were broad and encouraged participants to discuss how they felt about the various changes occurring since the city's inscription in 2017, to reflect on the value of the UNESCO inscription, how it affected their everyday lives and their well-being, and to express how involved they felt in the overall WHS inscription process and the resulting changes to historic Yazd. Specifically, the first five questions pertained to evoking reflection on the equitable distribution of outcomes while the remainder of the questions related to diversity and recognition as well as fairness in the inscription and subsequent WHS management processes.

All interviews were conducted by either the first or the second author through imo (free video call and chat app) and were recorded with the permission of the participants. Both interviewers (one male, one female) are very familiar with the culture and geography of Yazd, with the second author having been born and raised in the city. This insider perspective not only allowed us as a research team to effectively recruit interviewees but also to establish rapport quickly and to interpret statements that pertained to specific locations in Yazd or required knowledge of the general cultural or political contexts. The interviews took between 30 to 45 minutes each and were conducted in Persian.

Data gathering and analysis occurred simultaneously and jointly for both the social media data and the interviews. The first and second author, who are bilingual in English and Persian, transcribed the interview data and translated key comments from both data sets to facilitate data analysis which involved the third author, who was not familiar with Yazd before embarking on the research and therefore was in the position to question implicit assumptions and encourage a more reflexive research narrative. The analysis involved continuous reflection on the preliminary findings and discussion of emerging themes. To establish further credibility, key quotes, themes and concepts were checked by one additional bilingual speaker, who is originally from Yazd and has tourism development expertise. The final coding was discussed with two academic experts.

Findings

The findings concern the distribution of impacts among different resident groups, the recognition of heterogeneous values, and the fairness of decision-making processes. To ensure anonymity, no identifying information is given for social media quotes and interviewee names appearing in the result section are pseudonyms (see Table 1).

Table 2. Interview guide.

Interview Questions
1. How do you feel about tourism in your city? a. What are the impacts of tourism on your city (both positive and negative)? b. In what way do you personally benefit or suffer from tourism in your city?
2. Has tourism changed since the city's inscription as a WHS? a. If yes, in what ways? If not, why do you think it has not changed? b. How do you feel about the change/lack of change?
3. What do you think about the heritage site preservation post inscription? a. Do you think it affects the traditions and cultural identity of the community? b. Do you think it affects the resources available for its protection?
4. Have the value and meaning of the heritage site (city) changed for you personally after the inscription? Why? Why not?
5. Has the inscription affected your livelihood and/or quality of life in any way?
6. Did you know your city was nominated for the UNESCO world heritage list? a. If No, Next question b. If Yes, How did you receive the information? Do you know who initiated the nomination process? What did you initially think about the nomination? Were you officially involved in the nomination process? Did you have any opportunity to have your say? What did you think about the nomination process?
7. Since the inscription as a WHS, how would you describe the WHS management process? a. Would you describe it as fair and just? b. If yes, why; if no, why not?
8. In your opinion, who does currently benefit the most from the WHS inscription?
9. Do you think future generations of residents will benefit from the inscription? If yes, how; if no, why not?
10. Do you think that there are now more opportunities for diverse groups in the community (such as women or different ethnic groups) to benefit from tourism? If no, why not? If yes, what are they?
11. Please tell me about how tourism development in your city is managed post inscription. a. Whose opinion counts? b. Who decides? c. Do you see any change from before the inscription? What changes would you like to see?
12. In general, what do you think about the involvement of residents in tourism planning and decision-making processes? What are the advantages? What are disadvantages?
13. What do you think about the role of residents in future tourism development of your city? a. How could/should residents be integrated into the process? b. What could motivate them to actively participate?

Community pride

The UNESCO inscription was a big step for Yazd and its residents. Recognition of the city, its historic monuments, and its traditions by others brought a sense of pride and belonging to residents. The number of posts and videos shared by residents on social media increased dramatically after the inscription. Post-inscription, residents shared videos speaking about history, culture, attractions and historic places in Yazd. The posts include residents' selfies with historic monuments and with tourists, especially international tourists, with the captions "Happy to see tourists" or "Happy to see many people travelling to Yazd". These social media posts mirror residents' eagerness to display the rich cultural heritage visible when strolling through the old part of the city (Figure 1).

In these social media posts, Yazd residents, male and female alike, expressed a strong sense of pride and ownership. There are lots of videos captioned "My beautiful city" in Aparat, and local social media users described the city in their videos as "My historic city". A resident proclaimed in one of the YouTube videos how proud he was to call Yazd "My City" and added: "my city has been inscribed as UNESCO WHS!" He continued to express his happiness by saying "my ancestors had something to make us proud today" (YouTube video, 2017). In many other posts residents strongly conveyed their pride in their traditions and heritage. For example, in another YouTube video, when a resident is asked about Yazd winning the sustainable energy prize, he speaks about the city's history and how what his ancestors had done resulted in many achievements today (YouTube video, 2018). Similarly, in his interview, Mehran expressed his

Figure 1. Traditional handicraft demonstrations in Yazd; Photo credit: Author 1.

sense of pride by saying: "I am so proud that my city has been inscribed as a heritage site and now everyone knows my city".

Responsibility to conserve and sustain for future generations

The inscription of the city not only made the community proud but also brought the community together with respect to their responsibility to sustain their cultural heritage. Interviewees use the personal pronouns "we" and "our" to express the need to act as a community. For example, Neda stated: "we have to pass our traditions to the next generations as our ancestors did, it is our responsibility to make sure it happens". Likewise, Mahtab explained: "we [the residents of Yazd] have realised the importance of our cultural heritage and everyone agrees that we need to do everything to protect it". Ava adds: "My son has recently joined … a group of young people trying to promote the culture and heritage of Yazd. They even try to find old buildings and record their stories. I am really happy to see youth value their culture and traditions". This post inscription sentiment also indicates that residents feel personally responsible for protecting their cultural assets. A video posted on YouTube in 2017 shows that locals volunteered to rebuild heritage sites in Yazd. In the same video, residents expressed their happiness to play a part in protecting their culture by saying: "we are really happy to be involved in this project, we can conserve our heritage and inform others about our culture". These posts suggest that the inscription has raised the community's awareness, interest and desire to conserve as well as share their rich cultural heritage with visitors.

Concerns about changes to "authentic" Yazd

The above descriptions of pride and ownership only refer to those residents who belong to the traditional culture the WHS seeks to protect and do not apply to residents who have recently moved to the city. The traditional Yazdi culture is one of peaceful coexistence of religious groups, of emphasis on traditional religious festivals and ceremonies (e.g. Nakhl Gardani, an important mourning ceremony), and of limited resource use and sharing of resources, with special types of social organization including Waqf (endowment) benefitting public buildings. Despite being the capital of the province and therefore being prone to outside influences, Yazd has succeeded in protecting its unique culture over several thousand years. The UNESCO inscription (https://whc.unesco.org/en/list/1544) specifically highlights the city's continuity in architecture as well as in social organization. The same notion of traditional culture was emphasized by the UNESCO representative in one of the visits to assess the city pre-inscription (YouTube video, 2016).

Several interviewees similarly felt that the culture of Yazd has not changed much from previous generations and that it should be preserved. For instance, Baran stated: "our city is well known in Iran for preserving its culture, we would like to keep it this way", and Ehsan emphasizes that preserving the "authentic" culture of Yazd is important for tourism: "when you go to other cities, you see many of them are becoming very similar, but Yazd is different! … tourists love to see our heritage". Comments posted by residents on social media platforms like TripAdvisor after the inscription also promote "old Yazd" and refer frequently to UNESCO. For example:

1. *The old city of Yazd which is listed on UNESCO world heritage is a Must see in Yazd.*
2. *If you walk in the old city of Yazd, soon you will realise why UNESCO registered the city as a World Heritage City, in each alley you can see some unexpected place.* (TripAdvisor comments, 2017, 2018).

This focus on the pure, ancient heritage of Yazd is now causing some conflict. As most buildings in the heritage part of the city were old and received minimum attention before the WHS inscription, the houses were usually much cheaper compared to those located in other parts of the city and popular for low-income families or in-migrants moving to Yazd. After the inscription, some residents started to feel that people living in the area were not "true" representatives of the local culture and that they threatened the cultural values protected by the WHS. This was evident in the conversations by residents on social media. For example, a local expressed his unhappiness by saying; "we need to show tourists the real culture and history of the city, people living in the area have totally different cultures" (Instagram, 2018). Some residents even recommended that the government needed to relocate "these people" (resident quote) to other parts of the city to maintain the real (in the sense of "authentic") culture of Yazd. Interviewees had similar concerns, with Ehsan, for example, stating: "some people living in the old city are not true locals and do not represent our culture". There is a clear sense of new in-migrants being problematic to protecting the architecture and culture reflected in the WHS inscription.

Tourists' rights versus residents' cultural heritage rights

The numbers of domestic travelers increased rapidly after the inscription, especially during special occasions such as the Persian New Year holiday. Nevertheless, social media conversations evidence support for further tourism development, even at the cost of some residents. As mentioned in the previous section, people living in and around the old city are from low income families. Businesses such as shops in the area have been operated to serve the same community. However, in the last few years, and particularly after the inscription, many old houses and shops have been transformed into hotels and souvenir shops to serve tourists. There is a general sentiment among residents that the remaining old shops must be renovated and be ready to serve tourists as they may otherwise negatively affect the image of the destination. For example, a male resident expressed his idea by saying: "Shops around the Amir Chagmagh (an attraction) should change their products and goods from common daily products to tourist related products such as handicraft from Yazd" (Instagram post).

A generational divide was noticeable when residents talked about the influence of tourism on their culture. Some locals, usually younger residents who enjoy being exposed to other cultures, believed residents needed to be more flexible to welcome more tourists. For example, Arman mentioned that: "We need to be less strict about our culture if we want to attract more tourists". In contrast, the older generation tends to be dissatisfied with the influence of foreign visitors, who they believe negatively affect their traditions. For example, Mehran expressed his unhappiness by saying: "Our culture is changing fast as more and more tourists visit the city". These

Figure 2. Renovated old houses; Photo credit: Author 1.

residents are happy about the inscription and both national and international recognition of their culture but not happy to witness the influence of tourists on their culture, which they feel is losing its authenticity. Concerns relate to new dress codes promoted by tourists and a noticeable decrease in residents who speak with a local accent. For example, Jafar said; "we are losing our cultural authenticity, there are so many tourists in the city. They have massive impact on our special local accent, even my grandchildren do not want to talk with a Yazdi accent." These concerns reflect the earlier discussion about cultural heritage rights – who has the right to a destination community's cultural heritage when it is a world heritage; they also raise the question of what constitutes "world heritage" and who decides what it is.

Tensions over tourism and the role of tourists continue to play out in different ways. For instance, many residents participated in a discussion on Instagram about a proposal to change the name of a historic monument called Alexander's Prison, a 15th century building that some believe was built as a prison by Alexander the Great, while others consider it to be an old religious school and call it Ziaiyeh School. The historical evidence showed that the monument had different names in the past. Residents were divided into two groups, for and against the name change. Both groups considered the impact a name change could have on tourists' perception and some residents grappled with its meaning to their own culture and tradition. For example, one resident said: "If the government changed the name of the monument, tourists may not be interested in visiting the place anymore". However, other residents felt it more important to focus on their cultural heritage saying, for instance: "We need to rely on historical evidence rather than being worried about tourists" (Instagram, 2018).

Community well-being

The process of rebuilding and maintaining the historic monuments in the city generates entertainment and socialization opportunities, safety, security and beautification for residents. Residents believe the face of the city and, particularly, the old part suddenly changed in the last

Figure 3. Old houses transformed into traditional hotels and coffee shops; Photo credit: Author 1.

few years leading up to the inscription in 2017 (Figure 2). Mehran mentioned: "The government started to preserve, maintain and in some cases rebuild the old, traditional and historic buildings just a year before the inscription". Another resident commented that "many of these buildings were converted to museums, traditional restaurants and hotels" (TripAdvisor comment, 2018) which are not only used by tourists but also provided an opportunity for residents to visit these traditional houses and buildings. Ava in her interview said; "city council started to maintain and preserve old houses and heritage sites which give us an opportunity to visit those places and learn more about our culture and history". The government effort to preserve the cultural heritage did not stop after the inscription. Many more old buildings have been renovated since the inscription. For example, a representative from ICHHTO reported about 2000 cultural assets being repaired to preserve the local culture and meet the needs of the increasing number of domestic tourists (YouTube video, 2019). He further remarked that additional efforts are required to preserve the heritage of the city and ensure that it stays on the UNESCO World Heritage list.

Preservation efforts have included installing streetlights in dark and narrow alleys of the city, which transformed the area into a more secure and safe neighborhood for residents, especially women commuting at night (Instagram, 2018). City council also started installing lights around and behind monuments to enhance the view, particularly at night. As a result, residents began describing the city as safe and beautiful (TripAdvisor comments, 2017). Post inscription, the City Council has also paid more attention to the condition of roads particularly in tourism areas. Ava summarizes: "when I think about the changes due to the inscription, I believe residents are benefiting from the changes".

Increasing gender equity and opportunities for women

Inscription created job opportunities, particularly tourism related jobs such as service personnel in restaurants, coffee shops, hotels, souvenir shops and tour guides. Many of these jobs have been created in the old part of the city where old, and in most cases previously vacant properties, have now been transformed into traditional hotels, restaurants or coffee shops (Figure 3). Most owners had inherited the properties and were not actively using them before the inscription. Opening new businesses in these families required labour. As men already had paying jobs, in many cases women became responsible for the new business. Women started to run the businesses as managers, with the income from the business representing extra spending money for the family. Thus, tourism development provided opportunities for women who had previously stayed at home or worked in low paid jobs. Many more women are now working as chefs, shopkeepers or make handicrafts. Neda confirmed: "after the inscription, tourism has provided opportunities that were not available before, now many women are engaged in handicraft making for tourists or started to work as a chef in traditional hotels and restaurants". Mahtab also described the direct impact of the inscription on her life: "I was looking for a job for a long time, it was difficult for me to find a job as I have two little children. After a new hotel opened close to my house, I got a before and after-hours job. My husband looks after the kids when I am at work".

Inequitable distribution of benefits

Yazd was one of the top tourist destinations in Iran even before the inscription. Social media comments from before the inscription expressed residents' happiness about tourists and the income resulting from tourism (YouTube video, 2016). However, after the inscription there was a dramatic increase in the number of domestic travelers. It was even noticed by residents living far from the tourist attractions and mentioned in the interviews; "even as a someone not living in the old part of the city, I can see the increasing number of tourists" (Ava). At the same time, residents believe that due to political issues they never saw the same increase in the number of international tourists. Saman stated: "we are really disappointed to see that the number of international tourists has not increased much due to the political issues". But it is also evident in the social media and interview comments that the growth in domestic tourism was enough to stimulate demand for converting old houses into traditional restaurants and hotels, resulting in increased property values and commodity prices. Surprisingly, many residents are happy about this trend as they consider it a boost to the local economy. As Saman explained: "tourists and the tourist industry have impacts on local life, it does not matter whether we are directly involved in tourism or not. It will affect people even who are not directly involved".

Many residents living in the old part of the city used to pay extremely low rent, but are now no longer able to afford the rent. Arman complained: "living in the same area is now much more expensive, you need to pay more for everything". This resident believed that the benefits of the inscription and tourism development are only for certain groups who either owned property and benefited from the price rise or have the skills to run a tourism business. He mentioned that many people like him are thinking about moving away from the old part of the city to places they can afford.

A sense of disagreement is also observed among residents who believe businesses should be run by locals. For example, after news of a foreign company investing in the hotel sector in Yazd became public, locals started expressing their dissatisfaction on social media. Almost all the respective comments on Instagram were negative. Locals expressed their feelings by saying: "locals need to keep tourist businesses for themselves", "why are foreigners allowed to buy old houses and a hotel in the city?" or "one can be followed by many more, after a few years we will see many foreigners running the businesses" (Instagram, 2018).

Adverse influences on quality of life

The increasing number of tourists and local people visiting the monuments and heritage sites has increased traffic in the narrow streets of the old part of the city. One of the main attractions

Figure 4. Stone-paved street in popular tourist area. Photo credit: Author 1.

situated in the middle of the city, the Amir Chagmagh monument and its surrounding area is a good example. There are lots of shops around the square and many people walk from there to visit the sites and buy souvenirs. The lack of parking created a problem for people visiting the area. The transport department installed traffic signs but there is no regular police control. One of the videos in Aparat interviewed the shopkeepers and drivers to highlight the issue; a shopkeeper from the old part of Yazd commented: "there is no traffic control around the heritage site Amir Chagmagh monument. While there are many traffic signs, people ignore the signs, which leads to traffic congestion around the monument" (Aparat video, 2018). Shopkeepers were dissatisfied as they felt the traffic congestion disrupted their business. Increased traffic has also negatively affected the quality of life of residents who either live or work in the area or need to visit one of the many government buildings in the area. Baran stated: "locals usually try to avoid the area due to its traffic congestion; it is also not a great place to work or live".

Participation and control over heritage decision-making

Although the ICHHTO started the process of preparing the nomination document long before the inscription and finally submitted it to UNESCO in 2016, our research found no evidence of community consultation or formal resident involvement in the process. The interviewees claimed that they were informed about the nomination after the government had started renovating the old area a year before the inscription. Ava stated: "we realised this when we saw the news just a few months before the inscription". Even after the inscription, there was no conversation on social media that suggested the engagement of residents in designing, planning or other decision making regarding the old city and the heritage monuments. Some social media conversations clearly show residents' dissatisfaction. On Instagram, several residents voiced their disappointment with the way the government is planning and designing the old city. A concrete example pertains to streets in popular tourist areas. In the process of beautifying the old city, the government used stone to pave some of the streets (Figure 4). Residents feel the changes do not reflect their culture. Farzad commented on this: "it looks beautiful, I have seen this in other cities, but it is not our culture and not the way it looked in the past".

There were some attempts by the government to inform the residents about the process. A pre-inscription YouTube video shows the UNESCO representative meeting with a few residents. However, the interviewees asserted that they lacked knowledge of any formal processes. They indicated that, besides starting a tourism-related business, they were not aware of any roles they could play in shaping tourism development in Yazd. The interviewees stressed that tourism-related development is firmly in the hands of government officials and that they did not expect any changes to this in the future.

Several residents mentioned the need for more information and the desire for opportunities to participate in heritage-related decisions. One was Arman, who stated: "we really appreciate the opportunity to be informed about the changes and involvement in the decision-making process. We know the decisions are made by experts in the government organisations, but we need the opportunity to hear them first". Similarly, Saman expressed a need for more involvement in processes: "sometimes we might have different ideas, even between different groups of people. But when people feel valued by being informed about the detailed plans, I am sure they will want the best for their city. Everyone wants to contribute and they feel responsible". Farzad noted a clear desire to exercise more control: "I think we should be given the right to express our ideas and also contribute to the decisions on the type of development, it is our heritage and culture!".

WHS inscription has heightened residents' interest in heritage-related issues. Social media comments showed that, pre-inscription, there was minimum attention given to the old city or preserving the cultural heritage; however, post-inscription many residents actively participated in the discussions and expressed ideas related to aspects affecting the cultural heritage of the city. For example, a resident posted a photo of graffiti on an old clay wall showing her anger over this act of vandalism. The comments below the post exhibited similar reactions by other residents: "What do tourists think about our culture?" or "These actions will destroy the old walls". Since the inscription, locals have clearly felt more responsible for preserving old and historic buildings and pay much more attention to tourists' perceptions. An increase in posts and comments was found across all social media platforms; however, while social media clearly provide an important forum for residents to voice opinions, it is not clear whether government officials are listening.

Discussion and directions forward

The findings of this study show that, overall, residents have positive attitudes towards the inscription and tourism development in the city. Specifically, we identified positive developments in terms of community pride, sense of ownership and belonging, as well as eagerness to preserve Yazd's built and intangible heritage for future generations. Residents also mention tangible benefits derived from the WHS inscription, such as significant infrastructure improvements, job creation and increased entrepreneurship that positively affect their livelihoods and well-being. Notably, the WHS is also credited with improving the situation for local women who, out of necessity, now have opportunities they did not have before. The newfound enthusiasm for the city's heritage is striking and the general feeling of responsibility for its preservation is promising. There is real potential that this enthusiasm for protecting the common heritage will help the community come together despite religious and socio-economic differences.

At the same time, we uncovered several areas that are problematic from a social justice point of view. Not only are the positive economic impacts (e.g. increased property values) and the negative burdens (e.g. traffic congestions) unequally distributed, but some underprivileged community members are facing displacement. There is clear evidence of the lack of inclusion in decision-making processes and ability to assert cultural heritage rights. Residents indicated they neither had sufficient access to information before or after the inscription nor real participation opportunities. As well, they were unaware of ways in which they could instigate change. Their responses suggested that they were eager to participate but did not know how. Further, they reported feeling cheated of additional benefits because of the political decisions made at higher levels without their inputs.

Most importantly, the results show a clear tension between cultural heterogeneity and community as discussed by Fainstein (2014) in the context of ethnically diverse city neighbourhoods that do not value difference and therefore decrease trust, social exchanges, and community cooperation. The increased ethnic diversity in Yazd through in-migration seems to lead to less social solidarity. This hints at problems with recognizing and embracing diverse values and identities in the community and suggests that the inscription has opened chasms in the community that were not previously apparent. The focus on a particular community group's heritage and a very narrow notion of culture created negative sentiment towards those that allegedly threatened the purity of the protected heritage and traditions. Justice principles like benefiting the least well-off (in this case the economically disadvantaged in-migrants who had settled in the old city of Yazd) or minimally not harming them (e.g. through physical displacement) are clearly violated in the case of the Yazd WHS. This suggests that a nuanced notion of community is needed when judging the social justice effects of a WHS. While this specific WHS case increased community well-being overall, it did not lead to justice for all.

Fainstein (2017) calls for justice to be the first principle to guide planning and evaluate planning policy and defines diversity, equity and democracy as important values for just governance. As our case study illustrates, equity and fairness in the distribution of costs and benefits of heritage inscription is clearly an important principle. Additionally, WHS inscription processes should involve diverse groups of residents to ensure a fair distribution of benefits and costs, enable cultural heritage rights and recognize diversity in heritage. Following Meskell (2002), achieving social justice in relation to WHS inscription processes further requires making room for specific cultural, political or religious positions that might diverge from Western, secularist viewpoints and therefore do not necessarily represent "outstanding universal value". Similarly, Human (2015) highlights the need to democratize and vernacularize the WHS inscription process. Fainstein (2014) indicates that there is not a lot of research on justice in developing country contexts at the city level, despite the pressing need to understand it in these particular spatial and social contexts. This research therefore contributes to literature that seeks to promote the need for justice frameworks in urban development and, specifically, for tourism planning in developing countries (Jamal &0020Camargo, 2018). Fainstein's principles of democracy, diversity and equity seem to be useful to guide socially just decisions, but attention must be paid to how justice is enacted in non-Western countries and with attention to cultural heritage rights at the local level. Finally, it suggests that research on resident attitudes and community well-being in tourism might portray positive impacts, but that this does not mean they are just and increase equity.

From a practical point of view, like Seyfi et al. (2019) in their study of a tentatively listed WHS in Iran, our research finds that Iran currently lacks effective mechanisms for public participation in WHS inscription processes and subsequent tourism development. Our research indicated that top-down approaches prevail, and that no attention is given to the inclusion of residents in important decisions, confirming the findings of Nasrolahi et al. (2019). While promotional efforts might feature residents, their input outside of volunteer work is neither considered necessary nor appreciated. There are no public investments to specifically help the disadvantaged groups in Yazd, measures that Fainstein (2011) identifies as critical to fair and equitable development. Further, neo-liberal notions of fostering community well-being through the creation of entrepreneurship opportunities and attracting foreign investment seem to guide the thinking, at least in the case of Yazd. The desire of residents to keep businesses in local hands and the need for businesses that serve residents rather than tourists are ignored. That women benefited from the inscription was coincidental. There is a clear need to encourage and fund women entrepreneurship in Yazd. In addition, the focus of conservation is currently on the built heritage, with no concrete programs in place to protect the intangible heritage despite residents' worry that their language is threatened by demographic changes and the influx of tourists.

The research clearly demonstrates that residents are interested in preserving their local heritage and that social media provide them with a new channel to voice their opinions and organize to initiate change (Gretzel, 2017). This is especially true for women who, in the conservative city of Yazd, are not traditionally participating in public decision-making processes, but are now using social media extensively to comment on developments in the city. WHS managers and tourism developers need to realize that ignoring community sentiment is a dangerous proposition in the age of social media. At the same time, they should exploit opportunities for community engagement through mobile technologies and social media as residents are already engaged in lively discussions about their cultural heritage through these media. One promising example in this context is the personal Instagram page of one of Yazd's city councillors through which he presents council decisions and invites residents to engage in open dialogues. Because of the anonymity afforded by the platform, residents have actively shared ideas via the page but also voiced disapproval of political decisions. With democracy being a central tenet of realizing social justice in urban areas (Fainstein, 2011) and Yazd residents being eager to engage, such opportunities for open communication need to be further explored by local politicians in Yazd.

The story of the Yazd WHS is still unfolding. While the research captured resident perceptions before and after the inscription, it could only provide a glimpse into the longer development trajectory. Changes to residents' relationship with the WHS are likely if justice concerns are not resolved. Although we were able to gather a diversity of opinions by integrating personal interview data with social media posts, the voices of the most negatively affected residents (i.e. the international in-migrants) are still under-represented in the data. Research that specifically targets the most vulnerable groups living in and around the WHS is therefore urgently needed. In general, the findings suggest that the sustainability-social justice nexus needs to be further explored in future research to define the "just WHS" and to design principles and processes that can guide its sustainable development.

Acknowledgements

The authors would like to thank the three anonymous reviewers and Tazim Jamal for the invaluable feedback provided on earlier versions of the manuscript.

Disclosure statement

No potential conflict of interest was reported by the author(s).

References

Alavi, J., & Yasin, M. M. (2000). Iran's Tourism Potential, and Market Realities: An Empirical Approach to Closing the Gap. *Journal of Travel & Tourism Marketing*, *9*(3), 1–22. https://doi.org/10.1300/J073v09n03_01

Baral, N., Kaul, S., Heinen, J., & Ale, S. (2017). Estimating the value of the World Heritage Site designation: a case study from Sagarmatha (Mount Everest) National Park, Nepal. *Journal of Sustainable Tourism*, *25*(12), 1776–1791. https://doi.org/10.1080/09669582.2017.1310866

Blake, J. (2011). Taking a human rights approach to cultural heritage protection. *Heritage & Society*, *4*(2), 199–238. https://doi.org/10.1179/hso.2011.4.2.199

Bourdeau, L., Gravari-Barbas, M., & Robinson, M. (Eds.). (2017). *World Heritage Sites and tourism: Global and local relations*. Routledge.

Buckley, R. (2018). Tourism and natural world heritage: A complicated relationship. *Journal of Travel Research*, *57*(5), 563–578. https://doi.org/10.1177/0047287517713723

Camargo, B. A. (2019). Palenqueras in Colombia: A case of cultural injustice in tourism. In T. Jamal (Ed.), *Justice and Ethics in Tourism*. (pp. 72–77). Routledge.

Caust, J., & Vecco, M. (2017). Is UNESCO World Heritage recognition a blessing or burden? Evidence from developing Asian countries. *Journal of Cultural Heritage*, *27*, 1–9. https://doi.org/10.1016/j.culher.2017.02.004

Choi, H. C., & Sirakaya, E. (2006). Sustainability indicators for managing community tourism. *Tourism Management*, *27*(6), 1274–1289. https://doi.org/10.1016/j.tourman.2005.05.018

Cuccia, T., Guccio, C., & Rizzo, I. (2016). The effects of UNESCO World Heritage List inscription on tourism destinations performance in Italian regions. *Economic Modelling*, *53*, 494–508. https://doi.org/10.1016/j.econmod.2015.10.049

Dans, E., & Gonzalez, P. (2019). Sustainable tourism and social value at World Heritage Sites: Towards a conservation plan for Altamira. *Spain. Annals of Tourism Research*, *74*, 68–80.

Della Lucia, M., & Franch, M. (2017). The effects of local context on World Heritage Site management: the Dolomites Natural World Heritage Site. *Journal of Sustainable Tourism*, *25*(12), 1756–1775. https://doi.org/10.1080/09669582.2017.1316727

Díaz-Andreu, M. (2017). Heritage Values and the Public. *Journal of Community Archaeology & Heritage*, *4*(1), 2–6. https://doi.org/10.1080/20518196.2016.1228213

Drost, A. (1996). Developing sustainable tourism for world heritage sites. *Annals of Tourism Research*, *23*(2), 479–484. https://doi.org/10.1016/0160-7383(96)83345-7

Fainstein, S. S. (2011). *The just city*. Cornell University Press.

Fainstein, S. S. (2014). The just city. *International Journal of Urban Sciences*, *18*(1), 1–18. https://doi.org/10.1080/12265934.2013.834643

Fainstein, S. S. (2017). Urban planning and social justice. In Gunder, M., Madanipour, A., & Watson, V. (Eds.). *The Routledge Handbook of Planning Theory*. (pp. 130–142). Routledge.

Fathi, E. (2016). Religions in Iran in the last fifty years. *Amar*, *21*, 23–26.

Gretzel, U. (2017). Social media activism in tourism. *Journal of Hospitality and Tourism*, *15*(2), 1–14.

Gretzel, U. (2018). Tourism and Social Media. In Cooper, C., Gartner, W., Scott, N. & Volo, S. (Eds.), *The Sage Handbook of tourism management* (Vol. 2, pp. 415–432). Sage.

Hall, C. M., & Piggin, R. (2001). Tourism and world heritage in OECD countries. *Tourism Recreation Research*, *26*(1), 103–105. https://doi.org/10.1080/02508281.2001.11081182

Hodder, I. (2010). Cultural heritage rights: From ownership and descent to justice and well-being. *Anthropological Quarterly*, *83*(4), 861–882.

Human, H. (2015). Democratising world heritage: The policies and practices of community involvement in Turkey. *Journal of Social Archaeology*, *15*(2), 160–183. https://doi.org/10.1177/1469605314566557

ICHHTO (2016). Yazd Nomination Text (Deputy of Conservation, Revitalization and Inscription of Iranian Cultural Heritage, Handicrafts and Tourism Organization report). Retrieved August 10, 2019, fromhttps://whc.unesco.org/en/list/1544/documents/.

IRNA (2019). 10% of Yazd's population are foreign immigrants. Retrieved May 31, 2020, from https://www.irna.ir/news/83533935/.

Jaafar, M., Rasoolimanesh, S. M., & Ismail, S. (2017). Perceived sociocultural impacts of tourism and community participation: A case study of Langkawi Island. *Tourism and Hospitality Research*, *17*(2), 123–134. https://doi.org/10.1177/1467358415610373

Jafari, H. (2015). Even our president is more social than you! Retrieved March 1, 2018, from http://techrasa.com/2015/08/30/iran-even-president-digs-social-media/

Jamal, T. (2019). *Justice and ethics in tourism*. Earthscan.

Jamal, T., & Camargo, A. (2014). Sustainable tourism, justice and an ethic of care: Toward the just destination. *Journal of Sustainable Tourism*, *22*(1), 11–31. https://doi.org/10.1080/09669582.2013.786084

Jamal, T., & Camargo, B. (2018). Tourism governance and policy: Whither justice?. *Tourism Management Perspectives, 25*, 205–208. https://doi.org/10.1016/j.tmp.2017.11.009

Jimura, T. (2011). The impact of World Heritage Site designation on local communities – a case study of Ogimachi. *Tourism Management, 32*(2), 288–296. https://doi.org/10.1016/j.tourman.2010.02.005

Kozinets, R. V. (2020). *Netnography: The essential guide to qualitative social media research.* Sage.

Latkova, P., & Vogt, C. (2012). Residents' attitudes toward existing and future tourism development in rural communities. *Journal of Travel Research, 51*(1), 50–67. https://doi.org/10.1177/0047287510394193

Logan, W. (2014). Heritage rights—avoidance and reinforcement. *Heritage & Society, 7*(2), 156–169. https://doi.org/10.1179/2159032X14Z.00000000032

Macrotrends (2020). Yazd, Iran Metro Area Population 1950–2020. Retrieved May 31, 2020, from https://www.macrotrends.net/cities/21524/yazd/population.

Melubo, K., & Lovelock, B. (2019). Living Inside a UNESCO World Heritage Site: The Perspective of the Maasai Community in Tanzania. *Tourism Planning & Development, 16*(2), 197–216. https://doi.org/10.1080/21568316.2018.1561505

Meskell, L. (2002). Negative heritage and past mastering in archaeology. *Anthropological Quarterly, 75*(3), 557–574. https://doi.org/10.1353/anq.2002.0050

Nasrolahi, A., Roux, J. M., Ghasvarian Jahromi, L., & Khalili, M. (2019). Assessment of local people opinion after world heritage site designation. *Heritage, 2*(2), 1739–1747. https://doi.org/10.3390/heritage2020106

Nicholas, L. N., Thapa, B., & Ko, Y. J. (2009). Residents' perspectives of a world heritage site. *Annals of Tourism Research, 36*(3), 390–412. https://doi.org/10.1016/j.annals.2009.03.005

Noor, S., Rasoolimanesh, M., Jaafar, M., & Barghi, R. (2019). Inscription of a destination as a world heritage site and residents' perceptions. *Asia Pacific Journal of Tourism Research, 24*(1), 14–30. https://doi.org/10.1080/10941665.2018.1541183

Park, S., & Santos, C. A. (2017). From translocal to transnational: WHS articulations. *Annals of Tourism Research, 64*, 102–113. https://doi.org/10.1016/j.annals.2017.02.003

Patuelli, R., Mussoni, M., & Candela, G. (2013). The effects of world heritage sites on domestic tourism: A spatial interaction model for Italy. *Journal of Geographical Systems, 15*(3), 369–402. https://doi.org/10.1007/s10109-013-0184-5

Ramires, A., Brandão, F., & Cristina Sousa, A. (2018). Motivation-based cluster analysis of international tourists visiting a World Heritage City: The case of Porto. *Journal of Destination Marketing & Management, 8*, 49–60. https://doi.org/10.1016/j.jdmm.2016.12.001

Rasoolimanesh, M., & Jaafar, M. (2017). Sustainable tourism development and residents' perceptions in World Heritage Site destinations. *Asia Pacific Journal of Tourism Research, 22*(1), 34–48. https://doi.org/10.1080/10941665.2016.1175491

Rasoolimanesh, S. M., Taheri, B., Gannon, M., Vafaei-Zadeh, A., & Hanifah, H. (2019). Does living in the vicinity of heritage tourism sites influence residents' perceptions and attitudes? *Journal of Sustainable Tourism, 27*(9), 1295–1317. https://doi.org/10.1080/09669582.2019.1618863

Rasoolimanesh, S., Jaafar, M., Kock, N., & Ahmad, A. (2017). The effects of community factors on residents' perceptions toward world heritage site inscription and sustainable tourism development. *Journal of Sustainable Tourism, 25*(2), 198–216. https://doi.org/10.1080/09669582.2016.1195836

Rastegar, M., Hatami, H., & Mirjafari, R. (2017). Role of social capital in improving the quality of life and social justice in Mashhad, Iran. *Sustainable Cities and Society, 34*, 109–113. https://doi.org/10.1016/j.scs.2017.05.024

Rastegar, R. (2010). Tourism development and residents' attitudes: A case study of Yazd. *Iran. TOURISMOS, 5*(2), 203–211.

Rastegar, R. (2019). Tourism development and conservation, do local resident attitudes matter? *International Journal of Tourism Sciences, 19*(3), 181–191. https://doi.org/10.1080/15980634.2019.1663998

Rastegar, R. (2020). Tourism and justice: Rethinking the role of governments. *Annals of Tourism Research*, 102884. https://doi.org/10.1016/j.annals.2020.102884

Rezaei, N. (2017). Resident perceptions towards tourism impacts in historic centre of Yazd, Iran. *Tourism Geographies, 19*(5), 734–755. https://doi.org/10.1080/14616688.2017.1331261

Seyfi, S., Hall, C. M., & Fagnoni, E. (2019). Managing world heritage site stakeholders: a grounded theory paradigm model approach. *Journal of Heritage Tourism, 14*(4), 308–324. https://doi.org/10.1080/1743873X.2018.1527340

Shrestha, K. K., Bhattarai, B., Ojha, H. R., & Bajracharya, A. (2019). Disaster justice in Nepal's earthquake recovery. *International Journal of Disaster Risk Reduction, 33*, 207–216. https://doi.org/10.1016/j.ijdrr.2018.10.006

Smith, L. (2006). *Uses of heritage.* Routledge.

Sovacool, B. K., & Dworkin, M. H. (2015). Energy justice: Conceptual insights and practical applications. *Applied Energy, 142*, 435–444. https://doi.org/10.1016/j.apenergy.2015.01.002

Statistic Center of Iran (2020). Population of Yazd. Retrieved March 30, 2020, from https://www.amar.org.ir/english/Population-and-Housing-Censuses.

Su, L., Huang, S., & Nejati, M. (2019). Perceived justice, community support, community identity and residents' quality of life: Testing an integrative model. *Journal of Hospitality and Tourism Management*, *41*, 1–11. https://doi.org/10.1016/j.jhtm.2019.08.004

Svels, K. (2015). World heritage, tourism and community involvement: A comparative study of the High Coast (Sweden) and Kvarken Archipelago (Finland). *Scandinavian Journal of Hospitality and Tourism*, *15*(1-2), 183–201. https://doi.org/10.1080/15022250.2015.1009708

Tavakoli, R., & Mura, P. (2018). Netnography in tourism–Beyond web 2.0. *Annals of Tourism Research*, *73*(C), 190–192. https://doi.org/10.1016/j.annals.2018.06.002

Timothy, D. J., & Boyd, S. W. (2006). Heritage tourism in the 21st century: Valued traditions and new perspectives. *Journal of Heritage Tourism*, *1*(1), 1–16. https://doi.org/10.1080/17438730608668462

UNESCO (2019). World Heritage Centre. Retrieved September 15, 2019, from https://whc.unesco.org/.

Venn, A. (2019). 24 - Social justice and climate change. In T. M. Letcher (Ed.), *Managing global warming* (pp. 711–728). Academic Press.

Witcomb, A. (2012). Tensions between world heritage and local values: the case of Fremantle Prison (Australia). In M.-T. Albert, M. Richon, M.J. Viñals, & A. Witcomb (Eds.), *Community development through world heritage* (pp. 62–68). UNESCO.

YCHHTO (2019). Increasing number of tourists in Yazd. http://portal.yazdcity.ir

Young Journalist Club (YJC). (2019). Yazd received more than 470000 tourists. https://www.yjc.ir/fa/news/7081465.

Zhuang, X., Yao, Y., & Li, J. J. (2019). Sociocultural impacts of tourism on residents of world cultural heritage sites in China. *Sustainability*, *11*(3), 840. https://doi.org/10.3390/su11030840

Smart Korea: governance for smart justice during a global pandemic

Jiyoung Choi, Seunghoon Lee and Tazim Jamal

ABSTRACT

This research note offers preliminary insight into how one smart destination, South Korea, rapidly 'flattened the curve' during the global COVID-19 pandemic without shutting down its economy or its borders, by implementing strong measures with the help of smart technologies and rapid innovations. Proactive information-sharing enabled citizens to develop a shared understanding of the situation, comply with the newly adopted rules and safety measures, and build confidence in their government's abilities to manage the crisis. The case suggests that smart governance in South Korea can help to enable a smart form of justice that involves equity and fairness in information sharing and resource allocation among its residents and visitors, mitigates discrimination and exclusion, and facilitates democracy through an informed and involved citizenry.

The novel coronavirus arrives

The World Health Organization (WHO) declared a global pandemic on 11 March 2020. More than 211 countries, autonomous areas and territories have been affected by the highly contagious SARS-CoV-2 coronavirus to date. Tracing infections in different countries revealed that COVID-19 (the disease name) originated in travelers from Wuhan or from being in contact with an infected traveler. The US closed its borders to Chinese travelers at the end of January, one month after China informed WHO of the novel coronavirus outbreak in Wuhan (China). An infected bartender in an Austrian ski resort was identified to be a "super spreader" as hundreds of infected holidaymakers from different countries were traced to him (Hruby, 2020). Lombardi, Italy, became an epicenter in February from which the highly contagious virus migrated to other parts of Italy and beyond through regional and international travelers. Country-level lockdowns continued and New York became the next epicenter in March. Border closures, social isolation and shelter-in-place directives have resulted in travel and tourism being paralysed as airlines, hospitality services, tour operators, and the cruise industry have all been severely impacted.

Yet, cruise lines continued to operate into March even as the desperate circumstances of the Diamond Princess quarantined off Yokoyama (Japan) unfolded and public health warnings arose (see CDC, 2020). Questions have been raised as to levels of preparation to contain COVID-19 and curtail further spread from passengers returning home (Mallapaty, 2020; Rocklöv et al., 2020). The *Greg Mortimer* embarked from Ushuaia (Argentina) on 15 March 2020 on an Antarctica expedition and eventually ended up off Uruguay seeking help as 60% of its 217 passengers contracted the virus (Griffiths & Castillo, 2020). Some tourists intentionally took advantage of attractive travel packages during the pandemic and did not adhere to destination directives, such as the three Florida tourists who arrived in Hawai'i separately and flaunted self-quarantine measures, endangering island populations (Kelleher, 2020). First responders are also put at increased risk to rescue stranded tourists (Jamal & Budke, 2020). Many regions of the world have curtailed the freedom of movement rights of their citizens and subject vulnerable populations like migrants, refugees, and the poor to immense hardship and human rights violations (Roy, 2020).

Fortunately, glimmers of good news are emerging. Wuhan residents are finally able to travel out of their province after 76 days of stringent isolation, testing and hardships that commenced on 23 January 2020, two days before the Chinese New Year started (Gan, 2020). Although the WHO had counseled against closing borders, countries such as Australia and New Zealand locked down their borders shortly after the pandemic was announced and implemented social isolation and other virus suppression measures. New Zealand, for instance, started testing as early as 22 January, closed its borders on 19 March and committed to an elimination strategy on 23 March (a month after the first outbreak was detected), enabling it to swiftly "flatten the curve" (Cousins, 2020; see also Roy & Doherty, 2020). Other countries like Germany, Taiwan, Singapore and South Korea, also exercised quick responsibility to follow the science. The popular tourist destination of Iceland (pop. 364,134) had tested 10% of its population by early April, initiated rigorous contact tracing, and limited gatherings to twenty people (Hjelmgaard, 2020; Gudbjartsson et al., 2020).

Over 5.4 million coronavirus cases and greater than 340,000 deaths worldwide had been recorded as of 24 May 2020 (https://www.worldometers.info/coronavirus/). Overworked frontline health workers continue to battle desperately to save lives without adequate personal protective equipment. The pandemic has exposed and exacerbated economic inequalities and social injustices in the Global North and Global South, as over 100 million jobs face being lost at the rate of a million jobs a day (WTTC, 2020) and destinations seek to contain the virus, and restore trade and freedoms of movement. While awaiting a vaccine, technological innovations serve residents under limited mobilities and governments are implementing a range of rules and tools (including smart technologies) to contain or mitigate the disease. This Research Note draws on online and secondary literature to explore one such effort: South Korea's use of ICTs (Information and Communication Technologies) and other smart technologies to engage in smart governance of the COVID-19 outbreak within its borders. Our hope is to offer a timely snapshot that captures this moment in collective (tourism) memory and inspire future research and practice for fair, just and 'smart' responses to the pandemic as it continues to unfold and affect destinations worldwide.

Smart governance during a global pandemic

A growing literature on *smart tourism* and the *smart destination* is emerging with the development and application of Information and Communication Technologies (ICTs) to tourism and destination management (Buhalis & Law, 2008; Del Chiappa & Baggio, 2015; Ivars-Baidal et al., 2019; Jovicic, 2019). A smart tourism destination is "a knowledge-based destination, where ICTs are used to provide a technological platform on which information and knowledge relating to tourism could be instantly exchanged" among stakeholders so that they can innovate rapidly (Jovicic, 2019, p. 278). Boes et al. (2015) discuss the importance of information and an "infostructure" to facilitate smart innovation and governance. Saunders and Baeck (2015) discuss smart ways not just to collect data, but also to allocate resources and make decisions. ICTs have been studied with respect to public-private as well as consumer-related cooperation, collaboration and knowledge sharing (e.g., Del Chiappa & Baggio, 2015; Jovicic, 2019). Research on ICTs and smart mobile technologies is also growing in crisis communication and recovery, as well as emergency preparedness and management (Hu & Kapucu, 2016; Sigala, 2012), and more recently, mitigating biosecurity risk (Melly & Hanrahan, 2018). The South Korea government's efforts below offer some early insights and concerns for *smart governance* by mobilizing smart technologies to contain and mitigate the spread of COVID-19.

Smart Korea: governance during the COVID-19 outbreak

South Korea (pop. 51 million) is a popular tourism destination receiving around 15 million international tourists annually (Korea Tourism Organization, n.d.). It became the site of the first major COVID-19 outbreak outside China on 19 February with 11,122 confirmed coronavirus cases and 12 newly confirmed cases as of 21 May (KCDC, 2020). Fortunately, the country has been able to quickly flatten the curve through smart governance that actively involved its residents and tourists through innovative ICTs. The Korean government reacted immediately as soon as the first case was identified on 20 January, commencing daily briefing with the Korea Center for Disease Control and Prevention (KCDC). The KCDC acted swiftly to gather and disseminate data through public health communication on infections, address issues related to testing and contact tracing, and raise public awareness on necessary measures to prevent the spread of the coronavirus. Local residents have been able to easily browse local government websites for number of infections in their locality, obtain information on places infected people visited and the routes used, availability of testing stations in the area, and also location and availability of face masks. In addition, a number of other communication, monitoring and mitigation measures were undertaken:

Informational text messages

South Korea has the highest rate of cell phone/smart phone ownership in the world. A survey conducted by the Pew Research Center (2019) shows 100% cell phone ownership and 95% smart phone ownership rate in 2018. Given this domestic condition, its local governments have been sharing live updated information of confirmed cases in their area through mobile text messages. These texts are available to residents not just where they live, but also from the districts or provinces they visit, as their location data is collected in real-time using GPS. This helps individuals to plan their movements based on reliable information, thereby reducing infection risks. In addition, the texts encourage the informed individuals to be proactively involved in contact tracing and testing.

However, concerns have been raised about privacy violations relating to data being shared about location and places visited by infected travelers and residents. A number of infected people were subjected to anonymous criticisms, and the places they had visited suffered economic hardship due to the stigma associated with being discovered to be a COVID-19 infected area. Subsequent to objections by the National Human Rights Commission of Korea, the government changed its information disclosure guidelines. Now, it discloses information only when contact tracing shows the infected person has come in contact with one or more individuals who need to be advised. Information that can identify the location of infected person, such as apartments and office buildings, is no longer publicly available. Furthermore, local governments are restricting the duration of information disclosure to 14 days after the patient has fully recovered, saying they have a right to privacy and resuming normal life.

Smart mobile app

South Korea mobilized the country's advanced technological skills to rapidly develop a mobile application called the "self-quarantine safety protection" app (CDSCHQ, 2020). Every new arrival to South Korea (both Korean nationals and foreign nationals) and the close contacts of those infected are required officially to be quarantined and to report their health status twice a day in this app for two weeks. Based on the reports, the local health authorities inform them whether they should get tested and provide testing instructions as needed through a direct phone call. If the self-quarantined individuals do not submit their daily reports for three consecutive days, turn off the cell phone, or move out of their designated area, local government officials call or visit them to check their location and remind them to stay in quarantine. Thus, location and health data collected through the app helps the government to efficiently assist and regulate quarantine. For example, in late March, local, provincial, and national governments were informed through the app that a number of officially quarantined people who had arrived from other high-risk countries violated the quarantine rules. This led to imposing stricter penalties for quarantine violations (see Figure 1).

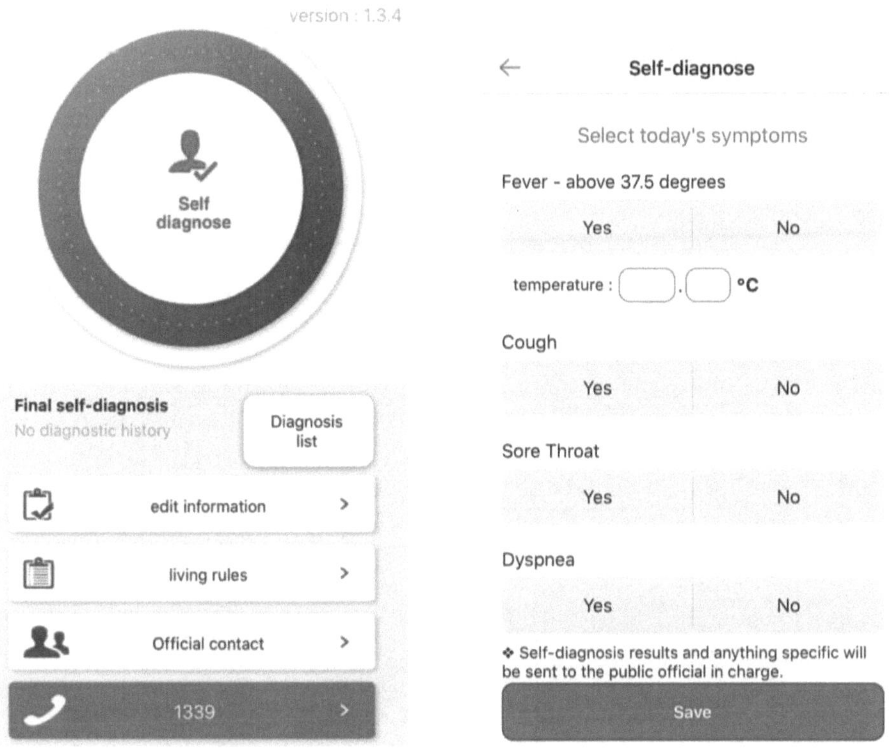

Figure 1. Self-quarantine safety protection app: (a) Main page (b) Self diagnosis.
Photo credit: author 1.

Open borders

Despite some public expressions of concern about the lack of travel bans, and worries over potential outbreaks from imported cases as well as the possible unfair burden on Korean citizens, the national government has maintained open borders in the interest of international solidarity and collaboration. Rather than blocking travel within and across borders through lockdowns, it implemented a special immigration procedure whereby international travelers can enter if they agree to comply with the quarantine rules and self-report through the above-mentioned mobile app (Ministry of Health & Welfare, n.d.). Those without a smart phone to download the app are required to submit their address and a contact number they can be reached at while in the country. Additionally, several local governments lease smart phones to visitors lacking one for free during the quarantine period. As of 16 May 2020, 93.1% of self-quarantined individuals are using the "self-quarantine safety protection" app (Ministry of Health and Welfare, May 17, 2020). In addition to this app, use of an electronic wristband was instituted to monitor passengers arriving from abroad who violate the 14-day self-quarantine order (AP News, April 11, 2020). When violation is detected, non-compliant visitors are requested to stay in their self-quarantine place wearing the electronic wristband for the rest of the quarantine period. If they do not consent to wearing the wristband, they are required to stay at a designated quarantine facility for the remainder of the quarantine period (Ministry of Health and Welfare, April 24, 2020).

Technological Infrastructure

Technological infrastructure such as the drive-thru and walk-thru testing stations were developed quickly to ramp up testing capacity and enable widespread, fair and free access to crucially vital testing for citizens, inbound passengers and foreign nationals alike. Testing at these stations was available immediately and test results were provided within 24 hours. Government authorities also allowed a highly vulnerable group–illegal migrant workers who were likely not enrolled in the national health insurance system–to get fair and free access to testing, and withdrew from cracking illegal migrants down. Measures to ensure equitable and effective access to testing encouraged the public to get tested and enabled the KCDC to trace the spread of coronavirus cases efficiently, contributing to making a safer public space for residents and tourists without a lockdown (see Figure 2).

Figure 2. Drive-through testing station in Daejeon, South Korea. As of April 8, there were over 600 testing sites in the country (including drive-through and walk-in).
Photo credit: author 2.

Distribution of food and essential goods

Distribution of food and essential goods (e.g., personal protective equipment) has been a critical issue that was well documented in the media (e.g., see Snider, March 5, 2020). Hoarding of toilet paper, hand sanitizers, disinfectants was prevalent in the US and Canada, and the shortage of face masks was ubiquitous. In such a situation, already-vulnerable groups including the disabled, the elderly, and pregnant women, have greater difficulty in getting essential goods and become even more susceptible to the virus. On 9 March the government started a 5-day rotation system for mask distribution, aimed to provide "public" masks at reasonable cost. Korean residents could buy two masks from local pharmacies on a designated day based on the last digit of their birth years for official identification documents (Mon: 1 and 6, Tues: 2 and 7, Wed: 3 and 8, Thurs: 4 and 9, Fri: 5 and 0). Selling locations and availability of "public" masks were made accessible through a citizen-developed private mobile app. Vulnerable social groups with a higher risk of contracting COVID-19 can purchase the masks by proxy. Foreign nationals could buy them by showing certificate of alien registration and enrollment in the Korean health insurance system. In another initiative, the district government or the district health authority has been distributing necessary goods to quarantined people with water, rice, food, toilet paper, face masks, and hand sanitizers delivered directly to their address. These efforts at fair distribution encouraged public confidence to comply with government-directed policy and practice citizenship.

Strategies to counter discrimination and exclusion

Several human rights organizations in Korea protested that illegal immigrants, unregistered foreigners, short-term visitors, and many international students were excluded from the above benefits because they were not enrolled in the national health insurance system (Kim, March 31, 2020). Some local governments therefore distributed locally available stockpiles through local centers for migrant workers as well as universities. Potential stigmatization of the country's minority LGBTQ+ population is also a worry. Widespread media reports on an outbreak in local nightclubs frequented by LGBTQ+ during the May golden week holidays increased concerns that Korea's testing and contact tracing method could lead to unfair outing of LGBTQ+ people as well as homophobic behaviors (Borowiec, May 14, 2020). In an attempt to mitigate potential discrimination while continuing the current prevention method, the Korean government has been offering anonymous testing where individuals can get tested without being required to provide personal information other than a contact phone number.

Smart justice for the smart destination?

South Korea has rapidly flattened the curve without shutting down its economy or its borders, by implementing strong measures with the help of smart technologies and rapid innovations. It acted quickly on available science and rapidly deployed its advanced technological capabilities, implementing practical and highly effective strategies to provide rapid testing (e.g., drive-in locations), and treatment locations away from main hospitals. Its well-educated resident population were kept updated and well-informed through ICTs and other media, and arriving visitors were provided fair and quality treatment, including rapid testing. Korean residents voluntarily complied with government policy and directives in part due to high trust in government, i.e., for being efficient, fair, and caring for the resident's welfare (Knowledge@Wharton, 2020).

Based on preliminary information gathered and concerns of discriminatory practices notwithstanding, it can be argued that the Korean government has engaged in *smart destination governance*. It acted rapidly and responsibly based on good scientific advice, ICTs and other innovations not only to its own citizens but also towards inbound visitors, ensuring their safety and protection while safe-guarding the rest of Korean society. Public health communication, awareness raising and stringent policies aided by smart technologies enable joint accountability and regulation by government, residents and visitors too. International and domestic travelers are being asked to take responsibility to comply with the destination's quarantine and social distancing rules through ICTs and other rapid smart innovations, health and safety procedures, joining its citizens in safeguarding the destination from further spread of COVID-19.

Transparency and proactive information-sharing has also enabled citizens to develop a shared understanding of the situation, comply with the newly adopted rules and safety measures, and build trust in their government's abilities to manage the crisis. As such, a smart form of justice is being enabled in the smart destination that involves empowering local residents to *co*-create smart governance to mitigate and manage the disease through an active, well informed and highly aware citizenry. Among other things, *smart justice* also involves providing equitable access to information and other necessary resources for its minority, diverse and poorer populations, countering discrimination and exclusion, facilitating freedom of movement and access to the destination through fair sharing of restrictions and risk (e.g., both visitors and Korean residents required to self-quarantine). Smart justice is also attentive to privacy infringement and issue of power and control, managing the tension between individual rights versus collective (societal) good in a democracy. Fairness in surveillance is an important principle as government authorities struggle to contain a public threat like COVID-19. The South Korean authorities have endeavored to protect the privacy and personal information of users while employing smart governance tools and processes during quarantine, contact tracking and social isolation.

Foucauldian biopolitics and governmentality will clearly continue to be a dominant issue in the smart pandemic movement (see, for instance, Lemm & Vatter, 2014). Two technology giants, Apple and Google have joined forces to develop contact tracing technology using a Smartphone's Bluetooth and state: "Privacy, transparency and consent are of utmost importance in this effort and we look forward to building this functionality in consultation with interested stakeholders" (Kelion, 2020). Health and other governance authorities are already recommending the use of coronavirus tracing apps elsewhere and see it as relevant for public safety and resuscitating tourism safety, while controversy over privacy abounds (e.g., Deutsche Welle, 2020; Yee, 2020; Zastrow, 2020) As to what exactly constitutes *smart tourism justice* or *smart destination justice*, conceptual and empirical work awaits future researchers as they define and redefine the meaning of 'smart tourism,' 'smart destinations' and 'smart governance' in relation to them (e.g., Fabry & Blanchet, 2019; Gretzel et al., 2015; Johnson & Samakovlis, 2019).

The lessons that South Korea and other proactive destinations offer may be informative in a 'smart' future where global disease outbreaks will be part of the 'new normal' (see Gardiner, 2020). What is clear is that we must begin now to identify ethical principles, create benchmarks and develop a (smart) knowledge base from which to tackle the upcoming challenges, identify and redress injustices, and contribute to destination resilience and societal well-being. Table 1 presents preliminary insights for smart justice in the case of the pandemic response in South Korea. A smart form of justice is emerging that involves equity and fairness in information sharing and resource allocation among its residents and visitors, mitigating discrimination and exclusion, and facilitating democracy through an informed and involved citizenry. However, future research is needed to explore and develop a comprehensive framework for smart justice in South Korea and elsewhere.

Table 1. Smart justice and smart governance: some principles and concepts.

Smart Justice	Smart governance	Examples from South Korea case
Citizen empowerment and social responsibility	Responsible actions by a well-informed and highly aware citizenry and visiting public	Residents and visitors receive updated information on cases, guidance on health precautions, and policy measures through government websites, text messages, and mobile apps; People follow social distancing measures and exercise precautionary behavior voluntarily as they are well-informed & included, which fosters trust and responsibility.
	Fostering trust between the public and government	Trust in government is strengthened by wide public health communication, swift policy responses and govt. action on challenges (e.g., making "public masks" available)
Transparency and inclusiveness	Co-creating smart governance	Using a smart mobile app, residents and tourists provide their data while the government manages its services and regulations accordingly.
	Enabling transparency in information sharing and public communication	Govt. and health authorities share COVID-19 information and policy measures through public websites, text messages, and mobile applications.
	Inclusive decision-making	The government makes timely policy decisions based on scientific data (e.g., from KCDC) and public opinions.
Equity and access	Equitable access to health- related and good (science-based) information	COVID-19 related information (statistics, contact tracing, testing stations, mask distribution pharmacies, and policies) is accessible through government websites and mobile text messages.
	Equitable access to resources	Both residents and visitors have access to free tests, face masks, and necessary goods.
	Facilitating freedom of movement and access to the destination	The government has maintained an open border policy by implementing the Special Entry Procedure using a mobile app.
Fairness in surveillance	Protecting privacy	The guidelines to restrict information disclosure has been established.
	Surveillance with consent	Quarantine measures require the consent of the quarantined person.
Mitigating exclusion and discrimination	Protecting minority and LGBTQ from discrimination	To protect the privacy of a minority group, the government offers anonymous testing.
	Including minority and vulnerable groups in the epidemic prevention system	Some measures are made to increase the accessibility of marginalized groups to testing and personal protective equipment.

Looking forward

The global crisis triggered by SARS-CoV-2 raises many ethical questions that need to be taken up, as well as challenging questions about the governance of travel and tourism, from the local to the global. Among the known and unknowns surrounding this coronavirus, one thing has become abundantly clear to policy makers, businesses, communities and residents. That is, the interdependencies of places and people in a highly mobile, "complex local-global system, where travel and tourism impacts are interrelated and can have local to global consequences" (Jamal, 2019, p. 15). *Action research* must (safely) follow so that the lessons learned from this global pandemic are well documented and shared to inform good practice—by destinations, policy makers, private and non-profit organizations, health practitioners, residents and tourists.

Action research may also be able to help identify injustices more quickly and facilitate *smart justice*. As South Korea's efforts to manage the spread of COVID-19 illustrate, it includes, among other things, rapid mobilization of smart technologies to share information, raise awareness and engage citizens and incoming visitors equitably in mitigation and management of the disease outbreak. An engaged, informed public aided by transparency in the government's use of smart technologies can call for accountability and, simultaneously, help to redress social injustices such as discriminatory practices and inequitable distribution of social goods and resources to vulnerable populations (e.g., see Rawls, 1971 on distributive justice and justice as fairness), as well as provide critical scrutiny to facilitate smart governance through unprecedented moments of pandemic crisis.

Disclosure statement

No potential conflict of interest was reported by the authors.

References

AP News. (2020, April 11). The latest: Electric wristbands for Koreans who break quarantine. AP News. https://apnews.com/b93ce6129f6e8f3d2307f3febf5a246d

Boes, K., Buhalis, D., & Inversini, A. (2015). Conceptualising smart tourism destination dimensions. In Tussyadiah I., Inversini A. (Eds.), *Information and communication technologies in tourism 2015*. Springer.

Borowiec, S. (2020, May, 14). How South Korea's nightclub outbreak is shining an unwelcome spotlight on the LGBTQ community, Time, https://time.com/5836699/south-korea-coronavirus-lgbtq-itaewon/

Buhalis, D., & Law, R. (2008). Progress in information technology and tourism management: 20 years on and 10 years after the Internet—The state of eTourism research. *Tourism Management*, *29*(4), 609–623. https://doi.org/10.1016/j.tourman.2008.01.005

CDC. (2020). Morbidity and mortality weekly report (MMWR) – Public health responses to COVID-19 outbreaks on cruise ships – worldwide, February-March 2020. Centers for Disease Control and Prevention. https://www.cdc.gov/mmwr/volumes/69/wr/mm6912e3.htm

CDSCHQ. (2020). Guide on the Installation of "Self-quarantine Safety Protection App." http://ncov.mohw.go.kr/upload/ncov/file/202004/1585732793827_20200401181953.pdf

Cousins, S. (2020). New Zealand eliminates COVID-19. World Report. www.thelancet.com, Vol. 345. Retrieved May 9, 2020, from https://www.thelancet.com/action/showPdf?pii=S0140-6736%2820%2931097-7

Del Chiappa, G., & Baggio, R. (2015). Knowledge transfer in smart tourism destinations: Analyzing the effects of a network structure. *Journal of Destination Marketing & Management*, *4*(3), 145–150. https://doi.org/10.1016/j.jdmm.2015.02.001

Deutsche Welle. (2020). DW.com. https://www.dw.com/en/eu-push-for-coronavirus-contact-tracing-suffers-setback/a-53219372

Fabry, N., & Blanchet, C. (2019). Monaco's struggle to become a smart destination. *International Journal of Tourism Cities*, *5*(4), 672–684. https://doi.org/10.1108/IJTC-02-2019-0034

Gan, N. (2020, April, 8). China lifts 76-day lockdown on Wuhan as city reemerges from coronavirus crisis. *CNN*. https://www.cnn.com/2020/04/07/asia/coronavirus-wuhan-lockdown-lifted-intl-hnk/index.html

Gardiner, B. (2020, March 23).: Coronavirus holds key lessons on how to fight climate change. Yale Environment 360. https://e360.yale.edu/features/coronavirus-holds-key-lessons-on-how-to-fight-climate-change

Gretzel, U., Sigala, M., Xiang, Z., & Koo, C. (2015). Smart tourism: foundations and developments. *Electronic Markets*, *25*(3), 179–188. https://doi.org/10.1007/s12525-015-0196-8

Griffiths, J., Castillo, J. (2020, April 8). Almost 60% of passengers on board Antarctic cruise ship have coronavirus. Retrieved April 14, 2020, from https://edition.cnn.com/2020/04/07/americas/greg-mortimer-cruise-ship-coronavirus-intl-hnk/index.html.

Gudbjartsson, D. F., Helgason, A., Jonsson, H., Magnusson, O. T., Melsted, P., Norddahl, G. L. , Saemundsdottir, J., Sigurdsson, A., Sulem, P., Agustsdottir, A. B., Eiriksdottir, B., Fridriksdottir, R., Gardarsdottir, E. E., Georgsson, G., Gretarsdottir, O. S., Gudmundsson, K. R., Gunnarsdottir, T. R., Gylfason, A., Holm, H., … Stefansson, K. (2020). Spread of SARS-CoV-2 in the icelandic population. *The New England Journal of Medicine*. https://doi.org/10.1056/NEJMoa2006100

Hjelmgaard, K. (2020, April 11). Iceland has tested more of its population for coronavirus than anywhere else. Here's what it learned. Retrieved April 14, 2020, from https://www.usatoday.com/story/news/world/2020/04/10/coronavirus-covid-19-small-nations-iceland-big-data/2959797001/

Hruby, D. (2020, March 24). How an Austrian ski resort helped coronavirus spread across Europe. CNN. https://www.cnn.com/2020/03/24/europe/austria-ski-resort-ischgl-coronavirus-intl/index.html

Hu, Q., & Kapucu, N. (2016). Information communication technology utilization for effective emergency management networks. *Public Management Review*, *18*(3), 323–348. https://doi.org/10.1080/14719037.2014.969762

Ivars-Baidal, J. A., Celdrán-Bernabeu, M. A., Mazón, J. N., & Perles-Ivars, Á. F. (2019). Smart destinations and the evolution of ICTs: a new scenario for destination management? *Current Issues in Tourism*, *22*(13), 1581–1600. https://doi.org/10.1080/13683500.2017.1388771

Jamal, T. (2019). *Justice and ethics in tourism*. Routledge.

Jamal, T., & Budke, C. (2020). Tourism in a world with pandemics: Local-global responsibility and action. *Journal of Tourism Futures*. https://doi.org/10.1108/JTF-02-2020-0014

Johnson, A. G., & Samakovlis, I. (2019). A bibliometric analysis of knowledge development in smart tourism research. *Journal of Hospitality and Tourism Technology*, *10*(4), 600–623. https://doi.org/10.1108/JHTT-07-2018-0065

Jovicic, D. Z. (2019). From the traditional understanding of tourism destination to the smart tourism destination. *Current Issues in Tourism*, *22*(3), 276–282. https://doi.org/10.1080/13683500.2017.1313203

Kelion, L. (2020, April 10). Coronavirus: Apple and Google team up to contact trace Covid-19. Retrieved April 14, 2020, from https://www.bbc.com/news/technology-52246319

Kelleher, J. S. (2020, April 8). Hawaii mayor: Florida man flouting quarantine was 'covidiot'. Retrieved April 14, 2020, from https://abcnews.go.com/Health/wireStory/hawaii-mayor-arrested-florida-man-covidiot-70035164

Kim, B. (2020, March 31). Seoul gives masks free to foreigners alienated in public rationing. The Korea Herald. http://www.koreaherald.com/view.php?ud=20200331000811

Knowledge@Wharton (2020). Combating COVID-19: Lessons from Singapore, South Korea and Taiwan. Knowledge@Wharton. Wharton University of Pennsylvania. Retrieved April 21, 2020, from https://knowledge.wharton.upenn.edu/article/singapore-south_korea-taiwan-used-technology-combat-covid-19/

Korea Centers for Disease Control and Prevention (KCDC). (2020). Press Release. Retrieved April 14, 2020, from https://is.cdc.go.kr/upload_comm/syview/doc.html?fn=158683989913200.pdf&rs=/upload_comm/docu/0030/

Korea Tourism Organization. (n.d.). Annual tourism statistics [data file]. https://kto.visitkorea.or.kr/kor/notice/data/statis/profit/board/view.kto?id=379522&isNotice=true&instanceId=294&rnum=0

Lemm, V., & Vatter, M. (Eds.). (2014). *The government of life: Foucault, biopolitics, and neoliberalism*. Fordham Univ Press.

Mallapaty, S. (2020, March 26). What the cruise-ship outbreaks reveal about COVID-19. Retrieved April 14, 2020, from https://www.nature.com/articles/d41586-020-00885-w

Melly, D., & Hanrahan, J. (2018). The potential role of smart mobile technology in mitigating Ireland's tourism biosecurity risk. *Journal of Tourism and Hospitality Management*, *6*(6), 264–280.

Ministry of Health and Welfare. (2020, April 24). Central disaster and safety countermeasures headquarters (CDSCHQ) briefing on COVID-19. Retrieved May 20, 2020, from http://ncov.mohw.go.kr/tcmBoardView.do?brdId=&brdGubun=&dataGubun=&ncvContSeq=354196&contSeq=354196&board_id=140&gubun=BDJ

Ministry of Health and Welfare. (2020, May 17). Central disaster and safety countermeasures headquarters (CDSCHQ) briefing on COVID-19. Retrieved May 20, 2020, from http://ncov.mohw.go.kr/tcmBoardView.do?brdId=&brdGubun=&dataGubun=&ncvContSeq=354587&contSeq=354587&board_id=140&gubun=BDJ

Ministry of Health and Welfare. (n.d.). Retrieved April 13, 2020, from http://ncov.mohw.go.kr/en/baroView.do?brdId=11&brdGubun=111&dataGubun=&ncvContSeq=&contSeq=&board_id=&gubun=

Pew Research Center. (2019, February). Smartphone ownership is growing rapidly around the world, but not always equally. https://www.pewresearch.org/global/wp-content/uploads/sites/2/2019/02/Pew-Research-Center_Global-Technology-Use-2018_2019-02-05.pdf

Rawls, J. (1971). *A theory of justice*. Harvard University Press.

Rocklöv, J., Sjödin, H., & Wilder-Smith, A. (2020). COVID-19 outbreak on the diamond princess cruise ship: estimating the epidemic potential and effectiveness of public health countermeasures. *Journal of Travel Medicine*, *27*(3), 1–7. https://doi.org/10.1093/jtm/taaa030

Roy, A. (2020, April 3). Arundhati Roy: 'The pandemic is a portal': Free to read. Retrieved April 14, 2020, from https://www.ft.com/content/10d8f5e8-74eb-11ea-95fe-fcd274e920ca

Roy, E. A., Doherty, B. (2020, April 9). Have Australia and New Zealand stopped Covid-19 in its tracks? https://www.theguardian.com/world/2020/apr/09/have-australia-new-zealand-stopped-covid-19-in-its-tracks-coronavirus

Saunders, T., Baeck, P. (2015). Rethinking smart cities from the ground up. Research report. Nesta. https://media.nesta.org.uk/documents/rethinking_smart_cities_from_the_ground_up_2015.pdf

Sigala, M. (2012). Social media and crisis management in tourism: Applications and implications for research. *Information Technology & Tourism*, *13*(4), 269–283. https://doi.org/10.3727/109830512X13364362859812

Snider, M. (2020, March 5). Coronavirus concerns lead to hoarding, panic buying to stock 'panic pantries,' Nielsen says. USA Today. https://www.usatoday.com/story/money/2020/03/04/coronavirus-us-spurs-panic-buying-stock-pandemic-pantries/4950835002/

WTTC. (2020). WTTC now estimates over 100 million job losses in the Travel and Tourism sector and alerts G20 countries to the scale of the crisis. World Travel and Tourism Council. May 24, 2020. https://wttc.org/News-Article/WTTC-now-estimates-over-100-million-jobs-losses-in-the-Travel-&-Tourism-sector-and-alerts-G20-countries-to-the-scale-of-the-crisis

Yee, F. Y. (2020). EU urges voluntary use of virus tracking apps to speed recovery from pandemic. https://www.reuters.com/article/us-health-coronavirus-eu-apps/eu-urges-voluntary-use-of-virus-tracing-apps-to-speed-recovery-from-pandemic-idUSKBN22P15H?il=0

Zastrow. (2020). Coronavirus contact-tracing apps. Can they slow the spread of COVID-19? *Nature*. Retrieved May 19, 2020. from https://www.nature.com/articles/d41586-020-01514-2.

Index

Note: Page numbers followed by 'n' refer to notes, those in **bold** refer to tables and those in *italics* refer to figures

Abdallah, J. E. 300, 309
Abraham, N. 185
academic works 254n3
accessibility 24, 154, 156, 163, 165, 167
actor-network theory (ANT) 10–11, 335
Adams, C. 181
advocacy 2, 11, 379–80
affirmative ethics 11, 13, 371–2, 375–8, 380, 382–4
African-American 281, 284, 291; and Blacks in American Southeastern tourism 166; exploration of 282; lens of Double Consciousness 288–9; societal racism and underrepresentation 290; travellers, lived experiences of 284; *vs.* White tourists 282
African diaspora 291
Agyeman, J. 336, 348
Aiken, Anne 198
Alderman, D. H. 301, 310
Alkon, A. H. 336, 348
Alston, Philip 195
alternative tourism 3, 11
Amazon region of Ecuador: Chakra Route 259–60 (*see also* Chakra Route); food sovereignty 260, 274; Indigenous food systems (*see* Indigenous food systems); Indigenous-led tourism 274; tourism 259
amenity-led migration 37
amenity migrants 35, 38
American Civil War 120
American Disabilities Act (ADA) 156, 160
American Southeastern tourism brochures 157–8
Americans with Disabilities Act (ADA) 163
animal ethics 173, 176, 177
animal justice 6, 9, 177–8, 181–9
animal rights 165, 173, 176, 177, 180, 188, 190
Antarctica 317, 408
Anthropocene 2, 12, 197, 200–1, 204–6, 212–14, 216n1, 336, 377, 385
anti-racism protests 1
anti-racism riots 12–13
apprenticeship 123
appropriation 7, 17, 18–22, 26, 28–30, 37, 73, 76, 82, 86, 88, 221, 225, 241
Archive of Indies 22
Arendt, H. 371
Argentina 408
Aristotelian eudemonia 190
Aristotle 5, 11
Ateljevic, I. 280
A Theory of Justice (Rawls) 6
atmospheres 197, 204, 205, 210–14, 216; and attractions 210–14; political crisis 215
attractions 210–14
Australia 315–17, 357
authenticity 7, 17, 19–20, 22, 23, 26, 28–30, 54, 60, 61, **61**, 62, 67, 82, 84

Bachrach, P. 297
Back to Africa movement 284
Bakhtin, M. M. 355, 356; carnivalesque 355; 'chronotope' 357; dialogic imagination 11, 356; dialogism 357
Barad, K. 106
Baratz, M. S. 297
Barcelona 16, 18, 114
Barker, A. J. 107n1
Barton, A. W. 289
Basarwa 3
Bauman, Z. 95
Beard, L. 338
Beeton, S. 245
Belhassen, Y. 96
Bell, M. M. 358
Bell, S. J. 101
Bengtsson, M. 302
Benjamin, Stefanie 8
Bennett, A. 318
Bennett, J. 318, 335, 338, 349
Benton, A. 282
Bermuda 112, *112*, 115, 117–18, 129, 130, 131n1
Bermuda National Trust 118, 120
Bermuda Tourism Authority (BTA) 115
Bianchi, Raoul 8–9

Biber, K. 325
Biddulph, R. 8, 156
biodiversity 253
Black Bermudian heritage 121
Black Diaspora 281–2
Black Lives Matter (BLM) 1
Black travelers 291; roots tourism trip to Ghana 280; United States 281
Black Travel Movement (BTM) 280, 281
Blake, J. 389
Boes, K. 409
Bottone, Ethan 8
Boyd, S. W. 149
Braidotti, R. 371
Bramwell, Bill 154
Brennan, T. 217n7
Bright, C. F. 8, 10
Brill, S. 103
Britton, S. 3
Broken Heel Festival 10, 316, 330; background and context 319–20; concept of 319; context 319–20; data 320–1; 'drag' 319; neo tribal festival behaviour and social justice 327–9; netnography approach 320; outback imagery 322–3; outback town 327; reflexivity and positionality of researchers 321; researchers, reflexivity and positionality 321; road trip 323–6, 329; rural stereotype 329
Broken Hill 10, 315, 316, 319–22, *325*, 325–7, 329, 330; *see also* Broken Heel Festival
Brundtland Commission 4, 196
BuenVivir (good living) ethos 274
Bunten, A. C. 74, 81
Burrows, Erskine 117
Buzinde, C. N. 138

Calakmul Biosphere Reserve 248
Callon, Michel 337
Camargo, B. A. 10, 66, 72, 179, 241, 249, 316, 336–7
Campbell, R. M. 310
Canada 202
Cañada, E. 262
Cape Town Declaration on Responsible Tourism 51
Capocchi, A. 16
carnivalesque 355
The Cathedral 22
Caton, K. 4, 8, 96, 105, 242, 249, 329
cautionary platform 2
centrality 18, 86, 266, 319
Chakra Mamas 264, *264*
Chakra Route 10; *Buen Vivir* (good living) ethos 274; community leaders 263; data analysis 266; definition 263; destination branding 271; food sovereignty, Kichwa perspectives 267–8; gender empowerment and traditional knowledge 271–2; Indigenous contexts 263; Kichwa and non-Kichwa people 274; Kichwa organizations **264**; participation 265–6; participatory governance and food sovereignty legislation 273; and relationships 268–71; setting and background 263–5; Six Pillars Framework 267, 273; trustworthiness 266–7
Chancellor, C. H. 286
Changing By Degrees: Steps to Reduce Greenhouse Gases 197
Chappell, Edward 125
Charlton, A. 176–7, 187
China 242, 408
Choi, Jiyoung 11
Christian humanism 242
'chronotope' 357
Cittaslow 334–6, 340, 341, 346–7, 349–51
civic disobedience 196–7
Civil War monuments 295
Clark, A. 217n11
Clark, E. 17
climate action 199–201
climate justice 9, 12, 195, 196, 199–201
climate litigation 195
Cloquet, I. 155
Cocola-Gant, A. 17
Coffin, Judge 198
Cohen, E. 105, 185
Cohen, J. J. 217n9
Colombia 202
colonial fantasies 139
colonialism 1, 4, 13, 96, 100, 113, 221, 270, 273, 274, 375; systemic injustices of 1
Comisariado de Bienes Comunales (Commissariat of Common Goods) 252
Commissariat of Common Goods (CCG) 252
commodification 7, 10, 18, 20, 25, 28–30, 72, 74, 75, 79, 84, 86, 130–1, 138, 190, 210, 221, 225, 273, 337, 349, 350, 371, 373, 374, 376, 383, 384
communitarianism 173
community-based tourism (CBT) 3, 10
community pride 395–6
commutative justice 179, 185, 187
compensation justice 87
conceive tourism 253
Confederate Museum 120
Congressional Office of Technology Assessment, 1991 197
Conran, M. 375
consumerism 374
contractarianism 173, 174–5, 181, 190
conviviality 11, 336, 337, 343, 344, 349–51
'Corredor Transístmico' megaproject 246
COVID-19 1, 5, 11, 190, 253, 385, 408–10, 412, 413, **414**, 415; crises 190; pandemic 1, 11; South Korea 410
critical disability theory 154–7, 160, 167; and inclusive tourism 155–6
critical multilogicality 96
Critical Race Theory (CRT) 282
Crow, Jim 281
Cruelty Prevention and Welfare of Animal Act 186
crystallization 115

CSD-NGO steering committee 222
cultural advocacy 379, 381
cultural discrimination 241
cultural heritage 9, 10, 13, 30, 52, **57**, **58**, 60, 65, 75, 83, 111–14, 118, 122, 124, 125, 128–30, 148, 296, 297, 337, 344, 348, 351, 388–91, 395–9, 402, 403
cultural imperialism 3, 137, 148
cultural injustices 3, 5, 239, 253
cultural integrity 4, 9, 85–6, 196, 371
cultural justice 73, 75, 81, 87, 88, 137, 241, 249
cultural landscapes 44, 52, 57, 59, 63, 65–8, 122; Honghe Hani Rice Terraces 54–6
Curry, P. 175

"dark" heritage 138
Davy, Humphrey 212
Dayutang 56
Dean, E. 103
decent work 9, 220–1, 225–33
decolonization 8, 93, 96–9, 138; and enacting responsibilities 96–8
Decolonizing Methodologies (Smith) 263
deep justice 181, 187–9
Deleuze, G. 371
De Line, Sebastian 106
de Man, Frans 8–9
democratic society 244
deontological ethics 376–7
Department of Justice (DOJ) 163
depoliticization 371, 373–5, 383
De Re Metallica 210
Destination Marketing Organization (DMO) 158
Detroit Zoo 186
Devlin, R. 156
dialogic imagination 11
dialogism 357–60
Diaz-Parra, I. 17
dignity 9, 10, 177, 184, 188, 190, 197, 242, 243–4; academic works 254n3; democratic society 244; human rights and humanism discourses 243; Indigenous groups 253; indigenous justice in tourism 249–51; individual and collective 249; International Labour Organization Constitution 243; Mayan residents 240; person's abilities 243; promotion of 252–3; protection of 252; PWDs 156; restoration 251–2; tourism employment 225; unconditional human 244; United Nations Declaration of Rights of Indigenous People 243; Universal Declaration of Human Rights 243
Dillette, Alana 10
direct participation 87
dis-citizenship 156, 165–7
disenfranchisement 117
distributive justice 8, 136–8, 174, 179, 241, 244
The Division of Labour in Society 319
Domaneschi, L. 338
Donovan, J. 181
Double Consciousness 10, 280, 282–4; African-American identity 288–9, 291; 'Black centric' 290; Du Bois' sociological theory 280, 282–6, 288; sociological theory of 282; sport tourism 284
Douglas, Frederick 281
Du Bois' theory 280, 282–6, 288
Du Bois, W. E. B. 282–4
Du Cros, H. 297
Duffy, Michelle 7
Duffy, R. 370
Dungog festival 35; data gathering and analysis 40; equitable distribution, costs and benefits 44–5; exclusion and "indifference" 47; facilitate connections between diverse groups 45–7; Film Festival 39; Koori 39; place frames 41–3; research methodology 40; rights claimants and their claims 43–4; trustworthiness 40
Durkheim, E. 298, 319
Dworkin, R. 177

ecocentrism 173, 175, **182**, 190
ecofeminism 12
ecological justice 195
economic injustice 3, 8
Edelheim, Johan 157
Ei Wada 140
Ekern, S. 53
elephants 172, **182–4**, 185–9
Elliot, Stephan 320
enacting responsibilities 96–8
environmental justice 3, 86, 137–8, 195, 206, 241
epistemic literalism 242
equity 6–8, 13
ethics of care 180–1, 189, 377
ethnic minority 54, 59, 67–8, 137, 226, 242
ethnography 77, 104, 284–6
European Commission 200
European Mediterranean 17
European Parliament 16
Ewart, William 296
existential authenticity 20
existentialist humanism 242
Experience Ghana: Journey to a Real Life Wakanda 284
Expo Company 57, 60

Fainstein, S. S. 5, 403
fairness 4, 7, 12, 79, 173, 175, 181, 185, 223, 239, 336, 348, 355–8, 367, 368, 370, 377, 390, 391, 403, 413, **414**
fair trade tourism 3
Fair Work Act, 2009 357
Falmouth: background and setting 123–4; challenges of place 124–8
Falmouth Heritage Renewal (FHR) 125–6, 131n2
Farm Animal Welfare Council, 2009 176
feminist ethics 12
Fennell, D. A. 9, 105, 188, 370
festivalization 344

festivals: the changing face of the rural 38–9; rights in places 37–8; the right to the city 36–7; and social justice 35–6
fetishization 26, 29
FIFO (fly-in fly-out) 319
Figueroa, R. 138, 241
flyskam (flight shaming) 200
Fondo Ecuatoriano de Cooperacion para el Desarrollo (FECD) 263
food sovereignty 256, 260, 261–2, 274; Kichwa perspectives 267–8
Forné, F. 336
Forrester, Bill 155
Fortenberry, B. R. 8
Foster, Kelly 8
foster settler accountability: common plantain 98–9; interrogation 99–100
Framework Convention on Tourism Ethics 222, 231, 233n1
Francione, G. L. 176–7, 187
Fraser, N. 137, 241
Friedmann, R. 316
Fromm, E. 242
Fyall, A. 54

Gao, J. 7, 54
Gardiner, M. 358
Garner, R. 188, 190
Gascón, J. 262
Gates Bay 131n3
Gear, C. 105
Gearey, A. 176
gender injustices 8, 136, 148, 149; mill girls and the TSM 149
gender representation 147, 148, 296
gentrification 7, 17, 20, 22–4, 27, 29, 30, 37, 230
geo-politics 117–18
Ghana: African-American, lens of Double Consciousness 288–9; African diaspora 291; Black travelers 291; data collection and analysis 285; during-trip 285–6; place, space and people 286–8; post trip 286; pre-trip 285; social justice 289–90
Gilligan, C. 3, 180
"glass ceiling" 135
global action and justice 199–201
Global Code of Ethics for Tourism 222, 233n1
global justice 174
global migrations 12
global mobilities 37
global pandemic 1, 253, 408–10, 415
Globe Hotels 118
Goodman, D. 339
Goodman, M. K. 339
Goodman, N. 98
Goodwin, H. 371
Gorman-Murray, A. 327, 329
Graham, Peter 9, 11
The Great Ape Project, 1993 189
Greater Caribbean 8, 10, 111–14, *112*, 114, 117, 120, 124, 129
The Great Global Warming Swindle 217n12
Greece 242
Grimwood, B. S. R. 8, 52, 96, 104
guanxi 60, 64
Guatemala *see* Tz'utujil Maya community

Haeckel, Ernst 212
Hahn, Harlan 164
Haiki, J. 371
Hales, R. 347
Hall, A. 310
Hall, C. M. 113
Hanna, S. P. 300
Harbor, Lucy 7
Hardy, A. 10, 318, 324
Harvey, David 19, 36
Hassanli, N. 316
Haussmann's Paris reforms 20
healing 8, 73, 94, 99, 102–4
Heidegger, M. 205, 216n3
Held, V. 180
heritage justice 8, 12–13, 112–14, 131; Caribbean heritage, tourism 113–14; facilitating 130; Greater Caribbean 114
heritage planning biases 118–19
heritage programming 8, 112, 115, 120–2, 128–31
heritage tourism 10, 52, 54, 62, 65–7, 111–15, 118, 122–5, 127–30, 138, 210, 297; *see also* Tennessee
heteroglossia 366–7
Hetherington, K. 328
Higgins-Desbiolles, F. 240, 371, 372
Historical Marker Database (HMDB) 301
Historic Falmouth Cruise Ship Port 127–8
historic roadside markers 8, 296, 302
Hobbesian social contract theory 174
Hodder, E. F. 300, 388
Hodge, D. 371, 372
Holifeld, R. 338
Hollinshead, K. 96, 138, 374, 375
Honghe Hani Rice Terraces 54–6
Hong Kong 18
Hosoi, Wakizo 140
Hughes, P. 179
Hultsman, J. 4, 240, 371
human dignity 9, 240, 243–5, 249, 253, 254n3
Human, H. 403
humanism 4, 10, 210, 240, 242–4, 254n2, 383
humanistic-management theory 240, 244–5, 249, 251
humanistic tourism **249**
humanitarian advocacy 379
humanitarianism 375–7, 381, 383
Human Research Ethics Committee 360
human rights 3, 5, 53, 54, 173, 243, 252, 280, 291, 317, 363, 366, 371, 374, 383, 389, 413
Hunt, Carter 7
Hunter, J. 136
Hunter Valley 39

inauthenticity 20, 286, 287, 291
inclusive green growth 9, 230–3
inclusiveness 8, 10, 13, 48, 130, 154, 167, 230–2, 298, 383, 390, 391, **414**
inclusive tourism 154, 163, 167; critical disability theory 155–6; PWDs 156–7
India 201, 242, 287
Indian Removal Act, 1830s 299
Indigenous community 61–2, 246
Indigenous environmental justice 8
Indigenous food systems: food sovereignty 261–2; Indigenous people 260; and tourism 261; traditional 260
Indigenous groups 239–40; dignity and wellbeing 253; economic and ecocultural well-being 253; restorative justice 241–2
Indigenous justice 240–2, 249; in tourism 249–51; in tourism studies 240–2
Indigenous-settler relations 94
Indigenous tourism 6, 72; communities 74–5; environmental justice 86, 241; and food sovereignty 256, 274; indigenous survival in globalized world 73; justice framework 253; promise of tourism and justice through tourism 75
Indigenous women 274
Information and Communication Technologies (ICTs) 409
Ingold, T. 213
Institutes of Justinian 173
Instone, L. 101
intermediate justice 181, 186–7
International Decade for People of African Descent 280
International Labour Organization, Convention 169 of 243, 248, 252
Iran's Cultural Heritage Handicrafts and Tourism Organization (ICHHTO) 392
Isaac, R. K. 371–2
Izmir *340*, 345, 348

Jackson, Andrew 299
Jamaica 112, *112*, 115, 130
Jamaica National Heritage Trust (JNHT) 115
Jamal, T. 11, 66, 72, 88n3, 106, 179, 260, 316, 336–7, 347, 356
Jansson, David 158
Japan 135, 408; gender inequality in 148; gender injustice in 136; mill girls, and TSM 141–2; Tomioka Silk Mill 139–40
Jenkins, J. 310
Jepson, A. S. 316
Jewish Americans 284
Johnson, M. 97
Joko Aishi (Hosoi) 140
Jones, J. B. 296, 299–302, 308, 311
Jourdan, Dawn 9
Journal of Sustainable Tourism 154
Jover, J. 17
Joyner, Andrew 8
Juliana *vs.* United States 197–9, 201
"Just Destination" 179
Justice and Ethics in Tourism (Jamal) 370–1
justice tourism 5, 11, 75, 371–2, 375, 384; concept of responsibility 377–8; concept of solidarity 378–80; dominant neo-liberal and liberal logics 383; elucidation 380–2; elucidation of 380–2; parameters of 377–80; tourism studies 372–3
Justice tourism (Scheyvens) 371, 372–3
just representation 8, 10, 296–8, 308–10; conceptualizion 296–8; historical roadside marker research, Tennessee 299–300; social importance 300–1; social representation theory 298

Karak**il**çik bread **("I" should be special character, please fix as per proof)**: actor-network theory 335, 337–8; discursive practices 342–3; fair food and slow food tourism 336–7; fair foods, gender equity and fairness 348; festivalization 344; heritage creation and conviviality 343–4; mutuality (co-creation) 345; production of 350; research approach 339–40; research methods 341–2; Seferihisar 334, 340–1, 349; slow food justice 346–8; slow movement 335–6; socio-ecological imagination 339, 348–51; socio-ecological-technological assemblages 345–6
Kateb, G. 243
Kathleen, J. M. 101
Kheel, M. 181
Kichwa community-based tourism associations 263
Kichwa Napo Runa people 265
Kichwa people; *see also* Chakra Route: Six Pillars of Food Sovereignty **269–70**
Kimmerer, R. W. 93, 98, 100, 101, 103
Kirtsoglou, E. 138
Kitchin, R. 156
Klu Klux Klan Founder Nathan Bedford Forrest 300
Knowledge Based Platform 3
Koensler, A. 374
Korea Center for Disease Control and Prevention (KCDC) 410, 412
Kriwoken, L. 318
Kuhnlein, H. V. 260
Kurtz, D. V. 137, 148
Kuznets, Simon 233n4

labor market 253
Laguna, D. 20
Lakoff, G. 97
land alienation 17
Lane, Bernard 154
Latour, B. 212, 337
Law, John 337
Lee, Miranda 8
Lee, Seunghoon 11
Lefebvre, H. 17–20, 30, 35, 37
legal equality 135

legal/general justice 180
Leonard, S. J. 289
Leopold, Aldo 175
Le Temps des Tribus (Maffesoli) 318
LGBTQI + communities: Broken Heel 315–16 (*see also* Broken Heel Festival); influx of non-heteronormative 330; marginalisation and discrimination 317; neo-tribal theory 317–18; neo-tribe 330; social justice, marginalisation and exclusion 316–17; spatiality, resistance and transformation 318–19
liberal egalitarianism 173
lifestyle migration 17
Lin, Hongxia 7
liquid modernity 95
Lisbon/Venice 18
litigation 195, 198, 200–1
Little, J. 329
Ljubici, G. J. 105
Lobo, M. 103
local government 58–60, 66
Loch and Standing Stones of Stenness 204–5, *205*
Locke, John 5
Lonely Planet 84
Loring, Philip 10
Lowman, E. B. 107n1
Lukes, S. 297
Lundholm, J. 105

Macbeth, J. 3
MacKay, W. 208, 216n4
Maffesoli, Michel 318, 319
Mainland Orkney 208–9
Mair, Judith 7
Maple 140
Marcuse, P. 27, 37
Mares, T. M. 338
marginalization 179, 239, 240, 316, 320–2, 327, 330; and discrimination 317; neo-tribal theory 317–18; social justice and exclusion 316–17
marker distribution: enhanced thematic categories **306**; Tennessee **306**; thematic categories **306**
Markwell, K. 329
Marobela, M. 3
Martin Luther King 5
Maruyama, Naho 8
Marwick, E. W. 208
Marxist humanism 242
Massey, D. 37
Material Engagement Theory (MET) 9, 213, 217n11
Mayan residents 240
Mayan Train project 240, 245, 254n6; controversies and conflict 248–9; emergence of 246–8; Indigenous groups and development, Mexico 245–6
May, Todd 92, 105
McCabe, S. 95
McGregor, D. 138
McGuire, M. 374
McIntosh, A. 209
Mead, George Herbert 282
megaliths 9, 205
Meiji government, 1872 139, 140, 147
Melé, D. 244
Mensah, I. 282
Meskell, L. 388, 403
metaphors 97, 106, 210, 324–5; and worldmaking 94–6
Mexico 76, 240, 241, 252, 261, 316; Indigenous groups and development 245–6
Mexico City International Airport 246
Miller, D. 114, 173
"mill girls" 136; cultural heritage 148; cultural imperialism 148; distribution and recognition injustice 148; in Japan and TSM 141–2; model recruiting factory 146–7; *moe-okoshi* and TSM 144–5; "retro boom" 146; tourism representation 142–4; and TSM 148
Mill, John Stuart 5
Mills, C. Wright 339
Mitchell House Museum 118
Mkono, M. 179
"model recruiting factory" 146–7
Molton Barrett House *126*
Monbiot, George 217n12
Montego Bay 127
Moore, A. 200, 217n6
Moralization of Tourism (Butcher) 215
Morant Bay rebellion 123
Morgan, N. 280
Morrison, Scott 194
Morton, T. 105
Moscardo, G. 4
Moscovici, S. 298
Mowatt, R. A. 286
Muldoon, M. 52
Murphy, L. 4
museumification 18

Nasrolahi, A. 389
National Disability Authority 156
National Fund for Tourism Promotion (FONATUR) 248
National Human Rights Commission of Korea 410
National Institute of Anthropology and History (INAH) 248
National Institute of Indigenous Peoples' Law 252
National Register of Historic Places 296
Nature Climate Change 2
Neilsen, N. 104
Nelson, Louis 125
Nelson Mandela 5
neo-colonialism 3, 113
neo-liberalism 374, 376
neo-lithic ethics 11, 205
neo-tribal festival behaviour 327–9
neo-tribal theory 317–18

netnography 320, 393
New Zealand 409
Nietzsche, F. 371
Noddings, N. 3
no justice 181, 184–5
Nomugi Pass (Yamamoto) 140
non-Aboriginal people 138
non-disabled hegemony 167
non-governmental organizations (NGOs) 5, 188, 222, 227, 254n6, 265, 271
Norway 202
Noy, C. 329
Nozick, R. 177, 188
Nussbaum, Martha 9, 12, 243
NVIVO program 266

Obama, Barak 195, 201
Oberndorfer, E. 105
Ocho Rios 127
The Odin Stone 207–10
oneness 289, 291
Open Doors Organization 153
Organisation for Economic Co-operation and Development (OCDE) 245
organizational justice 174
outside hostel operators 62–3
Outstanding Universal Value (OUV) 54
overtourism: alienation, appropriation and the right to the city 19–22; appropriation 26; authenticity 30; communities 29–30; displacement and place alienation 27–8; and gentrification 29; historic centres and urban tourism-led economies 28–9; historic city Centre of Seville 24–5; in Lefebvrés work 18–19, 30; perceptions of (over)tourism 28; place alienation and place authenticity 29; research approach and methods 22–3; scholars of 29; Seville's case, Spain 22, 29; socio-spatial justice 16–18
Oviedo, G. 52

Pakistan 202
Panelli, R. 329
Papa, C. 374
Pearce, P. L. 297
Peña, D. 338
people with disabilities (PWDs) 8, 156–7; ADA compliant 162–3; African-Americans/Blacks in American Southeastern tourism 166; American Southeastern tourism brochures 157–8; critical disability theory 167; critical disability theory and inclusive tourism 155–6; disabled *vs.* handicap language in tourism brochures *162*; "dis-citizenship" 166, 167; diversity 164–5; inclusive tourism framework 156–7; *Journal of Sustainable Tourism* 154; language of 'disability' in brochures *161*; non-disabled hegemony 167; Open Doors Organization 153; promotional tourism literature, 166; Southeastern tourism industry 162; state-level DMOs 167; tourism industry 154; within the tourism industry 155; travel brochures **161**; Travel Tourism Research Association Conference 154; visible physical disabilities 158–60; 'welcoming of pets' 165–6
perspective dualism 137
Pieper, J. 173, 178–80
Pierce, J. 7, 37, 39, 45, 47
Pirson, M. 245, 249
place alienation 17–18, 20, 22, 27, 29–30; in Seville 26–8
place authenticity 29
place frames **38**, 41–3, 45–8
place settlers: proximity and practices 101–2; sharing soils 100–1
Plantago major 8, 93, 98
Plato 5
Policy Options for Stabilizing Global Climate 197
political advocacy 379
Political Constitution of Mexico 252
political ecology 3, 12
political economy 3, 9, 221, 222–3, 225, 226, 230; sustainable tourism 222–3
political responsibility 11, 12, 371, 373, 375, 377–8, 381, 383–5
"politics of kindness" 11–12
polyphony 358, 363–6
posthumanism 4, 105, 375, 377, 383–4; affirmative ethics 377; promise of 383–4
posthumanistic justice 10–11
Pothier, D. 156
Prince, S. 96, 97
Priscilla, Queen of the Desert (Elliot) 320
Pritchard, A. 280, 291
procedural/participative justice 87, 174, 241
Proclamation, 1763 299
The Production of Space (Lefebvre) 18
pro-poor tourism 3, 372, 375, 381
public platform 3
Pugaolaozhai 56, 62, 64, 65
Purcell, M. 20, 36, 37
Puschkarsky, T. 52

Qingkou 56, **71**
Qiu, J. 52
Quetzaltenango 88n1

Rami, K. 371
Rawls, J. 5, 9, 175, 206, 241, 370
Receveur, O. 260
recognition justice 8, 13, 87, 138, 174, 254
reflexivity 94, 96, 97, 141, 303–4, 321
Regan, T. 174, 176, 177
Reid, S. 46
relational responsibility 377
Relph, E. 20
Ren, C. 338, 341
Renfrew, Colin 213
replica seed 345
resident perceptions 389, 391, 404

responsibility 377–8; Cape Town Declaration on Responsible Tourism 51; cultural sites, China 52; 'doing' 51–2, 67–8; Expo Company 57; heritage destination 54; indigenous people, decision making 66; lens of multi-stakeholder 67; multi-stakeholders' perspectives 58–63; protective titles and corresponding laws/regulations **58**; rights-based approaches to world heritage 52–3; semi-structured face-to-face interviews 57; stakeholder conflicts 63–5; world heritage designation process **57**
responsible tourism 3, 375
restorative justice 3, 174, 179, 185, 241–2
Restoring the Quality of the Environment 197
retributive justice 174, 356
retrospective analysis 139
Richards, C. 211
Rickly-Boyd, J. M. 357
rights-based approaches (RBAs) 53
rights claimants 43–4
rights in places 38, **38**
rights to the rural 7, 34, 35, 47
right to the city 17–18, 29, 30, 35–8; alienation and appropriation 19–22; Lefebvrés work 18–19
The Right to the City (Lefebvre) 17, 18, 30
Rio de Janeiro 18
Rio Earth Summit, 1992 220
Ritchie, J. 208
Rivera, L. A. 136, 138, 148
roadside markers 296
Robards, B. 318
Robinson, M. 241
Rogues and Runners 120
Roisek, J. L. 101
Rome 242
roots tourism 280, 282; Black Diaspora 281–2; Double Consciousness 282–4; Du Bois' theory 282–4
Rosenberg, R. 260
The Royal Palace 22
Rurubu 140

Saldana, J. 158
Sanchez Vazquez, A. 29
Sandel, Michael 6, 11
Sanguin, A. L. 20
Santafe-Troncoso, Veronica 10
Santiago Atitlan: cultural flourishing 79–81; ecological tourism 85–6; learning related reciprocities 83–4; mutual benefits 82–3; reciprocity and equity 83; tourism development and equitable access 84–5; Tz'utujil language 82
Santos, C. A. 138, 329
SARS-CoV-2 408, 415
scales of justice 9, 13, 173, 174, **182–4**; animals 173; theory and practice **182–4**
Schachter, O. 243
Schellhorn, M. 72
Scheyvens, R. 5, 8, 156, 240, 371, 372
Scotland 204
Sea Venture 116
second sight 283, 289–90
secular pilgrimage 330
Seed Exchange Festival 343–4
Seferihisar 334, 340–1, 349
segregation 117, 155, 281, 283, 299
self-determination 5, 12, 86, 104, 245–9, **249**, **250**, 252, 253, 261, 262, 273, 274, 336, 371, 372
self-quarantine safety protection app 410, *411*, 412
Sen, Amartya 9, 244
Senju, H. 138
service recovery justice 174, 179
Seyfi, S. 403
Shabazz, K. Z. 282
shallow justice 181, 185–6
sharing economy 174, 179, 230
Sheldon, Pauline 154
Sheppard, Valerie 9
Shields, R. 318
Shooster, J. 177
Silberberg, T. 296
Sinchi Warmi 271, 272
Singapore 138
Singer, P. 176
Sinnok *vs.* State of Alaska 201–2
Sirakaya, E. 139, 147, 149
Six Pillars Framework 267, 273
Skift 165
Slow City Movement (Cittaslow) 336
slow food justice 346–8; and tourism 335
Small Island Developing States (SIDS) 9, 199, 200
smart destination 409, 413–14
Smith, L. T. 101, 102, 263
Smith, M. 9, 11, 370
social-ecological system 2, 5, 10, 13
social justice 3, 6–8, 35, 136–7, 180, 244, 251, 289–90, 316, 327–30, 370–1; and festivals 35–6; sustainable WHS development and 390–1
social liberalism 376–7
social representation theory 298
social responsibility 377
societal racism 290
socio-demographic data 22
socio-spatial justice 16–18, 30
socio-technological seed 345
solidarity 378–9
Sonmez, S. 139, 147, 149
Sorenson, T. F. 211
The Souls of Black Folk (Du Bois) 283
Southeastern state-level DMOs 159–60
South Korea 11; COVID-19 outbreak 410–13; global pandemic 409–10; novel coronavirus 408–9; SARS-CoV-2 415; smart destination 413–14; smart justice and smart governance **414**, 415
Space and Politics (Lefebvre) 18
Spain 317
Speiran, Siobhan 9, 11
sport tourism 284

Stafford, Lady *207*
state-owned tourism enterprise 60–1
Steckler, E. 245, 249
St. George: aerial image 116, *116*; architectural character 118–19, *119*; cultural heritage tourism landscapes 122, 129; and Falmouth 129, 130–1; museum programming 121
Stinson, Michela 8
Strivings of the Negro People (Du Bois) 283
substantive citizenship 156
Suleman, R. 138, 374
Sundberg, J. 101
supremacism 374
sustainable development 9, 75, 173, 195, 200, 220, 222, 226–7, 232, 385, 390, 404
Sustainable Development Goals (SDGs) 75, 173, 220; application 220–1
sustainable tourism 371, 375; and labour 225–6; political economy 222–3; tourism capitalism and creation of value 224–5; tourism development and growth 223–4; UNWTO and 226–7

Tanny, Oceane 8
Tennes see:
community-based planning 310; data coding and analysis 302–3; data set 301; density of markers 306–7; funding 308; geographic distribution of markers 304; historical roadside markers 308–9; memorialization of history 311; memoryscape 308; Native American, women's and black history 307–8; racial distribution of markers 304–5; reflexivity and positionality 303–4; roadside historic markers 301; Shelby County 309; thematic distribution of markers 304; urban/non-urban locations *305*
"TESCO Family Dollar Superstore" 125
Teye, V. 287
Thailand 186
Theodossopoulos, D. 138
Thrift, N. 102
Tilley, C. 213
Timothy, D. J. 149
Tomioka Diary (Wada) 140
Tomioka Silk Mill (TSM) 136, 139–40, 147–9; Japan 8
Toronto Zoo 186
tourism: Caribbean heritage 113–14; commodification 373–4; degrowth 2; depoliticization 374–5; development 3, 239, 240–1, 253; environmentally and socially responsive 2; ethical approaches 375–7; gentrification 17; humanistic principles and government and industry responsibilities **250–1**; indigenous food systems 261; indigenous justice 249–51; justice *7*; and justice 137–9; justice and ethics 5; overtourism 2; people with disabilities 8; presage to the ends of 214–16; slow food justice and 335; tourismphobia 16; well-being 2
Tourism Enhancement Fund (TEF) 126
Tourism European Network (TEN) 222
Tourism for Development (Scheyvens) 279–80
Tourism Self-Management Committee (CAT) 78
traje (traditional Mayan vestments) 79, *80*
Travelers Today 165
Travel Tourism Research Association Conference 154
Tribe, J. 280, 291
TripAdvisor 179
Tuck, E. 96
Tucker, H. 113, 118
Turkey 188, 334, 335
twoness 283, 287, 289, 291
Tz'utujil Maya community 7; data collection and analysis 78–9; decolonizing methodology 77; ethnographic sensitivity 87; ethnographic work 77; Guatemala 76–7; Indigenous Tourism 87, 88; indigenous tourism and justice research 73–5; Quetzaltenango 88n1; Santiago Atitlan 79–86; Sustainable Tourism and Indigenous Peoples 72; in tourism 86

Uganda 202
unequal rights 12
UNESCO World Heritage site 7, 11, 22, 24, 122
Union of the Physically Impaired Against Segregation (UPIAS) 155
United Nations 155, 173, 200, 220
United Nations Commission on Sustainable Development (UNCSD) 222
United Nations Declaration on the Rights of Indigenous Peoples, Article 32 243, 248, 252
United Nations World Commission on Environment & Development (UNWCED) 195
United Nations World Tourism Organization (UNWTO) 4, 8–9, 196, 226–7, 279–80; embrace 220–1; SDG 8, 227–8; with SDG 221–2, 230, 232, 233; sustainable tourism 226–7; sustainable tourism and decent work 228–30
United States 8, 120, 154, 241, 285, 295, 297, 317
Universal Declaration of Human Rights 243
University of Technology (UTech) 115
unsustainability 4
UN World Tourism Organization 16
UNWTO Global Code of Ethics for Tourism 173
Urban European Initiative *(Ibidem)* 24
"urban neo-colonization" 18
The Urban Revolution 30
urban tourism 17–18
Urry, J. 95
utilitarian ethics 376
utilitarianism 173, 176–7, 186
Uzendoski, M. 268

Van der Duim, R. 338
Vázquez-Maguirre, Mario 10
the veil 283, 287, 291
virtue ethics 178, 181, 189
visible physical disabilities 158–60

volunteer tourism 3
Vorobjovas-Pinta, Oscar 10, 319

Wada, Ei 140
Waitt, G. 138, 241, 327, 329
Wakizo Hosoi 140
Walters, T. 316
Wang, H. 52
Wang, N. 20, 329
welfare 173, 176
Wertin, Jani 9
White, D. 335
White House Report, 1965 197
White tourists 282
WHS Management Committee 122
Whyte, K. P. 8, 72, 88, 88n4, 241, 254
William, James 282
Wilson, E. 96, 104
Wilson, S. 101, 103
Winchenbach, A. 240
Winters, N. 105
Witcomb, A. 388
Woodward, D. 233n3
Woosnam, Kyle 8
World Health Organization (WHO) 5, 408
World Heritage Site (WHS) 52, 136, 388, 404; inscription 389–90; sustainable development 390–1
worldmaking 93–6, 105–7, 136, 148, 374; and metaphors 94–6
Wu, B. 54
Wyckoff, J. 177

Yamamoto, Shigemi 140
Yang, L. 54
Yang, W. K. 96
Yazd: "authentic" 396–7; community pride 395–6; community well-being 398–9; heritage decision-making 401–2; inequitable distribution 400; netnography 393; quality of life 400–1; responsibility, future generations 396; semi-structured interviews 393–4; tourists' rights *vs.* residents' cultural heritage rights 397–8; women 399–400
Year of Return 284
Yin, R. 263
Young, I. M. 12, 137
youth activism 195
youth protesters 196–7
Yuanyang County, tourist arrival and tourism revenue of 56
Yucatán Peninsula 247
Yudina, O. 52
Yunseong, K. 137

Zhang, C. 7, 54
Zheng, Y. 52, 54